Dedicated to

My parents, Gopinath and Inchhamani, for their gift of life, love, and living examples

My wife, Jarana, for her life-long companionship, love, care, and support for following my dreams

My children, Megha, Sudhir and Subir, for their love and care

Contents

Preface

Toxicogenomics is the integration of genomics to toxicology. This technology is a powerful tool for collecting information from a large number of biological samples simultaneously; thus, it is very useful for large-scale screening of potential toxicants. This book provides up-to-date state-of-the-art information presented by the recognized experts and, therefore, it is an authoritative source of current knowledge in this field of research. The potential link between toxicology, genetics, and human diseases makes this book very useful to investigators in many and varied disciplines of science and toxicology.

The book is designed primarily for the research scientists currently engaged in toxicogenomics. However, it should be of interest to wide disciplines of science, including toxicology, genetics, medicine, and pharmacology, as well as in drug and food sciences. Also, it should be of interest to federal regulators and risk assessors of drug, food, environmental, and agricultural products. The tools used in the field of toxicogenomics can be used in humans as well as in human surrogate models, providing data that scientists and regulators would have more confidence in for making extrapolations from surrogates to humans.

Toxicogenomics is a rapidly developing field of science. There is a risk of being out of date in the not too distant future. However, neither the fundamental concepts and ideas, nor the experimental data are going to change. Therefore, for years to come, this book will be useful to the students and investigators in toxicogenomics, as well as in many disciplines of science to guide them in their future work.

Saura C. Sahu

List of Contributors

P. Ancian CIT, Evreux, France

S. Arthaud CIT, Evreux, France

Kirsten A. Baken Department of Health Risk Analysis and Toxicology (GRAT), Nutrition and Toxicology Research Institute Maastricht (NUTRIM), Maastricht University, PO Box 616, 6200 MD Maastricht, The Netherlands

Johanna M. Beekman Clinical Biomarkers, Clinical Pharmacology, Bayer Schering Pharma AG, Müllerstrasse 170–178, 13353 Berlin, Germany

Kathleen Boehme Merck Serono, Toxicology, 64271 Darmstadt, Germany

Daniel A. Casciano University of Arkansas for Medical Sciences, Department of Pharmacology and Toxicology, Little Rock, AR 72205, USA

Nabarun Chakraborti Department of Animal and Avian Sciences, University of Maryland, College Park, MD, USA

Leila Chekir-Ghedira Faculté de Médecine Dentaire de Monastir, Lab. de Biologie Moléculaire et Cellulaire, Rue Avicenne 5000, Monastir, Tunisie

Muireann Coen Department of Biomolecular Medicine, Sir Alexander Fleming Building, SORA Division, Faculty of Medicine, Imperial College London, SW7 2AZ, UK

Jennifer C. Davey Dartmouth Medical School, HB 7650 Remsen, Room 510, Hanover, NH 03755-3835, USA

Amy C. Ditewig   Abbott Laboratories, Department R4DA, Building AP9A, 100 Abbot
Park Rd, Abbott Park, IL 60064, USA

Bhalchandra A. Diwan   Basic Research Program, SAIC-Frederick, Inc., NCI Frederick,
MD, USA

Janine Ezendam   Laboratory for Health Protection Research, National Institute for Public Health and the Environment (RIVM), PO Box 1, 3720 BA Bilthoven, The Netherlands

R. Forster   CIT, Evreux, France

Brigitte Ganter   Ingenuity Systems, 1700 Seaport Blvd., Redwood City, CA 94063, USA

Elisavet T. Gatzidou   Department of Forensic Medicine and Toxicology, University of
Athens, Medical School, Athens, Greece

Julie A. Gosse   University of Maine, 5735 Hitchner, Room 108, Orono, ME 04469-5735,
USA

Karen Hamernik   Office of Science Coordination and Policy, Office of Prevention, Pesticides, and Toxic Substances, US Environmental Protection Agency, Washington, DC,
USA

Joshua W. Hamilton   Dartmouth Medical School, HB 7650 Remsen, Room 510,
Hanover, NH 03755-3835, USA

Thomas H. Hampton   Dartmouth Medical School, HB 7650 Remsen, Room 510,
Hanover, NH 03755-3835, USA

Kenneth Haymes   Office of Science Coordination and Policy, Office of Prevention,
Pesticides, and Toxic Substances, US Environmental Protection Agency, Washington, DC,
USA

Susan Hester   Office of Research and Development, National Health and Environmental
Effects Research Laboratory, US Environmental Protection Agency, Research Triangle
Park, NC, USA

Yoko Hirabayashi   Cellular & Molecular Toxicology Division, National Center for Biological Safety & Research, National Institute of Health Sciences, 1-18-1 Kamiyohga,
Setagayaku, Tokyo, Japan

Elaine Holmes   Department of Biomolecular Medicine, Sir Alexander Fleming Building,
SORA Division, Faculty of Medicine, Imperial College London, SW7 2AZ, UK

Taisen Iguchi   Division of Bio-Environmental Science, Department of Bio-
Environmental Science, Okazaki Institute for Integrative Bioscience, National Institute

for Basic Biology, National Institutes of Natural Sciences, 5-1 Higashiyama, Myodaiji, Okazaki Aichi 444-8787, Japan

Tohru Inoue National Center for Biological Safety & Research, National Institute of Health Sciences, 1-18-1 Kamiyohga, Setagayaku, Tokyo, Japan

Daehee Kang Molecular and Genomic Epidemiology Laboratory (MGEL), Seoul National University College of Medicine, 103 Daehangno, Chongnogu, Seoul 110-799, Korea

Yoshinao Katsu Division of Bio-Environmental Science, Department of Bio-Environmental Science, Okazaki Institute for Integrative Bioscience, National Institute for Basic Biology, National Institutes of Natural Sciences, 5-1 Higashiyama, Myodaiji, Okazaki Aichi 444-8787, Japan

J.J. Legrand CIT, Evreux, France

S. Leuillet CIT, Evreux, France

Lian Li Department of Preventive Medicine, Seoul National University College of Medicine, Seoul 110-799, Korea

Michael J. Liguori Abbott Laboratories, Department R4DA, Building AP9A, 100 Abbot Park Rd, Abbott Park, IL 60064, USA

John C. Lindon Department of Biomolecular Medicine, Sir Alexander Fleming Building, SORA Division, Faculty of Medicine, Imperial College London, SW7 2AZ, UK

Jie Liu Inorganic Carcinogenesis Section, Laboratory of Comparative Carcinogenesis, National Cancer Institute at NIEHS, Research Triangle Park, NC, USA

J. Thomas McClintock Office of Science Coordination and Policy, Office of Prevention, Pesticides, and Toxic Substances, US Environmental Protection Agency, Washington, DC, USA

Kirstin Meyer Early and Mechanistic Toxicology, Nonclinical Drug Safety, Bayer Schering Pharma AG, Müllerstrasse 170–178, 13353 Berlin, Germany

Arindam Mitra Virginia Maryland Regional College of Veterinary Medicine, University of Maryland, College Park, MD, USA

Stefan O. Mueller Merck Serono, Toxicology, 64271 Darmstadt, Germany

Suman Mukhopadhyay Center for Biosystems Research, University of Maryland Biotechnology Institute, College Park, MD, USA

Jeremy K. Nicholson Department of Biomolecular Medicine, Sir Alexander Fleming Building, SORA Division, Faculty of Medicine, Imperial College London, SW7 2AZ, UK

Paul A. Nony   Center for Toxicology and Environmental Health, LLC, North Little Rock, AR, USA

Woong-Yang Park   Department of Biomedical Sciences, Seoul National University College of Medicine, Seoul 110-799, Korea

Jeroen L.A. Pennings   Laboratory for Health Protection Research, National Institute for Public Health and the Environment (RIVM), PO Box 1, 3720 BA Bilthoven, The Netherlands

P. Scott Pine   Center for Drug Evaluation and Research, US Food and Drug Administration, Silver Spring, MD, USA

Barry Rosenzweig   Center for Drug Evaluation and Research, US Food and Drug Administration, Silver Spring, MD, USA

Saura C. Sahu   Division of Toxicology, Office of Applied Research and Safety Assessment, Center for Food Safety and Applied Nutrition, US Food and Drug Administration, Laurel, MD 20708, USA

Hiroshi Sawada   Development Research Center, Pharmaceutical Research Division, Takeda Pharmaceutical Company Limited, 2-17-85 Juso-Honmachi Yodogawa-ku, Osaka 532-8686, Japan

Susanne U. Schmidt   Merck Serono, Toxicology, 64271 Darmstadt, Germany

Susanne Schwenke   Global Drug Discovery Statistics, Global Biostatistics, Bayer Schering Pharma AG, Müllerstrasse 170-178, 13353 Berlin, Germany

Stephanie Simon   Merck Serono, Toxicology, 64271 Darmstadt, Germany

Laura Suter   F. Hoffmann-La Roche Ltd. Grenzacherstrasse, 90/505a, 4070 Basel, Switzerland

Raymond Tennant   National Institute of Environmental Health Sciences, Research Triangle Park, NC, USA

Stamatios E. Theocharis   Department of Forensic Medicine and Toxicology, University of Athens, Medical School, Athens, Greece

Karol Thompson   Center for Drug Evaluation and Research, US Food and Drug Administration, Silver Spring, MD, USA

Rob J. Vandebriel   Laboratory for Health Protection Research, National Institute for Public Health and the Environment (RIVM), PO Box 1, 3720 BA Bilthoven, The Netherlands

Henk van Loveren   Laboratory for Health Protection Research, National Institute for Public Health and the Environment (RIVM), PO Box 1, 3720 BA Bilthoven, The Netherlands

Antoaneta Vladimirova   Ingenuity Systems, 1700 Seaport Blvd., Redwood City, CA 94063, USA

Michael P. Waalkes   Inorganic Carcinogenesis Section, Laboratory of Comparative Carcinogenesis, National Cancer Institute at NIEHS, Research Triangle Park, NC, USA

Jeffrey F. Waring   Abbott Laboratories, Department R4DA, Building AP9A, 100 Abbot Park Rd, Abbott Park, IL 60064, USA

Hajime Watanabe   Division of Bio-Environmental Science, Department of Bio-Environmental Science, Okazaki Institute for Integrative Bioscience, National Institute for Basic Biology, National Institutes of Natural Sciences, 5-1 Higashiyama, Myodaiji, Okazaki Aichi 444-8787, Japan

Acknowledgments

I am deeply indebted to the following scientists who have influenced me directly or indirectly to write this book.

First, I must admit that writing and editing this book was a huge project. It took several contributors and I am but one of them. I am indebted to all these scientists for their excellent contributions in their own areas of expertise.

I thank Dr Thomas A. Cebula for his inspiration and leadership.

I thank Dr Joseph E. LeClerc for serving as a catalyst for my efforts and for his encouragement, guidance, and support.

I thank Dr Daniel A. Casciano for his professional advice, support, and contributed chapter.

I thank Dr Richard B. Raybourne for his encouragement and support.

I thank Dr Philip W. Harvey, Covance Laboratories, UK, for his professional advice, encouragement, and support.

Finally, I thank my publishers Martin Rothlisberger, Paul Deards, Jamie Summers, Rebecca Ralf, and Richard Davies of Wiley Publishing Company, who were instrumental in the publication of my previous book entitled 'Hepatotoxicity: From Genomics to *In Vitro* and *In Vivo* Models' published in 2007. They are directly instrumental in the publication of this book, too.

Section 1

Design, Analysis and Interpretation of Toxicogenomics

1

Mechanistic Toxicogenomics: Design and Analysis of Microarray Experiments

Kirstin Meyer, Susanne Schwenke and Johanna M. Beekman

1.1 Introduction

Mechanistic toxicogenomics, the elucidation of the underlying mechanisms of an observed toxicity at the transcriptional level, has become an accepted tool in exploratory toxicology. Especially since experimental procedures and data analysis are now more standardized, and quality parameters are established, toxicogenomics is frequently applied either as an integral part of toxicology studies or in separate mechanistic studies (Gant, 2003; Lord, 2004; Foster *et al.*, 2007). *In vivo* studies are clearly the main focus for mechanistic investigations, since comparative *in vitro/in vivo* investigations revealed fundamental differences in the gene expression pattern indicating limitations of *in vitro* systems (Boess *et al.*, 2003).

Mechanistic toxicogenomics is used to characterize toxicological findings at the molecular level, generate hypotheses about the mechanism of toxicity, identify potential safety biomarkers, contribute to the elucidation of species specificity and at the end also to support the risk assessment of new chemical entities (Boverhof and Zacharewski, 2006).

Nevertheless, until getting to the point of general acceptance, investigators had to put a lot of effort in optimizing the experimental design as well as in the sophistication of the data analysis.

In this chapter, we will describe the basic concepts in design and analysis of microarray experiments as we have developed them for use based on our experience. We will focus on the design of the underlying animal studies, the general performance of gene expression

Toxicogenomics: A Powerful Tool for Toxicity Assessment   Edited by S. C. Sahu
© 2008 John Wiley & Sons, Ltd

experiments and outline a strategy for data analysis. We describe the procedure using two of our first, not optimally designed, toxicogenomics experiments with the well-known compound acetaminophen (*N*-acetyl-*p*-aminophenol; APAP) in order to make caveats and pitfalls in the procedure obvious. APAP, which causes liver toxicity at high doses, is one of the most frequently studied compounds in toxicogenomics experiments, especially in rodent liver (Minami *et al.*, 2005; Powell *et al.*, 2006). Since these studies show that the current knowledge about the mechanism of APAP-induced toxicity is well reflected in the gene expression profile, APAP was an optimal example to verify the outcome of our own studies.

1.2   Experimental Design

1.2.1   Animals and Treatment

Male Wistar rats (Shoe:WIST; Tierzucht Schoenwalde, Germany), weighing 200–300 g, were used for the studies. The animals were allowed to adapt for at least 7 days prior to the start of treatment. They were housed under controlled conditions with a 12 h/12 h light/dark cycle and fed standard chow with water provided *ad libitum*. Acetaminophen (Sigma, Taufkirchen, Germany) was administered as suspension in Tween 80–sodium chloride by oral gavage of 800 mg kg^{-1} once with a volume of 8 ml kg^{-1}. The dosage was chosen based on literature data (Smith *et al.*, 1998) and in-house experience with the aim to induce overt hepatotoxicity after a single dose.

In study 1, groups of four animals were sacrificed at 24, 48 and 96 h after single oral administration of APAP. Four control animals, which received the vehicle only, were sacrificed after 96 h. In this study, histological findings were already observed after 24 h. Therefore, study 2 was performed, with an earlier time point of necropsy of 6 h ($n = 4$) and again at 24 h ($n = 4$). In study 2, the four control animals were sacrificed after 24 h.

At the end of the respective observation periods the animals were sacrificed under general anesthesia, and a full post mortem examination was performed. The left lateral loop of the liver was used for histological examination, the rest of the liver was snap-frozen in liquid nitrogen and stored at $-80\,°$C until processed for RNA extraction. The time of the day for necropsies varied between the studies: for study 1, necropsy was performed in the morning; for study 2, necropsy was performed in the early afternoon.

1.2.2   RNA Isolation

Total RNA was isolated from a randomly chosen piece of the snap-frozen liver using the Qiagen RNeasy Midi Kit (Qiagen, Hilden, Germany) according to the manufacturer's protocol. Integrity of RNA was assessed by denaturing agarose gel electrophoresis, and concentration of RNA was determined spectrophotometrically.

1.2.3   Affymetrix GeneChip Experiments

GeneChip Experiments were performed according to the recommendations of the manufacturer (Affymetrix, Santa Clara, USA). Briefly, 5 μg of total RNA were used for the synthesis of double-stranded cDNA with the Superscript Choice System (Invitrogen Life

Technologies, Karlsruhe, Germany) in the presence of a T7-(dT)$_{24}$ DNA oligonucleotide primer. The cDNA was purified using the Affymetrix GeneChip Sample Cleanup Module (Qiagen) and then transcribed *in vitro* using the BioArray High Yield RNA Transcript Labeling Kit (Enzo, Farmingdale, USA) in the presence of biotinylated ribonucleotides. The labeled cRNA was purified using the Affymetrix GeneChip Sample Cleanup Module (Qiagen) and quantified by optical density measurements. Quality was checked on an Agilent Bioanalyzer (Agilent, Waldbronn, Germany).

A 20 µg amount of labeled cRNA was fragmented and subsequently 8 µg fragmented cRNA was hybridized for approximately 16 h at 45 °C to an RG-U34A array. The array was then washed, stained with streptavidin-R-phycoerythrin (SAPE, Invitrogen) and the signal amplified using a biotinylated goat anti-streptavidin antibody (Vector Laboratories, Burlingame, USA) followed by a final staining with SAPE. Arrays were stained using the GeneChip Fluidics Workstation 400 (Affymetrix). The array was then scanned twice using a confocal laser scanner (GeneArrayScanner 2500, Affymetrix) resulting in one scanned image (*.DAT files). The scanner was adjusted to the highest sensitivity, giving approximately 10 times higher intensity than the standard settings. These files were quantitated using the MAS software (Affymetrix) resulting in *.CEL files. The CEL-files were condensed using the Affymetrix MAS 5.0 algorithm (Affymetrix, 2002).

1.2.4   GeneChip Quality Assessment

Selected quality criteria were taken from the MAS 5.0 reports and are summarized in Table 1.1.

Noise (Raw Q) is a measure of the pixel-to-pixel variation of probe cells on a GeneChip array. The two main factors that contribute to Noise are electrical noise of the scanner and sample quality. Each scanner has a unique inherent electrical noise associated with its operation. Since a significant portion of Noise (Raw Q) is electrical noise, values among scanners will vary. Array data (especially those of replicates) acquired from the same scanner should ideally have comparable Noise values below 40 (adjusted to high-intensity settings).

Background is a measure of the fluorescent signal on the array due to nonspecific binding and autofluorescence from the array surface and the scanning wavelength. Background values range from 200 to 1000 for GeneChips scanned with the GeneChip® Scanner 2500. We have set the threshold at 1000.

A scaling factor is used to make arrays comparable by adjusting the average intensity value of each array to a common value (target intensity). For replicates and comparisons involving a relatively small number of changes, the scaling factors should be comparable among arrays (within threefold).

Probe sets are called present when their detection p-value is smaller than 0.04. The number of probe sets called 'Present' relative to the total number of probe sets on the array is displayed as a percentage. Percentage Present (%P) values depend on multiple factors, including cell/tissue type, biological or environmental stimuli, probe array type and overall quality of RNA. GeneChip experiments within the same study should have comparable %P values of at least 25 %. Extremely low %P values are a possible indication of poor sample quality.

Table 1.1 *Summary of selected quality parameters from the MAS 5 reports[a]*

Experiment ID Recommended acceptable (adjusted) Measured range	Noise range <30 12.7–26.6	Background <1000 422–919	SF ratio <3 ratio = 1.3	%P >25% 26.8–43.6	GAPDH 3'/5' <3 0.9–9.1
study1_01_control	13.1	460	0.4	40.2	1.3
study1_02_control	14.7	509	0.4	38.5	0.9
study1_03_control	14.7	504	0.4	36.4	1.2
study1_04_control	14.5	494	0.4	37.3	1.4
study1_09_treated_24h	13.4	444	0.4	39.4	1.1
study1_10_treated_24h	26.6	919	0.7	26.8	1.2
study1_11_treated_24h	14.9	518	0.3	42.5	1.0
study1_12_treated_24h	14	477	0.4	42.7	1.3
study1_13_treated_48h	24.7	819	0.4	36.5	1.6
study1_14_treated_48h	14	476	0.3	43.6	1.8
study1_15_treated_48h	23.6	826	0.3	36.7	1.1
study1_16_treated_48h	12.7	432	0.5	39.2	1.3
study1_17_treated_96h	13.8	458	0.4	41.5	1.2
study1_18_treated_96h	13.8	457	0.4	39.3	1.1
study1_19_treated_96h	23.8	914	0.4	36.4	1.3
study1_20_treated_96h	23.4	934	0.3	37.0	1.2
study2_01_control	15.8	473	0.3	39.7	1.4
study2_02_control	14.8	459	0.3	39.9	1.9
study2_04_control	15.4	470	0.3	40.4	1.3
study2_03_control	15.3	476	0.3	40.7	1.4
study2_09_treated_6h	14.8	454	0.3	40.7	1.6
study2_09_treated_6h	14	435	0.4	37.9	1.8
study2_11_treated_6h	15.9	457	0.4	37.7	1.6
study2_12_treated_6h	15.4	478	0.3	40.0	1.6
study2_13_treated_24h	13	422	0.5	38.4	1.4
study2_14_treated_24h	15.5	490	0.3	40.7	1.5
study2_15_treated_24h	15.4	484	0.3	41.7	1.4
study2_16_treated_24h	17.2	552	0.3	37.9	**9.1**

[a] Italic indicates values within the acceptable range but close to the recommended limits. Bold indicates values outside the acceptable range.

The GAPDH transcript is measured using several probe sets. These probe sets have different locations on the transcript (3', middle and 5') and are used to get an indication of the quality of the RNA (integrity) and the efficiency of cDNA synthesis and/or *in vitro* transcription. The signal values of the 3' probe sets for GAPDH are compared with the signal values of the corresponding 5' probe sets. The ratio of the 3' probe set to the 5' probe set should be <3.

1.2.5   Statistical Analysis of Microarray Data and Pathway Analysis

The aim of mechanistic expression profiling using microarrays is to identify genes whose expression is altered by treatment with the toxicant of interest. A variety of different statistical methods is currently applied for such analyses. Depending on the specific objective of the analysis, one can use either class discovery and class comparison.

The goal for class discovery is to group experiments or genes with similar properties using unsupervised methods, such as principal component analysis (PCA) and hierarchical clustering (Eisen, 1998; Claverie, 1999). These methods identify patterns in data and depict them in such a way as to highlight their similarities and differences.

For investigating gene regulation in different classes (class comparison), supervised analysis can be used to find genes which are differentially expressed between two or more groups within a study to identify genes which characterize a specific experimental condition. Here, analyses like the *t*-test or analysis of variance (ANOVA) can be applied.

In the following sections, some of the approaches that we found useful for analysing microarray data are outlined in more detail.

1.2.5.1   Class Discovery

Principal Component Analysis. PCA is a projection method. It is a classical means of dimensionality reduction and visualization of multivariate data and involves a mathematical procedure that transforms a number of possibly correlated variables into uncorrelated variables called principal components. The first few principal components should contain most of the information, giving a less dimensional picture of hopefully interesting structures in the data. The first principal component accounts for as much of the variability in the data as possible. Each succeeding component accounts for as much of the remaining variability as possible and is chosen orthogonal to the preceding components, leading to uncorrelated variables. Restriction to the first two or three principal components often is sufficient for capturing dominant structures in the data and uncovering important relationships between individual samples or groups of samples. Using the covariance metric for computation of principal components leads to a stronger influence of variables on a higher scale (e.g. as genes on higher expression levels with larger variance), while using the correlation matrix leads to a scaling of variables and subsequent equal weighting of all variables.

Hierarchical Clustering. A hierarchical clustering builds up a so-called dendrogram, which represents a hierarchy with individual elements at one end and a single cluster containing all elements at the other end. In principal, hierarchical clustering starts by computing a pairwise distance matrix between all experiments, which are each considered as a cluster with one element in the first step. The distance metric can be based on Euclidean distance; however, in microarray studies, distance metrics involving the correlation (e.g., 1-Pearson correlation) are more frequently used. The distance matrix is explored for the two closest clusters (experiments), which are then merged to a new cluster, which now consists of two elements. The distance matrix is updated to reflect the distance of the cluster formed from all other clusters. Then the procedure searches for the nearest pair of clusters to agglomerate, and so on. This procedure leads to the dendrogram, which depicts the agglomeration distances in a hierarchical tree. The definition of distance between two clusters can, for example, be based on the distance between the two cluster means (average linkage) or the distance between the two closest elements of the two clusters (single linkage). In addition to average and single linkage, several other measures are available.

1.2.5.2   Class Comparison

Class comparison methods are used when genes are sought which differentiate between known groups in the data. The classical example compares treated versus untreated animals or multiple dose groups.

Several tests can be used to find these differentially expressed genes depending on whether two or more groups are compared, whether samples are paired (e.g. blood taken from an animal before and after treatment) and which distributional assumptions can be made.

A *t*-test, for example, tests for the difference in means and assumes that the distribution of values is normal within each group and is identical for both groups. If more than two groups are to be compared, then an ANOVA is the suitable statistical procedure if a normal distribution with identical variance can be assumed in all groups.

Both statistical tests consider two groups to be significantly different if the difference in means is large relative to the observed variance. In the case of groups with very small variances, therefore, quite small differences between means can become statistically significant, though they might not be considered as relevant by the analyst. To incorporate scientific relevance into the microarray analysis, genes are quite often required not only to be statistically significant, but also to display a minimum fold-change between the two groups. In the fold-change analysis, only the difference of log-transformed expression values (which corresponds to relative change in expression) is considered. A commonly used approach is to require a twofold change between two groups (MAQC Consortium, 2006).

However, since the GeneChips which we used are in reality many tests in one, a correction for multiple testing is needed. One quite well-known correction method is the Bonferroni correction (Abdi, 2007), which controls the so-called family-wise error (FWE), i.e. the probability of finding at most one false positive among all tests performed. In contrast to the FWE, the false discovery rate (FDR) estimates the number of false positive genes in the gene list generated. Control of the FDR constitutes a far less conservative approach than control of the FWE and is frequently used in microarray data analyses. Benjamini and Hochberg (1995) described a procedure for the calculation of the FDR.

1.2.5.3   Pathway Analysis

There are several commercially available software packages that provide varying degrees of capability with regard to the categorization of genes and assignment into metabolic or signal transduction pathways. We use MetaCore (GeneGo, San Diego, USA), which is based on a database of manually curated high-quality protein–protein and protein–DNA interactions, transcriptional factors, signaling, metabolism and bioactive molecules of human and rodent (mouse and rat). It allocates genes or proteins to their respective cellular pathways (provided as maps), which are then further condensed as subordinated map folders. The maps are constantly updated. In the current version, over 500 maps are available. The analytical package includes tools for data visualization, mapping and exchange, multiple networking algorithms and *in silico* filters. For any given list of interesting genes, MetaCore rates the pathways by comparing the number of genes in the pathway with the number of genes that would be expected just by chance when considering the length of the submitted gene list and the number of genes within a pathway. Pathways which contain more genes from the list than expected are considered the most interesting.

1.3 Results

1.3.1 Animal Experiments

Treatment-related lesions in the liver were histologically detected at all time points of necropsy, but with differences in the number of animals affected and grade of severity (Table 1.2). Centrilobular necrosis developed progressively up to 24 h. At 6 h, two out of four animals showed minimal to slight signs of necrosis. At 24 h, nearly all animals were affected with a moderate degree of damage. In addition, animals of study 1 showed a moderate infiltration of inflammatory cells, which usually accompanies necrotic lesions at 24 h after single application of APAP. In contrast, inflammatory cell infiltration was not reported for the 24 h animals of study 2. The severity of necrobiotic lesions decreased from 24 h onwards, reaching nearly the status of control animals at 96 h. The increased mitotic index after 48 and 96 h and the significant lower incidence and severity of hepatic lesions at 96 h clearly indicates replacement of necrotic cells and reversibility of the lesions. Formation of apoptotic bodies derived from hepatocytes, usually observed before or with the appearance of necrosis, was present at 24, 48 and 96 h, indicating the effort of the cells to avoid necrotic cell damage by initiating programmed cell death.

It is obvious that in both studies there is a considerable variation in the incidence and severity of observed histological findings between the rats in one group and between the two groups exposed for 24 h. This variability between animals following APAP treatment is known (Wells and To, 1986).

1.3.2 GeneChip Quality Assessment

RG-U34A arrays have 16 probe pairs (probe set) available to interrogate each gene. The intensity data for each probe set has to be condensed into one value, it has to be background corrected and the data of each array has to be normalized in such a way that arrays can be compared. Several algorithms, such as RMA (Irizarry *et al.*, 2003a,b), dCHIP (Li and Wong, 2001a,b), GC-RMA (Wu and Irizarry, 2004) and PDNN (Zhang *et al.*, 2003), are

Table 1.2 *Compound-related histological findings*[a]

| | Control | | APAP[b] | | | | |
| | | | Study 2 | | Study 1 | | |
	Study 2	Study 1	6 h	24 h	24 h	48 h	96 h
No. of animals	4	4	4	4	4	4	4
Histological findings							
centrilobular necrosis	–	–	2 (1.5)	3 (2.6)	4 (3)	3 (2)	1 (1)
inflammatory cell infiltration	–	–	–	–	4 (2.8)	3 (2)	1 (1)
hepatocellular apoptotic bodies	–	–	–	2 (1.0)	4 (2.0)	2 (2.5)	1 (2)
mitotic index increase	–	–	–	–	–	2 (1.5)	1 (2)

[a]Grading: 1 = very slight, minimal; 2 = slight; 3 = moderate; 4 = marked.
[b]No. of animals affected (average grading).

available to perform these tasks. We have chosen the Affymetrix MAS 5.0 algorithm, which uses the values of each probe pair and assumes that intensity differences between two or more arrays are linearly related and, therefore, normalizes the arrays by scaling them using the average intensity of each array.

To asses the quality of the arrays, Affymetrix provides a series of guidelines using the quality measures which are produced in the MAS reports. However, it is not really indicated what the user should do when the quality parameters fall outside the threshold and a definition of an outlier chip is not provided. Table 1.1 summarizes the quality parameters obtained from the MAS 5.0 software and shows that the GAPDH $3'/5'$ ratio is too high (9.1) for one chip out of 28, which indicates a low integrity of the RNA used for this experiment. Figure 1.1a–e shows these quality parameters as a scatter plot, giving an indication on the uniformity of the data. The scatter plots clearly show that five arrays have quality parameters which are deviant from the majority, even though the quality parameters are within the predefined thresholds.

Microarrays that are suspicious with respect to the quality properties compared with all other chips were further assessed in a second quality analysis step, in which the overall structure of the expression data was visualized by means of multivariate techniques (Expressionist Pro, version 4.5, Genedata, Basel, Switzerland). The dendrogram of hierarchical clustering revealed that two samples cluster apart from the majority, these being sample 16 of study 2 with an extreme high GAPDH $3'/5'$ ratio and sample 10 from study 1 which has high Noise, background, the highest scaling factor and is the outlier concerning percentage presence call (Figure 1.1f). The other four deviants seem to cluster within the vast majority, showing that their experimental variability is lower than the biological variability.

It was decided to omit the two outlier samples, the 24 h sample (No. 10) from study 1 and the 24 h sample (No. 16) from study 2.

1.3.3   Statistical Data Analysis

A workflow summarizing the major analysis steps we have applied is shown in Figure 1.2. Data analysis was performed using Expressionist (Genedata). As recommended by Simon *et al.* (2003), expression data were transformed by taking logarithms before analysis.

The RG-U-34A array contains 9762 probe sets in total. There was a 'present call' on at least two of the 26 arrays in 4327 probe sets. All statistical analyses were based on these 4327 probe sets.

1.3.3.1   Principal Component Analysis

PCA was initially performed with the datasets from both studies together based on all 4327 probe sets (Figure 1.3a). The first three principal components accounted for approximately 40 % of the variance in the dataset, indicating that a large share of variability could be captured in components 1–3. The clear separation of samples points to strong substructures in the data, especially between the two studies.

When investigating only the control and 24 h samples from both studies based on all 4327 probe sets, the process effect due to the two different experiments (difference in animals studies and sample processing) becomes obvious. Therefore, we decided to perform a study-specific statistical analysis.

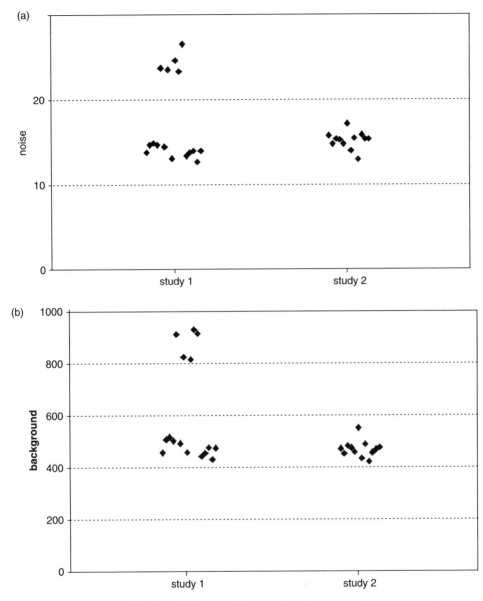

Figure 1.1 *Graphical display of the GeneChip quality parameters. (a) Noise (Raw Q) is a measure of the pixel-to-pixel variation of probe cells on a GeneChip array. Each dot represents one sample (=GeneChip) of the respective study. (b) Background is a measure of the fluorescent signal on the array due to non-specific binding and autofluorescence from the array surface and the scanning wavelength. Each dot represents one sample (=GeneChip) of the respective study.*

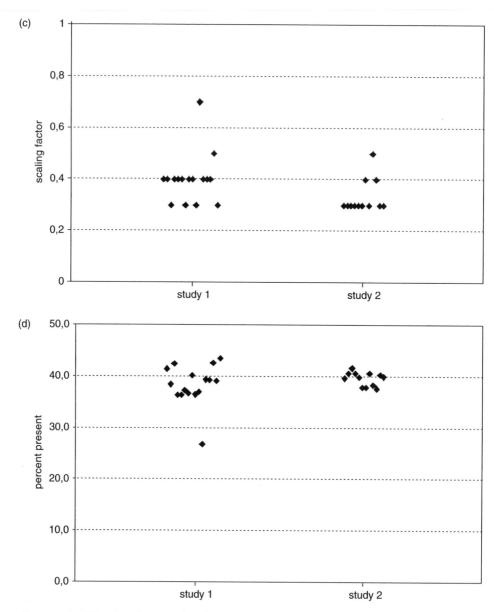

Figure 1.1 *(Continued) (c) Scaling factors are used to make arrays comparable by adjusting the average intensity value of each array to a common value (target intensity). Each dot represents one sample (=GeneChip) of the respective study. (d) Percentage presence is the number of probe sets called 'Present' relative to the total number of probe sets on the array. Each dot represents one sample (=GeneChip) of the respective study.*

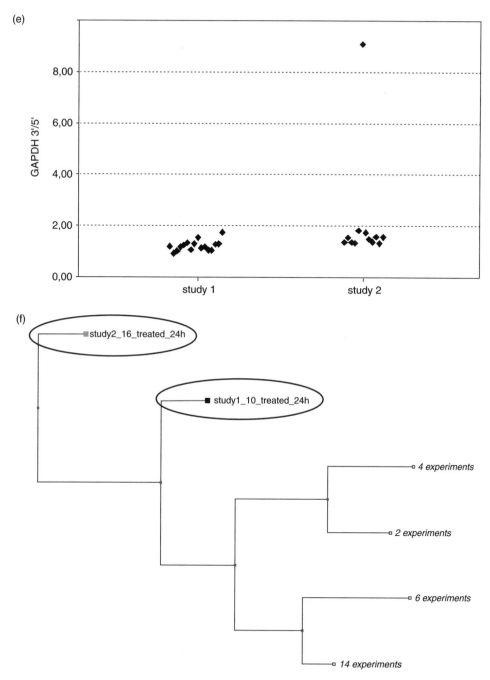

Figure 1.1 *(Continued) (e) GAPDH 3'/5' is a measure for the quality of the RNA (integrity) or the efficiency of cDNA synthesis and/or in vitro transcription. Each dot represents one sample (=GeneChip) of the respective study. (f) Hierarchical clustering of all samples from study 1 and study 2 (positive correlation with average linkage).*

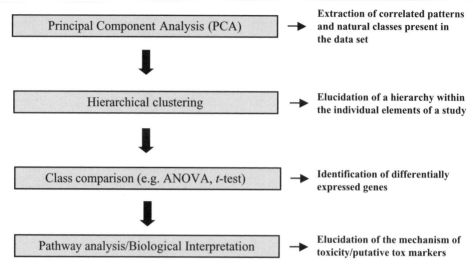

Figure 1.2   *Workflow of statistical data analysis.*

PCA of samples from study 1 shows a clear separation of controls and treated animals, where the strongest effects were seen for 24 h. Animals at 96 h after substance exposure are close to controls. In study 2, the strongest alterations in gene expression occurred after 6 h of substance exposure. In general, there was a high variability within the data for the 24 h time point.

1.3.3.2   Hierarchical Clustering

As an example, hierarchical clustering was performed with study 1 based on all 4327 probe sets (Figure 1.4). A similar picture as for the PCA can be observed, showing that the samples from the 24 h time point build a separate cluster. One sample from the 48 h time point clusters closest. All other samples form a separate big cluster, including controls, 48 and 96 h animals.

1.3.3.3   Class Comparison

Identification of significantly expressed genes was performed for each study individually. First, an ANOVA was performed and probe sets significantly expressed with a $p \leq 0.05$ were identified (study 1: 1055; FDR: 0.166, which means that approximately 17 % of the 1055 genes were detected falsely positive; study 2: 1222; FDR: 0.164). Those genes were then used for a two-group comparison by t-test ($p \leq 0.05$) and a twofold regulation change where the control group of each study was tested against the respective treatment groups. The number of differentially expressed genes for each study is summarized in Table 1.3 for each time point of necropsy. Changes in gene expression are most dramatic shortly after substance administration, which is especially obvious at 6 h of study 2. At the later time points the number of gene changes decreases.

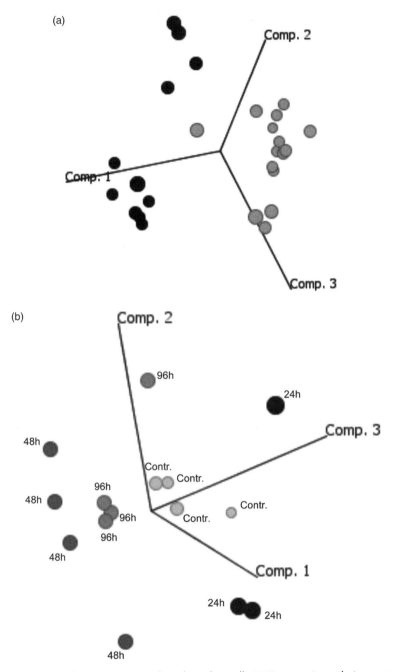

Figure 1.3 *Principal component analysis based on all 4327 genes (correlation matrix). (a) For all samples of study 1 (grey) and all samples of study 2 (black). PC1 captures 16.7 % of variation in the data, PC2 displays 12.6 % and PC3 contains 9.8 %. (b) For all samples of study 1 (Controls, 24 hours, 48 hours and 96 hours). (c) For all samples of study 2 (Control, 6 hours and 24 hours).*

(c)

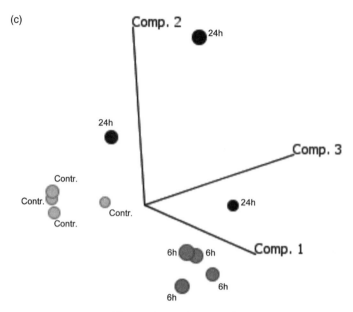

Figure 1.3   *(Continued)*

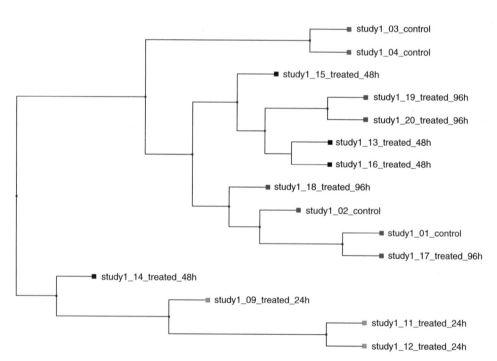

Figure 1.4   *Hierarchical clustering of all samples for study 1 (positive correlation with average linkage).*

Table 1.3   *Number of differentially expressed genes after statistical data analysis and affected pathways*

		No. of differentially expressed genes			
		6 h	24 h	48 h	96 h
Study 1	Upregulated		56	17	8
	Downregulated		29	13	4
	Affected pathways		30	22	10
Study 2	Upregulated	131	21		
	Downregulated	86	39		
	Affected pathways	96	31		

1.3.3.4   Pathway Analysis

The lists of differentially expressed genes were analyzed for the functions of their encoding gene products by MetaCore. This results in a listing of all pathways in which these genes play a role. In order to reduce complexity of the gene lists, a threshold of $p \leq 0.05$ was applied to identify most affected pathways. Only the significantly allocated genes were considered for more detailed evaluation (Table 1.3). Table 1.4 shows selected entries of such a pathway list. The number of differentially expressed genes correlates very well with the number of affected pathways.

1.3.4   Biological Interpretation

Patterns of gene expression clearly differ with time after single exposure to APAP, reflecting the differing phases of hepatocellular injury observed histologically at the individual time points of necropsy. Derived from the number of differentially expressed genes, the early time points of 6 and 24 h show the most dramatic changes in the liver, where the majority of animals developed centrilobular necrosis. Differentially expressed genes at these time points clearly reflect the postulated mode of action of APAP-induced liver toxicity, which is based on multiple cellular mechanisms (Figure 1.5). At low doses, APAP is primarily metabolized by sulfation and glucuronidation. To a lesser extent, oxidation with Cyp2E1 leads to the formation of the reactive metabolite *N*-acetyl-*p*-benzochinoneimine (NAPQI), which is detoxified by GSH conjugation. APAP induces hepatotoxicity when GSH reserves are exhausted. Then, NAPQI arylates a number of essential cellular proteins and induces massive oxidative stress and mitochondrial damage, which is enhanced by parallel depletion of glutathione. This leads to changes in energy metabolism, calcium homeostasis and induction of DNA damage. Since apoptosis is suppressed by the low levels of ATP and high levels of reactive oxygen species, necrotic cell death finally occurs (Mitchell *et al.*, 1973; Dahlin *et al.*, 1984).

Many genes involved in major energy biochemical pathways, like cholesterol biosynthesis (e.g. HMG-CoA synthase, 7-dehydrocholesterol reductase, squalene synthetase) and peroxisomal and mitochondrial fatty acid β-oxidation (e.g. 3-ketoacyl-CoA thiolase, carnitine-*O*-palmitoyltransferase, long-chain fatty acids ligases), are deregulated resulting in mitochondrial damage and depletion of energy. Damage to mitochondria is known to lead to the formation of reactive oxygen species (Mitchell *et al.*, 1973) which results in a

Table 1.4 *Selection of most affected pathways*

Map	Map folders	p-value[a]	Genes in pathway	Genes differentially regulated
Cholesterol biosynthesis	Metabolic maps/steroid metabolism	2.71×10^{-16}	21	Cytochrome P450 51A1 7-Dehydrocholesterol reductase 3-β-Hydroxysteroid-Delta(8),Delta(7)-isomerase Squalene synthetase Farnesyl pyrophosphate synthetase Hydroxymethylglutaryl-CoA synthase, cytoplasmic Isopentenyl-diphosphate Delta-isomerase 1 Lanosterol synthase Diphosphomevalonate decarboxylase Sterol-4-alpha-carboxylate 3-dehydrogenase, decarboxylating C-4 methylsterol oxidase Squalene monooxygenase
Glutathione metabolism	Metabolic maps/ vitamin and cofactor metabolism	3.32×10^{-6}	36	Aminopeptidase N Glutamate–cysteine ligase regulatory subunit Glutathione peroxidase 2 Glutathione reductase, mitochondrial precursor Glutathione synthetase Glutathione S-transferase A3 Glutathione S-transferase A5

Pathway	Function category	p-value	Count	Proteins
Mitochondrial unsaturated fatty acid β-oxidation	Metabolic maps/lipid metabolism	5.05×10^{-6}	15	Long-chain-fatty-acid–CoA ligase 1 Long-chain-fatty-acid–CoA ligase 5 Carnitine O-palmitoyltransferase I, liver isoform Carnitine O-palmitoyltransferase 2, mitochondrial precursor 3,2-trans-Enoyl-CoA isomerase, mitochondrial precursor
Peroxisomal branched chain fatty acid oxidation	Metabolic maps/ lipid metabolism	2.42×10^{-5}	20	Acyl-coenzyme A oxidase 2, peroxisomal Carnitine O-palmitoyltransferase I, liver isoform Carnitine O-palmitoyltransferase 2, mitochondrial precursor Peroxisomal carnitine O-octanoyltransferase 2-Hydroxyacyl-CoA lyase 1
IGF-RI signaling	Cell signaling/ growth and differentiation/growth and differentiation (common pathways) Function groups/ growth factors	2.00×10^{-3}	72	RAC-β serine/threonine-protein kinase Insulin-like growth factor IA precursor Insulin-like growth factor-binding protein 1 precursor Insulin-like growth factor-binding protein 2 precursor Myc proto-oncogene protein Phosphatidylinositol 3-kinase regulatory subunit alpha

(Continued)

Table 1.4 *(Continued)*

Map	Map folders	p-value[a]	Genes in pathway	Genes differentially regulated
Role of AP-1 in regulation of cellular metabolism	Function groups/transcription factors	7.67×10^{-3}	43	Glutamate–cysteine ligase regulatory subunit Hemoglobin subunit β Mitogen-activated protein kinase 9 NAD(P)H dehydrogenase [quinone] 1
Start of the mitosis	Cell signaling/ cell cycle control	1.13×10^{-2}	48	RAC-β serine/threonine-protein kinase G2/mitotic-specific cyclin-B1 Myc proto-oncogene protein Nucleolin
FAS signaling cascades	Cell signaling/cell death/apoptosis	4.61×10^{-2}	44	Heat-shock protein β-1 Mitogen-activated protein kinase 9 Poly [ADP-ribose] polymerase 1
Cytoskeleton remodeling	Cell signaling/ cell adhesion	1.04×10^{-1}	176	Fibronectin precursor Myc proto-oncogene protein Myosin-9 Nucleolin Phosphatidylinositol 3-kinase regulatory subunit α Vascular endothelial growth factor A precursor

[a]The *p*-value essentially represents the probability of the particular mapping arising by chance.

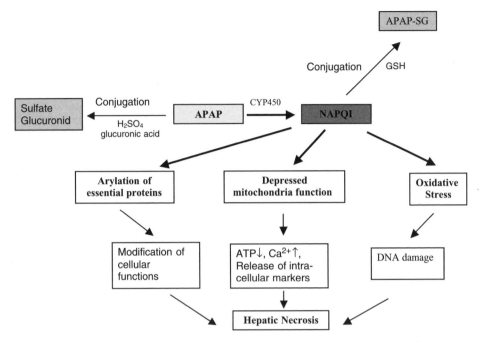

Figure 1.5 *Multiple pathways of liver cell injury induced by acetaminophen.*

specific gene signature (Heinloth *et al.*, 2004). With induction of cAMP-responsive element modulator, phospholipase D1 and glutathione peroxidase 2, we could detect regulation of several genes associated with oxidative stress supporting the literature data.

It is well known that the metabolic activation of APAP leading to GSH depletion is an important factor in APAP-induced hepatotoxicity. Several genes belonging to the glutathione metabolism pathway were found differentially expressed under APAP treatment in our studies, including aminopeptidase N, glutamate cysteine ligase regulatory subunit, glutathione peroxidase 2, glutathione reductase, glutathione synthetase and glutathione-*S*-transferases A3 and A5 (Table 1.4). Glutamate cysteine ligase, also known as gamma-glutamylcysteine synthetase, is the first and rate-limiting enzyme of glutathione synthesis. It is already upregulated at 6 h after dosing (1.5-fold, $p < 0.005$) and expression increases even more at 24 h, but is back to the status of control animals at 48 and 96 h (Figure 1.6). Also highly upregulated are the genes for the glutathione-*S*-transferase A3 and A5. GST A3 is upregulated up to 37-fold at 24 h after dosing. Both transferases play a role in the conjugation of reduced glutathione to various exogenous and endogenous metabolites. Increased expression of these transferases has been shown to protect mice treated with a toxic dose of APAP from hepatotoxicity (Dai *et al.*, 2005); upregulation can, therefore, be regarded as a protection mechanism of the cell against toxic damage.

In addition, changes are related to necrobiotic cell damage itself (e.g. mitogen-activated protein kinases, myosin) or reflect an attempt of the tissue for regeneration or compensation (e.g. upregulation of cell cycle genes, metabolic enzymes). At the later time points of 48 and 96 h, the regenerative processes are especially prominent, observable as upregulation

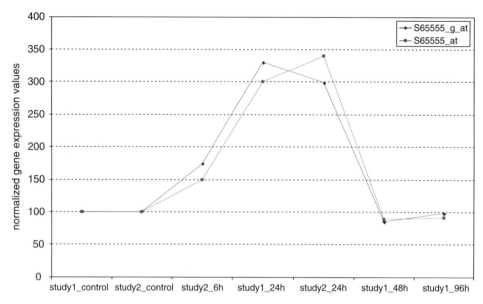

Figure 1.6   *Expression profile of glutamate cysteine ligase, modifier subunit. The gene expression values were normalized to the controls.*

of genes like early growth response 1 or insulin-like growth factor binding protein. At these time points the liver gets back to its normal status, which is also obvious after histopathological evaluation.

Individual genes between 6 h and 24 h in study 2 and the 24 h time points from study 1 and study 2 overlap to some extent but are not exactly the same. But, when comparing the affected pathways, the underlying mechanism of APAP toxicity is obvious.

1.4   Discussion

1.4.1   Design of Animal Experiments and Tissue Sampling

In order to interpret toxicogenomics data correctly, a properly designed animal experiment is a prerequisite. Since variability in biological samples, as seen in our studies, is well known, an appropriate number of animals (biological replicates) has to be chosen to allow an assessment of variation in responses of individuals and account for loss of animals during treatment. A group size of >5 is often suggested. In exceptional cases one has to cope with smaller numbers because of limited availability of tissue material, especially when using other toxicological-relevant species, like dog or monkey.

In most cases it is desirable to incorporate early time points of necropsy in order to see the immediate responses on the transcriptional level without manifest changes in histology or clinical parameters. This often requires prior knowledge, since these time points are not fixed and depend on the substance and dosing regimen. Times of 3, 6 and 24 h are often used for gene expression experiments, which is also on account of practicability. Time points where the toxicological finding (histological lesions, changes in biochemical parameters) becomes evident are important to account for the so-called phenotypic anchoring, the correlation of changes in gene expression to the observed lesion. In addition, doses should be chosen carefully. For single treatment as described here for APAP, a dose that does not cause toxicological observations at early time points but does at longer exposure is often suggested, as the early time point is especially not easy to determine. For repeated administration, at least one dose covering the pharmacological concentration or the low/no observed adverse effect level and one dose covering clear signs of toxicity are recommended for mechanistic investigations based on gene expression changes.

A number of genes show a circadian pattern in their expression. Therefore, timing of necropsies at the same time of the day and/or incorporation of time-matched controls is important (Desai *et al.*, 2004). In our experiment, the animals were sacrificed at different times of the day; therefore, differences in gene expression between the 24 h time points of both studies may not exclusively be related to substance treatment, but may also reflect circadian variability. For instance, genes involved in sugar (glucokinase, pyruvate dehydrogenase) and lipid (3 HMGCoA synthetase) metabolism, as well as several others, give a strong indication for that conclusion (Panda *et al.*, 2002).

Gene expression also varies depending on the part of the organ/tissue chosen for analysis. This is especially known for the liver, which exhibits a zone-specific expression pattern of a variety of genes, either increasing or decreasing from periportal to periveneous regions across the hepatic plate (Teutsch *et al.*, 1999). Therefore, it is advisable to sample the tissue in a way that covers most of the expected changes and use the same part within and between liver lobes to minimize the degree of variation between samples (Foley *et al.*, 2006). Since the liver tissue was randomly chosen in our experiment from animal to animal and study to study, this may be an additional point of variability.

1.4.2   Design and Performance of Microarray Experiments

Several issues in experimental design of microarray experiments should be considered to avoid generation of misleading effects causing wrong biological interpretation (Yang and Speed, 2002, Han *et al.*, 2004). Owing to the high number of variables under consideration and the low biological sample sizes, microarray studies are nearly always underpowered. This means that only large effects in the data (strong differences between the groups) can be observed. Whenever smaller effects are sought, it should be considered to reduce the number of groups in favor of increasing the number of animals per group. Dobbin and Simon (2005) give an example for sample size estimation based on the suspected effect size.

The introduction of process effects cannot always be avoided (e.g. additional groups have to be investigated). In those cases, great care should be taken not to introduce a process that has a so-called confounding effect, i.e. an effect that cannot be distinguished from the effects under investigation. One measure to uncover possible process effects is the introduction of identical groups in the two runs, as demonstrated here with the control and 24 h groups or the implementation of reference samples (e.g. Universal Reference RNA; commercially available for different species) implemented in every experiment. By this measure, possible process effects can be investigated in a first step and, if present, may be considered in the subsequent analysis (e.g. by performing separate analyses as demonstrated here) and in biological interpretation.

General guidelines for reporting of microarray results were developed by the Microarray Gene Expression Data Society (MGED: http://www.mged.org/). These Minimal Information About a Microarray Experiment (MIAME; Brazma *et al.*, 2001) include descriptions of experimental design (e.g. number of replicates, nature of biological variables) and experimental procedures (e.g. sample type, extraction, hybridization).

1.4.3   Statistical Data Evaluation and Biological Interpretation

The objective of transcriptional profiling using microarrays is to identify genes whose expression is altered by treatment with the compound under study. The task of data analysis is to organize the enormous amount of data into a format which covers the relevant changes and leads to a meaningful biological interpretation. The most obvious candidate genes are those whose expression exhibits the greatest increase or decrease between treated samples and controls. However, since changes in expression levels of smaller magnitude may be equally important for the underlying mechanism, and reliability of fold changes is influenced by other parameters (e.g. spot intensity), a straightforward approach of fold-change ranking plus statistical analysis based on a non-stringent p-cutoff from classical group comparison analysis has been proven to be successful in identifying reproducible gene lists (MAQC Consortium, 2006).

The computation of the false discovery rate for a used p-value cutoff provides a means to judge the presence of real effects in the data. High FDRs usually indicate that there are no strong effects in the data, whereas small FDRs point towards strong effects.

Visualization methods, like PCA plots and dendrograms from hierarchical clustering, can be used to get a first impression of the data and to uncover strong effects in the data, if present.

Regardless of the statistical methods applied, the identification of differentially expressed genes is only the first step of the biological data interpretation. With the availability of pathway analysis tools, the time to obtaining first biological insights has decreased. In addition, these tools offer a more unbiased categorization of genes in assigning them to their respective pathways and functions, including an estimation of significance of the pathways. In the final step of the analysis the investigator must fit all the information together into a biologically meaningful result, which most of the time involves working at the individual gene level. Several online resources for biological data (e.g. Kyoto Encyclopedia of Genes and Genomes (KEGG), TOXNET, GeneCards) and information for toxicogenomics studies, e.g. Chemical Effects in Biological Systems (CEBS), Toxicogenomics Research Consortium (TRC) can be referred to.

In conclusion, the experimental and design issues of microarray gene expressions experiments addressed in this chapter, including the performance of animal studies, tissue sampling, labeling and hybridization, quality controls and data analysis, show that each step of the procedure is important to get valid gene expression data. Taking all these thoughts into consideration, toxicogenomics can contribute to a greater understanding of the mechanisms of toxicity and can, along with the established evaluations, be a valuable tool in safety assessment.

References

Abdi H (2007). Bonferroni and Sidak corrections for multiple comparisons. In *Encyclopedia of Measurement and Statistics*, NJ Salkind (ed.). Sage: Thousand Oaks, CA.

Affymetrix (2002) *Statistical Algorithm Description Document* (after the Affymetrix analysis). Affymetrix Inc.

Benjamini Y, Hochberg Y (1995) Controlling the false discovery rate: a practical and powerful approach to multiple testing, *J. R. Statist. Soc. B* **57**, 289–300.

Boess F, Kamber M, Romer S, Gasser R, Mueller D, Albertinie S, Suter L (2003) Gene expression in two hepatic cell lines, cultured primary hepatocytes, and liver slices compared to the *in vivo* liver gene expression in rats: possible implications for toxicogenomics use of *in vitro* systems. *Toxicol. Sci.* **73**, 386–402.

Boverhof DR, Zacharewki TR (2006) Toxicogenomics in risk assessment: applications and needs. *Toxicol. Sci.* **89**, 352–360.

Brazma A, Hingamp P, Quackenbush J, Sherlock G, Spellman P, Stoeckert C, Aach J, Ansorge W, Ball CA, Causton HC, Gaasterland T, Glenisson P, Holstege FC, Kim JF, Markowitz V, Matese JC, Parkinson H, Robinson A, Sarkans U, Schulze-Kremer S, Stewart J, Taylor R, Vilo J, Vingron M (2001) Minimum information about a microarray experiment (MIAME) – toward standards for microarray data. *Nat. Gen.* **29**, 365–371.

Claverie JM (1999) Computational methods for identification of differential and coordinated gene expression. *Hum. Mol. Genet.* **8**, 1821–1832.

Dahlin DC, Miwa GT, Lu AY, Nelson SD (1984) *N*-Acetyl-*p*-benzoquinone imine: a cytochrome P-450-mediated oxidation product of acetaminophen. *Proc. Natl. Acad. Sci. U S A* **81**, 1327–1331.

Dai G, Chou N, He L, Gyamfi MA, Mendy AJ, Slitt AL, Klaassen CD, Wan YJ (2005) Retinoid X receptor alpha regulates the expression of glutathione *S*-transferase genes and modulates acetaminophen–glutathione conjugation in mouse liver. *Mol. Pharmacol.* **68**, 1590–1596.

Desai VG, Moland CL, Branham WS, Delongchamp RR, Fang H, Duffy PH, Peterson CA, Beggs ML, Fuscoe JC (2004) Changes in expression level of genes as a function of time of day in the liver of rats. *Mutat. Res.* **549**, 115–129.

Dobbin K, Simon R (2005) Sample size determination in microarray experiments for class comparison and prognostic classification. *Biostatistics* **6**, 27–38.

Eisen M, Spellmann PT, Botstein PO, Brown PO (1998) Cluster analysis and display of genome-wide expression patterns. *Proc. Natl. Acad. Sci. U S A* **95**, 14863–14867.

Foley JF, Collins JB, Umbach DM, Grissom S, Boorman GA, Heinloth AN (2006) Optimal sampling of rat liver tissue for toxicogenomic studies. *Toxicol. Pathol.* **34**, 795–801.

Foster WR, Chen SJ, He A, Truong A, Bhaskaran V, Nelson DM, Dambach DM, Lehman-McKeeman LD, Car BD (2007) A retrospective analysis of toxicogenomics in the safety assessment of drug candidates. *Toxicol Pathol.* **35**, 621–635.

Gant TW (2003) Application of toxicogenomics in drug development. *Drug News Perspect*, **16**, 217–221.

Han ES, Wu Y, McCarter R, Nelson JF, Richardson A, Hilsenbeck SG (2004) Reproducibility, sources of variability, pooling, and sample size: important considerations for the design of high-density oligonucleotides array experiments. *J. Gerontol. A Biol. Sci. Med. Sci.* **59**, 306–315.

Heinloth AN, Irwin RD, Boorman GA, Nettesheim P, Fannin RD, Sieber SO, Snell ML, Tucker CJ, Li L, Travlos GS. *et al.* (2004) Gene expression profiling of rat livers reveals indicators of potential adverse effects. *Toxicol. Sci.* **80**, 193–202.

Irizarry RA, Bolstad BM, Collin F, Cope LM, Hobbs B, Speed TP (2003a) Summaries of Affymetrix GeneChip probe level data. *Nucl. Acids Res.* **31**, e15.

Irizarry RA, Hobbs B, Collins F, Beazer-Barclay YD, Amtonellis KJ, Scherf U, Spees TP (2003b) Exploration, normalization, and summaries of high intensity oligonucleotide array probe level data. *Biostatistics* **4**, 249–264.

Li C, Wong WH (2001a) Model-based analysis of oligonucleotide arrays: expression index computation and outlier detection. *Proc. Natl. Acad. Sci. U S A* **98**, 31–36.

Li C, Wong WH (2001b) Model-based analysis of oligonucleotide arrays: model validation, design issues and standard error application. *Genome Biol.* **2**, 0032.1–0032.11.

Lord PG (2004) Progress in applying genomics in drug development. *Toxicol Lett.* **149**, 371–375.

MAQC Consortium (2006) The MicroArraz Quality (MAQC) project shows inter and intraplatform reproducibilitz of gene expression measurements. *Nat. Biotechnol.* **9**, 1151–1161.

Minami K, Saito T, Narahara M, Tomita H, Kato H, Sugiyama H, Katoh M, Nakajima M, Yokoi T (2005) Relationship between hepatic gene expression profiles and hepatotoxicity in five typical hepatotoxicant-administered rats. *Toxicol. Sci.* **87**, 296–305.

Mitchell JR, Jollow DJ, Potter WZ, Davis DC, Gillette JR, Brodie BB (1973) Acetaminophen-induced hepatic necrosis. I. Role of drug metabolism. *J. Pharmacol. Exp. Ther.* **187**, 185–194.

Panda S, Antoch MP, Miller BH, Su AI, Schook AB, Straume M, Schultz PG, Kay SA, Takahashi JS, Hogenesch JB (2002) Coordinated transcription of key pathways in the mouse by the circadian clock. *Cell* **109**, 307–320.

Powell CL, Kosyk O, Ross PK, Schoonhoven R, Boysen G, Swenberg JA, Heinloth AN, Boorman GA, Cunningham ML, Paules RS, Rusyn I (2006) Phenotypic anchoring of acetaminophen-induced oxidative stress with gene expression profiles in rat liver. *Toxicol. Sci.* **93**, 213–222.

Simon RM, Korn EL, McShane LM, Radmacher MD, Wright GW, Zhao Y (2003) *Design and Analysis of DNA Mircorarray Investigations.* Springer: New York.

Smith GS, Nadig DE, Kokosha ER, Solomon H, Tiniakos DG, Miller TA (1998) Role of neutrophils in hepatotoxicity induced by oral acetaminophen administration in rats. *J. Surgical Res.* **80**, 252–258.

Teutsch HF, Scheuerfeld D, Groezinger E (1999) Three-dimensional reconstruction of parenchymal units in the liver of the rat. *Hepatology* **29**, 494–505.

Wells PG, To ECA (1986) Murine acetaminophen hepatotoxicity: temporal interanimal variability in plasma glutamic–pyruvic transaminase profiles and relation to *in vivo* chemical covalent binding. *Fundam. Appl. Toxicol.* **7**; 17–25.

Wu Z, Irizarry RA (2004) Preprocessing of oligonucleotide array data. *Nat. Biotechnol.* **22**, 656–658.

Yang YH, Speed T (2002) Design issues for cDNA microarray experiments. *Nature Rev.* **3**, 579–588.

Zhang L, Miles MF, Aldape KD (2003) A model of molecular interactions on short oligonucleotide microarrays. *Nat. Biotechnol.* **21**, 818–821.

2

A New Approach to Analysis and Interpretation of Toxicogenomic Gene Expression Data and its Importance in Examining Biological Responses to Low, Environmentally Relevant Doses of Toxicants

Julie A. Gosse, Thomas H. Hampton, Jennifer C. Davey and Joshua W. Hamilton

2.1 Introduction

The use of microarrays to examine global changes in gene expression has evolved as a potentially powerful tool in toxicology. However, the basic nature of the assay and the standard analyses of data from such assays present some challenges as well. Currently, the most widespread analysis approaches have actually limited, in our view, the power of this tool for toxicology and for other biological inquiries, particularly with respect to low dose or similar experiments examining subtle changes in expression. We have developed an alternative approach to microarray analysis that allows one to obtain new insights into low-dose biological responses to toxicants, and which reveals robust biological patterns that were previously hidden from discovery by more traditional statistical approaches. The goal of this chapter is to present this approach in detail and discuss how it differs from other current approaches.

The first challenge of microarray data analysis stems from the semi-quantitative nature of the individual measurements, which are both less sensitive and less specific than quantitative

Toxicogenomics: A Powerful Tool for Toxicity Assessment   Edited by S. C. Sahu
© 2008 John Wiley & Sons, Ltd

real-time RT-PCR and similar approaches. The power of microarrays lies in the ability to analyze, simultaneously and in the same biological sample, hundreds, thousands or even tens of thousands of gene transcripts, virtually the entire expressed genome or 'transcriptome' of a system. However, this presents a second challenge, which is the multiple comparisons problem of statistics, i.e. the more things one measures, the greater likelihood that many of the statistically significant observations will be false positives (i.e. a type 1 error) that can occur by random chance. This 'false discovery rate' (FDR) has been a major concern and focus of statistical analyses of microarrays and, indeed, using techniques such as t-tests or analysis of variance (ANOVA) one would expect a large number of false positives on a gene list generated by these approaches. Thus, the statistical criteria for microarray analyses are generally very restrictive to begin with and often employ additional constraints, such as requiring that a statistically significant gene must also be altered by twofold or greater or have a certain mean-to-standard deviation ratio, to make the final gene list.

A third problem that is related to the comprehensive nature of microarrays is summarized by the old adage, 'be careful what you ask for, you might get it,' i.e. by probing the entire expressed genome one can still generate a list with hundreds of genes that appear to be significantly altered by a given treatment even after such rigorous statistics. A major challenge to genomics has been that the bioinformatics, particularly systems biology, has lagged considerably behind the technology that generates the microarray data. As a result, one was previously left with a long list of genes that literally had to be investigated one gene at a time, in order to begin to make sense of the biology underlying the response – or as some investigators in the field say, a week to do the experiment and a year to understand it. This combination of the statistical concern over false positives, and the practical concern over long, daunting gene lists, has, in our view, led to the development of an analytical paradigm that encourages more restrictive statistical and filtering criteria with the goal of generating the shortest gene list possible.

Although systems biology has advanced considerably over the past 2–3 years and now allows a more comprehensive analysis of genes at the pathways level, the approach outlined above remains the standard approach to analysis of microarrays. Our approach proposes that one should do the exact opposite, i.e. one should instead generate the longest, most catholic initial gene list possible and then use systems biology and an iterative exploratory data analysis approach to investigate pathways of interest rather than individual genes. This is best accomplished by using nonparametric and rank-based rather than conventional parametric statistical techniques – which we will argue is a superior approach for both statistical and biological reasons – and by using no restrictions or cutoffs or other data manipulation aside from basic background correction and normalization. We also propose using a combination of data graphing, visualization tools, pathway analysis programs, and more traditional statistical approaches both to identify candidate biological pathways from the initial gene list and then to reinvestigate the original data set for further gene observations based on the identified pathways. Ultimately, it is important to realize that a genomics experiment is best used as a hypothesis generator, not the end result in and of itself. Thus, the information obtained from a microarray experiment should be, at a minimum, confirmed by a complementary technique such as quantitative real-time RT-PCR, but also by other endpoints, such as alterations in protein expression (e.g. by western blot or proteomics approaches) or changes in cell biological or physiological responses that are linked to the pathways proposed by the microarray results.

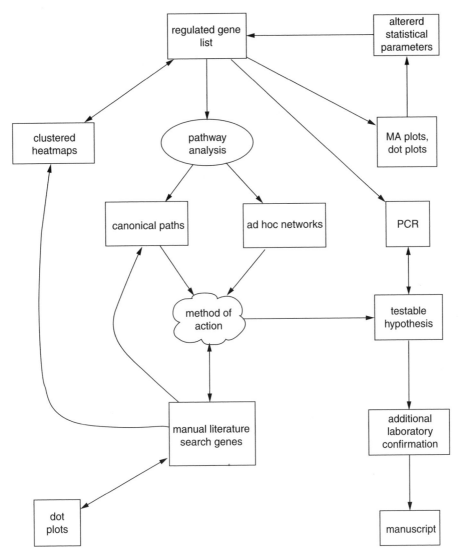

Figure 2.1 *An example of an iterative array analysis process (following data quality control and normalization). Information from one step helps to frame the question asked in the next, but that answer may also suggest new questions.*

Using the approaches we describe here (diagrammed in Figure 2.1), for example, we have discovered specific biological responses to arsenic and other toxic metals at extremely low doses that would otherwise have been undetectable using traditional analytical methods, and we have confirmed these results using various complementary experiments. In our hands, toxicogenomics has proven to be the most sensitive assay to date for investigating low-dose biological responses, and we believe that it will be an increasingly important tool in the toxicology arsenal. This new analytical approach will also have great utility in pharmacology (e.g. looking at low-dose drug effects), molecular ecology (e.g. looking at the effects of predation, dietary influences, and other 'low-dose' stressors) and many other biological disciplines where one may observe subtle responses that might otherwise be undetectable by traditional techniques. A fortunate circumstance of microarray experiments is that the initial raw data remain intact; thus, assuming that the basic procedures of proper experimental design and data collection were followed, previous experiments can be reanalyzed using newer analytical approaches such as this to glean additional information. Such reanalysis should yield important new insights in toxicology that will also allow more comprehensive comparison of microarray results across laboratories, model systems, experimental conditions and toxicants.

2.2   Experimental Design

Proper analysis and meaningful interpretation of microarray data, as in other areas of scientific inquiry, is crucially dependent upon sound experimental design (Ness, 2006). Almost any factor, such as laboratory diet, time of day, stress, degree of confluence of cells, or a minute contamination, can manifest itself as a change in gene expression, confounding a treatment effect unless the design controls or randomizes it. A critical factor in microarray experimental design is the number of biological replicates. Biological replicates are far more important than technical replicates (i.e. hybridizing aliquots of the same sample to multiple microarrays). There is surprisingly little technical variation among microarrays with the use of new-generation commercial microarrays and when hybridizations are performed by experienced staff (e.g. in a genomics core facility). Conversely, there is substantial biological variability in gene expression among individuals, even in highly inbred strains of animals and among littermates (discussed in Hamilton (2006)). Some of this relates to natural variation among individuals and some is attributable to the non-parametric nature of the regulation of many genes (see Hamilton (2006)). At least five or six biological replicates should be performed per treatment group in in vivo experiments with laboratory animals, and in human genomics experiments it has been recommended that 10–20 individuals is the minimum number required to see accurate patterns of gene expression changes. Cell culture experiments should have four or five replicate dishes or plates. There is an interesting debate as to whether individual cell culture plates represent true biological replication, since in most cases these replicates have been split out of the same parental plate and, therefore, are clonal. Nonetheless, biological replication in cell culture is more important than technical replication since, even in this highly controlled environment, there will be natural variation among replicate plates.

Treatment time and dose selection should be carefully considered. Given the transient nature of gene expression at the steady-state mRNA level, whatever dose and time chosen are merely a snapshot. But given the time and expense of microarray analysis, it is not normally possible to do extensive dose–response and time-course studies at the level of whole genome arrays, so key experimental conditions should be chosen, with subsequent follow-up using more traditional approaches, such as real-time RT-PCR, to do more extensive analysis of individual genes. As with any biological or toxicological experiment, it is very important to report all the pertinent technical details of the experiment. Such parameters as type of laboratory chow (Kozul *et al.*, 2007), animals per cage, etc. could turn out to have a significant influence on patterns of gene expression.

2.3 Microarray Platforms

A microarray platform describes a specifically designed array with a specific set and arrangement of probes (Ness, 2006; Wang and Cheng, 2006). We will focus on gene expression microarrays which examine the transcriptome or expressed mRNAs of the cell, but there are also array platforms for microRNA expression, comparative genomic hybridizations, DNA methylation analysis and other formats. The basic principle underlying gene expression microarray technology is common to most platforms, i.e. the hybridization between nucleic acids, one of which is immobilized on a solid chip or a glass slide.

Several features of current microarray platforms are important for understanding the analysis of microarray results. There are four general sources for microarrays:

1. pre-made, commercially available microarrays (e.g. from Affymetrix, Agilent);
2. custom-made, commercially available microarrays;
3. in-house spotting of commercially available cDNA or oligonucleotide collections;
4. in-house spotting of an in-house-prepared PCR-amplified clone collection.

There are two basic types of spotted nucleotides that are used for differential gene expression analyses: oligonucleotide-based and cDNA microarrays. Oligonucleotide microarrays (e.g. Affymetrix GeneChips, Agilent) are high-density arrays of short oligonucleotides (~20–80 bases each) that are synthesized in predefined positions onto quartz supports. Some custom glass slide arrays are also made with oligonucleotides. The oligonucleotides are computer designed in order to reduce cross-hybridization among spots. Oligonucleotide arrays require fewer production steps than cDNA arrays and are generally automated, leading to greater ease of production and fewer sequence errors. cDNA microarrays consist of PCR-amplified DNA sequences (200–2000 bases each), which correspond to unique expressed gene sequences, typically spotted onto the surface of treated glass slides using high-speed robotic printers. cDNAs can be derived from sequenced genes of known function or from collections of partially sequenced cDNA derived from expressed sequence tags (ESTs) corresponding to transcribed mRNAs. High-quality gene-specific probes for each gene of interest are important for any type of microarray.

Microarrays can be either one color or two colors. With one-color microarrays, each sample is hybridized to a separate, replicated microarray slide, the hybridization signal is scanned on a reader as a single color output and the results from the individual hybridizations are compared computationally following statistical normalization of signals from slide to slide. For two-color microarrays, samples from one treated and one control group are hybridized in competition with each other on the same chip. The sample from one group is labeled with one of the two colors (e.g. the Cy5 fluorophore) and the sample from the other group is labeled with the other color (e.g. Cy3 fluorophore). The labeled samples are then directly mixed together and are competitively hybridized to the microarray chip, and the two colors are analyzed and recorded on the reader. Theoretically, this at least partially solves the normalization problem, since the ratio of signals from the two different RNA samples are being compared on the same platform. However, there are some important normalization protocols that are required in two-color arrays, which both normalize the two signals on each slide and provide a means of normalizing among biological replicates on different slides. One important control is to do a 'dye-flip,' which is a technical replicate of each sample pair in which the sample that was first labeled with one color is now labeled with the other and vice versa. Any differences between the two dye-flip experiments with the exact same samples are due to bias in the labeling reaction or in the fluorescence yield on the reader.

Reproducibility across microarray platforms has been a concern. However, several recent studies have shown good agreement among various microarray platforms (Barnes *et al.*, 2005; Borovecki *et al.*, 2005; Irizarry *et al.*, 2005; Larkin *et al.*, 2005). Reproducibility appears to be improving because of advances in microarray technology, accumulation of experimental experience in core laboratories which perform the hybridizations and increased acceptance of the need for and use of multiple biological replicates. In one recent study, it was found that biological treatment had a much larger effect on measured gene expression effects than did platform for more than 90 % of genes (and this effect was validated by reverse transcription followed by quantitative real-time RT-PCR) (Larkin *et al.*, 2005). The Health and Environmental Science Institute (HESI) has focused on the sources of variability between microarray data sets from toxicogenomics experiments and has found that alterations in gene expression patterns, when placed in the context of the biological pathways in which they are involved, are robust across different platforms and laboratories (Pennie *et al.*, 2004). Similarly, the Toxicogenomics Research Consortium (http://www.niehs.nih.gov/research/supported/centers/trc/) has recently found that identification of differentially expressed genes is robust across microarray platforms but that variability arises from the technique of different workers in different laboratories (Mattes *et al.*, 2004), showing that it is human experimental technique rather than the microarray platform itself that is crucial to reproducibility. In fact, another manuscript reports a finding that there are relatively large differences in data from different laboratories utilizing the same microarray platform, but data from the best-performing laboratories agree well (Irizarry *et al.*, 2005). This finding suggests that the use of core facilities with experienced staff may be the most reliable way to perform the hybridizations, rather than having each research laboratory carry out its own microarray hybridizations. A recent review discusses many of the caveats in the use and standardization of different microarray processing and platforms (Verducci *et al.*, 2006).

Another consideration in choosing a microarray platform is the selection of genes represented on the array. For toxicogenomic studies, using whole-genome microarrays (which have only been available for a few years) will provide a complete and much richer dataset than that obtained with specialized microarrays. For example, there are several commercially available 'toxicology focused' arrays, which measure expression changes of genes that are known or suspected to be involved in toxicological responses in biology. However, many biological processes in addition to classical 'toxicology' responses – such as acute stress responses, DNA repair, xenobiotic metabolism, apoptosis and transport, among others – are now known to be affected by toxicants. Thus, the use of specialized 'toxicology' arrays may, in our view, potentially cause one to miss many of the biological responses that are characteristic for a given chemical, which is problematic in the screening of new toxicants and in investigating and comparing mechanisms of action among toxicants. However, such low-density arrays may be useful once the particular gene expression signature has been thoroughly characterized for a given type of exposure. For example, exposure of dendritic cells to chemical allergens leads consistently to expression changes in a battery of diagnostic genes, and a low-density array with these genes could be used as a relatively low-cost diagnostic tool for the identification of chemical allergens (Gildea *et al.*, 2006). Thus, identification, via whole-genome microarrays initially, of a smaller number of genes could be useful as predictive markers on specialized arrays for early screening tests with new molecules (some examples are given in de Longueville *et al.* (2004)).

2.4 Microarray Data Analysis Software

There are several steps in the analysis of microarray data, and, because of the complexity of the data sets, these are best performed using software that can automate these steps, perform these functions across tens of thousands of observations per sample, and compare multiple samples from the same experiment. The first step in microarray analysis is intended to ensure that labeling, hybridization and scanning of the array has produced distributions of raw intensity values that have been minimized for systematic errors (such as introduction of artifacts in the scanning process). The next step transforms raw intensity values from the arrays into 'expression values,' i.e. values that are background-corrected, scaled summaries of replicate determinations for each gene or gene probe. These expression values can then be used in statistical analyses and graphical representations to compare values among replicates and between treatments, ultimately with the aim of generating a subset of differentially expressed genes, or a gene list, which can be further analyzed and also confirmed with other complementary methods, such as quantitative real-time RT-PCR. Such gene lists would constitute the initial input for pathway analysis, clustering and manual literature searches, with the goal of determining information about a toxicant, such as its mechanism of action or comparing patterns among doses, time points, tissues or other toxicants. Figure 2.1 outlines this general data flow.

There are several software packages with which one can perform microarray analysis. Some of these packages are free and publicly available; others are commercially available. Free, publicly available software packages for the analysis of microarray data include several plug-ins for Microsoft Excel, including those for Cyber-T and Significance Analysis for Microarrays (SAM) statistical analyses (see below). BRB Array Tools (found at http://linus.nci.nih.gov/BRB-ArrayTools.html), which works with either one- or two-color arrays, is not restricted to any particular platform and is used within Excel as a plug-in for image analysis, cluster analysis and a limited number of normalization and statistical tests. Because most toxicologists are very familiar with the use of Excel, these applications are readily useable.

However, there are other software platforms that offer much greater flexibility and the ability to perform essentially all normalization, statistical, visual and clustering analyses within the same framework. One of the best tools currently available can be found in Bioconductor, an open-source software framework for the comprehensive analysis of microarray data. Bioconductor, which has been reviewed previously (Parmigiani *et al.*, 2003; Reimers and Carey, 2006), is comprised of various packages that perform a wide array of analyses. Information on downloading and use of Bioconductor is available at http://www.bioconductor.org/, and the 'package vignettes' provide usage manuals, including programming code. Although Bioconductor represents one of the best current microarray analysis tools because of its flexibility and comprehensiveness, it should be noted that it also has a steep learning curve because it runs as part of R, a statistical programming language. However, for those with some programming experience, there are workshops and on-line training materials (found at http://www.bioconductor.org/workshops/) which, in addition to the package vignettes, may be sufficient for getting started. In our experience, it requires as little as 10 lines of R code to read in data, normalize them, scale them and generate a list of genes at a particular significance level. Thus, while it is powerful, it is also relatively simple and transparent with respect to the operations it can perform.

However, because it requires learning the R language and programming experience, there have been recent attempts to build graphical user interfaces (GUIs) for Bioconductor. One example is CARMAweb (Comprehensive R- and Bioconductor-based Microarray Analysis web service), found at https://carmaweb.genome.tugraz.at/carma/ (Rainer *et al.*, 2006). CARMAweb works with several different microarray platforms and performs background correction, quality control, normalization, differential gene expression analysis, cluster analysis, data visualization and gene-ontology analysis. However, the number of different statistical analyses available is limited. Several companies do sell commercial GUIs that allow Bioconductor analyses, including GeneSpring (from Agilent; http://www.genespring.com), Spotfire DecisionSite (http://spotfire.tibco.com/products/decisionsite_microarray_analysis.cfm), and S+ArrayAnalyzer (from Insightful). There is also a web-based interface, called webbioc, for using Bioconductor without typing in code, and it can be obtained at http://bioconductor.org/packages/2.0/bioc/html/webbioc.html (further information about it can be found at http://cran.r-project.org/doc/Rnews/Rnews_2003–2.pdf). However, webbioc requires a Unix-based operating system and a dedicated webserver, and it does not currently allow the use of all of the available Bioconductor analysis packages. A recent online search indicates that several other public GUIs for Bioconductor are currently in development. One that is free and retains most of Bioconductor's capabilities is greatly needed by the toxicogenomics research community.

The Institute for Genomic Research (TIGR) offers the free, publicly available software suite TM4 (http://www.tigr.org/software/tm4), which does not require computer programming expertise. TM4 includes four applications: Madam for data entry and tracking, Spotfinder for image analysis, Midas for normalization and filtering and MultiExpression Viewer (MeV) for data analysis (e.g. SAM; see below), statistics, and visualization/clustering. MeV allows visualization of sample-by-sample data from individual probe sets/genes, using the GeneGraph function to see the normalized data in a line plot, across user-defined treatment groups. For a recent, thorough review of TIGR's TM4, see Saeed *et al.* (2006). TIGR MeV is limited in the different types of statistical analyses that it will perform – many of the techniques available in Bioconductor are, unfortunately, not available in TIGR's easily navigable GUI.

Also, at least as of mid 2007, TIGR MeV will not accept raw, Affymetrix CEL files as input. One must normalize Affymetrix data prior to loading into MeV. Robust multi-array (RMA) normalization can be done before going into TIGR MeV (or many other applications) by using the free, open-source RMAExpress (http://rmaexpress.bmbolstad.com/), a standalone GUI that takes in Affymetrix CDF and CEL files and outputs normalized expression values to a text file.

Some other free, publicly available programs for microarray analysis include PowerArray (found at http://www.niss.org/PowerArray/index.html), which performs data visualization, limited statistical analysis and some cluster analysis. Also, MA-Explorer (found at http://maexplorer.sourceforge.net) does many types of microarray processing and some clustering analysis. Cluster and TreeView and Mapletree (http://rana.lbl.gov/EisenSoftware.htm) perform several types of clustering analysis. Additional free, publicly available statistical packages for the assessment of differential gene expression have been previously reviewed (Steinhoff and Vingron, 2006).

Several companies sell commercial software for microarray analysis. The availability of this software has been previously reviewed (Olson, 2006). For commercial microarray platforms, the websites of the microarray chip manufacturers will often provide suggestions of software that is capable of handling and analyzing their microarray chips (e.g. the Affymetrix website provides this information about GeneChip-compatible software). These programs include Stratagene's ArrayAssist (http://www.stratagene.com), which performs differential expression statistical analysis, data visualization, and gene annotation via Affymetrix's NetAffx program. Stratagene's GeneTraffic (http://www.stratagene.com) is a web-based program for data management, normalization and filtration, differential expression statistical analysis, clustering and gene annotation; GeneTraffic works with multiple array platforms. Rosetta Biosoftware's Rosetta Resolver System (http://www.rosettabio.com/products/resolver/default.htm) is similar to Genetraffic, except that it is not web based. Though powerful, most of these programs typically do not have the complete functionality of Bioconductor in R.

One other significant consideration for preparing to do microarray analysis is computing power. Significant memory and computing power are needed for non-web- or server-based applications, such as Bioconductor in R on a desktop computer.

2.5 Normalization

Normalization facilitates direct comparisons between array measurements, removing systematic variation attributable to any factor other than differential gene expression. For one-color microarrays, normalization is used to remove obvious systematic differences (e.g. differences in labeling, array production, scanner specifications) between the microarrays that are to be compared and to deal with noise in the system. Normalization adjusts the data obtained from individual one-color arrays so that they can be more directly compared with one another. For two-color microarrays, because each individual microarray already contains ratio information, normalization is simply needed to adjust for labeling and photophysical differences between the Cy3 and Cy5 fluorophores in order to place all the arrays on a shared scale so that biological replicates are directly comparable. Many different types of microarray normalization have been developed (Bolstad *et al.*, 2003; Wu Z. *et al.*, 2004; Wu W. *et al.*, 2005; Ness, 2006; Kibriya *et al.*, 2007). Different normalization procedures use different *a priori* assumptions about an experiment, so failure to choose normalization contingent on the realities of an experiment can introduce artifacts that make downstream analysis problematic. For example, if one were to normalize to the expression of a set of housekeeping controls – whose expression levels often do change in response to various toxicants – downstream analysis would suggest a broad pattern of differential expression purely due to normalization errors. Thus, global normalization strategies are least biased and will produce greater consistency across experiments.

For one-color microarrays, it is important not to use the standard normalization procedure in Affymetrix MAS5/GCOS software, which has been shown to perform poorly. This deficiency has been discussed extensively in the literature (Irizarry *et al.*, 2003b, 2006; Cope *et al.*, 2004; Shedden *et al.*, 2005; Qin *et al.*, 2006). Instead, there is a growing consensus in the microarray community that RMA analysis is currently the best available normalization method for one-color microarrays. For example, RMA normalization resulted in more overlap between lists of genes generated by two different statistical analyses (SAM and Cyber-T) of a single dataset than did another commonly used normalization method, i.e. MBEI (Davey *et al.*, 2004). RMA (Bolstad *et al.*, 2003; Irizarry *et al.*, 2003a,b) uses quantile normalization, which adjusts the distributions of probe intensities to be the same across all microarrays in a given set (Bolstad *et al.*, 2003). The process of quantile normalization assumes that the signal distributions are similar across all arrays in the dataset and transforms all the array data in order to place it on a shared scale. Thus, when the altered genes are a small subset of the total genes and/or do not significantly shift the overall distribution of the data set, the approach works quite well. However, quantile normalization may amplify or reduce differences in the original data and may create artifacts which confound true toxicant effects if this similarity assumption is not basically true, e.g. if most genes are affected by a given treatment. Following quantile normalization, RMA uses median polish in order to summarize the probe sets. Median polish is an iterative process which is similar to taking the median value (across the replicate arrays) of each perfect-match probe within a probe set for a gene (e.g. for Affymetrix GeneChips) and combining these median values in order to create a single intensity value for each probe set. RMA uses only perfect match probe information, because many previous studies have shown that mismatch probes fail to account for nonspecific binding correctly as originally intended. Thus, with Affymetrix GeneChips, RMA takes the expression data from the perfect match oligonucleotides and creates one normalized expression value per probe set. Additionally, RMA includes a background subtraction component. RMA (as well as other types of microarray normalization) also uses $\log_2$ transformation of the normalized data for the purpose of treating up- and down-regulated genes equally. The $\log_2$ ratios for genes that are up or down by the same absolute fold change amount are equally distant from the $\log_2$ ratio, i.e. 0, that represents a lack of expression change; a difference of 1 or -1 in $\log_2$ values of the ratios is a twofold difference (up or down respectively) in gene expression. One other consideration is to be sure to perform global RMA – to load and RMA-normalize together every file to be compared – rather than performing RMA on arbitrary subsets of the total dataset, which can lead to artifactual differences in gene expression.

For two-color microarrays, various normalization procedures have been used. We recommend background subtraction followed by loess normalization, with accounting for duplicate spots (Shaw *et al.*, 2007). The purpose of loess normalization is to correct for data that are not median centered at $\log_2(\text{ratio}) = 0$. Loess normalizes the arrays by median centering them in order to adjust for dye effects, thus placing them on a shared scale. Loess is a transformation of two-color data to deal with changes in photophysical properties of the Cy3 and Cy5 dyes at high intensities and with unequal labeling efficiencies for Cy3 and Cy5. A plot of Cy3 intensity versus Cy5 intensity for all spots on a two-color array should theoretically be centered around a line with a slope of unity. However, the line will often deviate from the zero axis at high intensities due to experimental dye artifacts, such as fluorescence quenching, changes in photobleaching, etc. Thus, loess, a locally weighted linear regression method, is used to transform the data to fit a line with unit slope. The loess method is derived from the statistical function lowess (Cleveland, 1979), which uses a locally weighted least squares estimate of a regression fit. This method takes a section of the data and fits a regression line to that section; then the function is repeated to cover all of the data, resulting in a smoothed curvilinear regression line. The ratios obtainable from such a curve are representative of the fold-change ratios for the genes on the microarray; additional normalization is typically not necessary for comparing biological replicates of two-color arrays.

The Measurements for Biotechnology Program (http://www.mfbprog.org.uk) has examined the accuracy of microarray analysis and has found that the type of normalization used has a large impact on data comparability among analyses. Thus, the use of standard and reliable normalization methods, such as RMA for one-color microarrays, could go a long way in the quest to standardize microarray results and obtain the best biological data.

2.6 Data Filters

Current practices in data filtering prior to performing tests for differential gene expression generally reflect the aim of using an array analysis pipeline to justify the statistical significance of an individual gene: filtering reduces the number of parallel tests, and perhaps noise, directly improving significance estimates. However, arbitrary filtering may cut out genes that are truly affected by treatment and that may be important for the identification of biological patterns. Consistent with the suggestions of Breitling (2006), we recommend that arbitrary global filtering of microarray data should generally be avoided prior to assessment of differential gene expression analysis. Use of cutoffs is not only unnecessary, but is also likely to result in loss of important biological information. For example, systematically cutting out low-expressed genes should be avoided, in our opinion, because many biologically important genes are naturally expressed at low levels, and modest changes in their expression can be biologically meaningful and provide important insights into the overall biological response. Likewise, only selecting those genes whose expression changes by twofold or greater, or similar 'threshold' approaches, should also be avoided, since this will, in our experience, result in extensive loss of genes that may be critical for identifying and understanding a given pathway. Standard deviation filters are also generally unnecessary for similar reasons, and because many genes behave in a nonparametric manner and thus will appear to be 'statistically noisy' even though their qualitative response up or down is highly consistent (see GREB1 example below and discussion in Hamilton (2006)). Good statistics and/or visualization of the data should obviate the need for these types of filtration. More importantly, there is likely to be a biological basis for many genes responding nonparametrically, i.e. in ways that are not amenable to such filtering, which is based on the assumptions of parametric statistics and previous concerns about FDR and generation of overly long gene lists (Hamilton, 2006).

2.7   Analysis of Data for Differentially Expressed Genes

Identification of differentially expressed genes is obviously the overall goal of microarray analysis in toxicology or any other discipline. This phase of the analysis process is intended to answer four basic and related questions: how many genes respond to the toxicant treatment, by how much, which specific genes or pathways are altered, and how sure are we that they actually change? Various statistical methods in this phase can lead to very different answers, primarily because of differences in how methods estimate the probability that the observed difference was the result of chance. We will discuss some of the current methods for assessing differential expression in this section, but we begin by introducing four important conclusions we have drawn from our own analyses.

1. Overemphasizing specific expression levels, fold change cutoffs, or p values to generate lists of differentially expressed genes will eliminate many genes from the final gene list that may be critically important in the actual underlying biological response.
2. Emphasis on limiting the FDR, particularly in low-dose toxicogenomics, may severely limit the ability to identify the biologically relevant patterns in gene expression change.
3. As a consequence of 1 and 2 above, certain array statistics may grossly underestimate the number of differentially expressed genes in an array experiment, perhaps leading to the incorrect conclusion that the treatment does not alter gene expression or that the patterns of altered gene expression are not consistent across doses, times, tissues, models or replicate experiments.
4. Nonparametric statistics, in particular rank-based methods, that place proportionally greater emphasis on effect size and qualitative consistency across replicates, as opposed to statistical significance, will produce longer and more reliable gene lists. In our experience, longer lists are both richer in biological significance and show greater concordance among replicates.

When analyzing toxicogenomic data, particularly data from low-dose exposures, the goal is to obtain a biologically rich and meaningful dataset. During the first years of microarray experiments, most researchers used very stringent microarray analysis techniques, in large part because of the concern about FDR and the multiple comparisons problem inherent in large data sets, which was exacerbated by use of a low number of biological replicates (primarily driven by the cost of microarrays). There was also a common belief that high-dose toxicant experiments would produce a robust enough biological response that such stringent analyses would still reveal the primary response genes. However, the continued use of such overly stringent analysis techniques for more recent experiments that have adequate numbers of biological replicates and/or are investigating low-dose exposures has resulted, in our view, in a decreased ability to generate meaningful data. For example, recent reports in the toxicogenomics literature have suggested that there are very few genes altered significantly at low (and sometimes relatively high) toxicant doses (examples include Clewell *et al.* (2007) and Wei *et al.* (2005)), and that the genes vary by dose, time point, and experiment in a manner that suggests no underlying pattern. However, our own reanalysis of these data sets revealed substantial concordance and biologically interesting patterns (Hampton *et al.*, 2007).

The use of conservative, parametric analyses has been attractive because these techniques (such as SAM; see below) are readily available and easy to apply in many software applications. They have been rigorously justified from a statistical standpoint, in large part because they produce relatively short lists of genes. Obtaining a short gene list may seem desirable in order to reduce the complexity of the downstream analysis and to facilitate biological interpretation. A major challenge for biologists using microarrays has been the seemingly daunting task of understanding and interpreting the role of dozens or perhaps hundreds of genes in the overall response to a given experimental treatment, particularly prior to development of pathways software that allows for integration of genes into recognizable biological pathways. And for high-dose toxicant exposures, such stringent techniques or high fold-change cutoff hurdles (i.e. twofold or higher) may still produce relatively long gene lists that suggest a robust response even though many other interesting genes will fail to make such cutoffs. However, as toxicology moves to investigating much lower, environmentally relevant toxicant exposures, these exposures are likely to produce much more subtle changes in gene expression. Use of these rigorous statistical techniques with low-dose experiments typically results in extremely short gene lists which exclude large amounts of data, and these lists do not reproduce well across replicates or experimental treatments. However, many genes that do not make the list using such methods are changing in ways that are highly reproducible, not only among microarrays, but also as confirmed by complementary approaches such as real-time RT-PCR, suggesting that they are an important component of the underlying biological response. Thus, the methodology used to analyze microarrays in these experiments should be able to capture this information.

It was in response to this challenge that we and others have explored alternative ways to analyze toxicogenomic results, and it is clear that the more traditional and rigorous approaches previously used are missing important biological information, not only in low-dose experiments, but also in the higher dose experiments. However, the problem is most acute at the lower doses. For example, in some recent experiments in our laboratory we observed that analysis of toxicogenomic data from the lungs of mice exposed to a low dose of arsenic in drinking water (10 ppb arsenic, the current federal drinking water standard, for 5 weeks) resulted in just a few genes being identified as significantly affected according to a standard SAM, whereas less stringent but statistically justifiable analysis methods resulted in much longer, more meaningful lists of genes that were highly reproducible (see below). As a result of such analyses, we have developed an approach to microarrays that we feel is far superior and that, surprisingly, not only is simpler, but also is statistically quite rigorous. Most important, it produces the highest level of concordance among replicate experiments, the ultimate goal of any statistical approach and one which suggests that these approaches are also most accurately revealing the underlying biology. This methodological approach is described in more detail in the next section.

Toxicology researchers new to toxicogenomics and microarray analysis may find the range of possibilities for statistical tests to be daunting and may believe that one would need a high level of statistical training in order to perform microarray analysis adequately. However, we have found that a basic knowledge of statistics, which most toxicologists have, is sufficient for navigating currently available options. Moreover, the proposed approach we and others have taken is highly intuitive and is biologically based rather than based on pure or theoretical statistics. There are rigorous underlying statistical principles to this approach as well, and a rich theoretical framework within classical statistics that supports these approaches. The primary tools available for microarray analysis and their strengths and weaknesses are discussed below, focusing in particular on a comparison of parametric and nonparametric approaches, stringent versus nonstringent parameters, various methods for dealing with multiple testing problems, and their influence on interpretation of the biological result.

2.7.1   The *t*-test and Analysis of Variance

The *t*-test is the simplest, though not the best, statistical method for the analysis of microarrays. A *t*-test can be done in Excel, Bioconductor, TIGR MeV and many other software applications. Any basic statistics text describes the calculations being performed for a *t*-test. The *t* statistic essentially measures the signal:noise ratio, i.e. difference between the means divided by the variability of the groups. Then the calculated *t* value is compared with Student's *t* distribution charts, which indicate whether the data are statistically significant for a certain cutoff (e.g. 5 % significance or $p < 0.05$). However, *t*-tests assume that the distribution of the data is symmetrical and approximately normal, that the variances are equal (an *F*-test can be performed in order to determine whether the variances are equal) and that the number of replicates is large, which is often not the case with microarray data.

Classical *t*-tests are parametric tests. Parametric tests typically identify probe sets as being differentially expressed among treatment groups when there is a large fold-change difference as well as small standard deviation among the replicates for each treatment group. Parametric tests involve the calculation of parameters (e.g. mean, variance) and assume that the data follow a statistical distribution with known mathematical properties (typically the normal distribution). Also, parametric tests often require that the variances of all the samples being compared are the same. Conversely, nonparametric tests, and rank-based tests in particular, typically identify probe sets as being differentially expressed among treatment groups when all or most of the replicates in one treatment group have higher values than replicates in the other treatment group, without taking into account the variability among replicates within each treatment group (thus, nonparametric tests do not utilize 'parameters'). Think of a chi-square test, for example, that tests the difference between the number of observed and expected elements, which is independent of a quantitative value for each observation. Nonparametric tests also make fewer assumptions about the distribution of the data.

However, most importantly, the use of *t*-testing with datasets as large as those from microarrays, with perhaps thousands to tens of thousands of gene probes per chip, leads to multiple testing complications. For example, the use of a 5 % significance testing level (an $\alpha < 0.05$), with 10 000 genes tested in parallel on a microarray, would directly lead to the expectation that about 500 genes (that are not being differentially expressed) would reach the 5 % significance level just by chance and would, therefore, be false positives. The probability of yielding false positive data increases rapidly as the total number of genes on a microarray increases (Verducci *et al.*, 2006). Thus, a multiple testing correction is needed. One is the Bonferroni correction to the *t*-test (Lin, 2005), a much stricter significance threshold. The Bonferroni-corrected significance level α is the quotient of the original α (e.g. 0.05) divided by the total number of genes being tested on the microarray (e.g.

10 000). Thus, in this case, a p-value of 5×10^{-6} would need to be achieved in order to 'believe' that a tested gene is being differentially expressed. Therefore, this correction is clearly too severe, particularly for most low-dose toxicogenomic experiments, as we will show below. While statistically rigorous, Bonferroni-corrected t-tests have very low power: the probability of correctly identifying differentially expressed genes is very small, so many interesting, affected genes will be ignored in such an analysis. One alternative is the use of the 'adjusted Bonferroni correction,' which ranks genes by their t statistics. Then, increasingly less severe standards are applied to the ranked genes until a certain p-value threshold is obtained. Another alternative to contending with multiple testing problems is to reduce the number of statistical tests being performed. For example, if one is interested *a priori* in the effect of a toxicant on a particular cellular signaling pathway, then it is possible to gather a list of all the genes in that one pathway (e.g. 100 genes) and to perform t-tests on the microarray data for just those 100 genes. At a 5 % significance level, only five false positives would then be expected for this data subset.

When comparing more than two treatment conditions in a study, the statistical method ANOVA, a variation of the t-test, is useful, and ANOVA has been used to analyze microarray data. Like t-tests, ANOVA is a parametric method and assumes a normal distribution of the data, independent observations and equal variances for each gene/probe set on the microarray. A thorough review of the background and application of this method to microarray data is found in Ayroles and Gibson (2006). As with the t-test approaches described above, multiple testing issues are a problem with ANOVA and adjustments are available (Ayroles and Gibson, 2006).

Overall, conventional t-tests (and ANOVA) typically fail as tools for the global analysis of microarrays, largely due to the large weight given to small standard deviations among biological replicates, several assumptions (such as that of normally distributed data), the large number of replicates truly required and the multiple testing problem. Nonparametric versions of these tests alleviate some of these constraints, but they still fundamentally suffer from the multiple testing problem and, thus, are of limited utility for microarray analysis (see below).

2.7.2 Significance Analysis for Microarrays and Other Modified t-tests

Several modified t-tests have been created in an attempt to address several of the short-comings of conventional t-tests in microarray analysis. SAM (Tusher *et al.*, 2001), which is available through http://www-stat.stanford.edu/~tibs/SAM/ and executable in Excel, Bioconductor, TIGR MeV and many other software applications, is currently perhaps the most commonly used technique for statistical analysis of differential gene expression on microarrays (Breitling *et al.*, 2004). The basis for SAM is rooted in conventional t-tests (described above), in combination with randomization/permutation techniques for evaluating significance levels and controlling the false positive rate. However, SAM makes no distributional assumptions. Data must be normalized prior to SAM analysis.

SAM is essentially a modified t-test. The method combines test statistics with permutation procedures to assess significance and to control for multiple testing issues. SAM assigns a significance score for each gene based on changes in gene expression relative to the standard deviation of repeated measurements, and the method uses permutations of repeated measurements to estimate the FDR. Permutation analysis is a method of simulating data that satisfy a null hypothesis by shuffling around the observed data. SAM determines the FDR associated with an output gene list, which the user can control to be at a chosen level: FDRs of 1–10 % are considered reasonably strict. The process of determining the FDR involves adjustment of the *p*-value in order to reflect the frequency of false positives in the list. Thus, the use of an FDR controls the fraction of chosen differentially affected genes ('positive calls') that are false positives among all rejected hypotheses, rather than trying to avoid all false positives. It controls the expected number of false positives in the list of results, rather than the total number of experiments in which any error is made.

Although SAM has many advantages over conventional *t*-testing, particularly the use of permutation procedures to alleviate multiple testing problems, it is still a rather stringent analysis. For low-dose toxicogenomics experiments, much of the true data will be lost when SAM is the sole statistical analysis (see below).

Another statistical test based on the simple *t*-test is Cyber-T (Baldi and Long, 2001). Cyber-T can be accessed at http://cybert.microarray.ics.uci.edu/, where it can be performed using a web interface. Cyber-T uses a computational method, based on modeling of *p*-value distributions, for estimating and controlling false-positive and false-negative rates. Cyber-T does assume that the noise is normally distributed, and this is probably a reasonable assumption if the signals are log-transformed prior to analysis. Cyber-T performs better than conventional *t*-tests when there are few replicates (as is typically the case for microarray experiments) because it allows the user direct control over the estimated variance. This means that, in an experiment with few replicates and small fold changes, Cyber-T can detect significantly affected genes. For example, a comparative study found Cyber-T to provide a less stringent (but still rigorous) analysis than did SAM: when a low-dose toxicogenomics microarray dataset was normalized via different methods, Cyber-T produced a longer list of significantly affected genes overlapping the differently-normalized analyses than did SAM (Davey *et al.*, 2004).

These modified *t*-tests, SAM and Cyber-T, along with methods based upon the empirical Bayes method of Efron *et al.* (2001; Eckel *et al.*, 2004), are based upon Bayesian statistics. They have the following features in common:

1. They try to make a more stable estimate of the variance for individual probes (by adding a constant to the standard deviation used in the t statistic), which leads to a greater emphasis on fold change compared with standard deviation.
2. They also compute a more optimistic FDR (i.e. using a multiple testing correction that is much less harsh than a Bonferroni correction), using Bayesian methods (e.g. permutation tests or linear modeling) and the strength of observations across all genes.

These modified *t*-tests also tend to make fewer distributional assumptions and perform better with fewer replicates than conventional *t*-tests. For these reasons, they tend to generate longer lists of affected genes than, for example, a Bonferroni-corrected *t*-test. However, these modified *t*-tests tend to generate shorter, less reliable lists than rank-based methods (see Section 2.7.5 on concordance).

2.7.3   Rank-based Methods

A rank-based, nonparametric microarray analysis method, and one that we have found to work extremely well with low-dose toxicogenomics data, is called Rank Products

(RankProd) (Breitling *et al.*, 2004; Breitling and Herzyk, 2005). RankProd was designed to utilize biological reasoning rather than a sophisticated statistical model. It is based on the ranks of the fold changes of the probe sets on the microarray. The RankProd algorithm can be compared to a series (e.g. five replicates) of road races (e.g. microarrays), each road race with the same composition of runners (e.g. 10 000 genes). If runner X (e.g. gene X) places in the top ten in four of the races (microarrays) but loses a shoe and places much farther behind in the fifth race (microarray), a biologically based interpretation would argue that this runner X (gene X) is one of the fastest runners (most strongly affected genes) in the large group even though runner X's standard deviation among the replicate races (microarrays) is very high. In a simple two-color microarray experiment comparing two different conditions on a single slide, the genes that are most highly up- or down-regulated due to the treatment, on this single slide, can be ranked by their fold-change values. Because biological data are typically noisy, a single two-color microarray is not sufficient. However, if the same genes show up near the top of the fold-change list in several replicate sample arrays, as with the example of runner X above, then confidence in the data increases. For this reason, RankProd is very robust against outlier data.

This method can be performed in Bioconductor using the RankProd package (Hong *et al.*, 2006) and in Excel using the method outlined by Breitling *et al.* (2004: Supplement). RankProd works with any microarray platform, as long as the results can be expressed as ranked lists, and it can analyze both normalized Affymetrix GeneChip data and normalized cDNA microarray data. Whereas *t*-tests and SAM technically require a fairly large number of replicates in order to be statistically justified, RankProd was specifically designed for small numbers of replicates (e.g. three to five replicates) and, thus, is ideal for microarray analysis. Because RankProd is nonparametric, it does not assume that intensity measurements of normalized ratios are normally distributed. However, RankProd does assume that:

1. only a minority of genes are differentially expressed;
2. replicate arrays are independent measurements;
3. most individual gene changes are independent of each other; and
4. variance is approximately the same for all genes.

For a microarray consisting of data from n total genes, with k replicate microarrays, the random probability of a particular gene being at the top (rank #1) of *each* replicate list is $1/n^k$, which is clearly a very small probability, if none of the genes are actually differentially expressed. Then, for each gene g in k replicates i, with each replicate microarray consisting of n_i genes, the corresponding combined probability is calculated as the rank product:

$$\text{RP}_g^{\text{up}} = \prod_{i=1}^{k} = \frac{r_{i,g}^{\text{up}}}{n_i}$$

where $r_{i,g}^{\text{up}}$ is the rank of the gene g in the fold-change list of the ith replicate sorted by decreasing fold change. In this way, RankProd detects probe sets that are consistently highly ranked in several replicate experiments. Rank products for down-regulated genes ($\text{RP}_g^{\text{down}}$) are calculated similarly, from the fold-change list sorted by fold change. If the same number of genes (probe sets) is on each replicate microarray, then the geometric mean rank can be used for the RP calculation. The RP measurements are applied to create a list of significantly differentially expressed genes. For one-color arrays, in which fold-change values must be calculated *across* microarrays, the method is modified by calculating the RP values over all possible pairwise comparisons (which are not independent, and, thus, the significance analysis is adjusted to take this into account). For determination of significance levels to accompany the RP values, a straightforward permutation-based estimation procedure is used to determine how likely it is to observe particular RP values in a random experiment, thus generating an FDR similar to that used in SAM. As discussed above, the multiple testing problem is inherent to and plagues *t*-test analysis and other non-rank-based statistical analyses of microarrays. However, importantly, the RankProd method actually becomes more robust with increasing parallel testing (i.e. increasingly large data sets) because, the more items (probe sets, genes) that are available for ranking, the more unlikely it is for an individual gene to come in first by chance. In the road race analogy, if runner X consistently places in the top 10 out of 10 000 runners, then this is even more meaningful than placing in the top 10 out of 100 runners. Thus, coming in at or near the top or bottom of the fold-change list becomes more meaningful in large experiments, such as genome-wide microarrays. The mathematics behind the RP method is described comprehensively by Breitling *et al.* (2004). Other rank-based methods will also work well for analysis of microarray data because of their basic shared features.

RankProd has been shown to work robustly with noisy (e.g. biological) data (Breitling *et al.*, 2004). Importantly, RankProd is more sensitive and reliable and finds longer lists of differently expressed genes than does the more stringent SAM method. For example, Figure 2.2 shows a dot plot (categorical scatter plot) of RMA expression values (each dot representing an individual mouse) for a probe set for the gene GREB1 (Davey *et al.*, 2008). On the *x*-axis are four treatment groups, animals that were injected with either arsenic (As), arsenic and dexamethasone (As + Dex), dexamethasone (Dex), or a saline sham injection (control). On the *y*-axis are the $\log_2$ expression values (resulting from RMA normalization). From this dot plot, we would predict that this gene is being differentially affected in the

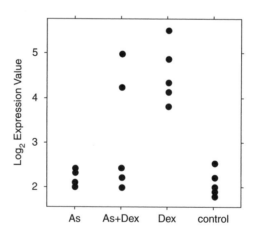

Figure 2.2 *Dot plot of individual-mouse RMA expression values of a probe set for the gene GREB1, which was found to be significantly affected by As + Dex versus Dex by visual analysis, RankProd (p < 0.01) and real-time RT-PCR confirmation, but which was not found by SAM.*

As + Dex group compared with the Dex group; analysis of the entire dataset by RankProd revealed this effect to be significant at $p < 0.01$. This result was then confirmed with real-time RT-PCR, which showed a similar change in expression as the dot plot. However, analysis by SAM did not reveal this probe set as being significant, even with the FDR pushed as high as 40 %. Breitling *et al.* (2004) found that the results of RankProd are often similar to data obtained with a simple fold-change analysis. Our laboratory has also found[1] that RankProd data tend to mirror fold-change-cutoff data and are typically much less stringent than the classical two-fold cutoff.

RankProd is, in our opinion, the best available method for generating an initial list of candidate genes. However, we have found that even the use of RankProd, which is the least stringent common microarray analysis method to our knowledge, does not reveal all the genes that are being differentially regulated in microarray experiments. Thus, even this method may sometimes be too stringent. RankProd operates with the underlying assumption that only a minority of genes are affected, which its creators clearly state is a conservative assumption that restricts the number of genes/probe sets that will be considered to be differentially expressed. This restriction means that the genes that are selected as being affected have been held to a high standard of statistical analysis. However, treatments that produce a large effect on many genes will be disadvantaged by this method. More importantly, it should not be assumed that any statistical method will, by itself, reveal all the important biological changes in an experiment since each method must balance rigorous statistics with inclusion, i.e. balance false positive and false negative discovery rates. Thus, we view RankProd as a useful starting place for further analysis using other approaches (see below).

2.7.4 Fold-change Cutoffs

Fold-change thresholding is generally believed to be an insufficient analysis method for microarrays because, without a statistical measurement and significance-level determination, it will fail to discriminate between spurious and true biological changes. However, as noted above, the RankProd method, which is statistically rigorous and contains a significance level determination, yields data which are very similar to a simple fold-change analysis with a cutoff value less rigorous than the classical two-fold cutoff. Also, simply selecting a fold-change cutoff is a straightforward analysis method. We have found that preparing an *M–A* plot, which is log_2(average OR median ratio) plotted against average log intensity, for all of the probe sets on a microarray will provide information about where an appropriate cutoff value might be for a given dataset. On an *M–A* plot, there is a 'swarm' of data close to the central line $M = 0$, which is the line for probe sets with no change in expression due to the treatment, and differentially expressed genes are typically outside of this central cluster.

2.7.5 Concordance and Reproducibility of Microarray Data

The best analytical method most accurately approximates what actually is happening biologically. Greater concordance among replicate experiments indicates increased accuracy.

---

[1] Andrew AS, Hampton T, Warnke LA, Nomikos AP, Kozul C, Waugh M, Davey JC, Gosse JA, Ihnat MA, Hamilton JW.

We have assessed the concordance, or degree of agreement, of microarray data from separate biological experiments (each experiment with multiple replicates per treatment group) using the different statistical approaches described above (Hampton *et al.*, 2007). In two separate experiments, mice ($n = 3$–5 per treatment group, per experiment) were given 0 or 10 ppb arsenic in their drinking water for 5 weeks, and lung RNA from each animal was extracted and analyzed separately on Affymetrix 430 2.0 whole-genome microarrays (Andrew *et al.*, unpublished data). Using open-source R software packages available as part of the BioConductor project (see above), the microarray data were normalized using RMA and three different statistical analyses were performed: SAM (parametric), RankProd (nonparametric), or simple mean fold-change. Figure 2.3 compares the lists of differentially expressed genes found in common between the two separate biological experiments (the number of probes concordant), at a range of list lengths. Analyzing the data with RankProd led to over 1632 genes in common within the top 5000 (about 33 %), almost twice the concordance found by SAM (887), which was slightly better than a simple *t*-test (806, data not shown) and which was much better than the 95th percentile expected by chance alone according to the hypergeometric distribution (589). Since RankProd gene lists are similar to mean fold-change lists, it was not surprising that mean fold resulted in 1444 concordant probes, a rate of 28 % at a depth of 5000. These data show that RankProd does a better job of identifying genes that behave consistently between experiments than SAM does and, therefore, provides a better estimate of the true biological response for this low-dose toxicological study. Similar results from a different study, showing that the simple fold-change method leads to greater concordance than either *t*-tests or SAM, have been reported by other researchers (Guo *et al.*, 2006). Thus, as discussed above, excessively rigid *p*-value

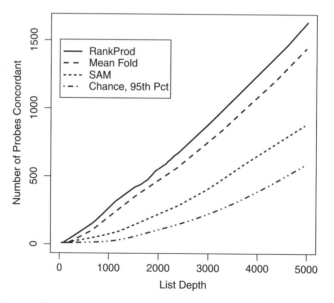

Figure 2.3 *Concordance of microarray data from two separate biological experiments, as a function of statistical approach: SAM (· · · ·), mean fold (– – –), or RankProd (——), compared with a random chance simulation (· – · –).*

and fold-change hurdles may generate misleading conclusions and gene lists that are too short and that do not reproduce well between experiments or treatments.

In light of these findings, we then reanalyzed several data sets from published low-dose toxicogenomic microarray experiments, using RankProd and other nonparametric methods. These analyses revealed that previously published toxicogenomics experiments are likely to contain useful information that was not previously identified because it was analyzed using parametric methods and/or it failed to meet excessively rigid significance criteria (Hampton *et al.*, 2007). Fortunately, this is a problem of analysis, not experimental methodology per se, i.e. the initial raw data files are still highly amenable to reanalysis assuming that the other basic parameters of a good toxicogenomics experiment have been met. Although we have found that RankProd works very well for the assessment of differential gene expression in many types of low-dose toxicogenomics experiments, the choice of statistical microarray test is highly dependent on the type of experiment being performed. We advocate an iterative process of exploratory and thorough analysis, including the use of nonparametric, rank-based tests and direct visualization of the data in dot plots of normalized expression values (discussed in more detail in subsequent sections; see Figure 2.1).

Another important aspect of concordance is to confirm key gene expression changes predicted from microarray data using complementary methods such as the quantitative and sensitive method of real-time RT-PCR. Recent reports have shown that the rate of confirmation of microarray data with real-time RT-PCR is very high ($\sim$90 %) for genes selected by standard microarray analysis procedures (Tusher *et al.*, 2001; Yao *et al.*, 2004). A paper looking specifically at differentially expressed genes that were affected by less than a twofold change also found a high rate of confirmation (using northern blot confirmation) for these small but biologically important changes in gene expression (Yao *et al.*, 2004). This finding bolsters the assertion that it is important to analyze microarray data in such a way so as to avoid exclusion of subtle but reproducible transcriptional effects due to low-dose toxicant exposures. Recent findings have shown that small changes in gene expression can be very meaningful biologically, e.g. in the process of tumorigenesis (Yan *et al.*, 2002). Also, microarrays are semi-quantitative at best and typically underestimate the magnitude of true gene expression changes (Yuen *et al.*, 2002). This underestimation, coupled with the potential importance of small gene changes, should be considered in the process of analyzing microarray data. For example, we were interested in the possible effect of arsenic on expression of DNA repair genes as a potential mechanism of arsenic's co-carcinogenicity (Andrew *et al.*, 2003, 2006). In data from whole-genome arrays of mice exposed to arsenic in drinking water, we analyzed the expression of all $\sim$150 known DNA repair genes in the exposed mouse livers. Liver RNA was analyzed using Affymetrix 430 2.0 whole-genome microarrays. Changes in expression of most of these genes due to the arsenic exposure were not detectable using several different statistical tests (although some were detected as differentially expressed by RankProd). In addition, it is well known that these genes are naturally expressed at very low levels even though they are extremely important biologically. Thus, we performed *t*-tests on this small subset of the entire genome data (to reduce the multiple testing problem with the small dataset) and also inspected the data with the use of dot plots of the RMA-normalized expression values, which revealed significant expression changes in a large number of DNA repair genes. These gene expression changes were then analyzed for a subset of these genes using real-time RT-PCR, all of which were confirmed. Thus, while much of the focus of microarray analysis has been on false positives, there are likely to be many false negatives as well using standard microarray analyses, requiring one to use additional analytical tools and complementary experimental approaches to reveal the underlying biology.

Many microarray platforms have multiple probe sets or cDNA spots representing the same gene. This repetition can be useful whenever a particularly interesting gene is extracted from a microarray analysis. For example, if all of the probe sets for a gene exhibit similar effects on expression due to a treatment, then this information bolsters the hypothesis that the gene of interest is truly being affected by the treatment. Conversely, it should not be assumed that discordance among replicate probes for a given gene means that there is not a significant effect. For example, if the different probes within a set represent different regions of the transcript, attenuation of transcription along the gene, which is common, could lead to higher levels of response for probes near the transcription start site than those more distal; likewise, some probes might represent alternatively spliced or alternatively transcribed regions of complex genes. Additionally, it has been reported that much of the disagreement between probe sets for certain genes in the Affymetrix GeneChips may be due to misannotation (Elbez *et al.*, 2006). However, the majority of probe sets on Affymetrix whole-genome GeneChips have been shown to be accurate (Alberts *et al.*, 2007).

2.8   Clustering Analysis of Microarray Data

Clustering is the statistically based process of organizing, grouping and visualizing gene patterns based upon gene expression profiles. Unsupervised, global clustering is done without *a priori* assumptions about how the data are grouped. Genes that are clustered in this way may have similar biological functions or other similar properties. For example, they may all be regulated by a common transcription factor, whose function has been altered by a toxicant, or part of a similar biological pathway. Clustering of microarray data can be performed in many different software applications, including Bioconductor and TIGR MeV and many others, including several with easy-to-use GUIs. A colored 'heat map' is a display of such clustering, and typically one in which probe sets (for particular individual genes) are on one axis and individual replicate microarray samples are on the other axis. The heat map is color coded to display the expression levels and direction of change of the genes (i.e. red–green in the original convention or blue–yellow for color-blind accessible displays). If the genes are randomly ordered in the heat map, then there will not be any obvious patterns, whereas treatment effects that produce consistent alterations in gene expression will produce readily visualized patterns that provide insights and lead to further analyses. A dendrogram plotted next to the heat map indicates relationships among groups, with short branches signifying close relationships. There are many different methods for clustering, which have been reviewed previously and will not be discussed in detail here (Sherlock, 2000; Butte, 2002; Chen *et al.*, 2002; Goldstein *et al.*, 2002; Verducci *et al.*, 2006; Lee and Saeed, 2007).

However, a major flaw of global clustering for analysis of microarrays is that, because the focus is on large-scale patterns, it may miss more subtle but important patterns in low-dose experiments. Therefore, global clustering analyses should always be done in conjunction with a comprehensive analysis of differential gene expression (see above) and pathways analysis (see below). Moreover, in our experience it is more informative to cluster genes following other analyses, in contrast to a priori global clustering. Also, it is most informative when smaller subsets of genes are clustered rather than clustering the entire array or even the entire list of significant genes from the initial analysis. These subsets might come from literature searches or suggestions from pathway analysis programs. Used this way, clustering is a nonparametric, outlier-resistant method for the visualization and analysis of toxicogenomic data. Heat maps can provide an overview of the degree and direction of differential expression due to a toxicant treatment, of a chosen subset of genes previously found to be grouped in a particular pathway (see Figure 2.1, connection between manual literature searches done after automated pathway analysis and clustered heat maps). This is also a powerful way to visualize data in a manner that can reveal important patterns of expression (i.e. are the genes in this subset behaving as a group?) across all groups and individuals.

2.9   Gene Ontology and Pathway Analysis

Once one has obtained an initial list of differentially expressed genes, the next goal is to determine their biological significance, including the functions of individual genes, the pathways into which they fit, the interactions among them and comparisons of different treatments (dose, time, tissue, etc.) (see Figure 2.1). The use and understanding of bioinformatics tools is necessary for this process, and it takes a considerable amount of time and effort to explore a data set fully, but fortunately these tools have dramatically and progressively improved over the past 2–3 years, and they are now readily accessible to biologists.

To begin, basic information on the molecular function, biological processes and cellular location of a particular gene product is particularly useful, and the Gene Ontology (GO) database (http://www.geneontology.org/) provides this information for all available genes across many organisms. Entrez Gene (http://www.ncbi.nlm.nih.gov/entrez/query.fcgi?db=gene) is another comprehensive source for the curation of sequence and functional information for all available genes. Affymetrix offers the free NetAffx program which can also provide an overview of some of the biological processes, cellular components or molecular functions being affected using the GO browser.

The next step is to integrate the expression data from a microarray dataset into biological pathways, and newly available pathway analysis tools allow one to rapidly move beyond the ontology of individual genes to more comprehensive biological analyses (Verducci *et al.*, 2006). Pathway analysis methods can also serve as a filter in order to make sense of a list of differentially expressed genes and both generate and narrow down specific hypotheses relating treatment to response. Many pathway analysis tools provide up-to-date information on the whole genome, including gene functions, literature references and various types of relationships among gene products in which a large amount of diverse information is collected into a single interface. One analytical approach is to overlay the gene expression data onto previously well-described ('canonical') biochemical pathways, e.g. glucose metabolism or signal transduction. A powerful alternative is to let the software develop ad hoc networks of genes that are not part of canonical pathways but that show known interactions among genes or gene products based on the published literature. Additional statistical power can be derived from either of these pathways representations, since one can then ask what the probability is of creating and populating such a diagram of related genes based on random chance, given the size of the starting list of candidate genes and the number of genes populating a specific pathway. For example, if a list of 250 candidate genes appear to respond to a given treatment, and within that list there are 25 genes that are all part of the same pathway, such as cell adhesion, then the software can calculate the probability of this result happening by chance. Using the Ingenuity Pathways software, we tested the validity of this analysis by inputting random lists of genes and asking whether it would find statistically significant networks or pathways, and this test resulted in no significant associations.

Such pathways software analysis also provides an additional layer of replication and concordance. For example, we have found that replicate experiments with the same toxicant, even at different doses or time points (and, in some cases, from different laboratories), often revealed the same pathways and/or networks even though the precise list of individually affected genes varied among experiments (Hampton *et al.*, 2007). This provides an additional level of confidence that there is a true biological response underlying the treatment, which is especially important in low-dose experiments at the very lowest end of the dose–response scale.

There are many free pathway analysis programs which can project gene expression data onto basic molecular interaction networks and pathways. The Kyoto Encyclopedia of Genes and Genomes (KEGG; http://www.genome.jp/kegg/pathway.html) is a collection of molecular interaction pathway maps for processes involved in metabolism, genetic information and processing, environmental information and processing, cellular processes, human diseases and drug development. MAPMAN (http://gabi. rzpd.de/projects/MapMan/#mapman_overview; Thimm *et al.*, 2004; Usadel *et al.*, 2005), Gene Microarray Pathway Profiler (GenMAPP, http://www.genmapp.org; Dahlquist *et al.*, 2002), DAVID (http://david.abcc.ncifcrf.gov/; Dennis *et al.*, 2003), and Cytoscape (http://www.cytoscape.org/; Shannon *et al.*, 2003) are also available free pathways software tools. Commercial packages include the web-based Ingenuity Pathways Analysis (http://www.ingenuity.com/) and Ariadne Pathway Studio (http://www.ariadnegenomics .com/products/pathway/), which both provide extensive gene annotation, pathway and function analysis and network analysis (protein interaction maps).

While pathways programs can be extremely helpful for placing a set of differentially expressed genes into a meaningful biological context, there are some concerns and caveats regarding their use. One major drawback with all currently available pathway analysis programs is that they are not comprehensive, i.e. they are missing many biochemical and physiological pathways, including many that are well described in the literature. This deficiency could lead a researcher to incorrect conclusions about a response, i.e. that certain biological processes were unaffected by a particular toxicant exposure, whereas the effects on them were simply not displayed by the software. Conversely, some of the pathways that are shown may be annotated based on a particular viewpoint of a field. For example, a cell signaling pathway may be labeled as being part of an immune response, whereas in reality that pathway may also be part of several other responses. But this may erroneously lead to the conclusion that the response is primarily an immune response. Also, genes that are peripherally (but importantly) involved in a particular pathway are often not included in the pathways software, which can also lead to incorrect conclusions. Thus, caution is required in using and interpreting such results; and, as in the case with the statistical analyses that precede this step, there is no pipeline that can substitute for comprehensive, hands-on human analysis of the results. As these tools and others in bioinformatics continue to expand and improve over time, many of these problems will be reduced. But, in any event, it is important to recognize that the results of such analyses are not a final endpoint; rather, they provide insights that can (and should) lead to further analysis and further experimentation.

2.10 Interpretation and Use of Toxicogenomics Results

Whether the specific goal is to understand mechanism, explore the low end of the dose–response curve, do comparative testing, or other endpoints, the overarching goal of toxicogenomics is to provide a highly sensitive, comprehensive, genome-wide snapshot of alterations in gene expression. Because of their sensitivity and specificity, well-designed toxicogenomics experiments hold the promise of detecting subtle and toxicant-specific effects due to low-dose, environmentally relevant exposures. Conversely, microarray experiments are a semi-quantitative technique, and are not the best means of determining the precise level of altered gene expression for individual genes, which is more appropriately done using a technique such as quantitative real-time RT-PCR. In our view there are four important applications of low-dose toxicogenomics results: the elucidation of mechanisms of toxicant action; the identification of potential biomarkers of exposure, susceptibility, or effect; the prediction of physiological or pathophysiological consequences of toxicant exposure; and the characterization of the low end of the biological dose–response to a given toxicant.

With the recent advances in pathway and network analysis tools, microarray data can now provide a rich starting point of testable hypotheses for determining the mode of action of a toxicant. Toxicogenomic gene expression data are particularly useful when anchored to a specific conventional toxic or pathological phenotype (similar to the conventional use of serum levels of liver enzymes that correlate with hepatic damage), and this process is referred to as 'phenotypic anchoring' (Tennant, 2002; Paules, 2003; Waters and Fostel, 2004). Anchoring the changes in molecular expression to such physiological phenotypes (e.g. tumorigenesis) can provide a context to alterations in gene expression and can help to define the sequence of key molecular events in a toxicant mechanism of action. Also, toxicogenomic data can be mined for the identification of biomarkers of both exposure and toxicity (reviewed in Merrick and Bruno (2004)).

One of the key goals for many in the field of toxicogenomics is to use microarray data for the prediction of consequences of exposure to new, not previously tested, toxicants, or identification of new consequences of concern for existing toxicants, with the goal of rapid screening for intelligent prioritization of lists of chemicals to be tested for further toxicity or human health effects. The field of predictive toxicogenomics is based on the assumption that chemicals with similar overt toxicology will induce similar patterns of gene expression; this assumption has been supported by several studies (reviewed in Hayes and Bradfield (2005)). The goal is to infer the likelihood of occurrence of a toxic event with exposure to a new agent based upon comparative responses within large databases of toxicogenomic data. For example, gene expression profiles obtained from hepatotoxin exposure were used in cluster analysis to compare known hepatotoxins in order to extract common features (Waring *et al.*, 2001a,b) (summarized in Hamadeh *et al.* (2002)) that could be used for future screening of potential but unknown hepatotoxins. Predictive toxicogenomics and its

use in risk assessment has been previously reviewed (Boverhof and Zacharewski, 2006; Fielden and Kolaja, 2006; Pognan, 2007), and analysis methods specifically for predictive toxicology have also been reviewed (Maggioli *et al.*, 2006).

Toxicogenomics has also been proposed as a way to predict or explain joint toxicant action of complex mixtures, a challenging but important issue in toxicology. Predictive toxicogenomics also may allow drug developers to identify potential adverse drug effects earlier in the drug development pipeline (reviewed in Searfoss *et al.* (2005)). Even though advances in drug development have led to an exponential increase in the number of potential drug targets, classical toxicity testing of drug candidates is slow, insensitive and expensive, and it is often a significant and rate-limiting step in the drug development process (de Longueville *et al.*, 2004). Toxicogenomics studies may allow for more accurate and sensitive detection of early stages of toxicity, seen in the gene expression signatures before traditional toxicological endpoints occur (e.g. cancer, death), which may take months or years of exposure or require large numbers of individuals. Thus, toxicogenomics – particularly its use in low-dose exposures – holds great promise for both toxicology and pharmacology.

2.11   Conclusions

Gene expression microarrays are sensitive tools for the detection of biological effects due to low-dose, environmentally relevant toxicant exposures. However, in order to maximize the usefulness of such microarray data in low-dose toxicology and risk assessment, several key points should be considered:

- As with any biological experiment, experimental design is crucial, and several biological replicates are necessary.
- There are many software options for the analysis of microarray data, but there is a need for a free, publicly available and readily useable GUI that contains much of the functionality of Bioconductor.
- The avoidance of arbitrary data filtering and the use of standard and reliable normalization techniques will help significantly in the standardization and optimization of microarray data.
- Until very recently, most toxicogenomic experiments used excessively rigid fold-change and significance criteria, resulting in the failure to observe what are likely to be highly relevant biological responses. Conventional *t*-tests place a large weight on standard deviation, make many assumptions about the structure of the data and suffer acutely from the multiple testing problem. Modified *t*-tests, such as SAM, emphasize fold change over standard deviation, make fewer assumptions about the data and use computational procedures to generate a more reasonable FDR. Rank-based methods, such as RankProd, place emphasis on effect size and qualitative consistency among replicates, rather than standard deviation, and actually become more robust with multiple testing. In our experience, rank-based methods generate the longest gene lists, which are the most concordant among replicate experiments. Thus, we recommend less rigid but scientifically justifiable analysis methods, such as the use of rank-based methods and also direct visualization of data in order to extract a more comprehensive and meaningful dataset.

- Small changes in gene expression are sometimes biologically meaningful and can be measured reliably (confirmable by other methods) on microarrays using the appropriate analysis techniques. Likewise, low-dose toxicant exposures produce biologically significant responses that are highly reproducible and of potentially great value to toxicology and pharmacology.
- Pathway analysis programs are very useful to integrate expression data from a candidate list of affected genes and to place the gene expression changes into the context of meaningful biological responses. However, it is important to remember that such programs are currently not comprehensive in the pathways that they represent, and so additional analysis is still required following identification of candidate pathways.
- Overall, we recommend an iterative process of exploratory data analysis of microarrays (illustrated in Figure 2.1), including further experimental confirmation of gene expression effects and downstream physiological effects. Such analyses can lead to better understanding of mechanisms of toxicant action, the identification of biomarkers, more robust predictive toxicology, and the ability to characterize the low end of dose-response curves.

Acknowledgments

We thank Courtney Kozul and Dr Michael Spinella for critical reading of the manuscript. We also thank Manida Wungjiranirun for experimental assistance.

References

Alberts R, Terpstra P, Hardonk M, Bystrykh LV, de Haan G, Breitling R, Nap JP, Jansen RC (2007) A verification protocol for the probe sequences of Affymetrix genome arrays reveals high probe accuracy for studies in mouse, human and rat. *BMC Bioinform* **8**, 132.

Andrew AS, Karagas MR, Hamilton JW (2003) Decreased DNA repair gene expression among individuals exposed to arsenic in United States drinking water. *Int J Cancer* **104**, 263–268.

Andrew AS, Burgess JL, Meza MM, Demidenko E, Waugh MG, Hamilton JW, Karagas MR (2006) Arsenic exposure is associated with decreased DNA repair *in vitro* and in individuals exposed to drinking water arsenic. *Environ Health Perspect* **114**, 1193–1198.

Ayroles JF, Gibson G (2006) Analysis of variance of microarray data. *Methods Enzymol* **411**, 214–233.

Baldi P, Long AD (2001) A Bayesian framework for the analysis of microarray expression data: regularized *t*-test and statistical inferences of gene changes. *Bioinformatics* **17**, 509–519.

Barnes M, Freudenberg J, Thompson S, Aronow B, Pavlidis P (2005) Experimental comparison and cross-validation of the Affymetrix and Illumina gene expression analysis platforms. *Nucleic Acids Res* **33**, 5914–5923.

Bolstad BM, Irizarry RA, Astrand M, Speed TP (2003) A comparison of normalization methods for high density oligonucleotide array data based on variance and bias. *Bioinformatics* **19**, 185–193.

Borovecki F, Lovrecic L, Zhou J, Jeong H, Then F, Rosas HD, Hersch SM, Hogarth P, Bouzou B, Jensen RV, Krainc D (2005) Genome-wide expression profiling of human blood reveals biomarkers for Huntington's disease. *Proc Natl Acad Sci U S A* **102**, 11023–11028 (2005).

Boverhof DR, Zacharewski TR (2006) Toxicogenomics in risk assessment: applications and needs. *Toxicol Sci* **89**, 352–360.

Breitling R (2006) Biological microarray interpretation: the rules of engagement. *Biochim Biophys Acta* **1759**, 319–327.

Breitling R, Herzyk P (2005) Rank-based methods as a non-parametric alternative of the *T*-statistic for the analysis of biological microarray data. *J Bioinform Comput Biol* **3**, 1171–1189.

Breitling R, Armengaud P, Amtmann A, Herzyk P (2004) Rank products: a simple, yet powerful, new method to detect differentially regulated genes in replicated microarray experiments. *FEBS Lett* **573**, 83–92.

Butte A (2002) The use and analysis of microarray data. *Nat Rev Drug Discov* **1**, 951–960.

Chen G, Jaradat SA, Banerjee N, Tanaka TS, Ko MSH, Zhang MQ (2002) Evaluation and comparison of clustering algorithms in analyzing ES cell gene expression data. *Stat Sinica* **12**, 241–262.

Cleveland WS (1979) Robust locally weighted regression and smoothing scatterplots. *J Am Stat Assoc* **74**, 829–836.

Clewell HJ, Thomas RS, Gentry PR, Crump KS, Kenyon EM, El-Masri HA, Yager JW (2007) Research toward the development of a biologically based dose response assessment for inorganic arsenic carcinogenicity: a progress report. *Toxicol Appl Pharmacol* **222**, 388–398.

Cope LM, Irizarry RA, Jaffee HA, Wu Z, Speed TP (2004) A benchmark for Affymetrix GeneChip expression measures. *Bioinformatics* **20**, 323–331.

Dahlquist KD, Salomonis N, Vranizan K, Lawlor SC, Conklin BR (2002) GenMAPP, a new tool for viewing and analyzing microarray data on biological pathways. *Nat Genet* **31**, 19–20.

Davey JC, Andrew AS, Barchowsky A, Soucy NV, Mayka DD, Lantz RC, Hays A, Hamilton JW (2004) Toxicogenomics of drinking water arsenic *in vivo*: effects of replicates on microarray analysis. *Toxicol Sci* **78** (Suppl. 1), 60.

Davey JC, Gosse JA, Kozul CD, Hampton TH, Nomikos A, Warnke LAT, Bodwell JE, Ihnat MA, Hamilton JW (2008) Microarray analysis of mouse liver transcripts following low dose arsenic/dexamethasone exposure. *Toxicologist* **102**, 293.

De Longueville F, Bertholet V, Remacle J (2004) DNA microarrays as a tool in toxicogenomics. *Comb Chem High Throughput Screen* **7**, 207–211.

Dennis Jr G, Sherman BT, Hosack DA, Yang J, Gao W, Lane HC, Lempicki RA (2003) DAVID: Database for Annotation, Visualization, and Integrated Discovery. *Genome Biol* **4**, R60.

Eckel JE, Gennings C, Chinchilli VM, Burgoon LD, Zacharewski TR (2004) Empirical Bayes gene screening tool for time-course or dose–response microarray data. *J Biopharm Stat* **14**, 647–670.

Efron B, Tibshirani R, Storey J, Tusher V (2001) Empirical Bayes analysis of a microarray experiment. *J Am Stat Assoc* **96**, 1151–1160.

Elbez Y, Farkash-Amar S, Simon I (2006) An analysis of intra array repeats: the good, the bad and the non informative. *BMC Genomics* **7**, 136.

Fielden MR, Kolaja KL (2006) The state-of-the-art in predictive toxicogenomics. *Curr Opin Drug Discov Dev* **9**, 84–91.

Gildea LA, Ryan CA, Foertsch LM, Kennedy JM, Dearman RJ, Kimber I, Gerberick GF (2006) Identification of gene expression changes induced by chemical allergens in dendritic cells: opportunities for skin sensitization testing. *J Invest Dermatol* **126**, 1813–1822.

Goldstein DR, Ghosh D, Conlon EM (2002) Statistical issues in the clustering of gene expression data. *Stat Sinica* **12**, 219–240.

Guo L, Lobenhofer EK, Wang C, Shippy R, Harris SC, Zhang L, Mei N, Chen T, Herman D, Goodsaid FM, Hurban P, Phillips KL, Xu J, Deng X, Sun YA, Tong W, Dragan YP, Shi L (2006) Rat toxicogenomic study reveals analytical consistency across microarray platforms. *Nat Biotechnol* **24**, 1162–1169.

Hamadeh HK, Amin RP, Paules RS, Afshari CA (2002) An overview of toxicogenomics. *Curr Issues Mol Biol* **4**, 45–56.

Hamilton JW (2006) Toxicogenomic and toxicoproteomic approaches for biomarkers. In *Toxicologic Biomarkers* AP DeCaprio (ed.). Marcel Dekker: New York.

Hampton T, Davey JC, Gosse JA, Warnke LA, Ihnat MA, Andrew AS, Shaw JR, Hamilton JW (2007) Nonparametric and graphical methods reveal new gene patterns in low dose arsenic microarray experiments. *Toxicol Sci* **96** (Suppl.), 252.

Hayes KR, Bradfield CA (2005) Advances in toxicogenomics. *Chem Res Toxicol* **18**, 403–414.

Hong F, Breitling R, McEntee CW, Wittner BS, Nemhauser JL, Chory J (2006) RankProd: a bioconductor package for detecting differentially expressed genes in meta-analysis. *Bioinformatics* **22**, 2825–2827.

Irizarry RA, Bolstad BM, Collin F, Cope LM, Hobbs B, Speed TP (2003a) Summaries of Affymetrix GeneChip probe level data. *Nucleic Acids Res* **31**, e15.

Irizarry RA, Hobbs B, Collin F, Beazer-Barclay YD, Antonellis KJ, Scherf U, Speed TP (2003b) Exploration, normalization, and summaries of high density oligonucleotide array probe level data. *Biostatistics* **4**, 249–264.

Irizarry RA, Warren D, Spencer F, Kim IF, Biswal S, Frank BC, Gabrielson E, Garcia JG, Geoghegan J, Germino G, Griffin C, Hilmer SC, Hoffman E, Jedlicka AE, Kawasaki E, Martinez-Murillo F, Morsberger L, Lee H, Petersen D, Quackenbush J, Scott A, Wilson M, Yang Y, Ye SQ, Yu W (2005) Multiple-laboratory comparison of microarray platforms. *Nat Methods* **2**, 345–350.

Irizarry RA, Wu Z, Jaffee HA (2006) Comparison of Affymetrix GeneChip expression measures. *Bioinformatics* **22**, 789–794.

Kibriya MG, Jasmine F, Argos M, Verret WJ, Rakibuz-Zaman M, Ahmed A, Parvez F, Ahsan H (2007) Changes in gene expression profiles in response to selenium supplementation among individuals with arsenic-induced pre-malignant skin lesions. *Toxicol Lett* **169**, 162–176.

Kozul CD, Nomikos AP, Hampton TH, Davey JC, Gosse JA, Warnke LA, Ihnat MA, Jackson BP, Hamilton JW (2007) Laboratory diet alters gene expression in mice. *Toxicol Sci*, **96** (Suppl.), 154.

Larkin JE, Frank BC, Gavras H, Sultana R, Quackenbush J (2005) Independence and reproducibility across microarray platforms. *Nat Methods* **2**, 337–344.

Lee NH, Saeed AI (2007) Microarrays: an overview. *Methods Mol Biol* **353**, 265–300.

Lin DY (2005) An efficient Monte Carlo approach to assessing statistical significance in genomic studies. *Bioinformatics* **21**, 781–787.

Maggioli J, Hoover A, Weng L (2006) Toxicogenomic analysis methods for predictive toxicology. *J Pharmacol Toxicol Methods* **53**, 31–37.

Mattes WB, Pettit SD, Sansone SA, Bushel PR, Waters MD (2004) Database development in toxicogenomics: issues and efforts. *Environ Health Perspect* **112**, 495–505.

Merrick BA, Bruno ME (2004) Genomic and proteomic profiling for biomarkers and signature profiles of toxicity. *Curr Opin Mol Ther* **6**, 600–607.

Ness SA (2006) Basic microarray analysis: strategies for successful experiments. *Methods Mol Biol* **316**, 13–33.

Olson NE (2006) The microarray data analysis process: from raw data to biological significance. *NeuroRx* **3**, 373–383.

Parmigiani G, Garett ES, Irizarry RA, Zeger SL (2003) *The Analysis of Gene Expression Data*. Springer: New York.

Paules R (2003) Phenotypic anchoring: linking cause and effect. *Environ Health Perspect* **111**, A338–A339.

Pennie W, Pettit SD, Lord PG (2004) Toxicogenomics in risk assessment: an overview of an HESI collaborative research program. *Environ Health Perspect* **112**, 417–419.

Pognan F (2007) Toxicogenomics applied to predictive and exploratory toxicology for the safety assessment of new chemical entities: a long road with deep potholes. *Prog Drug Res* **64**, 217–238.

Qin LX, Beyer RP, Hudson FN, Linford NJ, Morris DE, Kerr KF (2006) Evaluation of methods for oligonucleotide array data via quantitative real-time PCR. *BMC Bioinform* **7**, 23.

Rainer J, Sanchez-Cabo F, Stocker G, Sturn A, Trajanoski Z (2006) CARMAweb: comprehensive R- and bioconductor-based web service for microarray data analysis. *Nucleic Acids Res* **34**, W498–W503.

Reimers M, Carey VJ (2006) Bioconductor: an open source framework for bioinformatics and computational biology. *Methods Enzymol* **411**, 119–134.

Saeed AI, Bhagabati NK, Braisted JC, Liang W, Sharov V, Howe EA, Li J, Thiagarajan M, White JA, Quackenbush J (2006) TM4 microarray software suite. *Methods Enzymol* **411**, 134–193.

Searfoss GH, Ryan TP, Jolly RA (2005) The role of transcriptome analysis in pre-clinical toxicology,. *Curr Mol Med* **5**, 53–64.

Shannon P, Markiel A, Ozier O, Baliga NS, Wang JT, Ramage D, Amin N, Schwikowski B, Ideker T (2003) Cytoscape: a software environment for integrated models of biomolecular interaction networks. *Genome Res* **13**, 2498–2504.

Shaw JR, Colbourne JK, Davey JC, Glaholt SP, Hampton TH, Chen CY, Folt CL, Hamilton JW (2007) Gene response profiles for *Daphnia pulex* exposed to the environmental stressor cadmium reveals novel crustacean metallothioneins. *BMC Genomics* **8**, 477, DOI: 10.1186/1471-2164-8-477.

Shedden K, Chen W, Kuick R, Ghosh D, Macdonald J, Cho KR, Giordano TJ, Gruber SB, Fearon ER, Taylor JM, Hanash S (2005) Comparison of seven methods for producing Affymetrix expression scores based on false discovery rates in disease profiling data. *BMC Bioinform* **6**, 26.

Sherlock G (2000) Analysis of large-scale gene expression data. *Curr Opin Immunol* **12**, 201–205.

Steinhoff C, Vingron M (2006) Normalization and quantification of differential expression in gene expression microarrays. *Brief Bioinform* **7**, 166–177.

Tennant RW (2002) The National Center for Toxicogenomics: using new technologies to inform mechanistic toxicology. *Environ Health Perspect* **110**, A8–A10.

Thimm O, Blasing O, Gibon Y, Nagel A, Meyer S, Kruger P, Selbig J, Muller LA, Rhee SY, Stitt M (2004) MAPMAN: a user-driven tool to display genomics data sets onto diagrams of metabolic pathways and other biological processes. *Plant J* **37**, 914–939.

Tusher VG, Tibshirani R, Chu G (2001) Significance analysis of microarrays applied to the ionizing radiation response. *Proc Natl Acad Sci USA* **98**, 5116–5121.

Usadel B, Nagel A, Thimm O, Redestig H, Blaesing OE, Palacios-Rojas N, Selbig J, Hannemann J, Piques MC, Steinhauser D, Scheible WR, Gibon Y, Morcuende R, Weicht D, Meyer S, Stitt M (2005) Extension of the visualization tool MapMan to allow statistical analysis of arrays, display of corresponding genes, and comparison with known responses. *Plant Physiol* **138**, 1195–1204.

Verducci JS, Melfi VF, Lin S, Wang Z, Roy S, Sen CK (2006) Microarray analysis of gene expression: considerations in data mining and statistical treatment. *Physiol Genomics* **25**, 355–363.

Wang S, Cheng Q (2006) Microarray analysis in drug discovery and clinical applications. *Methods Mol Biol* **316**, 49–65.

Waring JF, Ciurlionis R, Jolly RA, Heindel M, Ulrich RG (2001a) Microarray analysis of hepatotoxins *in vitro* reveals a correlation between gene expression profiles and mechanisms of toxicity. *Toxicol Lett* **120**, 359–368.

Waring JF, Jolly RA, Ciurlionis R, Lum PY, Praestgaard JT, Morfitt DC, Buratto B, Roberts C, Schadt E, Ulrich RG (2001b) Clustering of hepatotoxins based on mechanism of toxicity using gene expression profiles. *Toxicol Appl Pharmacol* **175**, 28–42.

Waters MD, Fostel JM (2004) Toxicogenomics and systems toxicology: aims and prospects. *Nat Rev Genet* **5**, 936–948.

Wei M, Arnold L, Cano M, Cohen SM (2005) Effects of co-administration of antioxidants and arsenicals on the rat urinary bladder epithelium. *Toxicol Sci* **83**, 237–245.

Wu W, Dave N, Tseng GC, Richards T, Xing EP, Kaminski N (2005) Comparison of normalization methods for CodeLink Bioarray data. *BMC Bioinform* **6**, 309.

Wu Z, Irizarry RA, Gentleman R, Martinez-Murillo F, Spencer F (2004) A model based background adjustment for oligonucleotide expression arrays. Johns Hopkins University Department of Biostatistics Working Papers Series 1001, Berkeley Electronic Press, Berkeley, CA.

Yan H, Dobbie Z, Gruber SB, Markowitz S, Romans K, Giardiello FM, Kinzler KW, Vogelstein B (2002) Small changes in expression affect predisposition to tumorigenesis. *Nat Genet* **30**, 25–26.

Yao B, Rakhade SN, Li Q, Ahmed S, Krauss R, Draghici S, Loeb JA (2004) Accuracy of cDNA microarray methods to detect small gene expression changes induced by neuregulin on breast epithelial cells. *BMC Bioinform* **5**, 99.

Yuen T, Wurmbach E, Pfeffer RL, Ebersole BJ, Sealfon SC (2002) Accuracy and calibration of commercial oligonucleotide and custom cDNA microarrays. *Nucleic Acids Res* **30**, e48.

3

Principles of Data Mining in Toxicogenomics

Yoko Hirabayashi and Tohru Inoue

3.1 Introduction

When animals are exposed to ionizing radiation, the radioactivity induces a probabilistic quantum effect based on the uncertainty principle followed by sequential early events in physicochemical processes, which is in the realm of subatomic particle physics (Czapski and Peled, 1973; *ibid*, 1975; Hunt, 1976). Because the initial radiation hit may be probabilistic and random with respect to radiation beams, radiation-induced damage may be random and stochastic from one molecule to another, from one cell to another, from one tissue to another and, furthermore, from one individual to another. Consequently, data should not be statistically focused but must reveal a limited probabilistic divergence, including divergence from one cluster to another. Recent developments of DNA chips, photolithographic microchip analysis and integrated microarray methods (Brown and Botstein, 1999) have enabled the analysis of the global expressions of more than 34 000 genes as ultimate biological responses (Lovett, 2000; Hamadeh *et al.*, 2001; Storck *et al.*, 2002). The method is effective in not only providing deterministic genetic information, but also nondeterministic epigenetic information. Nondeterministic epigenetic information, together with computational toxicology, can be used to elucidate the mechanism underlying the above-mentioned background and to identify genes including responsible gene ontologies (GOs).

It is also of interest to compare gene expression profiles between radiation-induced myelogenous leukemias and spontaneous myelogenous leukemias as a sample analytical model for computational toxicology; principal component analysis (PCA) primarily differentiates the expression profiles between radiation-induced and spontaneous myelogenous leukemias. Although radiation-induced myelogenous leukemias are expected to show a convergent gene expression pattern because of random but limited damage induced by radiation exposure linked to, for example, radiation-fragile sites (Hastie and Allshire, 1989; Yunis *et al.*, 1987), spontaneous myelogenous leukemias are associated with more highly stochastic diversities in gene expression patterns. The ordered list generated by PCA provides the contributing genes that differentiate the expressions associated with radiation-induced myelogenous leukemias from those associated with spontaneous myelogenous leukemias. These genes are assumed to provide useful biomarkers for differentiating these myelogenous leukemias that have different characteristics, although additional computational treatment is required to elucidate the radiation-specific gene expression profiles (see Section 3.8).

3.2   From gene Expression to Toxicological Application

Genetic information is carried by chromosomes, the major composition of which is nucleic acid, i.e. DNA, consisting of four different bases, namely guanine, adenine, thymine, and cytosine (Watson and Crick, 1953). Such genetic information is transcribed into messenger RNA (Brenner *et al.*, 1961), triplets of which, 'codons,' are translated into the 20 different amino acids (Crick *et al.*, 1961). On the basis of the translated information on the amino acid sequence, the protein, a source of life, is synthesized (Crick, 1958). A full set of chromosomes is called the genome, and its composition (DNA sequence) is replicated semiconservatively as genetic information (a master plan), which is transferred to descendant DNA in the case of germ cells or contributes to the expression of various biological activities, including cellular proliferation, apoptosis or metabolic activities, in the case of somatic cells (Singer and Berg, 1991; Bloomfield *et al.*, 2000; Lewin, 2004; Lodish *et al.*, 2004; Watson *et al.*, 2007). From the above-described background, the use of 'toxicogenomics' is aimed at collecting such information, which is assumed to reflect all the information about life, including on the one hand intrinsic biological activities and on the other hand xenobiotic responses (Inoue, 2003).

The sequencing of the genomes of *C. elegans* (*C. elegans* Sequencing Consortium, 1998), *Drosophila* (Adams *et al.*, 2000), mouse (Waterston *et al.*, 2002) and human (Lander *et al.*, 2001) was completed around the year 2000. This development makes it possible to establish a method of detecting global gene expression as macroscientific information, including the determination of xenobiotic hazardous effects (Inoue, 2003). Such information can be utilized for many different purposes, and its toxicological application, particularly for predicting toxicological gene expression, is called toxicogenomics (Borlak, 2005).

3.3   Toxicogenomics

Toxicogenomics can be assumed as an 'expression gene microscope' because the method focuses on patterns of global gene expression as a whole, rather than on individual gene expressions (Figure 3.1). Original data obtained and examined by gene chip or microarray technologies provide mathematical values based on expression intensities. These values have essentially no biological meaning; rather, they reveal only a pattern. The patterned expression intensities obtained and those of the corresponding reference findings are

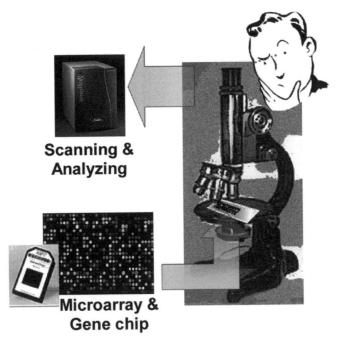

Figure 3.1 *Toxicogenomics methodology can be assumed as an 'expression gene micro-scope' because the method focuses on patterns of global gene expression as a whole rather than on individual gene expressions. See text. Two different methods are utilized: one using DNA microarray invented by Brown and Botstein (1999) and one using photolithographic gene chips invented by Fodor et al. (1993). The patterns of gene expression of targeted groups and the control are analyzed computationally using various pieces of external information (supervised analysis) or solely with internal data (unsupervised analysis).*

mathematically analyzed, which is considered 'transcriptomics analysis.' One can find similarities in the histopathological examination; there are different findings between un-treated and treated histological specimens, and the gaps in the findings can be identified and considered as a histological alteration induced by the treatment. Toxicogenomics also requires an established database as well as an analytical computational methodology; this is similar to histopathology, which was established using the histopathological disease entities established during a hundred years of pathological history (Henke and Lubarsch, 1924–1952).

3.4   Mining of Information Provided by Toxicogenomics

Categorically, two different types of information are provided by gene expression profiles: 'genetic information' and 'epigenetic information.' The former is based on the geneti-cally defined deterministic information. However, the latter is based on the developmental tissue-specific gene expressions, for example, or based on the consequences of xenobiotic responses with modified gene expressions by methylation, acetylation, or phosphorylation of the genome; thus, the alteration would be probabilistic and plastic. The former, for example, includes single-individual gene-specific nucleotide polymorphism (SNP), which may be within a category of a new type of genomic bioassay. On the other hand, the latter seems to be analogous to conventional microscopy, which is considered to provide analytical information related to xenobiotic responses when one compares altered gene expressions after xenobiotic responses with an appropriate control database. Either type of information on gene expressions obtained from the former and the latter materials is useful in the field of toxicology; both are managed similarly from the technological viewpoint; thus, the latter is mainly discussed in the text.

Gene expression profiles from mice after whole-body irradiation were used as experimental models. In this experiment, mice were exposed to a single dose of radiation at 0.6, 1, and 3 Gy and the gene expression profile in the bone marrow of the exposed mice compared with that of the nonirradiated control. From the results, one can find gene expression profiles with sequential changes in intensity with increasing radiation dose (*dose-dependent profiling*). On the other hand, one can also find a dose-independent expression pattern among the groups (*dose-specific profiling*). The former profiling is considered to be a useful tool for comparing dose–response relationships from the outcome of various testing protocols. The latter, on the other hand, is not only dose specific, but, interestingly, also possibly valid for a wide range of interspecies extrapolation; thus, the gene expression profile of the latter can be another potentially useful biomarker (data not shown).

3.5   Points to Consider in Data Mining Using Gene Chip and Microarray Technologies

Gene chip and microarray technologies provide a large amount of genetic information distributed in the over 34 000-dimensional Euclidian universe, because such data are based on the independently regulated expression of 34 000 genes (GeneChip® Mouse Genome 430 2.0 Array; Affymetrix, Santa Clara, CA). Various computational aspects of trials have been made available to reduce such a large number of dimensions (Zhang and Shmulevich, 2006; Albanese *et al.*, 2007).

One piece of software provides a single-dimensional gene matrix based on each gene expression intensity, which is clusterized by Euclidean distance on the basis of gene expression intensity, and links them to each other according to their vector values (GeneSpring GX 7.3.1; Agilent Technologies Inc., Santa Clara, CA). A sample dendrogram is shown in Figure 3.2. These expression data can be clusterized and linked on the basis of two factors, namely gene expression intensity and groups of conditional trees. Two-dimensional and higher multiphasic dendrographic analyses can be carried out. These processes are carried out unsupervised, namely statistically autogenerated.

In this dendrographic analytical method, the arrangement of each gene is linked solely on the basis of expression intensity; thus, the relationships of genes are solely based on a single-dimensional diagram. Specifically, in this dendrographic analytical method, the statistical structure allows deterministic alterations, such as mutations induced directly by a genotoxic carcinogenic compound, the mechanism of which needs to be elucidated more clearly. Unpublished data, as an example, provided by Professor T. Iguchi, Okazaki

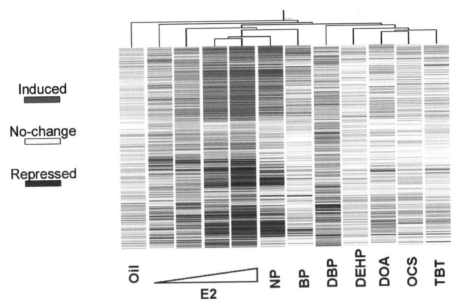

Induced

No-change

Repressed

Oil

E2

NP
BP
DBP
DEHP
DOA
OCS
TBT

Figure 3.2   *Sample data for dendrographic analysis. Four columns, next to oil on the left, are from mice treated with a graded increase in dose of 17-beta-estradiol (E2). NP: nonylphenol; BP: benzophenone; DBP: dibutylphthalate; DEHP: diethylhexyl phthalate; DOA: dioctyl adipate; OCS: octachlorostyrene; TBT: tributyltin. (Unpublished data provided by Professor T. Iguchi, Okazaki National Biology Research Institute.)*

National Biology Research Institute, show a characteristic analytical function of the dendrographic analysis introduced above (Figure 3.2). In this figure, the gene expression profiles associated with estradiols at increasing reference doses and those associated with several endocrine-disrupting chemicals are shown. From the comparative gene expressions, the expression profiles associated with the reference estradiols, i.e. nonylphenol (NP) and benzophenone (BP), are clustered on the left half of the figure. On the other hand, the unsupervised autogenerated dendrogram does not show any representative profiles related to the role of endocrine-disrupting chemicals in the expression profiles associated with other known endocrine-disrupting chemicals, such as dibutylphthalate (DBP), diethylhexyl phthalate (DEHP) and dioctyl adipate (DOA). Specifically, dendrographic analysis is solely based on the expression intensity of each gene, and these autogenerated vectors clustered by approximation along the single-dimensional analysis cannot be compared with other dimensional data. Thus, multiphasic data profiling and multidimensional data analysis are based on different functions. That is, when one plots the expression data of each gene along a fixed dimension, it may be difficult to compare these data with the gene expression data along a different multidimensional axis having a pleiotropic interrelationship.

Furthermore, not all gene functions are important when genes are strongly expressed. Similarly, not all gene functions are of less importance when genes are weakly expressed, although the functions of weakly expressed genes are sometimes difficult to analyze. It is more difficult to analyze the interrelationship of genes on the basis of transcriptional activation and to elucidate the consequent gene expression pathway using this methodology.

(1)   (2)   (3)

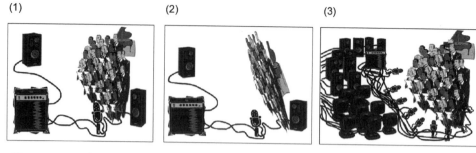

Figure 3.3 *Information from gene chip and/or microarray analyses is multidimensional. A microarray consisted of 34 000 genes behaving in the 30 000-dimensional Euclidian universe; thus, proper computational analysis requires appropriate multidimensional analytical power. See text.*

Figure 3.3a shows a cartoon where 30 000 persons are expressing (dispatching) their individual information. Depending on how a receiver receives the information, such information might be distorted because of the low-dimensional accepting system; the information might be distorted solely by expression intensities, as shown in Figure 3.3b for example. The possible reason why the intended information is not obtained from the microarray data is not always because of the poorly normalized raw data or the lack of significant data. One may have to realize that the reason may be the insufficient computational incorporation of multidimensional aspects of microarray data. Consequently, it may be appropriate when a possible multidimensional analysis of microarray data can be carried out to recover the data multidimensionally, as shown in Figure 3.3c.

3.6   Reproducibility of Microarray Data and Comparison among the Data Obtained from Different Platforms

Reproducibility, homogeneity and stability during data sampling are critical so that global gene expressions can be compared, because RNAs sampled from tissues are labile in general. Thus, the proper robotic administration of test compounds and the standardization of test materials are of special importance, the technical advancement of which has contributed markedly to the present microarray technologies (Schadt *et al.*, 2001; Churchill, 2002; Kroll and Wolfl, 2002; Astrand, 2003; Stoyanova *et al.*, 2004). Furthermore, spotting and spike gene incorporations also contributed to the development of a hardware system that supports qualitative data evaluation and semiquantitative gene expression profiles (Hill *et al.*, 2001). Consequently, newly introduced journals in the field of computational sciences publish a number of informational studies. Data from different platforms are also actively compared and reported because of minimum requirements in the field of microarray study (Brazma *et al.*, 2001). Actual sample evaluation data are not presented in this paper; however, the similarities of those obtained data with those obtained from different platforms are often confirmed by PCA. This is not only because of equivalences at each technical level of data processing, but also because it is more likely that those gene expression data are sequentially linked together. Current gene expression data obtained by microarray and gene chip technologies are largely comparable, unless specific spike genes or other additional modifiers are incorporated in the system (Petersen *et al.*, 2005).

3.7 Pathological and Toxicological Endpoints

In this section, let us consider the general data mining of observed outcomes obtained by gene chip and microarray technologies. From Figure 3.4, which shows different routes from the bottom to the mountain top, one can assume that the top where the routes merge is a pathological endpoint where common genes are mostly expressed, whereas the other enclosed area at the foot of the mountain can be considered a toxicological endpoint, which is assumed to be based on the different routes to the summit represented by stochastic and probabilistic gene expression clusters. The former is considered to include a group of specific and deterministic gene expression profiles, i.e. an 'essential leukemogenic gene' profile, which could be used as possible biomarker genes for diagnosis, whereas the latter clusters represent various probabilistic uncertainties whose profiles are considered to be different from one cluster (route) to another, i.e. 'stochastically necessary genes.' In this regard, toxicological endpoints do not seem to provide definitive information of gene expression; however, toxicological endpoints represented by 'stochastically necessary genes'

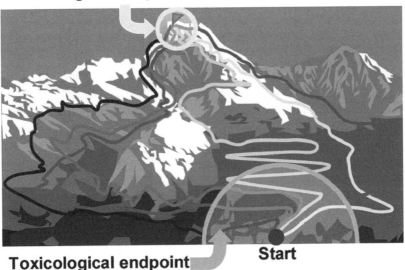

Pathological endpoint

Toxicological endpoint   **Start**

Figure 3.4 *'Pathological endpoint' and 'toxicological endpoint'. Each route from the start to the summit represents an individual probabilistic variety of different gene expression profiles. Depending on the characteristics of toxicological impacts, responders may show different toxicological endpoints in different clusters, even if one uses an identical and homogeneous experimental protocol with a highly purified inbred strain (a probabilistic quantum effect based on the uncertainty principle; see text).*

can provide probabilistic but cluster-specific predictability among various independent clusters. The latter clusters may not be statistically determined unless hundreds of a relatively large number of quantum cases are examined. Both an 'essential leukemogenic gene' profile and a 'stochastically necessary gene' profile are required for the early prediction of radiation-induced myelogenous leukemogenesis.

3.7.1 Pathological Endpoints

The strain C3H/He mouse develops a similar myelogenous leukemia both spontaneously and upon irradiation. The average incidence of the latter is 35% after 3 Gy irradiation, whereas that of the former is 1% (Seki *et al.*, 1991; Yoshida *et al.*, 1997). The line configuration in Figure 3.5 shows six cases of spontaneous leukemias on the left and six cases of radiation-induced myelogenous leukemias on the right. In this figure, the line configuration

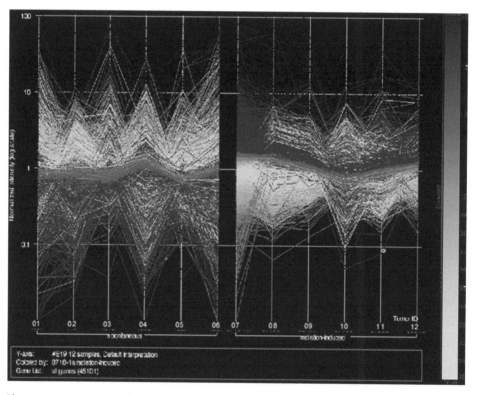

Figure 3.5 *Linear configurations of spontaneous and radiation-induced myelogenous leukemias. Six individual data on the left are from spontaneously developed myelogenous leukemias in C3H/He mice. The other six individual cases on the right are from radiation-induced myelogenous leukemias in C3H/He mice after 3 Gy X-ray exposure (Seki et al. 1991; Yoshida et al. 1997). Along the gene expression intensity from the highest (red) to the lowest (green) of group #07, the same gene in the other groups was connected and designated with the same color. Accordingly, overexpressed genes in the radiation-induced myelogenous leukemia groups are largely repressed in the spontaneous myelogenous leukemias. See text.*

associated with the spontaneous leukemias on the left shows very prominent and wide divergence in expression intensities from one individual mouse to another, and individual differences in gene ordering are also significant among each other. In contrast, each case of radiation-induced myelogenous leukemia on the right shows relatively homogeneous expression intensities compared with the former cases of spontaneous leukemias. When one compares the linear configurations of spontaneous and radiation-induced myelogenous leukemias, genes associated with radiation-induced myelogenous leukemias are not expressed in the spontaneous leukemias similarly but rather diversely. These findings are compatible with the observation by real time RT-PCR (Applied Biosystems 7900 Sequence Detection System, ABI, Foster, CA) shown in Figure 3.6, in which the former shows diverged *Sh2d1a* expressions from spontaneous myelogenous leukemias (*Sh2d1a* top) and the latter, relatively homogeneous ones from radiation-induced myelogenous leukemias (*Sh2d1a* bottom). Similarly, another gene, *Gngt1*, depicted from radiation-induced myelogenous leukemia shows relatively homogeneous expressions in most of radiation-induced myelogenous leukemias (*Gngt1* bottom), whereas the expressions of *Gngt1* were not detected in spontaneous myelogenous leukemias (*Gngt1* top). These expression profiles are further analyzed by PCA and the gene cluster associated with radiation-induced myelogenous leukemia can be discriminated from that associated with spontaneous leukemia (Figure 3.7). Representative genes that can be used to differentiate between the two types of myelogenous leukemia can be determined by PCA. The list of genes that differentiate both leukemias include *Met* (met proto-oncogene), *Fosl2* (fos-like antigen2), *Fancd2* (Fanconi anemia, complementation group D2), and *Fmr2* (fragile X mental retardation 2 homolog), among others. It is important that the expression intensities of these discriminant genes described above are not always high in all radiation-induced myelogenous leukemias, but sometimes low in a stochastic manner. Therefore, it is assumed that these genes cannot be the 'essential leukemogenic genes,' but 'stochastically necessary genes' for radiation-induced myelogenous leukemias. As described later, these genes, however, still seem to be less discriminant for radiation-induced myelogenous leukemias, although the function of each gene is linked to radiation injury. Why are radiation-induced leukemia-specific gene expression profiles not determined by the above computational analysis?

Regardless of the difference between spontaneous and radiation-induced myelogenous leukemias, it took nearly a lifetime for both myelogenous leukemias to develop fatally. Thus, the profile of each type of myelogenous leukemia may be associated with an age-related gene expression profile. Such an age-related gene expression profile may overlap to some extent with the gene expression profiles associated with spontaneous and radiation-induced myelogenous leukemias. The relationship of this factor with the above-mentioned differentiation will be considered later.

3.7.2 Toxicological Endpoints

Toxicological endpoints are different from pathological endpoints in terms of the nature of probabilistic clusters, as discussed above. One such frequent epigenetic modification, e.g. DNA methylation, constitutes a post-replicative modification, in which a methyl group is added covalently to a DNA residue. As it is known that cancer diagnosis before the cancer reaches the 'point of no return' is considered 'logically' difficult, toxicologists consider the possibility of identifying gene repertoires that are possibly associated with

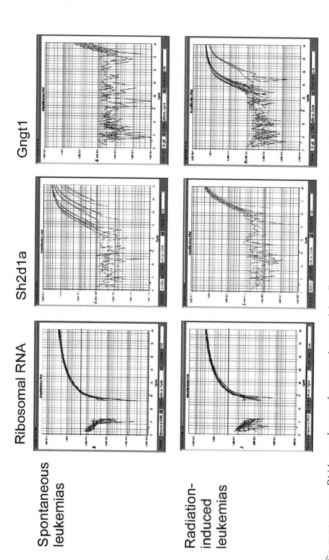

Figure 3.6 Specific messenger RNA products of genes detected by fluorescent probes (TaqMan™ probes, ABI). Ordinate axes represent relative fluorescence (VIC™, 6-FAM™) and horizontal axes the number of cycles amplified. Ribosomal RNAs as control for RT-PCR on the left; a sample gene from spontaneous leukemia, Sh2d1a, in the middle; and a sample gene from radiation-induced leukemia Gngt, on the right are evaluated in both spontaneous and radiation-induced myelogenous leukemias (six cases each, upper and lower row respectively). Sh2d1a expressions in spontaneous leukemias (Sh2d1a top) are diverged in expression intensities, whereas those from radiation-induced leukemias are relatively homogeneous (Sh2d1a bottom). Similarly, another gene, Gngt1, shows relatively homogeneous expressions in most of radiation-induced leukemias (Gngt1 bottom), whereas the expressions were not detected in spontaneous leukemias (Gngt1 top). Quantitative real-time PCR triplicates are shown in each figure.

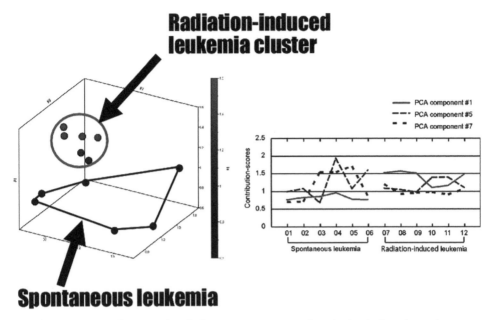

Figure 3.7 *PCA of six each of the spontaneous and radiation-induced myelogenous leukemias is shown in the three-dimensional contribution scores for components #1, #5 and #7, which discriminate the radiation-induced myelogenous leukemia cluster from the spontaneous myelogenous leukemia cluster. The line graph on the right shows actual contribution scores converted from each eigenvector value, which were used for the three-dimensional expression on the left. Note that the contribution scores of the spontaneous leukemias, except for component #1, are relatively divergent in comparison with those of radiation-induced leukemias.*

carcinogenesis at the early stage. There seems to be no definitive theoretical understanding of this issue. The most discouraging reason concerning this issue is that previous results obtained by the study of the International Life Science Institute (ILSI) consortium showed that the gene expression profiles associated with genophilic and nondirect genophilic pharmaceutical compounds were clearly differentiated, but both expression profiles from mice 4 h and 24 h after treatment showed up-regulation and down-regulation respectively (data not shown; Hu *et al.*, 2004). The possible diagnostic profiles, therefore, are considered to change with observation time or depending on the strain or treatment dose. Where can one find an appropriate discriminant axis? This is the question that needs to be answered.

The experimental results obtained by the ILSI consortium, however, can be interpreted differently. Those profiles may change with observation time, dose and strain; however, those profiles associated with genophilic compounds are always found to change in a direction opposite to that of the change of profiles associated with nondirect genophilic compounds; it is plausible that discriminant biomarkers for both genophilic and nondirect genophilic compounds would show this trend in both groups. In this regard, some sample trials carried out to determine these discriminant biomarker genes are discussed next.

The effect of age on experimental animals during toxicological experiments is one of the interesting factors associated with toxicological endpoints. Two groups of linear configurations represented by five mice each for the 2-month-old and 21-month-old groups are shown in Figure 3.8a. These two different age groups show different expression diversities; the aged group (21 months old) shows a much wider range of expression intensities than the younger (2 months old) group. Furthermore, the order of gene expression intensity between the two groups is reversed. The wider expression intensities of the aged group suggest that senescent changes appear stochastically and are probabilistically different from one individual to another. Note that the wider expression intensities do not indicate differences in the quality of individual animals but the different probabilistic responses of individual animals during aging. These individual differences are also observed at the cellular level, if each cellular function is affected by gene expression (Bahar *et al.*, 2006).

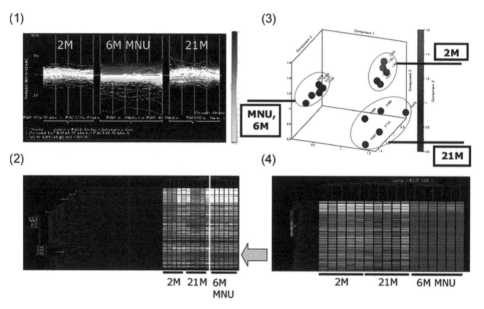

Figure 3.8 (a) Age-related gene expression profiles determined from the bone marrow of 2-month-old and 21-month-old mice are shown in the line configuration. See text. (Six gene expression profiles of bone marrow obtained from mice 6 months after treatment with a single dose of methyl-nitroso-urea (MNU) at 50 mg kg^{-1} b.w.) (b) Two-dimensional dendrographic diagram of the same expressed genes from the bone marrow of 2-month-old and 21-month-old mice. (Six gene expression profiles of bone marrow obtained from mice 6 months after treatment with a single dose of MNU. Arrow indicates the gene cluster specifically up-regulated in the MNU-treated groups; an expanded view of the profiles is shown in (d).) (c) PCA of age control group, five mice each for the 2-month-old and 21-month-old groups, is shown in the three-dimensional contribution scores for components #1, #2 and #3, which discriminate between the clusters from the 2-month-old and 21-month-old groups. Six gene expression profiles of bone marrow obtained from mice 6 months after MNU treatment belong to another separate cluster. (d) Expanded view of profiles indicated by the arrow in (b) showing the gene cluster specifically up-regulated in the MNU-treated groups.

Such differences between the two age groups cannot be defined using a dendrogram, as observed in the 10 columns on the left of Figure 3.8b, in which the aged group and the young group are not clearly separated. The reason that these different age groups are not clearly separated in the dendrogram is because of the relatively small number of genes with weak expressions that may define these different age groups. To pinpoint the possible responsible genes that define both groups, PCA was applied. Results show that there are discriminant components defined by components #1, #2 and #3, as shown in Figure 3.8c.

When one compares the gene expression profiles of these two age groups with those of mice treated with methyl-nitrosourea (MNU), a direct genophilic leukemogenic compound, the expression profiles of six bone marrow tissues from MNU-treated mice are clearly defined in the six columns on the right in Figure 3.8d. The discrimination of the clusters shown in the three-dimensional expressions is clearly separated in Figure 3.8c. The genes responsible for these discriminations can be autogenerated (data not shown).

3.8   Radiation-specific Probabilistic Union Genes after Subtracting Age-specific Gene Expressions

As mentioned previously, regardless of the difference between spontaneous and radiation-induced myelogenous leukemias, it took nearly a lifetime for both leukemias to develop fatally. Furthermore, radiation-induced myelogenous leukemias express stochastically divergent profiles, as observed in the line configuration in Figure 3.5. Therefore, each gene expression profile from a total of six radiation-induced myelogenous leukemia cases was analyzed by PCA to elucidate each unique gene list for nontreated mice and compared with the gene expression profiles of bone marrow cells from five 21-month-old nontreated mice. To obtain the union gene list, the gene expression profile of the 21-month-old group was compared separately with each individual expression profile of radiation-induced myelogenous leukemias by PCA followed by the selection of genes with a contribution score of over 1.0 from components #2 or #3 and #6 from the result of PCA (Table 3.1). In this PCA, the 287 union genes obtained were subtracted by 45 genes that overlapped with another 249 union genes analyzed by the PCA combination between the profile of the 21-month-old bone marrow group and that of spontaneous myelogenous leukemias (data not shown). Consequently, the final number of union genes obtained was 128 genes after the subtraction of 114 expression sequence tags (ESTs), which were generated in an unsupervised manner; however, most of them showed gene functions consistent with the response to radiation exposure. Specifically, five radiation-damage-related genes, including *Hus1*, *Eef1a2*, *Vegfc*, 13 cell cycle/cell-growth-related genes, and 12 apoptosis/cell-death-related genes were observed. Notably, 42 tumorigenesis-related genes (i.e. 33 %) were observed in which the down-regulation of tumor suppressor genes and the up-regulation of tumor promoter genes were commonly observed. The expressions of genes for cytoskeleton and cell adhesion molecules and those of genes for oxidative stress and inflammatory cytokines were also observed and included in the list; the expressions of these genes are also considered consequences of the xenobiotic responses to radiation exposure, because these genes were not included in the list of genes associated with spontaneous myelogenous leukemia (data not shown). As a reference to the functions of the autogenerated genes,

Table 3.1   Union gene list for radiation-induced leukemias

Affymetrix systemic name		Common name	Genbank ID	Description
1415874_at		Spry1	NM_011896	sprouty homolog 1 (Drosophila)
1416001_a_at		Cotl1	NM_028071	coactosin-like 1 (Dictyostelium)
1417097_at		Nrbf1	NM_025297	nuclear receptor binding factor 1
1417194_at		Sod2	NM_013671	superoxide dismutase 2, mitochondrial
1417602_at		Per2	AF035830	period homolog 2 (Drosophila)
1417623_at	*	Slc12a2	BG069505	solute carrier family 12, member 2
1417851_at		Cxcl13	AF030636	chemokine (C–X–C motif) ligand 13
1418062_at		Eef1a2	NM_007906	eukaryotic translation elongation factor 1 alpha 2
1418094_s_at		Car4	NM_007607	carbonic anhydrase 4
1418450_at		Islr	NM_012043	immunoglobulin superfamily containing leucine-rich repeat
1418547_at		Tfpi2	NM_009364	tissue factor pathway inhibitor 2
1418597_at		Top3a	NM_009410	topoisomerase (DNA) III alpha
1418666_at		Ptx3	NM_008987	pentaxin related gene
1418697_at	*	Temt	NM_009349	thioether S-methyltransferase
1418712_at		Cdc42ep5	NM_021454	CDC42 effector protein (Rho GTPase binding) 5
1418713_at		Pcbd	NM_025273	6-pyruvoyl-tetrahydropterin synthase/dimerization cofactor of hepatocyte nuclear factor 1 alpha (TCF1)
1418764_a_at		Bpnt1	BB412311	bisphosphate 3'-nucleotidase 1, metal-dependent lithium-inhibited phosphomonoesterase protein family
1419196_at		Hamp1/Hepc	NM_032541	hepcidin antimicrobial peptide
1419353_at		Dpm1	NM_010072	dolichol-phosphate (beta-D) mannosyltransferase 1
1419365_at		Pex11a	NM_011068	adaptor-related protein complex 3, sigma 2 subunit
1419417_at		Vegfc	NM_009506	vascular endothelial growth factor C
1419561_at	*	Ccl3	NM_011337	chemokine (C–C motif) ligand 3

Table 3.1   *(Continued)*

Affymetrix systemic name		Common name	Genbank ID	Description
1419664_at		*Srr*	BC011164	serine racemase
1419714_at		*Pdcd1lg1*	NM_021893	programmed cell death 1 ligand 1
1419967_at		*Seh1l*	AW540070	SEH1-like
1419970_at		*Slc35a5*	C86506	solute carrier family 35, member A5
1420034_at		*Ppp2r2d*	AU019644	protein phosphatase 2, regulatory subunit B, delta isoform (AU019644 Mouse eight-cell stage embryo cDNA *Mus musculus* cDNA clone J0520E06 3-, mRNA sequence)
1420052_x_at	*	*Psmb1*	C81484	proteasome (prosome, macropain) subunit, beta type 1
1420090_at	*	*Raf1*	AA990557	v-raf-1 leukemia viral oncogene 1
1420688_a_at		*Sgce*	NM_011360	sarcoglycan, epsilon
1420843_at	*	*Ptprf*	BF235516	protein tyrosine phosphatase, receptor type, F
1420872_at	*	*Gucy1b3*	BF472806	guanylate cyclase 1, soluble, beta 3
1421251_at		*Zfp40*	NM_009555	zinc finger protein 40
1421462_a_at		*Lepre1*	NM_019783	leprecan 1
1421619_at	*	*Kcnh3*	NM_010601	potassium voltage-gated channel, subfamily H (eag-related), member 3
1422025_at	*	*Mitf*	NM_008601	microphthalmia-associated transcription factor
1422218_at		*P2rx7*	NM_011027	purinergic receptor P2X, ligand-gated ion channel, 7
1423070_at		*Rpl21*	BG922742	general transcription factor III A
1423259_at		*Idb4*	BB121406	inhibitor of DNA binding 4
1423499_at		*Sncaip*	AK017012	synuclein, alpha interacting protein (synphilin)
1423677_at		*Fkbp9*	AF279263	FK506 binding protein 9
1424041_s_at		*C1s*	BC022123	complement component 1, s subcomponent
1424228_at		*Polr3h*	AK019868	polymerase (RNA) III (DNA directed) polypeptide H
1424295_at		*Dppa3*	AY082485	*Mus musculus* stella mRNA, complete cds
1424322_at		*Apex2*	AB072498	apurinic/apyrimidinic endonuclease 2

(Continued)

Table 3.1   (Continued)

Affymetrix systemic name		Common name	Genbank ID	Description
1424586_at		Ehbp1	AF424697	EH domain binding protein 1
1424651_at		BC021611	BC021611	hypothetical protein LOC257633
1424893_at		Ndel1	BC021434	nuclear distribution gene E-like homolog 1 (A. nidulans)
1425198_at		Ptpn2	BG076152	protein tyrosine phosphatase, non-receptor type 2
1425278_at		Ube4a	BC021406	ubiquitination factor E4A, UFD2 homolog (S. cerevisiae)
1425366_a_at		Hus1	AF076845	Hus1 homolog (S. pombe)
1425555_at	*	Crk7/Crkrs	BG070845	Cdc2-related kinase, arginine/serine-rich (RIKEN cDNA 1810022J16 gene)
1425597_a_at		Qk	AW060288	quaking
1425608_at		Dusp3/VHR	BC016269	dual specificity phosphatase 3 (vaccinia virus phosphatase VH1-related)
1425750_a_at		Jak3	L40172	Janus kinase 3
1425865_a_at		Lig3	U66057	ligase III, DNA, ATP-dependent
1425918_at		Egln3	BC022961	EGL nine homolog 3 (C. elegans)
1427558_s_at	*	Alg12	AJ429133	asparagine-linked glycosylation 12 homolog (yeast, alpha-1,6-mannosyltransferase)
1427595_at		Acac	BE650741	acetyl-Coenzyme A carboxylase alpha
1427833_at		Spi16/mBM17	U96702	serine protease inhibitor 16
1427843_at		Cebpb	AB012278	CCAAT/enhancer binding protein (C/EBP), beta
1428386_at		Acsl3	AK012088	acyl-CoA synthetase long-chain family member 3
1430148_at		Rab19	BM241400	RAB19, member RAS oncogene family
1430391_a_at		Siat8d	AK003690	sialyltransferase 8 (alpha-2,8-sialyltransferase) D
1430483_a_at		Tmem79	AK010144	transmembrane protein 79 (RIKEN cDNA 2310042N02 gene)
1430651_s_at		Zfp191	AI504586	zinc finger protein 191
1431066_at		Fut11	BB626220	fucosyltransferase 11

Table 3.1 *(Continued)*

Affymetrix systemic name		Common name	Genbank ID	Description
1432072_at	*	*Kif2a*	AK016720	kinesin family member 2A
1432115_a_at		*Pign*	AK014165	phosphatidylinositol glycan, class N
1433509_s_at		*Reep1*	BQ174328	receptor accessory protein 1(D6Ertd253e)
1433992_at	*	*Apxl*	BQ176992	apical protein, *Xenopus laevis*-like
1434349_at		*Vars2l*	AV258022	valyl-tRNA synthetase 2-like
1434369_a_at		*Cryab*	AV016515	crystallin, alpha B
1435132_at		*Disp1*	AI505698	dispatched homolog 1 (*Drosophila*)
1435557_at	*	*Fhod1*	AV298805	formin homology 2 domain containing 1
1435962_at		*Rps6*	BG089974	ribosomal protein S6 (Transcribed sequence with strong similarity to protein sp:P10660 (*H. sapiens*) RS6_HUMAN 40S ribosomal protein S6)
1436429_at		*Zfp606*	BB198855	zinc finger protein 606 (BB198855 RIKEN full-length enriched, 0 day neonate thymus *Mus musculus* cDNA clone A430007N09 3-, mRNA sequence)
1436521_at		*Slc36a2*	AI596194	solute carrier family 36 (proton/amino acid symporter), member 2
1436623_at		*Entpd7*	AV381133	ectonucleoside triphosphate diphosphohydrolase 7 (RIKEN cDNA 2900026G05 gene)
1436682_at		*Tmsb10*	AW259435	thymosin, beta 10 (up29e07.x1 NCI_CGAP_Mam2 *Mus musculus* cDNA clone IMAGE:2655780 3' similar to gb:S54005 THYMOSIN BETA-10 (HUMAN), mRNA sequence)
1436895_at		*Centd1*	BB182934	centaurin, delta 1
1436904_at		*Thrap1*	BB667559	hypothetical protein D030023K18

(Continued)

Table 3.1   (Continued)

Affymetrix systemic name	Common name	Genbank ID	Description
1436993_x_at	Pfn2	BB560492	profilin 2 (BB560492 RIKEN full-length enriched, 10 days neonate olfactory brain Mus musculus cDNA clone E530111B09 3' similar to AL096719 Homo sapiens mRNA; cDNA DKFZp566N043 (from clone DKFZp566N043), mRNA sequence.)
1437059_at	Sox21	BB046776	SRY-box containing gene 21 (BB046776 RIKEN full-length enriched, 11 days embryo Mus musculus cDNA clone 6230417M22 3-, mRNA sequence)
1437106_at	Jarid1a	BM246184	jumonji, AT rich interactive domain 1A (Rbp2 like) (K0734F05–3 NIA Mouse Hematopoietic Stem Cell (Lin-/c-Kit-/Sca-1-) cDNA Library (Long) Mus musculus cDNA clone NIA:K0734F05 IMAGE:30076864 3-, mRNA sequence)
1437123_at	Mmrn2	BB038352	multimerin 2
1437307_at	Senp8	BG069815	SUMO/sentrin specific protease family member 8
1437473_at	Maf	AV284857	avian musculoaponeurotic fibrosarcoma (v-maf) AS42 oncogene homolog (RIKEN cDNA A230108G15 gene)
1437789_at	Birc6	BB527646	baculoviral IAP repeat-containing 6 (BB527646 RIKEN full-length enriched, 15 days embryo head Mus musculus cDNA clone D930041P14 3-, mRNA sequence)
1437863_at	Bche	BB667762	butyrylcholinesterase (BB667762 RIKEN full-length enriched, adult male liver tumor Mus musculus cDNA clone C730038G20 3-, mRNA sequence)

Table 3.1 *(Continued)*

Affymetrix systemic name		Common name	Genbank ID	Description
1438463_x_at		Zdhhc6	AV142865	AV142865 *Mus musculus* C57BL/6J 10–11 day embryo *Mus musculus* cDNA clone 2810427C08, mRNA sequence.
1438825_at		Calm3	AV047570	calmodulin 3 (Similar to calmodulin – rabbit (tentative sequence) (LOC384465), mRNA (AV047570 *Mus musculus* adult C57BL/6J testis *Mus musculus* cDNA clone 1700069D17, mRNA sequence))
1438857_x_at		Irak1/pelle-like	BB058253	Irak1(interleukin-1 receptor-associated kinase 1)/pelle-like (BB058253 RIKEN full-length enriched, 2 days neonate sympathetic ganglion *Mus musculus* cDNA clone 7120478B17 3- similar to U56773 *Mus musculus*
1439247_at		Dock10	BB763030	dedicator of cytokinesis 10 (BB763030 RIKEN full-length enriched, B16 F10Y cells *Mus musculus* cDNA clone G370018M23 3-, mRNA sequence)
1440180_x_at		Zbtb3	AV258279	zinc finger and BTB domain containing 3 (AV258279 RIKEN full-length enriched, adult male testis (BNN132) *Mus musculus* cDNA clone 4923101A10 3′, mRNA sequence)
1440871_at		Baiap1	AI835038	BAI1-associated protein 1
1441272_at	*	Matr3	BI249188	matrin 3 (602994742F1 NCI_CGAP_Mam5 *Mus musculus* cDNA clone IMAGE:5150530 5-, mRNA sequence)
1442100_at		Inpp5f	BB619843	inositol polyphosphate-5-phosphatase F

(Continued)

Table 3.1   (Continued)

Affymetrix systemic name	Common name	Genbank ID	Description
1443229_at	Atad2	AV319821	ATPase family, AAA domain containing 2 (RIKEN cDNA 2610509G12 gene (AV319821 RIKEN full-length enriched mouse cDNA library, C57BL/6J testis male 13 days embryo *Mus musculus* cDNA clone 6030413I17 3-, mRNA))
1443493_at	Dhx37	BB766805	DEAH (Asp–Glu–Ala–His) box polypeptide 37
1443952_at	Nr1d1	BI525006	nuclear receptor subfamily 1, group D, member 1 (602924093F1 NCI_CGAP_Lu33 *Mus musculus* cDNA clone IMAGE:5056607 5-, mRNA sequence)
1445195_at	C77631	C77631	expressed sequence C77631 (Mouse 3.5-dpc blastocyst cDNA *Mus musculus* cDNA clone J0035A08 3′ similar to Mouse T-cell receptor (TCR V-alpha 16.1) gene exons 1–2, mRNA, mRNA sequence)
1447753_at	Cdc37l	BB391093	cell division cycle 37 homolog (*S. cerevisiae*)-like (BB391093 RIKEN full-length enriched, 0 day neonate cerebellum *Mus musculus* cDNA clone C230073C03 3-, mRNA sequence)
1447897_x_at	Anapc11	AV019615	AV019615 *Mus musculus* 18-day embryo C57BL/6J *Mus musculus* cDNA clone 1190010L24, mRNA sequence.
1448169_at	Krt1-18	NM_010664	keratin complex 1, acidic, gene 18
1448443_at	Serpini1	NM_009250	serine (or cysteine) proteinase inhibitor, clade I, member 1
1448986_x_at *	Dnase2a	NM_010062	deoxyribonuclease II alpha

Table 3.1   *(Continued)*

Affymetrix systemic name	Common name	Genbank ID	Description
1449481_at	Slc25a13	BC016571	solute carrier family 25 (mitochondrial carrier; adenine nucleotide translocator), member 13
1449493_at	Insl5	NM_011831	insulin-like 5
1449700_at	Igbp1	C81413	immunoglobulin (CD79A) binding protein 1
1449789_x_at	Ly6g6c	AV088850	lymphocyte antigen 6 complex, locus G6C (*Mus musculus* tongue C57BL/6J adult *Mus musculus* cDNA clone 2310040E07, mRNA sequence)
1449851_at *	Per1	AF022992	period homolog 1 (*Drosophila*)
1450046_at	Tmem59/ORF18	NM_019801	transmembrane protein 59 thymic dendritic cell-derived factor 1
1450135_at	Fzd3	AU043193	frizzled homolog 3 (*Drosophila*)
1450173_at	Ripk2	NM_138952	receptor (TNFRSF)-interacting serine-threonine kinase 2
1450199_a_at	Stab1	NM_138672	stabilin 1
1450208_a_at	Elmo1	NM_080288	engulfment and cell motility 1, ced-12 homolog (*C. elegans*)
1450296_at *	Klrb1a	NM_010737	killer cell lectin-like receptor subfamily B member 1A
1450297_at	Il6	NM_031168	interleukin 6
1450424_a_at	Il18bp	AF110803	interleukin 18 binding protein
1451541_at	Bcs1l	BC019781	RIKEN cDNA 1700112N14 gene
1451583_a_at	BC025076	BC025076	hypothetical protein LOC216829 membrane magnesium transporter 2
1451592_at	P42pop	AF364868	Myb protein P42POP
1451768_a_at	Slc20a2	AF196476	solute carrier family 20, member 2
1451950_a_at	Cd80	D16220	CD80 antigen
1451996_at	Bbp	AF353993	beta-amyloid binding protein precursor
1452253_at	Crim1	AK018666	cysteine-rich motor neuron 1
1452905_at	Gtl2	AV015833	GTL2, imprinted maternally expressed untranslated mRNA

(Continued)

Table 3.1   *(Continued)*

Affymetrix systemic name		Common name	Genbank ID	Description
1453055_at		Sema6d	BB462688	sema domain, transmembrane domain (TM), and cytoplasmic domain, (semaphorin) 6D
1453227_at		Rhobtb3	BG801497	Rho-related BTB domain containing 3
1453481_at		Zdhhc2	BB342242	zinc finger, DHHC domain containing 2
1453690_at	*	Mpp7	AV292557	membrane protein, palmitoylated 7 (MAGUK p55 subfamily member 7) (RIKEN full-length enriched, 6 days neonate head *Mus musculus* cDNA clone 5430426E14 3′, mRNA sequence)
1454414_at	*	Btbd7	AK017755	BTB (POZ) domain containing 7
1455158_at		Itga3	BI664675	integrin alpha 3
1455297_at		SPIN-2	BG070258	Similar to Spindlin-like protein 2 (SPIN-2) (LOC278240), mRNA
1455404_at		Jph2	BG870711	junctophilin 2
1455717_s_at	*	Daam2	BM206030	dishevelled associated activator of morphogenesis 2
1455985_x_at		Shmt2	AV213251	serine hydroxymethyltransferase 2 (mitochondrial) (AV213251 RIKEN full-length enriched, ES cells *Mus musculus* cDNA clone 2410126G07 3′, mRNA sequence)
1456975_at		Taok1	BM238077	TAO kinase 1(RIKEN cDNA 2810468K05 gene)
1457040_at		Lgi2	BE947711	leucine-rich repeat LGI family, member 2
1457311_at		Camk2a	AW490258	calcium/calmodulin-dependent protein kinase II alpha
1457451_at		Acvr2	BB199213	activin receptor IIA
1458047_at		Tnfsf13b	BB667811	tumor necrosis factor (ligand) superfamily, member 13b
1458381_at		Clic5	BB028501	chloride intracellular channel 5
1458641_at		Braf	BM217816	Braf transforming gene

Table 3.1   *(Continued)*

Affymetrix systemic name	Common name	Genbank ID	Description
1459597_at	Mtpn	BG074849	myotrophin
1459868_x_at	Il11ra1	AV313111	interleukin 11 receptor, alpha chain 1 (RIKEN full-length enriched, adult male thymus *Mus musculus* cDNA clone 5830408C01 3' similar to X74953 *M. musculus* ETL-2 mRNA, mRNA sequence.)
1460170_at	Ext2	NM_010163	exostoses (multiple) 2
1460666_a_at	Ebf3	NM_010096	early B-cell factor 3

*: overlapped in both lists of union genes for radiation-induced and spontaneous leukemias.

10 genes that represent the 128 genes mentioned above are described as follows: *Eef1a2* (eukaryotic translation elongation factor 1 alpha 2), expressed in tumors of the ovary, breast and lung (Amiri *et al.*, 2007; Tomlinson *et al.*, 2007) and plays a role in the resistance to apoptosis induced by oxidative stress (Chang and Wang, 2007); *Top3a* (topoisomerase (DNA) III alpha), required for accurate DNA replication (Oh *et al.*, 2002) and related to telomere–telomere recombination (Tsai *et al.*, 2006), cancer, and aging (Laursen *et al.*, 2003); *Ppp2r2d* (protein phosphatase 2, regulatory subunit B, delta isoform), expression biomarker for blast crisis (Liu *et al.*, 2007; Neviani *et al.*, 2007); *Leprel* (leprecan 1), basement-membrane-associated proteoglycan that functions in growth suppression and is a potential suppressor gene (Wassenhove-McCarthy and McCarthy, 1999); *Idb4* (inhibitor of DNA binding4), promotes neuronal stem cell proliferation (Yun *et al.*, 2004) and induction of leukemia cell apoptosis (Yu *et al.*, 2005); *Hus1* (hydroxyurea sensitive1), DNA damage checkpoint (Harris *et al.*, 2006), required for telomere maintenance and functions as Rad9–Hus1–Rad1 checkpoint; *Dusp3/VHR* (dual-specificity phosphatase 3 (vaccinia virus phosphatase VH1-related)), induces expression of cyclin D1 in breast cancer (Hao and ElShamy 2007) and arrests the cell cycle in VHR (Rahmouni *et al.*, 2006); *Igbp1/alpha 4* (immunoglobulin (CD79A) binding protein 1), apoptosis inhibitor via dephosphorylation of c-Jun and p53 (Kong *et al.*, 2004) and biomarker for acute myelogenous leukemia (Cruse *et al.*, 2005; Bhargava *et al.*, 2007); *Itga3* (integrin alpha 3), inhibitor of caspase 3 activity (Manohar *et al.*, 2004), which is up-regulated in adenocarcinoma of the lung (Boelens *et al.*, 2007), esophagus (Hourihan *et al.*, 2003) and stomach (Varis *et al.*, 2002); *Il11ral* (Interleukin 11 receptor alpha 1), functions in carcinogenesis associated with up-regulation of PI3K and p44/p42 MAPK in gastric cancer (Nakayama *et al.*, 2007) and colon cancer (Yoshizaki *et al.*, 2006) associated with STAT3 in the prostate cancer (Zurita *et al.*, 2004), and constitutively activated in myeloma/B-CLL (Tsimanis *et al.*, 2001). The functions of these genes may satisfy the characteristics of cluster-specific gene expression profiles linked to radiation-induced leukemias.

Among the 149 genes, 21 genes, including *raf1*, *Mitf*, and *Crkrs*, overlap in both lists of union genes for both radiation-induced and spontaneous leukemias (see asterisks in

Table 3.1). These overlapped genes observed in both union gene lists imply that not all the radiation-specific union genes (i.e. 'stochastically necessary genes') are always required for the development of radiation-induced myelogenous leukemias, but a combination of the 'stochastically necessary genes' in the list is required in addition to the common leukemogenic genes. An essential rule for the combination of the 'stochastically necessary genes' for radiation-induced leukemogenicity is not yet identified.

3.9   New Risk Evaluation Strategy Using Gene Expression Profiles

These toxicological endpoints obtained by gene chip and microarray technologies can be used to evaluate their quantitative and qualitative differences in gene expression for risk evaluation by PCA. Figure 3.9a shows sample expression data between the two groups, one for bone marrow tissues exposed to 0.6 Gy and the other for nonirradiated controls, which is shown by two-dimensional expressions from PCA components #1 and #2. No observed effect level (NOEL) or no observed adverse effect level (NOAEL) can be statistically calculated by the 95 % confidence areas of the two clusters, so that the differences between the two groups can be considered as the risk evaluation parameters. Furthermore, when one attempts to evaluate the possible differences between the two groups by gene chip and microarray methods, one can use any component(s) with low differential contribution. However, when one attempts to evaluate homogeneity, one should evaluate possible

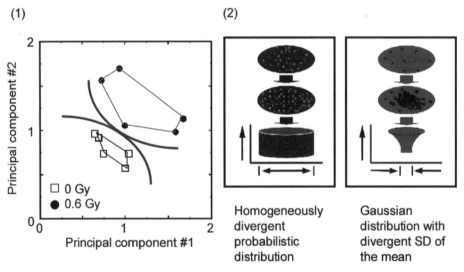

Figure 3.9   *(a) Sample gene expression clusters associated with the bone marrow after 0.6 Gy whole-body irradiation and those associated with the control are shown in the two-dimensional principal component diagram plotted along the contribution factors. Areas of each cluster are statistically defined along with the confidence levels of NOEL and/or NOAEL. See text. (b) Two types of data distribution pattern: homogeneously divergent probabilistic distribution on the left and divergent distribution due to Gaussian distribution with the error of the mean on the right.*

consistencies not only for the major component, but also for other components with low contribution, and determine whether the differences can be ignored. Because minor components may sometimes play an important toxicological role, careful examination of the genomic repertoire is required to determine whether the expressions of responsible genes are identical in a case-by-case manner.

Lastly, it is very important to recognize the characteristics of each data point. When one calculates NOEL or NOAEL, one should note the different characteristics and diversity of data between the two types, such as whether the data show probabilistic homogeneously diverse distribution or a Gaussian normal equivalent distribution due to error/deviation, as shown on the left and right sides of Figure 3.9b. Interestingly, data from developmental toxicology and the growth parameters tend to show the latter convergent distribution. However, data from toxicological changes in relation to senescence tend to show the former scattered distribution.

In this chapter, murine spontaneous and radiation-induced myelogenous leukemias were used as experimental models to discuss an essential principle of data-mining in toxicogenomics. The examinations were focused only on the bone marrow. A possible reason for the predictability observed in tissues other than the bone marrow may be attributable to the characteristics of radiation-induced tissue injury, of which the general rule may be stochastic but generally applicable to other tissues. In the case of chemicals, the predictability of the results of a single tissue examination may be limited owing to the possible tissue-specific interactions of such test chemicals.

References

Adams MD, Celniker SE, Holt RA, Evans CA, Gocayne JD, Amanatides PG, Scherer SE, Li PW, Hoskins RA, Galle RF *et al.* (2000) The genome sequence of *Drosophila melanogaster*. *Science* **287**, 2185–2195.

Albanese J, Martens K, Karkanitsa LV, Dainiak N (2007) Multivariate analysis of low-dose radiation-associated changes in cytokine gene expression profiles using microarray technology. *Exp Hematol* **35**, 47–54.

Amiri A, Noei F, Jeganathan S, Kulkarni G, Pinke DE, Lee JM (2007) eEF1A2 activates Akt and stimulates Akt-dependent actin remodeling, invasion and migration. *Oncogene* **26**, 3027–3040.

Astrand M (2003) Contrast normalization of oligonucleotide arrays. *J Comput Biol* **10**, 95–102.

Bahar R, Hartmann CH, Rodriguez KA, Denny AD, Busuttil RA, Dolle ME, Calder RB, Chisholm GB, Pollock BH, Klein CA, Vijg J (2006) Increased cell-to-cell variation in gene expression in ageing mouse heart. *Nature* **441**, 1011–1014.

Bhargava P, Kallakury BV, Ross JS, Azumi N, Bagg A (2007) CD79a is heterogeneously expressed in neoplastic and normal myeloid precursors and megakaryocytes in an antibody clone-dependent manner. *Am J Clin Pathol* **128**, 306–313.

Bloomfield VA, Crothers DM, Tinoco I (2000) Nucleic Acids: Structures, Properties, and Functions. University Science Books: Sausalito, CA.

Boelens MC, van den Berg A, Vogelzang I, Wesseling J, Postma DS, Timens W, Groen HJ (2007) Differential expression and distribution of epithelial adhesion molecules in non-small cell lung cancer and normal bronchus. *J Clin Pathol* **60**, 608–614.

Borlak J (2005) *Handbook of Toxicogenomics*. Wiley–VCH Verlag: Weinheim.

Brazma A, Hingamp P, Quackenbush J, Sherlock G, Spellman P, Stoeckert C, Aach J, Ansorge W, Ball CA, Causton HC *et al.* (2001) Minimum information about a microarray experiment (MIAME) – toward standards for microarray data. *Nat Genet* **29**, 365–371.

Brenner S, Jacob F, Meselson M (1961) An unstable intermediate carrying information from genes to ribosomes for protein synthesis. *Nature* **190**, 576–581.

Brown PO, Botstein D (1999) Exploring the new world of the genome with DNA microarrays. *Nat Genet* **21**, 33–37.

C. elegans Sequencing Consortium (1998) Genome sequence of the nematode *C. elegans*: a platform for investigating biology. *Science* **282**, 2012–2018.

Chang R, Wang E (2007) Mouse translation elongation factor eEF1A-2 interacts with Prdx-I to protect cells against apoptotic death induced by oxidative stress. *J Cell Biochem* **100**, 267–278.

Churchill GA (2002) Fundamentals of experimental design for cDNA microarrays. *Nat Genet* **32** (Suppl.), 490–495.

Crick FH (1958) On protein synthesis. *Symp Soc Exp Biol* **12**, 138–163.

Crick FH, Barnett L, Brenner S, Watts-Tobin RJ (1961) General nature of the genetic code for proteins. *Nature* **192**, 1227–1232.

Cruse JM, Lewis RE, Pierce S, Lam J, Tadros Y (2005) Aberrant expression of CD7, CD56, and CD79a antigens in acute myeloid leukemias. *Exp Mol Pathol* **79**, 39–41.

Czapski G, Peled E (1973) Reaction rates of electrons at short times. *J Phys Chem* **77**, 893.

Czapski G, Peled E (1975) Reaction rates of electrons at short times. In *Radiation Research: Biomedical, Chemical and Physical Perspectives*, OF Nygaard, HI Adler, WK Sinclair (eds). Academic Press: New York.

Fodor SP, Rava RP, Huang XC, Pease AC, Holmes CP, Adams CL (1993) Multiplexed biochemical assays with biological chips. *Nature* **364**, 555–556.

Hamadeh HK, Bushel P, Paules R, Afshari CA (2001) Discovery in toxicology: mediation by gene expression array technology. *J Biochem Mol Toxicol* **15**, 231–242.

Hao L, ElShamy WM (2007) BRCA1-IRIS activates cyclin D1 expression in breast cancer cells by downregulating the JNK phosphatase DUSP3/VHR. *Int J Cancer* **121**, 39–46.

Harris J, Lowden M, Clejan I, Tzoneva M, Thomas JH, Hodgkin J, Ahmed S (2006) Mutator phenotype of *Caenorhabditis elegans* DNA damage checkpoint mutants. *Genetics* **174**, 601–616.

Hastie ND, Allshire RC (1989) Human telomeres: fusion and interstitial sites. *Trends Genet* **5**, 326–331.

Henke FU, Lubarsch O (1924–1952). *Handbuch des speziellen pathologischen Anatomie und Histologie*, vols **1–12**. Springer: Berlin.

Hill AA, Brown EL, Whitley MZ, Tucker-Kellogg G, Hunter CP, Slonim DK (2001) Evaluation of normalization procedures for oligonucleotide array data based on spiked cRNA controls. *Genome Biol* **2**, RESEARCH0055.

Hourihan RN, O'Sullivan GC, Morgan JG (2003) Transcriptional gene expression profiles of oesophageal adenocarcinoma and normal oesophageal tissues. *Anticancer Res* **23**, 161–165.

Hu T, Gibson DP, Carr GJ, Torontali SM, Tiesman JP, Chaney JG, Aardema MJ (2004) Identification of a gene expression profile that discriminates indirect-acting genotoxins from direct-acting genotoxins. *Mutat Res* **549**, 5–27.

Hunt JW (1976) Early events in radiation chemistry. In *Advances in Radiation Chemistry*, vol. **5**, M Burton, JL Magee (eds). John Wiley and Sons, Ltd: New York, p. 185.

Inoue T (2003) Introduction: Toxicogenomics – a New Paradigm of Toxocology. In *Toxicogenomics*, T Inoue, WD Pennie (eds). Springer-Verlag: Tokyo, pp. 3–11.

Kong M, Fox CJ, Mu J, Solt L, Xu A, Cinalli RM, Birnbaum MJ, Lindsten T, Thompson CB (2004) The PP2A-associated protein alpha4 is an essential inhibitor of apoptosis. *Science* **306**, 695–698.

Kroll TC, Wolfl S (2002) Ranking: a closer look on globalisation methods for normalisation of gene expression arrays. *Nucleic Acids Res* **30**, e50.

Lander ES, Linton LM, Birren B, Nusbaum C, Zody MC, Baldwin J, Devon K, Dewar K, Doyle M, FitzHugh W *et al*. (2001) Initial sequencing and analysis of the human genome. *Nature* **409**, 860–921.

Laursen LV, Bjergbaek L, Murray JM, Andersen AH (2003) RecQ helicases and topoisomerase III in cancer and aging. *Biogerontology* **4**, 275–87.

Lewin, B (2004) *Genes VIII*. Pearson Prentice Hall, Upper Saddle River, NJ.

Liu Q, Zhao X, Frissora F, Ma Y, Santhanam R, Jarjoura D, Lehman A, Perrotti D, Chen CS, Dalton JT, Muthusamy N, Byrd JC (2007) FTY720 demonstrates promising pre-clinical activity for chronic lymphocytic leukemia and lymphoblastic leukemia/lymphoma. *Blood* **111**, 275–284.

Lodish H, Berk A, Matsudaira P, Kaiser C, Krieger M, Scott M, Zipursky SL, Darnell J (2004) *Molecular Cell Biology*. WH Freeman: New York.

Lovett RA (2000) Toxicogenomics. Toxicologists brace for genomics revolution. *Science* **289**, 536–537.

Manohar A, Shome SG, Lamar J, Stirling L, Iyer V, Pumiglia K, DiPersio CM (2004) Alpha 3 beta 1 integrin promotes keratinocyte cell survival through activation of a MEK/ERK signaling pathway. *J Cell Sci* **117**, 4043–4054.

Nakayama T, Yoshizaki A, Izumida S, Suehiro T, Miura S, Uemura T, Yakata Y, Shichijo K, Yamashita S, Sekin I (2007) Expression of interleukin-11 (IL-11) and IL-11 receptor alpha in human gastric carcinoma and IL-11 upregulates the invasive activity of human gastric carcinoma cells. *Int J Oncol* **30**, 825–833.

Neviani P, Santhanam R, Oaks JJ, Eiring AM, Notari M, Blaser BW, Liu S, Trotta R, Muthusamy N, Gambacorti-Passerini C, Druker BJ, Cortes J, Marcucci G, Chen CS, Verrills NM, Roy DC, Caligiuri MA, Bloomfield CD, Byrd JC, Perrotti D (2007) FTY720, a new alternative for treating blast crisis chronic myelogenous leukemia and Philadelphia chromosome-positive acute lymphocytic leukemia. *J Clin Invest* **117**, 2408–2421.

Oh M, Choi IS, Park SD (2002) Topoisomerase III is required for accurate DNA replication and chromosome segregation in *Schizosaccharomyces pombe*. *Nucleic Acids Res* **30**, 4022–4031.

Petersen D, Chandramouli GV, Geoghegan J, Hilburn J, Paarlberg J, Kim CH, Munroe D, Gangi L, Han J, Puri R, Staudt L, Weinstein J, Barrett JC, Green J, Kawasaki ES (2005) Three microarray platforms: an analysis of their concordance in profiling gene expression. *BMC Genomics* **6**, 63.

Rahmouni S, Cerignoli F, Alonso A, Tsutji T, Henkens R, Zhu C, Louis-dit-Sully C, Moutschen M, Jiang W, Mustelin T (2006) Loss of the VHR dual-specific phosphatase causes cell-cycle arrest and senescence. *Nat Cell Biol* **8**, 524–531.

Schadt EE, Li C, Ellis B, Wong WH (2001) Feature extraction and normalization algorithms for high-density oligonucleotide gene expression array data. *J Cell Biochem Suppl* **37**, 120–125.

Seki M, Yoshida K, Nishimura M, Nemoto K (1991) Radiation-induced myeloid leukemia in C3H/He mice and the effect of prednisolone acetate on leukemogenesis. *Radiat Res* **127**, 146–149.

Singer M, Berg P (1991) *Genes & Genomes: A Changing Perspective*. University Science Books: Mill Valley, CA.

Storck T, von Brevern MC, Behrens CK, Scheel J, Bach A (2002) Transcriptomics in predictive toxicology. *Curr Opin Drug Discov Dev* **5**, 90–97.

Stoyanova R, Querec TD, Brown TR, Patriotis C (2004) Normalization of single-channel DNA array data by principal component analysis. *Bioinformatics* **20**, 1772–1784.

Tomlinson VA, Newbery HJ, Bergmann JH, Boyd J, Scott D, Wray NR, Sellar GC, Gabra H, Graham A, Williams AR, Abbott CM (2007) Expression of eEF1A2 is associated with clear cell histology in ovarian carcinomas: overexpression of the gene is not dependent on modifications at the EEF1A2 locus. *Br J Cancer* **96**, 1613–1620.

Tsai HJ, Huang WH, Li TK, Tsai YL, Wu KJ, Tseng SF, Teng SC (2006) Involvement of topoisomerase III in telomere–telomere recombination. *J Biol Chem* **281**, 13717–13723.

Tsimanis A, Shvidel L, Klepfish A, Shtalrid M, Kalinkovich A, Berrebi A (2001) Over-expression of the functional interleukin-11 alpha receptor in the development of B-cell chronic lymphocytic leukemia. *Leuk Lymphoma* **42**, 195–205.

Varis A, Wolf M, Monni O, Vakkari ML, Kokkola A, Moskaluk C, Frierson Jr H, Powell SM, Knuutila S, Kallioniemi A, El-Rifai W (2002) Targets of gene amplification and overexpression at 17q in gastric cancer. *Cancer Res* **62**, 2625–2629.

Wassenhove-McCarthy DJ, McCarthy KJ (1999) Molecular characterization of a novel basement membrane-associated proteoglycan, leprecan. *J Biol Chem* **274**, 25004–25017.

Waterston RH, Lindblad-Toh K, Birney E, Rogers J, Abril JF, Agarwal P, Agarwala R, Ainscough R, Alexandersson M, An P *et al.* (2002) Initial sequencing and comparative analysis of the mouse genome. *Nature* **420**, 520–562.

Watson JD, Baker TA, Bell SP, Gann A, Levine M, Losick T (2007) *Molecular Biology of the Gene*. Pearson/Benjamin Cummings: San Francisco, CA.

Watson JD, Crick FH (1953) Molecular structure of nucleic acids; a structure for deoxyribose nucleic acid. *Nature* **171**, 737–738.

Yoshida K, Inoue T, Nojima K, Hirabayashi Y, Sado T (1997) Calorie restriction reduces the incidence of myeloid leukemia induced by a single whole-body radiation in C3H/He mice. *Proc Natl Acad Sci U S A* **94**, 2615–2619.

Yoshizaki A, Nakayama T, Yamazumi K, Yakata Y, Taba M, Sekine I (2006). Expression of interleukin (IL)-11 and IL-11 receptor in human colorectal adenocarcinoma: IL-11 up-regulation of the invasive and proliferative activity of human colorectal carcinoma cells. *Int J Oncol* **29**, 869–876.

Yu L, Liu C, Vandeusen J, Becknell B, Dai Z, Wu YZ, Raval A, Liu TH, Ding W, Mao C, Liu S, Smith LT, Lee S, Rassenti L, Marcucci G, Byrd J, Caligiuri MA, Plass C (2005) Global assessment of promoter methylation in a mouse model of cancer identifies ID4 as a putative tumor-suppressor gene in human leukemia. *Nat Genet* **37**, 265–274.

Yun K, Mantani A, Garel S, Rubenstein J, Israel MA (2004) Id4 regulates neural progenitor proliferation and differentiation *in vivo*. *Development* **131**, 5441–5448.

Yunis JJ, Soreng AL, Bowe AE (1987) Fragile sites are targets of diverse mutagens and carcinogens. *Oncogene* **1**, 59–69.

Zhang W, Shmulevich I (2006) *Computational and Statistical Approaches to Genomics*. Springer Science+Business Media: New York.

Zurita AJ, Troncoso P, Cardo-Vila M, Logothetis CJ, Pasqualini R, Arap W (2004) Combinatorial screenings in patients: the interleukin-11 receptor alpha as a candidate target in the progression of human prostate cancer. *Cancer Res* **64**, 435–439.

4

Design Issues in Toxicogenomics: The Application of Genomic Technologies for Mechanistic and Predictive Research

Woong-Yang Park, Lian Li and Daehee Kang

4.1 Introduction

The genetic sequence of the human genome is now completely determined (International Human Genome Sequencing Consortium, 2004), and the whole genome sequences of more than 1500 organisms are open to the public. Based on this sequence information, major advances in genomic technology to analyze the genome-wide expression and genotype of cellular genes in human and model organisms have been achieved. Toxicogenomics is the analysis that measures the activity of particular chemicals on cells according to the genomic information generated by profiling alterations in DNA, RNA, and protein levels.

DNA microarrays are commonly used by biomedical researchers and in genetic toxicology studies. Microarray technologies have enabled us to monitor tens of thousands of nucleic acid sequences simultaneously by quantifying the amount of specific transcripts (Brown and Botstein, 1999). If we can fully list the species of cellular mRNA by full-genome microarray, it will represent the cellular status in terms of intracellular constituents. Such a global analysis of gene expression has the potential to provide a more comprehensive view of toxicity than has been previously possible, since toxicity not only generally involves change in a single gene or in a few genes, but also rather is a series of gene–gene interactions.

Toxicogenomics: A Powerful Tool for Toxicity Assessment   Edited by S. C. Sahu
© 2008 John Wiley & Sons, Ltd

Table 4.1   *Public databases related to toxicogenomics studies*

Database	URL
CTD (Comparative Toxicogenomics Database)	http://ctd.mdibl.org/
ILSI	http://rsi.ilsi.org/
NPH	http://ntp-apps.niehs.nih.gov/ntp_tox/index.cfm
dbZach	http://dbzach.fst.msu.edu
TOXNET	http://toxnet.nlm.nih.gov/
NIEHS NCT (National Center for Toxicology)	http://www.niehs.nih.gov/nct/
TRC (Toxicogenomics Research Consortium)	http://www.niehs.nih.gov/nct/trc.htm
CEBS (Chemical Effects in Biological Systems)	http://www.niehs.nih.gov/nct/cebs.htm

Using toxicogemomic analysis, we can investigate the complex interaction between genetic variability and environmental exposure with regard to toxicological effects (Olden *et al.*, 2004). Gene expression changes can possibly provide more sensitive, immediate, and comprehensive markers of toxicity than typical toxicological endpoints, such as morphological change, carcinogenicity and reproductive toxicity (Marchant, 2003). Since the US National Institutes of Environmental Health Sciences Microarray Group started to analyze changing patterns of gene expression across the entire genome (The National Center for Toxicogenomics, 2004), there have been many efforts to compose huge datasets on toxicogenomics (Table 4.1).

Epidemiological studies focusing on genetic risk factors and environmental factors (i.e. exposure to toxic elements) can also be applied to toxicogenomic research. Epidemiological study design has been recently re-emphasized as a basis for effective application of new technologies such as genomics, proteomics and metabolomics (Potter, 2003). The association studies evaluating the interactive effects between genetic factors (e.g. single nucleotide polymorphisms) and environmental exposure might be considered one type of toxicogenomic research. Thus, the perspectives of epidemiological studies need to be supplemented or blended with those of '-omics' studies.

Toxicogenomics may provide us with a powerful measure to predict adverse toxic effects of pharmaceutical drugs on susceptible individuals. Genomic techniques such as gene expression level profiling and genotyping analysis of genetic variation of individuals can reveal the characteristics of those persons by genetic testing. Studies of those genotypes can then be correlated to adverse toxicological effects in clinical trials so that suitable diagnostic markers for these adverse effects can be developed.

Although nucleic acid microarray technologies have received much attention recently, other powerful new tools for global analysis of cellular constituents are already available and will also have a major impact on the field of toxicology. These include technologies for global analysis of proteins and peptides (proteomics) and of cellular metabolites (metabolomics). Among these advances are: improvements in classical two-dimensional (2D) gel electrophoresis; the introduction of multidimensional liquid chromatography, tandem mass spectrometry, and database-searching technologies (termed multidimensional protein identification technology, or MudPIT); and improved mass spectroscopic identification of protein sequences using matrix- or surface-enhanced laser desorption ionization (MALDI, SELDI), which are techniques that allow rapid characterization of proteins or protein fragments (Yates, 2000). These proteomic methods allow for the analysis of the functional and structural proteins in a sample.

Methods for simultaneously monitoring small molecules involved in intermediary metabolic pathways (metabolomics) are also at hand (Holmes *et al.*, 2001). The ability to monitor defense responses via proteomics or metabolomics in humans at subpathological doses is of particular importance, because it will make possible human studies that could not be carried out at overtly toxic exposures. The use of all of these tools will be important for obtaining a comprehensive picture of toxicological change in cellular constituents.

There is every reason to expect major change during the next decade, as new technologies and knowledge become incorporated into regulatory and industrial practice. Unlike other new approaches or methods in toxicology that have been adopted slowly, genomic, proteomic and metabolomic methods are being evaluated and adopted rapidly by industry, academia and regulatory agencies. As with most new toxicological methods and approaches, collaboration will be required to develop the data and approaches necessary to achieve worldwide acceptance and use. Rather, the broad impact and application of the field of molecular genetics will transform the field of toxicology into a science based on molecular biochemical knowledge and techniques. In this chapter, we summarize new approaches in genomic technologies, with special emphasis placed on genetic toxicology.

4.2   Genomic Technologies

4.2.1   Functional Annotation of the Human Genome

Sequencing of the human genome has provided us with valuable information of detailed DNA sequences of 3 billion base pairs (International Human Genome Sequencing Consortium, 2004). Protein-coding regions have been mapped to reveal ~40 000 genes, even though half of them are not yet fully annotated (Carninci, 2007). Even noncoding transcripts, which might regulate gene expression, have been reported (Rane *et al.*, 2007). The Encyclopedia of DNA elements (ENCODE) projects were launched to reveal biological information of the human genome by high-throughput techniques to identify and annotate functional DNA elements encoded by the human genome. In the first report of this effort, analysis data on 1 % of the human genome were published (The ENCODE Project Consortium, 2007), which will provide a 'dictionary' with growing annotations about biochemical functions.

4.2.2   Expression Analysis

The application of technologies for genome-wide analysis of all cellular components such as DNA, proteins and metabolites is not possible, with the exception of RNAs. Transcriptomes of cells or tissues can be easily analyzed by microarray. Application of microarray technology can provide a more comprehensive view of cellular and molecular responses than ever before. Gene expression is altered either directly or indirectly as a result of toxicant exposures such as ionizing radiation (Park *et al.*, 2002).

The expression of certain categories of genes is required to achieve pathological outcomes such as cell death or proliferation. Changes in gene expression associated with toxicity are often more sensitive and characteristic of the toxic response than currently employed endpoints of pathology. An ability to monitor the pre-pathological cellular responses to toxic damage can provide useful biomarkers. Such biomarkers will be of particular importance, because human studies that could not be carried out at overtly toxic exposures will be possible. In particular, those biomarkers can be assayed easily using high-throughput screening tools such as DNA chips (ToxChip, etc.). DNA chips from many different species that permit thousands of genes to be monitored simultaneously are already available. This capability of 'global' monitoring of essentially all expressed genes in any model organism provides the opportunity to characterize patterns of gene expression associated with specific types of damage and specific classes of chemicals.

4.2.3   Polymorphisms

In addition to the study of gene expression/proteins/metabolites in response to exposure to toxicants, evaluation of gene sequence variation (polymorphisms) is also available in microarray format. Such methods are enabling the systematic evaluation of the effects of variant genetic sequences in response to toxicants. Examples of genetic variants that affect the sensitivity to chemicals have long been known. Understanding the relationship between this genetic variability and the sensitivity to chemical exposure will greatly facilitate individual health risk assessment and extrapolation of findings from laboratory models to human risk. Mapping of sequence differences between individuals with and without diseases (most often single-base differences known as single nucleotide polymorphisms (SNPs)) is now revealing a growing number of disease susceptibility genes (Choi *et al.*, 2005), as well as polymorphisms that determine individual diversity in drug responses (Uetrecht, 2007).

4.2.4   Personal Genome Sequencing

In 2004, the National Human Genome Research Institute of the National Institutes of Health awarded $32 million in grants to enable full genome sequencing technology for $100 000 in 5 years and $1000 in 10 years (http://www.nih.gov/news/pr/aug2005/nhgri-08.htm). One of the fastest technologies, provided by 454 Life Sciences, succeeded in uncovering a genome sequence of an individual human. These data depict a definitive molecular portrait of a diploid human genome that provides a starting point for future genome comparisons and enables an era of individualized genomic information. Initially, personal genotyping might focus on hypothesis generation and cell-based testing. However, epidemiological and genetic association studies, such as genomic cohort studies, will benefit from thousands or even millions of participants. Biological and medical research needs not only new '-omics' technologies and systems models, but also new ways to assess technology and to obtain low-risk, 'open,' integrated datasets focused on interindividual variation. Personal genome sequencing is a work-in-progress and is likely to change and diversify in response to a variety of inputs.

4.2.5 Noncoding RNAs

In addition to its entirety, the whole genome can be fractioned into just the protein-coding or the euchromatic regions. At our current rate of discovering new genetic elements and phenomena that can affect human health, how close we can get to a whole genome sequence is what we want to measure. At present, microRNAs (miRNAs) are well-known species of noncoding RNAs, but Piwi-interacting RNA (piRNA) and variable lengths of noncoding RNAs are actively transcribed in cells. In mammals, miRNA has been known to suppress the translation of target RNAs. An individual miRNA can be targeted to more than 200 target RNAs related to carcinogenesis, development and immune function. Now, more than 600 miRNAs have been registered in miRbase, but they await functional annotations. In pancreatic tumors, a subset of miRNAs was specifically expressed higher than in normal or inflamed pancreas (Bloomston *et al.*, 2007).

4.2.6 Copy Number Variation

The copy number variant (CNV) is the variation of number of copies of a particular gene in the genotype of an individual (Redon *et al.*, 2006). Recent evidence shows that the gene copy number can be elevated in cancer cells. Studies of the human genome have revealed an unknown amount of variation between people with respect to the number of copies of genes they have. Such variations appear to involve as much as 12 % of our DNA and raise questions about what constitutes a 'normal' genome. People also differ in the number of copies of genes they have, with large portions of one person's DNA being duplicated or deleted when compared with another's. Extra copies of identical genes can even cause disease with no mutation involved. Now, when seeking genetic causes for responses to chemicals, scientists will have to consider CNVs together with mutations or SNPs.

4.2.7 Epigenomics

Methylation of cytosine in CpG dinucleotides of the mammalian genome has been implicated to be a key mechanism of gene expression control. Many technologies have provided systems to monitor the methylation status of certain regions in various physiological and pathological conditions. Array-based technologies offer genome-wide methylation analysis (Shi *et al.*, 2002; Ching *et al.*, 2005). Recently, it has been found that methylated DNA immunoprecipitation (MeDIP) can directly isolate methylcytosine-containing DNA fragments to analyze in genomic DNA arrays.

Although nearly every cell type has essentially identical sequences in their genomes, the different cell types in a multicellular organism maintain markedly different behaviors against various chemicals. Considerable evidence suggests that cellular responses may be closely related to the 'chromatin state' – that is, modifications to histones and other proteins that package the genome. Mikkelsen *et al.* (2007) constructed 'chromatin-state maps' for a wide variety of cell types, showing the genome-wide distribution of important chromatin modifications. They used the ChIP-seq method, in which methylated histones were immunoprecipitated with specific antibodies, and the individual sequences were analyzed by the 'Giga' sequencing method.

4.2.8   Proteomics

Proteomics is a field that studies the full set of proteins (the proteome) of a tissue or cell type encoded by a genome and deals with global separation, quantitation and functional characterization of expressed proteins in whole tissue samples (Anderson *et al.*, 1996). The most commonly used techniques for studying the proteome are 2D gel electrophoresis and mass spectrometry. Recently, promising new proteomic technologies such as MudPIT (Washburn *et al.*, 2001), ICAT (Gygi *et al.*, 1999) or iTRAQ methods (Ross *et al.*, 2004) and specific protein and antibody arrays (Cutler, 2003) allow the genome-wide functional and structural characterization of proteins. Profiling of global protein expression patterns provides important complementary information to genomics, as RNAs are subject to post-transcriptional control to regulate the amount and quality of proteins in cells, tissues and biological fluids (Klein and Thongboonkerd, 2004). Proteomic technology offers greater knowledge of cellular processes and complementary information to genomics (Heijne *et al.*, 2003). Nonetheless, the innate properties of proteins, such as size, solubility and isoelectric point, would restrict the high-throughput measurement of tens of thousands of proteins together. In addition, posttranslational modification, protein–protein interactions and membrane proteins are limiting the analysis of the proteome at present.

4.2.9   Metabolomics

In the 1990s, the concept of comprehensive analysis of the metabolic components in biological systems, termed metabolomics (alternately metabonomics), was established (Robertson, 2005). Metabolomics combines the application of analytical technologies such as nuclear magnetic resonance with statistics. Metabolomic studies allow the quantitative measurement of the levels of cellular metabolites from samples such as tissues or fluids. They can directly reflect the amount of toxins and their metabolism together with the physiological change of cells. The application of metabolomics to tissues and fluids is a nondestructive and quick procedure. While it is quite difficult to extract cellular mRNA and protein from limited amounts of tissues, metabolites can be easily purified from tissue, as well as from fluids such as serum, plasma, cerebrospinal fluid and urine. Another gain for metabolomics, especially in toxicogenomics, is the acquisition of a metabolic signature of an organism in a pathological or altered physiological state.

4.3   Hierachy of Toxicogenomic Studies

To design the optimal toxicogenomic studies, conventional toxicity tests should be considered together with genomic techniques (Pesch *et al.*, 2004). For *in vitro* or *in vivo* experiments, researchers should decide all the factors influencing the final outcome, such as type of cell line, culture methods, model animals, methods of treatment, route of administration, dose, endpoint, groups and sample size. For epidemiological studies, cohort and case–control studies can be applied according to the purpose of the experiments (Pesch *et al.*, 2004). In addition, publicly available databases in toxicogenomics can be used to obtain information on biomarkers, genomic analysis and characteristics of chemical compounds (Lee *et al.*, 2004; Mattes *et al.*, 2004).

The process of designing toxicogenomic studies can be grouped into six steps (Figure 4.1). First, search for information on the tested agent in the toxicology database.

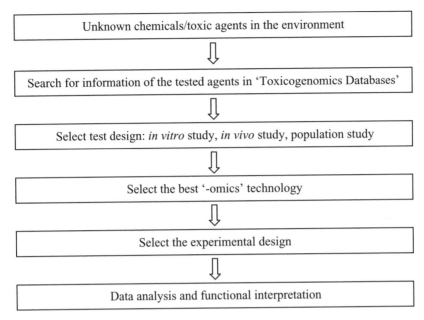

Figure 4.1 *Six steps for toxicogenomics study design.*

Next, select the test platform (e.g. *in vitro* or *in vivo* study, population study). Based on those platforms, the appropriate '-omics' technology should be selected, and then the toxicogenomic experimental design also should be determined according to appropriate genomic techniques. Finally, huge amounts of genomic data should be analyzed to provide the mechanistic and predictive aspects of agents tested.

4.3.1 *In Vitro* Study Design

For *in vitro* tests, researchers need to select the appropriate cell types according to the study design. Basal cytotoxicity in reaction to chemicals is a common phenomenon in almost all types of cell. However, there is a specificity of the toxic effect in certain cell types or tissues, and one needs to select the cell types carefully by considering the following:

1. cells that are more sensitive than others to the chemicals;
2. cells that show functional impairment by treatment with chemicals (functional toxicity);
3. cells that are major targets of specific usages of the chemicals (Eisenbrand *et al.*, 2002).

Next step would be the selection of the endpoint of the *in vitro* study. The endpoint of the toxicity test can be dictated by the toxic features and alteration of functions. Toxicity endpoints are usually determined by the breakdown of cell permeability barriers, reduced mitochondrial function, changes in cell morphology and changes in cell replication. These endpoints have been widely used in many cell types (Eisenbrand *et al.*, 2002; Kier *et al.*, 2004; Baken *et al.*, 2007).

4.3.2 *In Vivo* Study

The animal model for toxicogenomic study should be relevant to humans. Dosage, duration and route of administration must take the nature and extent of human exposure into account.

Usually, researchers choose either a high dose for a short duration or a low dose for a long duration (Barlow *et al.*, 2002). The sample size of the toxicogenomic in vivo study is much smaller than traditional toxicity tests. Usually, five animals of each case and control group in triplicated experiments are generally accepted (Kier *et al.*, 2004; Dere *et al.*, 2006).

Although genome-wide analysis is now available, subsets related to several categories, such as regulation of cell division, cell signaling, cell structure, apoptosis, and metabolism (Lee *et al.*, 2005), can be selected according to an *a priori* hypothesis and known mechanism through which a toxicant is working (Pennie *et al.*, 2000). For example, ToxBlot arrays include gene classes related to cancer, immunology, endocrinology and neurobiology, to investigative toxicology, to predevelopment toxicology, and to safety assessment. ToxBlot arrays are composed of approximately 2400 cDNA sequences, spanning about 600 genes of the relevant species. Four individual spots containing two nonoverlapping cDNAs on each array represent one gene. When selecting genes or proteins for knowledge-based '-omics' studies, several online resources for biological data and information for toxicogenomics study, including CTD (Mattingly *et al.*, 2006), KEGG, NTC, TRC, and CEBS, can be referred to, and a substantial matrix of data on chemicals with known exposure–disease outcomes needs to be obtained (Table 4.1)

4.3.3   Epidemiological Study

Epidemiology is an observational science that describes the patterns of diseases and their determinants in the human population. The epidemiological study of gene–environment interactions can evaluate the relationships between chemical exposure and disease status in the exposed population at the genetic level (Sellers, 2003). Because the study groups according to disease status or exposure status in epidemiological studies are not the same as experimental units, direct comparisons of different groups might not be possible. Therefore, study design is the most critical issue for the best epidemiological practices and, ultimately, for the application of new technologies.

The two major types of epidemiological study design are cohort and case–control studies (Boyes *et al.*, 2007). Cohort studies, either a prospective or a retrospective cohort study, can evaluate the occurrence of disease in individuals who presently have a certain condition. Case–control studies evaluate the history of exposure in individuals with and without disease (Joo *et al.*, 2004; Forrest *et al.*, 2005). Both types of study design have strengths and limitations. Case–control studies require a smaller sample size, but it is difficult to choose appropriate controls and to present clear relationships between chemical exposure and disease. On the other hand, cohort studies require long follow-up times and large sample sizes. Because many subjects are studied for long times, follow-up loss can occur and the information may not be detailed enough (Stovner, 2006).

Selection bias occurs when a systematic error from the failure to recruit study subject. In the toxicogenomics study, people can be exposed to low levels of multiple potentially toxic substances, and the exposures may last for long periods. In this sense, it is quite difficult to measure the exact level of chemicals for the characteristics and duration of exposure. It is important to collect all or well-defined random subject from a source population to avoid this kind of bias.

Information bias includes recall bias, outcome identification bias. It may occur when there is misclassification of exposure and/or outcome in the data collection procedure and experimental measurement error. The study must be conducted on the SOP (standard operating protocol) to avoid information bias. The standard operation protocol must include questionnaire design and administration, sample collection, storage, experimental genetic information measurement, data analysis.

In epidemiological studies, people may be exposed to low levels of multiple potentially toxic agents, even for years (Boyes *et al.*, 2007; Knudsen and Hansen, 2007). For example, painters are exposed not only to benzene, but also may be exposed to other toxic agents (Smith *et al.*, 2005; Vermeulen *et al.*, 2005). It is very difficult to establish the dose level in epidemiological studies, and also very difficult to link observed outcomes to one particular substance.

Confounding occurs when the exposed and nonexposed groups in the source population are not comparable because of inherent differences in background disease risk. The method to prevent confounding in advance of data analysis in population studies is randomization and matching the potential confounding variables, such as age, sex, etc. Confounding bias should also be controlled for with stratified or multiple regression analysis after data collection. If there is the potential for uncontrolled confounding, then it is important to assess its likely strength and direction (Lee KM *et al.*, 2005).

Toxicogenomics study aims to study the complex interaction between genome and environmental stressors and toxicants. Optimal sample size will be selected to the population based toxicogenomics study. It dependent on the statistical test for gene-to environment or environmental exposure (Gauderman, 2002a,b; Wang and Chen, 2004). The number of factors includes the specified significance level α, the desired statistical power $(1 - \beta)$, the fraction η of truly altered genes out of the total g genes studied, and the effect size Δ for the altered genes (Wang and Chen, 2004). Therefore, pooling samples might be a good choice in the application of '-omics' approaches to population studies under the condition that epidemiologically sound strategies, such as randomization and matching, to control potential confounders are adopted (Churchill, 2002; Page *et al.*, 2003).

4.4 Application of Toxicogenomic Research

4.4.1 Mechanistic Toxicology

The pattern of gene expression is a key feature of cellular responses to stimuli. Toxicogenomic technologies are very useful for analysis to identify and elucidate fundamental mechanisms of the actions of known compounds. Analysis of the transcriptome offers a preliminary indication of the biochemical or biological mechanism that could possibly be affected by external stimuli. In Jurkat cells devoid of p53, ionizing radiation could elicit NFκB-dependent transcription in response (Park *et al.*, 2002). Changes in gene expression can provide a more sensitive and specific endpoint than the toxicity itself. Gene expression profiles provide an improved understanding of the mechanisms of toxicity, identifying the relationships between compound exposure and alterations in genome-wide gene expression patterns. However, not all alterations in gene expression should be viewed as responses to toxins. Accurate and exhaustive analysis of the type of gene expression pattern is crucial in order to distinguish between harmful toxic responses and responses representing benign homeostatic adjustments (Irwin *et al.*, 2004). Gene expression results obtained from exposure of model systems to known compounds are analyzed and a common set of changes in gene expression in cells and tissues unique to the specific class of compound (molecular signature) is identified. This molecular signature is characteristic of a specific mechanism of induced toxicity.

4.4.2 Biomarkers for Predictive Toxicology

One major application of toxicogenomics to predict exposure of toxic materials is to identify genes predictive of toxicity, i.e. biomarkers. The ideal biomarker provides a sensitive and informative indicator of possible adverse effects at times or doses preceding tissue damage, toxicity or disease (Gatzidou *et al.*, 2007). Toxicogenomics through microarray technology would provide a powerful and relatively inexpensive tool to isolate biomarkers from gene expression profiles showing altered expression levels among tens of thousands of genes. The discovery and validation of these biomarkers is essential for the application in high-throughput experimental systems to characterize target organ-specific effects and to detect specific toxicity endpoints. The identification and utilization of biomarkers for toxicogenomics can be applied to many fields. First, we can use them in preclinical toxicology analysis for the risk characterization and risk assessment of chemicals. Second, in clinical stages of drug development, monitoring these markers will be essential to detect adverse effects in earlier stages. Lastly, in daily clinical diagnosis, these genes can be used for disease classification and therapeutic monitoring.

4.5 Conclusion

Genome-wide analysis using microarrays has enabled us to understand the landscape of physiological and pathological changes at the RNA or DNA level. However, we still need extensive follow-up work using more traditional approaches examining multiple endpoints at the molecular, cellular, physiological and histopathological levels (Smith, 2001). Most chemicals work through various mechanisms depending on dose, time and duration. In addition, the responses to chemicals are quite variable between different cell types. Moreover, level of exposure, age, gender, diet and hormonal status possibly affect the response to chemicals *in vivo*. Thus, differences in measurements of parameters such as gene expression occurring at the transcriptional level among different animal or cell model systems prohibit us from extrapolating one result of toxicogenomic studies to other models.

Genomics technologies have been used to mine information on the mechanisms of action of toxic substances. In particular, microarray experiments could be the best example of a successful application of toxicogenomic studies (Waring *et al.*, 2001; Lee WS *et al.*, 2004; Lee KM *et al.*, 2005). In addition, different levels of genomic research, such as DNA methylation, posttranscriptional regulation and small RNAs, are now emerging in technology as well as in quality of information. Proteomics also helps us understand the toxicity mechanism of chemicals that induce alterations in the level and modification of cellular proteins (Amacher *et al.*, 2005; Gibbs, 2005). Results from metabolomics studies on chemicals have now been presented to reveal the effect of chemicals on cellular metabolism. Each platform might be powerful enough to understand the effect of toxicity induced by certain chemicals, but the integration of these techniques will deliver a greater impact on toxicogenomic research.

Acknowledgments

This work was supported by a grant from the Korea National Institute of Health to D. Kang (2007).

References

Amacher DE, Adler R, Herath A, Townsend RR (2005) Use of proteomic methods to identify serum biomarkers associated with rat liver toxicity or hypertrophy. *Clin Chem* **51**: 1796–1803.

Anderson NL, Taylor J, Hofmann JP, Esquer-Blasco R, Swift S, Anderson NG (1996) Simultaneous measurement of hundreds of liver proteins: application in assessment of liver function. *Toxicol Pathol* **24**: 72–76.

Baken KA, Arkusz J, Pennings JLA, Vandebriel RJ, van Loveren H (2007) *In vitro* immunotoxicity of bis(tri-*n*-butyltin)oxide (TBTO) studied by toxicogenomics. *Toxicology* **237**: 35–48.

Barlow SM, Greig JB, Bridges JW, Carere A, Carpy AJM, Galli CL, Kleiner J, Knudsen I, Koeter HBWM, Levy LS, *et al.* (2002). Hazard identification by methods of animal-based toxicology. *Food and Chemical Toxicology* **40**: 145–191.

Bloomston M, Frankel WL, Petrocca F, Volinia S, Alder H, Hagan JP, Liu CG, Bhatt D, Taccioli C, Croce CM (2007) MicroRNA expression patterns to differentiate pancreatic adenocarcinoma from normal pancreas and chronic pancreatitis. *J Am Med Assoc* **297**: 1901–1908.

Boyes WK., Moser VC, Geller AM, Benignus VA, Bushnell PJ, Kamel F (2007) Integrating epidemiology and toxicology in neurotoxicity risk assessment. *Hum Exp Toxicol* **26**: 283–293.

Brown PO, Botstein D (1999) Exploring the new world of the genome with DNA microarrays. *Nat. Genet* **21**: 33–37.

Carninci P, Hayashizaki Y (2007) Noncoding RNA transcription beyond annotated genes. *Curr Opin Genet Dev* **17**: 139–144.

Ching TT, Maunakea AK, Jun P, Hong C, Zardo G, Pinkel D, Albertson DG, Fridlyand J, Mao JH, Shchors K, Weiss WA, Costello JF (2005) Epigenome analyses using BAC microarrays identify evolutionary conservation of tissue-specific methylation of SHANK3. *Nat Genet* **37**: 645–651.

Choi JY, Lee KM, Park SK, Noh DY, Ahn SH, Chung HW, Han W, Kim JS, Shin SG, Jang IJ, Yoo KY, Hirvonen A, Kang D (2005) Genetic polymorphisms of SULT1A1 and SULT1E1 and the risk and survival of breast cancer. *Cancer Epidemiol Biomarkers Prev* **14**: 1090–1095.

Churchill GA (2002) Fundamentals of experimental design for cDNA microarrays. *Nat Genet* **32**(Suppl): 490–495.

Cutler P (2003) Protein arrays: the current state-of-the-art. *Proteomics* **3**: 3–18.

Dere E, Boverhof DR, Burgoon LD, Zacharewski TR (2006) *In vivo–in vitro* toxicogenomic comparison of TCDD-elicited gene expression in Hepa1c1c7 mouse hepatoma cells and C57BL/6 hepatic tissue. *BMC Genomics* **7**: 80.

Eisenbrand G, Pool-Zobel B, Baker V, Balls M, Blaauboer BJ, Boobis A, Carere A, Kevekordes S, Lhuguenot JC, Pieters R, Kleiner J (2002) Methods of *in vitro* toxicology. *Food Chem Toxicol* **40**: 193–236.

Forrest MS, Lan Q, Hubbard AE, Zhang L, Vermeulen R, Zhao X, Li G, Wu YY, Shen M, Yin S, Chanock SJ, Rothman N, Smith MT (2005) Discovery of novel biomarkers by microarray analysis of peripheral blood mononuclear cell gene expression in benzene-exposed workers. *Environ Health Perspect* **113**(6): 801–807.

Gatzidou ET, Zira AN, Theocharis SE (2007) Toxicogenomics: a pivotal piece in the puzzle of toxicological research. *J Appl Toxicol* **27**: 302–309.

Gauderman WJ (2002a) Sample size requirements for association studies of gene–gene interaction. *Am J Epidemiol* **155**: 478–484.

Gauderman WJ (2002b) Sample size requirements for matched case–control studies of gene–environment interaction. *Stat Med* **21**: 35–50.

Gibbs A (2005) Comparison of the specificity and sensitivity of traditional methods for assessment of nephrotoxicity in the rat with metabonomic and proteomic methodologies. *J Appl Toxicol* **25**: 277–295.

Gygi SP, Rist B, Gerber SA, Turecek F, Gelb MH, Aebersold R (1999) Quantitative analysis of complex protein mixtures using isotope-coded affinity tags. *Nat Biotechnol* **17**: 994–999.

Heijne WH, Stierum RH, Slijper M, van Bladeren PJ, van Ommen B (2003) Toxicogenomics of bromobenzene hepatotoxicity: a combined transcriptomics and proteomics approach. *Biochem Pharmacol* **65**: 857–875.

Holmes E, Nicholson JK, Tranter G (2001) Metabonomic characterization of genetic variations in toxicological and metabolic responses using probabilistic neural networks. *Chem Res Toxicol* **14**: 182–191.

International Human Genome Sequencing Consortium (2001) Initial sequencing and analysis of the human genome. *Nature* **409**: 860–921.

Irwin RD, Boorman GA, Cunningham ML, Heinloth AN, Malarkey DE, Paules RS (2004) Application of toxicogenomics to toxicology: basic concepts in the analysis of microarray data. *Toxicol Pathol* **32**(Suppl 1): 72–83.

Joo W-A, Sul D, Lee D-Y, Lee E, Kim C-W (2004) Proteomic analysis of plasma proteins of workers exposed to benzene. *Mut Res Gen Toxicol Environ Mutagen* **558**: 35–44.

Kier LD, Neft R, Tang L, Suizu R, Cook T, Onsurez K, Tiegler K, Sakai Y, Ortiz M, Nolan T. *et al.* (2004) Applications of microarrays with toxicologically relevant genes (*tox* genes) for the evaluation of chemical toxicants in Sprague Dawley rats *in vivo* and human hepatocytes *in vitro*. *Mut Res Fundam Mol Mech Mutagen* **549**: 101–113.

Klein JB, Thongboonkerd V (2004) Overview of proteomics. *Contrib Nephrol* **141**: 1–10.

Knudsen LE, Hansen AM (2007) Biomarkers of intermediate endpoints in environmental and occupational health. *Int J Hyg Environ Health* **210**: 461–470.

Lee KM, Kim JH, Kang D (2005) Design issues in toxicogenomics using DNA microarray experiment. *Toxicol Appl Pharmacol* **207**: 200–208.

Lee WS, Lee GJ, Lee GJ, Yeo CD, Kang JS, Kim YH, Kim SJ, Kim JS, Hwang SY, Park JS *et al.* (2004) The intelligent data management system for toxicogenomics. *J Vet Med Sci* **66**: 1335–1338.

Marchant GE (2003) Toxicogenomics and toxic torts. *Trends Biotechnol.* **20**(8): 329–332.

Mattes WB, Pettit SD, Sansone SA, Bushel PR, Waters MD (2004) Database development in toxicogenomics: issues and efforts. *Environ Health Perspect* **112**: 495–505.

Mattingly CJ, Rosenstein MC, Colby GT, Forrest Jr JN, Boyer JL (2006) The Comparative Toxicogenomics Database (CTD): a resource for comparative toxicological studies. *J Exp Zool A Comp Exp Biol* **305**: 689–692.

Mikkelsen TS, Ku M, Jaffe DB, Issac B, Lieberman E, Giannoukos G, Alvarez P, Brockman W, Kim TK, Koche RP, Lee W, Mendenhall E, O'Donovan A, Presser A, Russ C, Xie X, Meissner A, Wernig M, Jaenisch R, Nusbaum C, Lander ES, Bernstein BE (2007) Genome-wide maps of chromatin state in pluripotent and lineage-committed cells. *Nature* **448**: 553–560.

Olden K, Call N, Sobral B, Oakes R (2004) Toxicogenomics through the eyes of informatics: conference overview and recommendations. *Environ Health Perspect* **112**(7): 805–807.

Page GP, Edwards JW, Barnes S, Weindruch R, Allison DB (2003) A design and statistical perspective on microarray gene expression studies in nutrition: the need for playful creativity and scientific hard-mindedness. *Nutrition* **19**: 997–1000.

Park WY, Hwang CI, Im CN, Kang MJ, Woo JH, Kim JH, Kim YS, Kim JH, Kim H, Kim KA, Yu HJ, Lee SJ, Lee YS, Seo JS (2002) Identification of radiation-specific responses from gene expression profile. *Oncogene* **21**: 8521–8528.

Parker BJ, Günter S, Bedo J (2007) Stratification bias in low signal microarray studies. *BMC Bioinform* **8**: 326.

Pennie WD, Tugwood JD, Oliver GJA, Kimber I (2000) The principles and practice of toxicogenomics: applications and opportunities. *Toxicol Sci* **54**: 277–283.

Pesch B, Bruning T, Frentzel-Beyme R, Johnen G, Harth V, Hoffmann W, Ko Y, Ranft U, Traugott UG, Thier R *et al.* (2004) Challenges to environmental toxicology and epidemiology: where do we stand and which way do we go? *Toxicol Lett* **151**: 255–266.

Potter JD (2003) Epidemiology, cancer genetics and microarrays: making correct inferences, using appropriate designs. *Trends Genet* **19**(12): 690–695.

Rane S, Sayed D, Abdellatif M (2007) MicroRNA with a MacroFunction. *Cell Cycle* **6**: 1850–1855.

Redon R, Ishikawa S, Fitch KR, Feuk L, Perry GH, Andrews TD, Fiegler H, Shapero MH, Carson AR, Chen W *et al.* (2006) Global variation in copy number in the human genome. *Nature* **444**: 444–454.

Robertson DG (2005) Metabonomics in toxicology: a review. *Toxicol Sci.* **85**: 809–822.

Ross PL, Huang YN, Marchese JN, Williamson B, Parker K, Hattan S, Khainovski N, Pillai S, Dey S, Daniels S, Purkayastha S, Juhasz P, Martin S, Bartlet-Jones M, He F, Jacobson A, Pappin DJ (2004) Multiplexed protein quantitation in *Saccharomyces cerevisiae* using amine-reactive isobaric tagging reagents. *Mol Cell Proteomics* **3**: 1154–1169.

Sellers TA (2003) Review of proteomics with applications to genetic epidemiology. *Gen Epidemiol* **24**: 83–98.

Shi H, Yan PS, Chen CM, Rahmatpanah F, Lofton-Day C, Caldwell CW, Huang TH (2002) Expressed CpG island sequence tag microarray for dual screening of DNA hypermethylation and gene silencing in cancer cells. *Cancer Res* **62**: 3214–3220.

Smith LL (2001) Key challenges for toxicologists in the 21st century. *Trends Pharmacol Sci* **22**: 281–285.

Smith MT, Vermeulen R, Li G, Zhang L, Lan Q, Hubbard AE, Forrest MS, McHale C, Zhao X, Gunn L. *et al.* (2005) Use of 'omic' technologies to study humans exposed to benzene. *Chemico-Biol Interact* **153–154**: 123–127.

Stovner LJ (2006) Headache epidemiology: how and why? *J Headache Pain* **7**: 141–144.

The ENCODE Project Consortium (2007) Identification and analysis of functional elements in 1 % of the human genome by the ENCODE pilot project. *Nature* **447**: 799–816.

The National Center for Toxicogenomics (2004) NCT update, microarray demystified. *Environ Health Perspect* **112**(4): A222–A223.

Uetrecht J (2007) Idiosyncratic drug reactions: current understanding. *Annu Rev Pharmacol Toxicol* **47**: 513–539.

Vermeulen R, Lan Q, Zhang L, Gunn L, McCarthy D, Woodbury RL, McGuire M, Podust VN, Li G, Chatterjee N *et al.* (2005) Decreased levels of CXC-chemokines in serum of benzene-exposed workers identified by array-based proteomics. *Proc Natl Acad Sci U S A* **102**: 17041–17046.

Wang SJ, Chen JJ (2004) Sample size for identifying differentially expressed genes in microarray experiments. *J Comput Biol* **11**: 714–726.

Waring JF, Ciurlionis R, Jolly RA, Heindel M, Ulrich RG (2001) Microarray analysis of hepatotoxins *in vitro* reveals a correlation between gene expression profiles and mechanisms of toxicity. *Toxicol Lett* **120**: 359–368.

Washburn MP, Wolters D, Yates III JR (2001) Large-scale analysis of the yeast proteome by multi-dimensional protein identification technology. *Nat Biotechnol* **19**: 242–247.

Yates III JR (2000) Mass spectrometry: from genomics to proteomics. *Trends Genet* **16**: 5–8.

5

Sources of Variability in Toxicogenomic Assays

Karol Thompson, P. Scott Pine and Barry Rosenzweig

5.1 Introduction

New technologies for simultaneous high-throughput measurement of the relative levels of expressed transcripts provide increased molecular endpoints for toxicity assessments. However, toxicogenomic data will be of limited utility unless the important sources of biological and technical variability are understood and controlled. Standard methods for measuring assay and laboratory performance cannot be easily applied to the multiple endpoints measured by microarray technology. An additional challenge in formulating guidelines of best practices for conducting microarray assays is the need to address the diversity of platforms, instrumentation, reagents and protocols that are used to generate toxicogenomic data.

The reproducibility of microarray data across platform formats is an important factor in the validation of this technology as a reliable and accurate measure of relative gene expression. A critique of cross-platform validation studies highlights some of the factors that have contributed to increases in correlation between laboratories and platforms that are reported in more recently published studies (2004 and beyond) (Yauk and Berndt, 2007). The increased availability of qualified reagent kits and of automated systems for sample and array processing have been important factors in reducing technical variability. Although incorrect annotation was an issue with cDNA-based arrays due to incorrect or contaminated clone sets, oligonucleotide probes have since become the *de facto* standard for arrays from both commercial and noncommercial sources. For the field of toxicogenomics, an important development was the complete sequencing of the rat genome in 2004, which has led to improvements in probe annotation and microarray design. A comprehensive study recently conducted by the Microarray Quality Control (MAQC) Consortium to examine the comparability of microarray data between Affymetrix, Agilent, GE Healthcare Codelink, Illumina, and Applied Biosystems microarray platforms showed high overall reproducibility within and between experienced laboratories using standardized protocols, and between different platforms using probes that were mapped to curated cDNA sequence databases (MAQC Consortium, 2006). Although microarray technology can achieve reproducible results, there are other factors involved in performing toxicogenomic studies, such as study design, sample preparation or protocol choice that were not variables in the MAQC study, but which can contribute to discrepancies in the data. This chapter highlights some recent studies on major sources of variability in toxicogenomic study data that were conducted in US Food and Drug Administration laboratories and at other institutions.

5.2   The Importance of Study Design

Toxicant-specific effects on gene expression can be obscured or misidentified when nonuniform handling of animals or samples introduces significant differences in expression levels within or between treatment groups. A good experimental design minimizes the introduction of bias in the data due to nonrandomized processing of treatment groups during steps such as necropsy, RNA isolation, sample labeling and hybridization. For example, variability in toxicogenomic data can be introduced through unequal sampling of tissue for RNA extraction. Careful and consistent sectioning between individual animals is critical to reduced biological variation which may arise from regional differences in gene expression. Several recent studies that have examined the comparability of gene expression in different regions of the kidney have found that the medulla and cortex are highly variant in expression (Tamura *et al.*, 2006; Boedigheimer *et al.*, 2008). Analysis of control animal data from different liver lobes found tissue section to be a much less prominent source of variance in baseline gene expression for liver than kidney (Boedigheimer *et al.*, 2008). However, differences in transcriptional responses to liver toxicants have been observed between different liver lobes. The left and median lobes of F344 rat liver exhibited differences in acetaminophen-induced gene changes (Irwin *et al.*, 2005). Furan induced an increased severity of lesions and a greater number of gene expression changes in the right and caudate lobes of the liver, compared with the left and medial lobes (Hamadeh *et al.*, 2004). These results suggest that a recommended best practice for toxicogenomics might include performing histopathology, gene expression and other toxicology endpoints on the same region of a given tissue, where practical.

Comparisons of gene expression levels in tissues from rats in control (untreated and vehicle-treated) groups between differently designed toxicogenomic studies could provide some additional insight on the impact of variations in study protocol on gene expression. The Health and Environmental Sciences Institute (HESI) Technical Committee on Genomics has compiled a publicly accessible dataset of over 500 arrays from control animal liver and kidney that has served as a resource to identify and analyze baseline fluctuations in gene expression that are due to biological or technical variables (Boedigheimer *et al.*, 2008). The data were collected from rat toxicogenomic studies conducted at 16 different institutions using Affymetrix arrays. Each array was annotated with descriptors of experimental protocol variables like diet, vehicle type, route, and frequency of administration, method of sacrifice and the use of anesthetics during the study. These studies encompassed eight different routes of vehicle administration, 13 different vehicles, and six sacrifice methods. The lengths of the studies ranged from less than 1 day to 90 days, although the most common study length (22 % of total studies) was 24 h in duration. Because of the wide range in study design for the collected data, most study factors could not be analyzed independent of confounding factors. Nonetheless, gender, organ section, strain and fasting state were identified as some of the key factors associated with signal variability.

The HESI study of control animal data also identified genes that have high baseline variance, which was estimated from the level of residual variance not associated with any identified study factor. The inclusion of genes with high inherent variability in signatures of toxicity or pharmacology could have a negative impact on signature reproducibility across laboratories or studies. We found that several of the genes associated with renal injury by cisplatin (*Spp1, Lcn2, Egf, Slc34a2, Col1a1, G6pc* and *Igfbp3*) in an earlier study conducted through HESI (Thompson *et al.*, 2004) were also identified as having high baseline variance. High baseline variability in control liver, kidney or both tissues was also associated with genes which are components of curated pathways (Kyoto Encyclopedia of Genes and Genomes (KEGG) pathways) of relevance to toxicology or drug disposition. The KEGG pathways associated with genes of high variance included androgen and estrogen metabolism, biosynthesis of steroids, insulin signaling pathway, mitogen-activated protein kinase signaling pathway and xenobiotic metabolism.

In toxicogenomic studies of subchronic or chronic duration, organ weight and total body weight are important overall indicators of drug toxicity. Little information is publicly available on the effect of organ weight reductions on gene expression, the tissue specificity of the effect in type or magnitude and the threshold of response. A working group in the HESI Technical Committee on Genomics that is focused on mechanism-based markers of toxicity is conducting an extensive toxicogenomics study of doxorubicin cardiotoxicity that may provide the level of detail needed to fill gaps in information on the effect of heart weight on cardiac gene expression.

Circadian variations in gene expression are important factors for consideration in toxicogenomic study design and interpretation. Circadian regulation of metabolic programs in tissues is controlled by peripheral oscillators under the coordination of the master clock in the suprachiasmatic nuclei of the hypothalamus. It has been estimated that 10 % of genes expressed in liver and 8 % of genes expressed in heart show circadian regulation in the mouse (Storch *et al.*, 2002). Of direct relevance to toxicogenomics, transcript levels of certain genes involved in metabolism and cholesterol synthesis exhibit a diurnal pattern of expression in liver and kidney in rats acclimated to a light/dark cycle (Kita *et al.*, 2002; Boorman *et al.*, 2005). Circadian variations in gene expression have been observed in xenobiotic detoxification enzymes, including CYP2a family members, glutathione *S*-transferases, *Abcb1a* (P-glycoprotein), *Nr1i3* (CAR), *Ppara* (PPAR-alpha) and *Arnt* (Lim *et al.*, 2006).

Steroid hormones are known to affect circadian rhythm in different tissues in mammals (Lim *et al.*, 2006). Food restriction has also been shown to have an interacting effect on circadian rhythms of gene expression. Restricting access to food to 4 or 8 h during the light portion of a 12 h/12 h light/dark cycle has been shown to entrain the rhythm of expression of the clock gene *Per1* in the liver of rats (Stokkan *et al.*, 2001). In another study, subsets of liver genes were identified that exhibited a significant difference in day to night expression ratios only in fasted animals, only in nonfasted animals or irrespective of feeding status (Kita *et al.*, 2002). For example, genes involved in cholesterol synthesis and glucose utilization are diurnally regulated in the liver but differ in their sensitivity to fasting. Additionally, genes identified as diurnally expressed genes in kidney tended to be distinct from those identified in liver and were less affected by fasting.

The mechanisms responsible for the circadian rhythm of expression of xenobiotic metabolism genes includes direct regulation by circadian 'clock' genes or activation by glucocorticoids, which are secreted in circadian rhythms. One study found that the action of glucocorticoids accounted for 60 % of circadian-regulated expression in mouse liver (Reddy *et al.*, 2007). The transcription factor HNF4α, which is a central regulator of lipid metabolism in the liver (Naiki *et al.*, 2002), has been identified as a key mediator of circadian- and glucocorticoid-regulated transcription (Reddy *et al.*, 2007). Design and analysis of toxicogenomic studies should include awareness of potential confounding effects on genes in pathways regulated or influenced by circadian rhythms.

5.3 The Effect of RNA Quality on Toxicogenomic Data

At necropsy, tissues are commonly preserved for subsequent RNA isolation either through immediate freezing in liquid nitrogen or by immersion in tissue-collection solutions which stabilize RNA in tissues at ambient temperatures. The methods used to harvest and preserve source tissue for gene expression analyses can impact the quality of isolated mRNA. A standard practice for classifying the level of RNA integrity is through automated assignment of an RNA integrity number (RIN) to each sample based on microcapillary electrophoresis tracings. The RIN software algorithm classifies the integrity of eukaryotic total RNA on a scale of 1 to 10 (most to least degraded) based on the most informative features of an electropherogram, including the 18s and 28s rRNA peaks (Schroeder *et al.*, 2006).

We recently compared the kinetics of RNA degradation in fresh rat liver tissue with thawed frozen rat liver sections (Figure 5.1) (Thompson *et al.*, 2007). No measurable effect on RNA integrity was observed in fresh liver tissue incubated up to 6 h at room temperature. RNA degradation occurred most rapidly in frozen tissue upon thawing. The incubation times required to generate poor-quality RNA (RIN $\leq$ 7) were about 30 min for frozen tissue at room temperature and about 3 h for fresh tissue at 37 °C. These results suggest that rapid homogenization of frozen tissue for RNA is important to preserve RNA integrity. Whether tissue is stored at -70 °C or in stabilization solutions prior to RNA extraction, in our experience it is important to dissect tissue into small pieces prior to storage to facilitate the rapid homogenization of tissue.

We examined the effect of RNA quality on microarray assay performance through analysis of a set of rat liver RNA samples with a progressive change in RNA integrity (Thompson *et al.*, 2007). Select samples of defined RNA integrity levels generated by thawing frozen tissue over a time course were run on Affymetrix RAE230A arrays. Probe sets on Affymetrix arrays are designed to reside within 600 nucleotides of the 3′ end of a transcript so that signal levels are relatively tolerant to moderate RNA degradation (Schoor *et al.*, 2003). An array-based metric for Affymetrix GeneChip® arrays measures the ratio of probe signals that hybridize to 5′ or 3′ regions of the universally expressed β-actin and glyceraldehyde-3-phosphate dehydrogenase (GAPDH) gene transcripts. Target length will be affected by both the level of RNA degradation in the sample and the progressivity

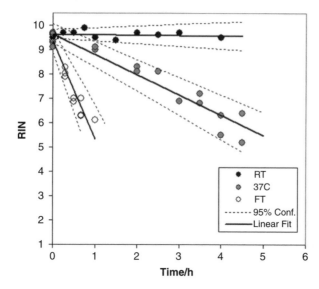

Figure 5.1   *Effect of tissue handling on RNA degradation in rat liver. Liver tissue sections were incubated at room temperature (RT), in a 37 °C water bath (37C), or flash frozen and thawed at room temperature (FT). At each time point, liver was immediately homogenized and RNA extracted as described (Thompson et al., 2005). The RIN for each sample was measured using an Agilent 2100 Bioanalyzer.*

of cDNA and cRNA synthesis reactions. For RNA degraded by a freeze–thaw cycle, we found that an RIN ≤ 7 corresponded to a $3'/5'$ GAPDH ratio ≥ 3. The use of RNA with a $3'/5'$ GAPDH ratio <3 has been recommended as a best practice in conducting microarray experiments (Hoffman *et al.*, 2004).

The sensitivity of individual probe sets to RNA degradation was characterized using the dataset of rat liver samples with a progressive loss in RNA quality. Eighty-six probe sets were identified that were as sensitive to RNA degradation as the $5'$ GAPDH and $5'$ β-actin probe sets used to assess sample quality on Affymetrix GeneChip® arrays. Most of these probe sets mapped to locations more than 1000 nucleotides upstream of their transcription termini, similar to the positioning of the $5'$ control gene probe sets.

The effect of RNA integrity on the sensitivity and specificity of microarray data generated on Affymetrix GeneChip arrays was also assessed with this dataset. Detection of false positives was significantly increased in comparisons of array data from undegraded liver RNA with RNA of RIN ≤ 7. Known changes in signal between samples were modeled *in silico* based on a four-tissue RNA design previously described for measurements of assay performance (Thompson *et al.*, 2005). Detection of true positives was reduced using target signals from RNA with RIN ≤ 7 for probe sets identified as most sensitive to sample integrity. Quantitative reverse transcription polymerase chain reaction (qRT-PCR) experiments are more tolerant to RNA degradation. The level of RNA integrity found to impact the accuracy and precision of the results negatively was RIN < 6 (Schroeder *et al.*, 2006) or RIN < 5 (Fleige and Pfaffl, 2006). It is important to define the threshold of RNA quality that impacts the sensitivity and specificity of toxicogenomic data, and for studies to be annotated with this information.

5.4 Optimizing Laboratory Protocols

Target preparation usually involves a two-step protocol of cDNA synthesis, followed by linear amplification with T7 RNA polymerase to produce cRNA. The efficiency of these reactions can be monitored by the yield and the size of products generated. The expected yield will be dependent on the protocol and reagents used. In pilot experiments for the Affymetrix sites in the MAQC study (MAQC Consortium, 2006), certain variations in the standardized protocol, such as the model of thermocycler used (e.g. heated versus nonheated lid), were found to impact cRNA yield. If long overnight transcription reactions are specified in a protocol, then they can be prematurely terminated by condensation on the reaction tube lid. Examination of the intersite comparison of results from the main MAQC study that were based on the titration response between a universal human reference RNA, human brain RNA and two different mixtures of these samples (Shippy *et al.*, 2006) suggests that cRNA yield might impact performance. The Affymetrix site with the lowest overall cRNA yields (see Supplemental Table 1 in MAQC Consortium (2006)) also had the lowest percentages of titrating genes, although other contributing factors are possible.

Two-color array designs incorporate different fluorescently labeled nucleotides into targets for treated and reference or control samples to allow simultaneous hybridization of both samples to the same array. There are differences between dyes (usually Cy3 and Cy5) in the rate of incorporation into the nucleotide target and in quantum efficiency that can introduce bias into the data. Indirect labeling methods first introduce reactive nucleotides (e.g. amino-allyl and amino-hexyl modified) into the target during cDNA synthesis, followed by coupling with mono-reactive fluorescent derivatives. This approach can correct for dye bias that results from differential dye incorporation. Dye bias can also result from the differential sensitivity of Cy5 and Cy3 dyes to quenching, photobleaching or degradation. The environmental ozone sensitivity of certain cyanine dyes (i.e. Cy5, Alexa 647) may track as a seasonal or diurnal effect on microarray data quality (Fare *et al.*, 2003). The installation of carbon filters in laboratory air-handling systems can effectively control the decay in Cy5 signal due to ozone (Branham *et al.*, 2007).

The low concordance exhibited by some inter-platform studies has been attributed to the use of suboptimal protocols (Yauk and Berndt, 2007). Reference standards are common components of quality assessment programs that allow researchers to test the performance of their laboratory methods and technique. We have developed a material for toxicogenomics laboratories to use for proficiency testing, process drift monitoring, protocol optimization (Han *et al.*, 2006) and other performance assessments on rat whole-genome expression microarrays (Thompson *et al.*, 2005). The material can be created by researchers from rat tissue readily available from animal studies and is composed of a set of two mixed-tissue RNA samples formulated to contain different known ratios of RNA for each of four rat tissues (brain, liver, kidney and testes). The two mixes form the basis of an assay that measures the ratios of a set of reference probes that bind tissue-selective gene transcripts that we have identified for Affymetrix, GE Healthcare CodeLink and Agilent arrays. Mix 1 consists of 10 % testis, 40 % brain, 30 % liver and 20 % kidney RNA and Mix2 consists of 40 % testis, 20 % brain, 20 % liver and 20 % kidney RNA (Figure 5.2). The reference material is regenerable; it can be prepared from RNA pooled across multiple animals that is available from in-house or commercial sources.

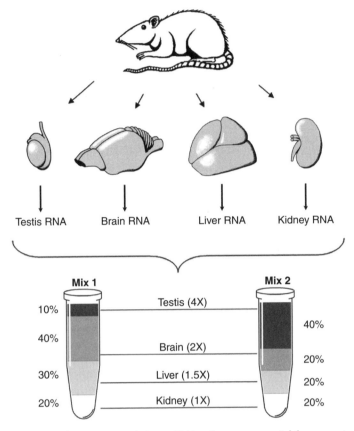

Figure 5.2 *Design for a rat mixed-tissue RNA reference material for assessing performance on microarrays.*

A reliable measure of performance using the rat mixed-tissue RNA reference material is the area under the curve (AUC) from receiver-operating characteristic (ROC) plots (Pine *et al.*, submitted). An ROC plot is generated by fitting a curve to a plot of the true positive fraction versus false positive fraction along a continuum of decision thresholds based on *P*-values or log ratios. Using 100 tissue-selective analytes per target ratio, the diagnostic accuracy of detection of true positive changes of 4-, 2-, and 1.5-fold fold-change ratios provides a quantitative assessment of assay sensitivity that can be compared internally or against an external reference datasets.

External controls are essential reagents for use in the qualitative assessment of different aspects of technical performance involved in generating microarray data (i.e. sample labeling, hybridization and grid alignment). These controls are provided by many array manufacturers and consist of nonmammalian sequences that are spiked into samples and hybridize to corresponding control probe sequences on their arrays. External controls tend to be selected from prokaryotic or plant gene sequences that do not cross-hybridize to mammalian transcripts or probes and are usually not interchangeable between platforms. Labeling controls are polyadenylated prokaryotic RNAs that are spiked into RNA samples before the reverse transcription step to monitor labeling and subsequent processes. Some platforms also contain probes for negative controls that can be used to calculate the noise threshold and to flag probes with signals below the lower limit of detection. Signal uniformity across arrays can be assessed if the array design includes multiple replicate spot locations for control sequences that are distributed across the array grid. External controls that are added after target synthesis are used to troubleshoot the array hybridization and staining steps.

5.5   Summary

Variation in toxicogenomic data can be reduced if attention is paid to biological sources of variation in designing toxicogenomic studies, if RNA quality thresholds are used and if laboratory performance is monitored using appropriate standards.

Disclaimer

The views presented in this chapter do not necessarily reflect official agency policy.

References

Boedigheimer MJ, Wolfinger RD, Bass MB, Bushel PR, Chou JW, Cooper M, Corton JC, Fostel J, Hester S, Lee JS, Liu F, Liu J, Qian H-R, Quackenbush J, Pettit S, Thompson KL (2008) Sources of variation in baseline gene expression levels from toxicogenomic study control animals across multiple laboratories. *BMC Genomics* **9**, 285.
Boorman GA, Blackshear PE, Parker JS, Lobenhofer EK, Malarkey DE, Vallant MK, Gerken DK, Irwin RD (2005) Hepatic gene expression changes throughout the day in the Fischer rat: implications for toxicogenomic experiments. *Toxicol Sci* **86**, 185–193.
Branham WS, Melvin CD, Han T, Desai VG, Moland CL, Scully AT, Fuscoe JC (2007) Elimination of laboratory ozone leads to a dramatic improvement in the reproducibility of microarray gene expression measurements. *BMC Biotechnol* **7**, 8.
Fare TL, Coffey EM, Dai H, He YD, Kessler DA, Kilian KA, Koch JE, LeProust E, Marton MJ, Meyer MR, Stoughton RB, Tokiwa GY, Wang Y (2003) Effects of atmospheric ozone on microarray data quality. *Anal Chem* **75**, 4672–4675.
Fleige S, Pfaffl MW (2006) RNA integrity and the effect on the real-time qRT-PCR performance. *Mol Aspects Med* **27**(2–3), 126–139.
Hamadeh HK, Jayadev S, Gaillard ET, Huang Q, Stoll R, Blanchard K, Chou J, Tucker CJ, Collins J, Maronpot R, Bushel P, Afshari CA (2004) Integration of clinical and gene expression endpoints to explore furan-mediated hepatotoxicity. *Mutat Res* **549**, 169–183.
Han T, Melvin CD, Shi L, Branham WS, Moland CL, Pine PS, Thompson KL, Fuscoe JC (2006) Improvement in the reproducibility and accuracy of DNA microarray quantification by optimizing hybridization conditions, *BMC Bioinform* **7**(Suppl 2), S17.
Hoffman EP, Awad T, Palma J, Webster T, Hubbell E, Warrington JA, Spira A, Wright G, Buckley J, Triche T *et al.* (2004) Expression profiling – best practices for data generation and interpretation in clinical trials. *Nat Rev Genetics* **5**, 229–237.

Irwin RD, Parker JS, Lobenhofer EK, Burka LT, Blackshear PE, Vallan MK, Lebetkin EH, Gerken DF, Boorman GA (2005) Transcriptional profiling of the left and median liver lobes of male f344/n rats following exposure to acetaminophen. *Toxicol Pathol* **33**, 111–117.

Kita Y, Shiozawa M, Jin W, Majewski RR, Besharse JC, Greene AS, Jacob HJ (2002) Implications of circadian gene expression in kidney, liver and the effects of fasting on pharmacogenomic studies. *Pharmacogenetics* **12**, 55–65.

Lim FL, Currie RA, Orphanides G, Moggs JG (2006) Emerging evidence for the interrelationship of xenobiotic exposure and circadian rhythms: a review. *Xenobiotica* **36**(10–11), 1140–1151.

MAQC Consortium (2006) The MicroArray Quality Control (MAQC) project shows inter- and intraplatform reproducibility of gene expression measurements. *Nat Biotechnol* **24**, 1151–1161.

Naiki T, Nagaki M, Shidoji Y, Kojima H, Imose M, Kato T, Ohishi N, Yagi K, Moriwaki H (2002) Analysis of gene expression profile induced by hepatocyte nuclear factor 4alpha in hepatoma cells using an oligonucleotide microarray. *J Biol Chem* **277**(16), 4011–4019.

Reddy AB, Maywood ES, Karp NA, King VM, Inoue Y, Gonzalez FJ, Lilley KS, Kyriacou CP, Hastings MH (2007) Glucocorticoid signaling synchronizes the liver circadian transcriptome. *Hepatology* **45**(6), 1478–1488.

Schoor O, Weinschenk T, Hennenlotter J, Corvin S, Stenzl A, Rammensee HG, Stevanovic S (2003) Moderate degradation does not preclude microarray analysis of small amounts of RNA. *Biotechniques* **35**, 1192–1201.

Schroeder A, Mueller O, Stocker S, Salowsky R, Leiber M, Gassmann M, Lightfoot S, Menzel W, Granzow M, Ragg T (2006) The RIN: an RNA integrity number for assigning integrity values to RNA measurements. *BMC Mol Biol* **7**, 3.

Shippy R, Fulmer-Smentek S, Jensen RV, Jones WD, Wolber PK, Johnson CD, Pine PS, Boysen C, Guo X, Chudin E, Sun YA, Willey JC, Thierry-Mieg J, Thierry-Mieg D, Setterquist RA, Wilson M, Lucas AB, Novoradovskaya N, Papallo A, Turpaz Y, Baker SC, Warrington JA, Shi L, Herman D (2006) Using RNA sample titrations to assess microarray platform performance and normalization techniques. *Nat Biotechnol* **24**, 1123–1131.

Stokkan KA, Yamazaki S, Tei H, Sakaki Y, Menaker M (2001) Entrainment of the circadian clock in the liver by feeding. *Science* **291**, 490–493.

Storch KF, Lipan O, Leykin I, Viswanathan N, Davis FC, Wong WH, Weitz CJ (2002) Extensive and divergent circadian gene expression in liver and heart. *Nature* **417**, 78–83.

Tamura K, Ono A, Miyagishima T, Nagao T, Urushidani T (2006) Comparison of gene expression profiles among papilla, medulla and cortex in rat kidney. *J Toxicol Sci* **31**(5), 449–469.

Thompson KL, Afshari CA, Amin RP, Bertram TA, Car B, Cunningham M, Kind C, Kramer JA, Lawton M, Mirsky M, Naciff JM, Oreffo V, Pine PS, Sistare FD (2004) Identification of platform-independent gene expression markers of cisplatin nephrotoxicity. *Environ Health Perspect* **112**, 488–494.

Thompson KL, Rosenzweig BA, Pine PS, Retief J, Turpaz Y, Afshari CA, Hamadeh HK, Damore MA, Boedigheimer M, Blomme E, Ciurlionis R, Waring JF, Fuscoe JC, Paules R, Tucker CJ, Fare T, Coffey EM, He Y, Collins PJ, Jarnagin K, Fujimoto S, Ganter B, Kiser G, Kaysser-Kranich T, Sina J, Sistare FD (2005) Use of a mixed tissue RNA design for performance assessments on multiple microarray formats. *Nucleic Acids Res* **33**, e187.

Thompson KL, Pine PS, Rosenzweig BA, Turpaz Y, Retief J (2007) Characterization of the effect of sample quality on high density oligonucleotide microarray data using progressively degraded rat liver RNA. *BMC Biotechnol* **7**, 57.

Yauk C.L, Berndt ML (2007) Review of the literature examining the correlation among DNA microarray technologies. *Environ Mol Mutagen* **48**(5), 380–394.

6

Key Aspects of Toxicogenomic Data Analysis and Interpretation as a Safety Assessment Tool to Identify and Understand Drug-Induced Toxicity

Antoaneta Vladimirova and Brigitte Ganter

6.1 Introduction

The drug discovery and development process has been shown to be very costly, time consuming and inefficient. Bringing a drug to the market can cost up to $900M and can take 10–12 years (DiMasi *et al.*, 2003; Kola and Landis, 2004) mainly because only about one out of nine clinical candidates successfully reaches the market. The two primary reasons for drug candidate failure are the lack of efficacy, estimated to attribute to about 30 % of the overall failure rate, and toxic liabilities accounting also for about 30 % of the attrition rate (DiMasi *et al.*, 2003; Kola and Landis, 2004). The majority of drug candidates fail in Phase II and III of clinical development, after significant funds and resources have been invested (DiMasi *et al.*, 2003; Kola and Landis, 2004). Better and more efficient methods are needed to aid the drug development process and help eliminate problematic compounds at much earlier stages. There is a crucial need to improve the *status quo* by allowing new technologies to contribute and improve the safety assessment by speeding up and facilitating the selection of successful drug candidates.

Toxicogenomics, the discipline of gaining and applying knowledge based on gene expression data of chemicals in model systems, offers a great promise. It utilizes the wealth

Toxicogenomics: A Powerful Tool for Toxicity Assessment   Edited by S. C. Sahu
© 2008 John Wiley & Sons, Ltd

of transcriptional profile information and is currently being applied in two main modes: (1) developing and applying predictive biomarkers of toxicity and/or mechanism of action (Moggs *et al.*, 2004; Fielden *et al.*, 2005; Fielden and Kolaja, 2006; Ganter *et al.*, 2005, 2006; Powell *et al.*, 2006) and (2) elucidating compound molecular mechanism(s) of action and/or toxicity (Debouck and Goodfellow, 1999; Nuwaysir *et al.*, 1999; Burczynski *et al.*, 2000; Harris *et al.*, 2001; Waring *et al.*, 2001a,b, 2006; Hamadeh *et al.*, 2002; Bailey *et al.*, 2003; Peterson *et al.*, 2004; Elrick *et al.*, 2005; Liguori *et al.*, 2005; Sawada *et al.*, 2005; Tugendreich *et al.*, 2006). In recent years this new molecular approach has been adopted with increasing rates as more researchers gain expertise and demonstrate success in generating, analyzing and interpreting toxicogenomic data as a result of incorporating toxicogenomics as an additional tool for safety assessment.

6.2 Expectations and Applications of Toxicogenomics in Risk Assessment

Traditional toxicological approaches have been applied in the drug discovery and development process for decades. Clinical chemistry and histopathology findings are still considered the 'gold standards' for detecting toxic liabilities in animal models. A main limitation of this approach is that the current assays are primarily used for diagnostic purposes, as none of these traditional measurements can predict the future onset of toxicity. The inability to predict potential toxic liabilities earlier in the process in the absence of detectable changes in the organs of model animals is the main driver to develop alternative methods. The main challenge that toxicogenomics addresses is to bring safer drugs to the market faster and at reduced costs.

The premise of toxicogenomics is that drug treatments cause gene expression changes that not only can coincide with toxicity manifestations, but also can precede these events and can constitute surrogate biomarkers of future toxic outcomes (Suter *et al.*, 2004; Fielden *et al.*, 2005, 2007; Nie *et al.*, 2006; Tugendreich *et al.*, 2006). When anchored to specific traditional standard toxicology endpoints (Moggs *et al.*, 2004; Powell *et al.*, 2006; Beyer *et al.*, 2007), such as clinical chemistry measurements or histopathology findings, gene expression profiles can form the basis of biomarkers that are predictive or diagnostic of the toxicological outcome studied. On the other hand, one of the main strengths of toxicogenomics is the ability to investigate a toxic outcome retrospectively and help decipher the molecular mechanism(s) of a specific drug-induced organ toxicity to streamline the selection of safe alternative drug candidates. Toxicogenomics opens the door to focused screening of back-up compounds and allows the selection of molecules devoid of toxicity in cases where the toxicity is not target related. As scientists get better at gathering and analyzing transcriptional data, the application of toxicogenomic approaches as a complementary tool to currently established practices will help improve the decision-making process and will bring the scientific community closer to the goal of developing safer and more efficacious drugs while saving significant amounts of time and resources in the drug discovery and development cycle.

6.3 Setting up the Stage

Formulating the specific question that a toxicogenomic study addresses determines the type of data that needs to be collected and analyzed. Ideally, toxicogenomic applications will be ultimately targeted to predict human toxicity and risk. Although there have been some studies addressing the prediction of human toxic liabilities or understanding pharmacological response(s) in humans (DePrimo *et al.*, 2003; Horvath *et al.*, 2006), toxicogenomics is currently applied mostly to predict, diagnose or aid the understanding of animal toxicity. One of the major limitations in assessing human risk is the availability of human biological material amenable to gene expression or genomics studies in general. For example, although limited biopsy tissue samples are available through extractions from human surgery materials, the types and sizes of the tissue samples, their heterogeneity and variability of the collection methods often limit their practical utility. In contrast, multiple variability factors in humans, such as diet, age and health state, can be largely eliminated in animal studies where enough samples are available to allow for better statistical analyses. The most accessible human biological material is blood (Burczynski and Dorner, 2006), and while toxicogenomics applications in blood are still in their early stages and the types of response that can be studied in this system are mainly limited to immune reactions, some promising results have recently been described (Golub *et al.*, 1999; Burczynski *et al.*, 2000; Moos *et al.*, 2002; Feezor *et al.*, 2004; Baranzini *et al.*, 2005; Pearson *et al.*, 2006).

To date, animal models still remain the most suitable study systems for toxicogenomics applications, followed recently by encouraging results obtained by using various *in vitro* model systems, such as drug treatment of primary rat hepatocytes (Day *et al.*, 2006) or precision-cut liver slices (Zidek *et al.*, 2007). Although *ex vivo* drug administration has apparent limitations and cannot be reflective of the systemic body response achieved *in vivo*, it still can provide valuable insights into specific biological pathways and drug metabolism processes that operate in a specific cell type or tissue. The knowledge gained from this reductionistic approach can be complementary to solving the complex puzzle of interrelationships between different cell types, tissues and organs and help explain the multidimensional systemic response to a drug molecule.

Selecting the proper animal study is a key prerequisite for the generation of meaningful experimental data and the basis for successful delineation of drug mechanism(s) of action or toxicity. Ideally, a study design should include analysis of a candidate compound administered at multiple doses and time points, which allows the measurement of temporal and dose responses and adaptive and progressive processes to better understand the progression of drug-induced toxicity. Administration of pharmacologically active doses with no toxic findings along with doses leading to toxic outcomes allows researchers to better classify gene expression changes associated with pharmacological responses from those linked to toxicity and vice versa. Inclusion of multiple time points helps establishing consistent temporal patterns and eliminating sporadic and nonsignificant gene changes from consideration. As the length of the study is proportional to cost and the amount of compound is largely limited, the duration of the studies is typically short, ranging from a few days to a few weeks. This period may allow collecting samples while the toxicity is not yet manifested in conjunction with samples gathered at time points when the toxicity is present. This approach helps elucidate gene responses predictive of injury and also helps define biomarkers based on diagnostic gene changes or elucidate the mechanism of toxicity.

Another important component of the study design is the choice of controls and the number of replicates to run for each drug–dose–time combination. To enable statistical assessment of the data, each experiment should to a minimum be run as a biological triplicate (Ganter *et al.*, 2005). This allows for the identification of genes that are significantly perturbed and associated with the toxicity under investigation. Frequently, not all animals from the same drug–dose–time treatment group show identical or even similar toxic responses in terms of clinical chemistry of histopathology findings. This can be explained, in part, by the different exposure rates of the chemical in individual animals. However, even if the bioavailability of the drug is similar across the animals in the same experimental group, biological variability may account for the observed differences. To understand the toxicity induced as a result of the drug treatment, a researcher would be faced with two possible analysis approaches: (1) study the toxicity at the molecular level only within the animal subgroup that exhibits the toxicity or (2) analyze samples across all the animals in the study group including those that do not show the toxic outcome. Using only the animals within the study group that have similar toxicity findings to generate gene expression data and the analysis of those assures that the statistically significant gene expression responses are very tightly anchored to the toxic endpoint in the target tissue, which can be a powerful method to delineate the molecular events that accompany the toxicity manifested. The limitation of this approach is that it normally requires a larger animal study to ensure that the number of animals with the expected toxicity for phenotypic anchoring will be sufficient to allow statistical analysis of the data, e.g. to ensure a minimum of a biological triplicate within the group to be analyzed. Of course, larger animal groups translate into higher expenses due to increased animal and handling costs and, at the same time, larger amounts of the test compound administered. The most important drawback of this approach, however, is that it may limit the understanding of the mechanism(s) of toxicity, as it only zeroes in on the biological perturbations that accompany the toxicity at the time it is manifested and does not reflect changes that uncover the development of the adverse effect(s). Therefore, this analysis approach is limited in scope to diagnosing and not predicting molecular patterns of toxicity before it occurs. Thus, this type of study design captures predominantly the final stages of the development of the toxicity when overt changes are readily detectable.

Considering that one of the main goal of toxicogenomics is to be able to predict the toxic liability before it is manifested, including all animals of the same treatment group with and without exhibited toxicity may be more beneficial, as it better reflects the ultimately desired functionality of the application. Using averages of gene expression values across all animals from the study group, including animals not exhibiting comparable signs of toxicity, allows the detection of specific gene changes that precede the gross outcome and can, therefore, become a valuable tool for prediction of compound toxicity before it occurs. This latter approach results in gene expression changes that form the basis of a successful predictive biomarker identification, validation and application (Fielden *et al.*, 2005; Natsoulis *et al.*, 2005; Waring *et al.*, 2006; Fielden *et al.*, 2007).

The type and number of controls to be included along with the compound(s) of interest are other important considerations when planning an experiment. Ideally, each study group representing an experimental condition (drug–dose–time combination) should be accompanied by a control group of untreated animals to allow statistical analyses and to increase the signal-to-noise ratio in the expression data. In addition to the untreated control animals, inclusion of negative and positive controls that represent animals treated with various other comparator molecules allows for benchmarking of the test compound to other drugs or chemicals known to be associated with specific therapeutic or toxic outcome(s), as they permit both qualitative and quantitative evaluation of the compound of interest. Assessing such contextual compound information in parallel to the test compound data serves as a point of reference and helps confirm or reject a hypothesis that would otherwise be done solely based on test compound results. Without benchmarking, the mountain of test molecule data may become uninterpretable and too complex. Even with sufficient positive and negative control compound data, it can still be challenging to select informative and specific gene sets representative of a specific toxicity or compound class effect.

6.4 Analysis Approaches

Instead of focusing on technical details and specific tools of analysis, such as hierarchical clustering, principal component analysis (PCA), analysis of variance (ANOVA) or other statistical methods, we will focus on some important conceptual approaches instead. Details and descriptions on available methods of analysis can be found elsewhere (Castle *et al.*, 2002; Natsoulis *et al.*, 2005).

Transcriptional data can be analyzed in two ways: (1) in a compound-centric way; or (2) in a contextual way, as outlined in Figure 6.1. There is a fundamental difference between these two approaches, whereby the compound-specific approach could be compared with a 'tunnel vision' where the test data are being analyzed in isolation without including data obtained using other molecules. This compound-specific approach evaluates the data in terms of impact on well-known biological processes or pathways or against various biomarkers. Although valuable information can be obtained when applying this approach, no major conclusions can be drawn without proper benchmarking as to whether the gene expression findings for the compound under investigation are common or untypical relative to a diversified molecule reference set. This method can also be used for hypothesis building, where the investigator builds a model to correlate gene changes to an observed phenotype. However, without additional contextual data, the results cannot be tested, validated or rejected *in silico*.

In contrast, the contextual approach allows the analyst to reference and benchmark the test compound data against gene expression data derived from pharmacologically relevant molecules, such as approved and marketed drugs or toxicity-inducing molecules, which include biochemicals, toxicants or withdrawn drugs. This second approach allows the researcher quickly to distinguish the real signal from the noise and focus on the key compound gene expression findings that are tightly linked to the endpoint of interest. This results in a more educated and precise hypothesis-building process.

Analysis of gene expression data requires expertise to understand gene expression changes induced by the drug molecule. There are multiple branching points where the analyst may ask 'what if' type of questions, each of which can lead to an explosion of possibilities that can help explain the experimental results. To simplify the decision-making process and limit the time spent on manual tasks, automated analysis presents a viable and time-saving option and can form the basis of a deeper analysis. An example of an automated toxicogenomic analysis suite is offered through ToxFX™ (http://www.toxfx.com). ToxFX scores gene expression data against a battery of potential toxic endpoints via validated biomarker sets (Drug Signature®) (Fielden *et al.*, 2005, 2007; Fielden and Halbert, 2007). This analytical suite also provides information on highly impacted biological pathways relative to comparator molecules and highly up- or down-regulated gene sets in the dataset. Although gene expression profiling is still a relatively expensive undertaking, this automated analysis approach allows for a comprehensive view at the data and helps

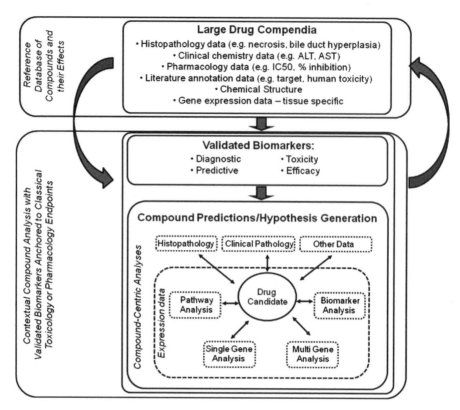

Figure 6.1 *Compound-centric and contextual compound analysis. A compound-centric analysis begins with gene expression data from several experiments representing different compound dose and time combinations and focuses on statistically significantly perturbed genes. Initial steps in the analysis include assessment of compound impact on biological processes such as evaluations of canonical pathways and gene ontology categories, selected literature-based gene sets associated with toxicological or pharmacological outcomes, and investigation of genes with most significant perturbations. The process involves multiple iterations of hypothesis building and adjustments to explain the experimental results. Gene expression data are interpreted in the context of other types of data from the same study, such as histopathology, clinical chemistry and hematology findings. To fully understand the meaning of the gene changes in the experiments though, the gene expression trends of the compound under investigation have to be benchmarked against reference molecules representing similar toxicity or pharmacology endpoints allowing qualitative and quantitative comparisons. A large reference database provides the additional experimental data to evaluate the compound-centric hypotheses and to put the findings in perspective of the results of other marketed or withdrawn drugs or biochemical or toxicological standards. This second layer of analysis also may include multiple iterations and compounds comparisons. Applying validated diagnostic or predictive biomarkers is one of the key steps in the analysis and is dependent on a rigorous process of biomarker derivation and validation. The same reference database, which includes gene expression and other domains of compound data such as classical toxicological, pharmacological or literature findings, is used as a basis for biomarker development. The biomarker(s) can be applied to evaluate expression data, broaden the contextual analysis and score the test compound relative to known standards. Depending on the algorithm used to derive the genes in the biomarker, the biomarker genes may also be useful in the process of mechanism of toxicity (MOT) or action (MOA) delineation.*

uncover multiple potential liabilities along with their potential mechanisms, thus saving time and resources, and sometimes helping the researchers focus their attention on unexpected outcomes while broadening the range of the anticipated analysis. The ToxFX results can serve as a foundation of any additional in-depth toxicogenomic analysis.

6.5   Biomarker Derivation and Validation

One way of addressing the analysis of an overwhelming number of gene changes is to look only at a select subset of gene changes, or to apply biomarkers proven to be statistically associated with a specific toxicity or mechanism of action. Pre-established validated biomarker gene sets consist of a limited number of genes where the contribution of each gene to the toxic outcome is firmly established and statistically validated. This approach focuses the analysis on a smaller subset of gene changes that are predictive or diagnostic of a specific phenotype of interest. The real value of toxicogenomics lies in the derivation and validation of surrogate biomarkers of toxicity that can predict toxic liabilities much earlier and with greater precision than the traditional measurements of toxicological endpoints. This has not been an easy undertaking and has proven to be more complex, time-consuming and expensive than originally anticipated. Derivation of such biomarkers requires highly standardized datasets containing gene expression data in specific tissues and clinical pathology (clinical chemistry, hematology and histopathology) findings from the same tissues and animals which then can be anchored to the transcriptomic data for the successful derivation of a good biomarker. Similarly, other data domains, such as drug annotation information, pharmacology or literature data domains can be used to group compounds into classes or to distinguish those classes by selecting the best biomarker genes.

The successful process of biomarker development builds on a well annotated and large reference toxicogenomic database that contains gene expression data from a variety of different molecules, including marketed and withdrawn drugs, biochemical standards and standard toxicants. There are only a few examples of comprehensive reference toxicogenomics databases that couple gene expression profiles with classical toxicological findings from the same animals. These include databases from commercial sources, such as Gene-Logic's ToxExpress® System (Castle *et al.*, 2002) (http://www.genelogic.com/genomics/toxexpress/system.cfm), and Entelos (formerly Iconix Biosciences) DrugMatrix® (http://www.iconixbiosciences.com) (Ganter *et al.*, 2005). In addition, there is the Chemical Effects of Biological Systems (CEBS) database (http://cebs.niehs.nih.gov/cebs-browser/cebsHome.do;jsessionid=75F2DABB5163FDA37 80C89BF5FF21808), which is a publicly available database and integrated with ArrayTrack (http://www.fda.gov/nctr/science/centers/toxicoinformatics/ArrayTrack/), a microarray data analysis solution for data managing, analysis and interpretation. Within this chapter we will focus mainly on the content and details of the DrugMatrix database.

Derivation of biomarkers is largely dependent on the availability of large positive and negative drug-treatment sets representing different compound–dose–time combinations. DrugMatrix contains multiple tissue expression profiling information on over 630 compounds from short-term repeat dose studies (toxic and pharmacologically nontoxic doses). This extensive reference database contains not only gene expression profiling information, but also extensive blood chemistry assay panels, clinical chemistry, hematology and histopathology profiles, all gathered from the same animals within a study. In addition it contains extensive pharmacology, structure and literature information domains collected for each compound. Such a large collection of compounds and associated gene expression linked to ancillary information allows for an extensive and comparative analysis of multiple features of the candidate drug. For biomarker derivation, assignment of the experiments to a positive or negative class is guided primarily based on specific phenotypes, which can include different type(s) of tissue injury (e.g. bile duct hyperplasia), elevation of certain clinical chemistry parameters (e.g. ALT or AST increase), specific pharmacological targets (e.g. HMG-CoA reductase inhibitors, or statin drugs), or can be based on literature information (e.g. genotoxic carcinogens). The biomarker generation and validation process is outlined in Figure 6.2. In this process, it is very important to diversify the composition especially of the positive class of experiments, or drug–dose–time combinations, which is typically a much smaller set than the negative class of experiments. Furthermore, when deriving toxicity-related biomarkers it is important that the resulting gene sets do not reflect a specific structure–activity relationship (SAR), but rather cover a variety of mechanisms from a variety of unrelated molecules that all induce the observed toxicity.

Genes that can best distinguish the positive from the negative classes of experimental drug treatments form biomarkers (see Figure 6.2). These biomarkers have to be validated further through a variety of follow-up methods, including split-sample and forward validation methods. The *in silico* split-sample validation method allows training of the model on part of the experimental data while testing it on the rest. The forward validation process includes testing the model on compounds that have not been part of the original data generation and offers a rigorous biomarker performance test.

The power and precision of the biomarker sets developed from such large reference databases stems from the standardized and carefully designed collection of aggregate data from a variety of data domains that allow mining of the gene expression data to uncover gene associations with various other parameters recoded along with the gene perturbations. Once the biomarkers have been validated and accepted across the scientific community, they can be applied as acceptable surrogates for toxicity manifestations. Especially promising are the predictive biomarkers where gene expression findings reveal changes preceding the actual toxic outcome and offer an opportunity to shorten the drug developmental timeline. The predictive, as well as diagnostic, biomarkers can be used to evaluate individual test molecules, to prioritize series of related molecules or provide back-up screens for failed compounds. Depending on the method of derivation, the sets of genes forming the biomarker may also be used to gain insight into the mechanism of the phenomenon studied.

6.6   Mechanistic Understanding as the Major Challenge to Toxicogenomics

Toxicogenomics has been applied to aid the understanding of molecular mechanisms of action and toxicity of preclinical drug molecules with toxic liabilities and of failed clinical candidates. Solving the complex puzzle of intricate biological interrelationships to uncover specific genes and pathways that explain the mechanisms of toxicity is a nontrivial task. It not only involves proper experimental design as addressed earlier, but also requires a

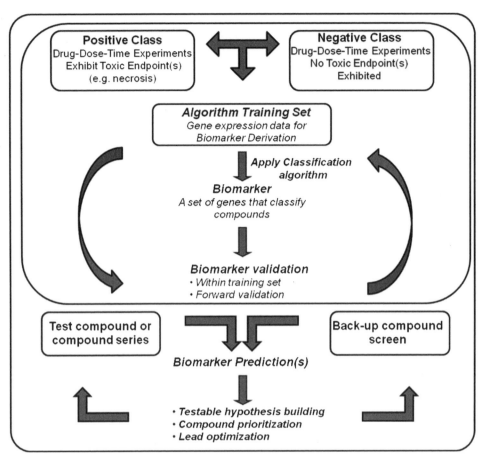

Figure 6.2 *Biomarker derivation, validation and application using a contextual reference database. The process of biomarker derivation starts with a large collection of gene expression and standard toxicology data from a large and a diverse collection of chemicals. Classical toxicology or pharmacology endpoints are used to assign the experiments into positive and negative classes depending on whether an experiment is annotated to have a given phenotype (positive class) or not (negative class). The same experiment can be used to derive multiple biomarkers and a given experiment can belong to the positive class of one biomarker and a negative class for another biomarker, such as when it causes necrosis (positive class for 'necrosis' biomarker) and does not cause proliferation (negative class for 'proliferation' biomarker). A classification algorithm is then applied to the gene expression data of both classes to select the genes and their direction of change that best distinguish the behavior of the two groups. To validate the biomarker, tests that evaluate its accuracy of the predictions are applied. This can be done by scoring either a reserved part of the experiments in the database that have not been originally used in the biomarker derivation, or making predictions based on gene expression data of chemicals that are completely independent and evaluating the accuracy of their predictions. Once the biomarker is validated, it can be used routinely to predict the phenotype that it is built to classify by evaluating individual compounds, compound series or applied in the lead optimization process. As more data become available, biomarkers may be revisited and rebuilt to achieve higher accuracy of predictions.*

solid understanding of the pathophysiology and an expertise in molecular toxicology for successful data interpretation. The mechanistic analysis, even if done by an expert with a lot of experience, is a very time-consuming undertaking. This process involves iterative steps where hypotheses that explain the results are generated. After evaluating various aspects of the test data, the process of analysis moves from an early phase of decomposing the complex gene expression picture into simpler components, to a more complex and difficult synthesis phase, where all the individual knowledge fragments or hypotheses need to be put together to build a master hypothesis of the mechanism to be explained. The whole process is very difficult, as in many cases it is not possible to interpret the data unambiguously or to form a single all-encompassing hypothesis. It is very important to emphasize again that reference databases representing expression profiles of a variety of pharmacological and toxicological endpoints are really crucial for delineation and understanding of mechanism.

Deciphering a specific toxicity mechanism is a task that is largely dependent on the organ studied and the state of the knowledge for a certain toxic liability. For example, liver and kidney toxicities are traditionally the ones that have been observed and studied most often in the past, mostly due to the fact that those are the organs that primarily metabolize the compounds and are exposed to the highest concentrations of the toxic agent, drug or metabolite. Examples of toxicants studied extensively throughout the years in liver and kidney are acetaminophen or carbon tetrachloride (Hardman *et al.*, 2001). As a result, molecular mechanisms of some of these well-studied compound toxicities are better understood than others. Consequently, toxicogenomic reference databases composed of drugs or toxicants with relatively well understood hepato-, nephro- or cardio-toxicity mechanisms provide a valuable tool to compare and contrast with the test molecule data. The test drug evaluation typically focuses on comparing and contrasting to drugs that impact specific biological processes or perturb specific genes representing key steps in the development of toxicities induced by the test drug treatment. For example, in the recent decade, many of the drugs in development, including some marketed drugs, have shown the potential to induce cardiotoxic liabilities (Herman *et al.*, 1998; Kerkelä *et al.*, 2006), which has created a new challenge for the pharmaceutical industry. As a result, more compounds have become available that serve as a reference for cardiotoxicities, which, in turn, allow the investigation of the basis of these toxicities. Detecting and eliminating toxicities in these three major organs (liver, kidney and heart) has, therefore, become more successful. This is further exemplified with the multiple cell-type-specific *in vitro* assays developed that aid the drug selection process. Yet, even in these tissues, understanding the causes and mechanism(s) of toxicity is not a trivial task and requires a detailed understanding of the toxicity of the reference compounds. Lastly, there are test drug-manifested toxicities that are completely based on novel mechanisms of drug-induced toxicity. They represent the most challenging cases, as the compound toxicity forms a precedent, where the observed toxicological endpoint may be similar to other observed phenotypes, yet the underlying molecular mechanism and networks of interactions are unique to the compound under study, and the reference database can only serve to eliminate certain hypotheses, but not confirm them.

Another scenario where toxicogenomics can aid the selection of drug candidates is the evaluation of the extent to which the observed toxicity in animal models is relevant to humans. There are cases where it has been clearly established that some toxicities are species specific and do not pose a risk to humans, such as those of PPAR alpha agonists

(Guo *et al.*, 2007) which have been shown to not pose a risk to humans but are toxic to rodents.

Confirming lack of toxicity is another application of toxicogenomic data analysis, which can help improve the drug candidate selection process. Demonstrating the safety of a compound and advancing the best candidate(s) devoid of detectable toxicities can be equally important if not more valuable than raising a concern about a molecule. By the same token, establishing that a mechanism of toxicity has a conserved origin across species has very high value, as it underscores the likelihood of risk for humans.

Very challenging are the cases of idiosyncratic toxicities in well-studied organs or the cases where rare toxicities are observed in tissues that have not been well studied or understood, such as toxicities in muscle, bone or testis. Increasingly, toxicities in organs different than liver, kidney or heart have become the road block, partially because the pharmaceutical industry has gained knowledge and experience by focusing mainly on eliminating the most frequent liver, kidney and heart toxic liabilities. It becomes increasingly important to provide toxicogenomic reference data in those less-studied organs or tissues to simplify the molecule evaluation process.

Last but not least, mechanistic studies are very important to understand whether the toxicity is target related, and considered to be an on- or off-target effect. Even today for the majority of drugs on the market or those in development this is not well established. There is still a limited understanding of whether an observed toxicity stems from the intended target of the drug or not. A recent concern was voiced over the safety of the 'wonder drug' Gleevec (imatinib mesilate), currently used to treat myelogenous leukemia (Kerkelä *et al.*, 2006). Gleevec is among the few cases where the target of the drug is known. Gleevec inhibits the kinase activity of the oncogenic mutant protein Bcr–Abl, resulting from a fusion of two gene products (Van Etten, 2004). Yet the recently reported cardiac toxicity associated with the anticancer agent might well be due to an inhibition of a cardio-protective function played by Abl in cardiomyocytes, as suggested by mouse and *in vitro* experiments (Kerkelä *et al.*, 2006). Getting a full understanding of the molecular mechanisms at play will allow researchers to best appreciate the efficacy as well as toxicity potential of promising drug candidates or drugs already on the market.

Toxicogenomics is especially valuable in predicting dose-related risks, as it evaluates the gene responses of the entire genome resulting from treatments with subtoxic and toxic doses. The transcriptomic approach can, therefore, be used to define similarities and dissimilarities in the biological networks affected by the two different drug dose conditions. For example, in the cases where the target for inhibition by a drug candidate serves key physiological role(s) which are nonredundant with the functions of other gene products, the development of this drug may be doomed from its inception, as it inhibits indispensible function(s) in the cell.

Knowing and understanding the cause of the toxicity can help eliminate problematic molecules and can help advance back-up candidates with better chances of success earlier in the drug development process. The real benefit of mechanistic understanding of the toxicological outcome comes in the shape of a testable hypothesis. Typically, the end result of a toxicogenomic analysis boils down to a set of interrelated genes that are believed to have a causal effect or are simply associated with the toxicity and are believed to be predictive of the toxicological outcome or the toxicity mechanism at play. To be able to really advance the drug selection process, newly developed hypotheses need to be tested in the laboratory and, if validated, shared with the scientific community so that they can be translated into valuable assays that allow 'screening out' of molecules with undesirable adverse effects and focusing on candidates devoid of these toxic liabilities. A good example of this approach is the work from Sawada and co-workers (Sawada *et al.*, 2005, 2006). This team evaluated phospholipidosis-inducing compounds, and proposed a set of phospholipidosis predictive genes. Translating this knowledge into a PCR-based devise allows tracking of these specific gene changes and, later, a translation of the approach into a high-throughput method of detection to predict toxicity (Sawada *et al.*, 2006).

6.7   Summary and Conclusions

This overview has highlighted recent advances in the area of toxicogenomics, as well as the requirements, limitations and hurdles this new technology faces for a widespread adoption across the industry. This review demonstrated the advantages of utilizing contextual databases in various applications of toxicogenomics, such as mechanistic analysis, generation of validated biomarkers or benchmarking of a drug candidate. Integrating toxicogenomic data with proteomic and metabolomic data and linking them to the corresponding clinical endpoints will provide a powerful systems-wide approach for the drug discovery and development. Toxicogenomics has a head start compared with other genomic technologies, in that the technology to measure RNA expression changes has been fairly well developed over the last decade. The future systems biology perspectives gained from toxicogenomics, proteomics and metabolomics platforms combined with traditional toxicological studies will enable rational drug development and facilitate risk assessment even further.

References

Bailey NJC, Oven M, Holmes E, Nicholson JK, Zenk MH (2003) Metabolomic analysis of the consequences of cadmium exposure in *Silene cucubalus* cell cultures via ^{1}H NMR spectroscopy and chemometrics. *Phytochemistry* **62**(6), 851–858.

Baranzini SE, Mousavi P, Rio J, Caillier SJ, Stillman A, Villoslada P, Wyatt MM, Comabella M, Greller LD, Somogyi R, Montalban X, Oksenberg JR (2005) Transcription-based prediction of response to IFNbeta using supervised computational methods. *PLoS Biol* **3**(1), e2.

Beyer RP, Fry RC, Lasarev MR, McConnachie LA, Meira LB, Palmer VS, Powell CL, Ross PK, Bammler TK, Bradford BU *et al.* (2007) Multi-center study of acetaminophen hepatotoxicity reveals the importance of biological endpoints in genomic analyses. *Toxicol Sci* **99**(1), 326–337.

Burczynski ME, Dorner AJ (2006) Transcriptional profiling of peripheral blood cells in clinical pharmacogenomic studies. *Pharmacogenomics* **7**(2), 187–202.

Burczynski ME, McMillian M, Ciervo J, Li L, Parker JB, Dunn II RT, Hicken S, Farr S, Johnson MD (2000) Toxicogenomics-based discrimination of toxic mechanism in HepG2 human hepatoma cells. *Toxicol Sci* **58**(2), 399–415.

Castle AL, Carver MP, Mendrick DL (2002) Toxicogenomics: a new revolution in drug safety. *Drug Discov Today* **7**(13), 728–736.

Day G, Brady L, Ryan TP *et al.* (2006) Characterization of compound-induced haptotoxicity using gene expression profiling in rat primary hepatocytes. Itinerary Planner. Abstract No. 2392. Society of Toxicology, San Diego, CA.

Debouck C, Goodfellow PN (1999) DNA microarrays in drug discovery and development. *Nat Genet* **21**, 48–50.

DePrimo SE, Wong LM, Khatry DB, Nicholas SL, Manning WC, Smolich BD, O'Farrell A-M, Cherrington JM (2003) Expression profiling of blood samples from an SU5416 Phase III metastatic colorectal cancer clinical trial: a novel strategy for biomarker identification. *BMC Cancer* **3**, 3.

DiMasi JA, Hansen RW, Grabowski HG (2003) The price of innovation: new estimates of drug development costs. *J Health Econ* **22**(2), 151–185.

Elrick MM, Kramer JA, Alden CL, Blomme EAG, Bunch RT, Cabonce MA, Curtiss SW, Kier LD, Kolaja KL, Rodi CP, Morris DL (2005) Differential display in rat livers treated for 13 weeks with phenobarbital implicates a role for metabolic and oxidative stress in nongenotoxic carcinogenicity. *Toxicol Pathol* **33**(1), 118–126.

Feezor RJ, Baker HV, Xiao W, Lee WA, Huber TS, Mindrinos M, Kim RA, Ruiz-Taylor L, Moldawer LL, Davis RW, Seeger JM (2004) Genomic and proteomic determinants of outcome in patients undergoing thoracoabdominal aortic aneurysm repair. *J Immunol* **172**(11), 7103–7109.

Fielden MR, Halbert DN (2007) Iconix Biosciences, Inc. *Pharmacogenomics* **8**(4), 401–405.

Fielden MR, Kolaja KL (2006) The state-of-the-art in predictive toxicogenomics. *Curr Opin Drug Discov Dev* **9**(1), 84–91.

Fielden MR, Eynon BP, Natsoulis G, Jarnagin K, Banas D, Kolaja KL (2005) A gene expression signature that predicts the future onset of drug-induced renal tubular toxicity. *Toxicol Pathol* **33**(6), 675–683.

Fielden MR, Brennan R, Gollub J (2007) A gene expression biomarker provides early prediction and mechanistic assessment of hepatic tumor induction by non-genotoxic chemicals. *Toxicol Sci* **99**(1), 90–100.

Ganter B, Snyder RD, Halbert DN, Lee MD (2006) Toxicogenomics in drug discovery and development: mechanistic analysis of compound/class-dependent effects using the DrugMatrix database. *Pharmacogenomics* **7**(7), 1025–1044.

Ganter B, Tugendreich S, Pearson CI, Ayanoglu E, Baumhueter S, Bostian KA, Brady L, Browne LJ, Calvin JT, Day GJ *et al.* (2005) Development of a large-scale chemogenomics database to improve drug candidate selection and to understand mechanisms of chemical toxicity and action. *J Biotechnol* **119**(3), 219–244.

Golub TR, Slonim DK, Tamayo P, Huard C, Gaasenbeek M, Mesirov JP, Coller H, Loh ML, Downing JR, Caligiuri MA, Bloomfield CD, Lander ES (1999) Molecular classification of cancer: class discovery and class prediction by gene expression monitoring. *Science* **286**(5439), 531–537.

Guo Y, Jolly RA, Halstead BW, Baker TK, Stutz JP, Huffman M, Calley JN, West A, Gao H, Searfoss GH, Li S, Irizarry AR, Qian HR, Stevens JL, Ryan TP (2007) Underlying mechanisms of pharmacology and toxicity of a novel PPAR agonist revealed using rodent and canine hepatocytes. *Toxicol Sci* **96**(2), 294–309.

Hamadeh HK, Amin RP, Paules RS, Afshari CA (2002) An overview of toxicogenomics. *Curr Issues Mol Biol* **4**(2), 45–56.

Hardman JG, Limbird LE, Gilman AG (2001) *Goodman & Gilman's The Pharmacological Basis of Therapeutics*. McGraw-Hill.

Harris K, Lamson RE, Nelson B, Hughes TR, Marton MJ, Roberts CJ, Boone C, Pryciak PM (2001) Role of scaffolds in MAP kinase pathway specificity revealed by custom design of pathway-dedicated signaling proteins. *Curr Biol* **11**(23), 1815–1824.

Herman EH, Lipshultz SE, Rifai N, Zhang J, Papoian T, Yu ZX, Takeda K, Ferrans VJ (1998) Use of cardiac troponin T levels as an indicator of doxorubicin-induced cardiotoxicity. *Cancer Res* **58**(2), 195–197.

Horvath S, Zhang B, Carlson M, Lu KV, Zhu S, Felciano RM, Laurance MF, Zhao W, Qi S, Chen Z *et al.* (2006) Analysis of oncogenic signaling networks in glioblastoma identifies ASPM as a molecular target. *Proc Natl Acad Sci U S A* **103**(46), 17402–17407.

Kerkelä R, Grazette L, Yacobi R, Iliescu C, Patten R, Beahm C, Walters B, Shevtsov S, Pesant S, Clubb FJ *et al.* (2006) Cardiotoxicity of the cancer therapeutic agent imatinib mesylate. *Nat Med* **12**(8), 908–916.

Kola I, Landis J (2004) Can the pharmaceutical industry reduce attrition rates? *Nat Rev Drug Discov* **3**(8), 711–715.

Liguori MJ, Anderson MG, Bukofzer S, McKim J, Pregenzer JF, Retief J, Spear BB, Waring JF (2005) Microarray analysis in human hepatocytes suggests a mechanism for hepatotoxicity induced by trovafloxacin. *Hepatology* **41**(1), 177–186.

Moggs JG, Tinwell H, Spurway T, Chang HS, Pate I, Lim FL, Moore DJ, Soames A, Stuckey R, Currie R, Zhu T, Kimber I, Ashby J, Orphanides G (2004) Phenotypic anchoring of gene expression changes during estrogen-induced uterine growth. *Environ Health Perspect* **112**(16), 1589–1606.

Moos PJ, Raetz EA, Carlson MA, Szabo A, Smith FE, Willman C, Wei Q, Hunger SP, Carroll WL (2002) Identification of gene expression profiles that segregate patients with childhood leukemia. *Clin Cancer Res* **8**(10), 3118–3130.

Natsoulis G, El Ghaoui L, Lanckriet GR, Tolley AM, Leroy F, Dunlea S, Eynon BP, Pearson CI, Tugendreich S, Jarnagin K (2005) Classification of a large microarray data set: algorithm comparison and analysis of drug signatures. *Genome Res* **15**(5), 724–736.

Nie AY, McMillian M, Parker JB, Leone A, Bryant S, Yieh L, Bittner A, Nelson J, Carmen A, Wan J, Lord PG (2006) Predictive toxicogenomics approaches reveal underlying molecular mechanisms of nongenotoxic carcinogenicity. *Mol Carcinog* **45**(12), 914–933.

Nuwaysir EF, Bittner M, Trent J, Barrett JC, Afshari CA (1999) Microarrays and toxicology: the advent of toxicogenomics. *Mol Carcinog* **24**(3), 153–159.

Pearson C, Fujimoto S, Judo M *et al.* (2006) Blood as a toxicogenomic tissue. Itinerary Planner. Abstract No. 182, Society of Toxicology: San Diego, CA.

Peterson RL, Casciotti L, Block L, Goad ME, Tong Z, Meehan JT, Jordan RA, Vinlove MP, Markiewicz VR, Weed CA, Dorner AJ (2004) Mechanistic toxicogenomic analysis of WAY-144122 administration in Sprague-Dawley rats. *Toxicol Appl Pharmacol.* **196**(1), 80–94.

Powell CL, Kosyk O, Ross PK, Schoonhoven R, Boysen G, Swenberg JA, Heinloth AN, Boorman GA, Cunningham ML, Paules RS, Rusyn I (2006) Phenotypic anchoring of acetaminophen-induced oxidative stress with gene expression profiles in rat liver. *Toxicol Sci* **93**(1), 213–222.

Sawada H, Takami K, Asahi S (2005) A toxicogenomic approach to drug-induced phospholipidosis: analysis of its induction mechanism and establishment of a novel in vitro screening system. *Toxicol Sci* **83**(2), 282–292.

Sawada H, Taniguchi K, Takami K (2006) Improved toxicogenomic screening for drug-induced phospholipidosis using a multiplexed quantitative gene expression ArrayPlate assay. *Toxicol In Vitro* **20**(8), 1506–1513.

Suter L, Babiss LE, Wheeldon EB (2004) Toxicogenomics in predictive toxicology in drug development. *Chem Biol* **11**(2), 161–171.

Tugendreich S, Pearson CI, Sagartz J, Jarnagin K, Kolaja K (2006) NSAID-induced acute phase response is due to increased intestinal permeability and characterized by early and consistent alterations in hepatic gene expression. *Toxicol Pathol* **34**(2), 168–179.

Van Etten RA (2004) Mechanisms of transformation by the BCR–ABL oncogene: new perspectives in the post-imatinib era. *Leuk Res* **28**(Suppl 1), S21–S28.

Waring JF, Ciurlionis R, Jolly RA, Heindel M, Ulrich RG (2001) Microarray analysis of hepatotoxins *in vitro* reveals a correlation between gene expression profiles and mechanisms of toxicity. *Toxicol Lett* **120**(1–3), 359–368.

Waring JF, Jolly RA, Ciurlionis R, Lum PY, Praestgaard JT, Morfitt DC, Buratto B, Roberts C, Schadt E, Ulrich RG (2001) Clustering of hepatotoxins based on mechanism of toxicity using gene expression profiles. *Toxicol Appl Pharmacol* **175**(1), 28–42.

Waring JF, Liguori MJ, Luyendyk JP, Maddox JF, Ganey PE, Stachlewitz RF, North C, Blomme EA, Roth RA (2006) Microarray analysis of lipopolysaccharide potentiation of trovafloxacin-induced liver injury in rats suggests a role for proinflammatory chemokines and neutrophils. *J Pharmacol Exp Ther* **316**(3), 1080–1087.

Zidek N, Hellmann J, Kramer PJ, Hewitt PG (2007) Acute hepatotoxicity: a predictive model based on focused illumina microarrays. *Toxicol Sci* **99**(1), 289–302.

Section 2

Applications of Toxicogenomics

7

Toxicogenomics as a Tool to Assess Immunotoxicity

Kirsten A. Baken, Janine Ezendam, Jeroen L.A. Pennings, Rob J. Vandebriel
and Henk van Loveren

7.1 Introduction

Immunotoxicology is the study of adverse effects of substances on the immune system. Immunotoxicity has two main domains: direct toxicity to components of the immune system and indirect toxicity due to immune reactions provoked by chemicals. Direct toxicity of agents causes malfunctioning of the immune system, with consequences such as reduced resistance to infections or decreased surveillance of arising neoplasms. In addition, direct immunotoxicity may result in loss of regulatory activity. This may have consequences for the expression of allergic or autoimmune phenomena, which by themselves were induced for other reasons, such as pollen or house dust mite in the case of allergy or idiosyncratic in the case of autoimmunity. Indirect toxicity can be described as the adverse side effects of immune responses to the agent itself, such as is the case with haptens or certain proteins, or to self-components altered by chemicals (de Jong and van Loveren, 2007).

Numerous experimental methods are available to detect and study immunomodulatory effects, most of which make use of rodent models. Non-functional parameters of immunotoxicity include weight of lymphoid organs, histopathology, white blood cell counts and differentiation, and immunoglobulin levels in serum. In addition, functional assays are frequently used to assess specific immunotoxic effects. Functional effects are often assessed in host resistance models (bacterial, viral, and parasitic), tumor models, autoimmunity models, and allergy models. In addition, activity of immune cells can be examined, e.g. macrophage activity, natural killer cell activity, and mitogen responses of spleen cells (de Jong and van Loveren, 2007). Assessment of gene expression changes induced by immunomodulators is a relatively new approach to the study of immunotoxic processes.

7.2   Applications of Gene Expression Profiling in Immunotoxicology

Microarray analysis is increasingly used to study alterations in gene expression profiles after immunotoxicant exposure (reviewed by Burns-Naas *et al.* (2006) and Baken *et al.* (2007a)). This technique is expected to provide insight into the underlying mechanisms of action of toxicants. Furthermore, identification of genes that are regulated specifically by immunotoxicants could be of use in hazard identification. Some examples of immunomodulating agents that have been studied by toxicogenomics are discussed below.

7.2.1   Hexachlorobenzene

Hexachlorobenzene (HCB) is a persistent environmental pollutant that has several toxic effects in man and rat, including immunotoxic effects. Mechanisms of immunotoxicity have been studied in rodent models, particularly in brown Norway (BN) rats, which are very susceptible to HCB-induced immunopathology. Oral exposure to HCB induces, among other things, enlargement of liver, spleen and lymph nodes, increased serum IgM, IgG and IgE levels, and inflammatory skin and lung lesions. The mechanisms of HCB-induced immunopathology are not yet fully understood; they are very complex and involve multiple factors (Ezendam *et al.*, 2005).

To gain more insight into the molecular mechanisms of HCB-induced toxicity, gene expression profiling was performed (Ezendam *et al.*, 2004a). BN rats were exposed to a diet supplemented with 0, 150, or 450 mg HCB per kilogram of food for 4 weeks. Microarray analysis was subsequently performed in blood, thymus, kidney, liver, spleen, and mesenteric lymph nodes. Principal component analysis showed that the effects of HCB on transcript abundance were clearly dose related (Figure 7.1). The alterations of gene expression profiles were tissue specific, except for genes related to stress responses, which were affected in all organs assessed. As expected, the thymus was only weakly affected, since it is not a target organ of HCB. Microarray analysis proved to be a suitable tool to reveal changes in gene expression that are consistent with a number of the known (immuno)toxicological effects of HCB in the BN rat and its induction of enzymes involved in metabolism and reproduction. Novel findings included increased gene expression of pro-inflammatory cytokines, chemokines, complement components, cell adhesion molecules, antioxidants, and acute-phase proteins. These results are indicative for the involvement of macrophages and granulocytes and the mediators they release in the inflammatory response to HCB, which is accompanied by oxidative stress and an acute-phase response. This corresponds with previous findings (Ezendam *et al.*, 2004b, 2005). The study thus revealed the complexity of cells and mediators that participate in the response to HCB and provided more insight into the mechanisms of HCB toxicity. Assessment of gene expression at more than a single time point could, however, have yielded even more mechanistic information.

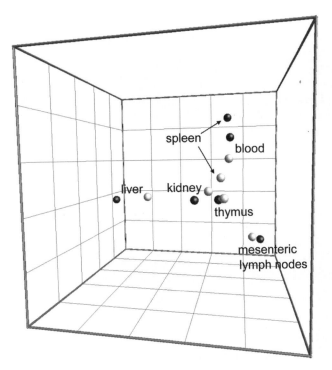

Figure 7.1 *Principal component analysis plot of gene expression data acquired for different organs of BN rats exposed to HCB for 28 days. The data points represent the gene expression levels per organ of the low-dose group (white dots) or high-dose group (black dots) compared with average control levels in the same organs. (Ezendam J., Staedtler F., Pennings J., Vandebriel RJ., Pieters R., Harleman JH. and Vos JG. Environ Health Perspect 2004, 112(7), 782–791. Reproduced with permission from Environmental Health Perspectives.)*

7.2.2   2,3,7,8-Tetrachlorodibenzo-*p*-dioxin

The xenobiotic 2,3,7,8-tetrachlorodibenzo-*p*-dioxin (TCDD) produces a variety of toxic effects, one of the targets being the immune system. Immunotoxic effects, such as thymus atrophy, suppression of cytotoxic T cell activity, and reduction of humoral immunity, have often been shown to be mediated through the transcription factor aryl hydrocarbon receptor (reviewed by Inadera (2006)). Zeytun *et al.* (2002), therefore, studied the effects of TCDD *in vivo* in mice at the transcriptome level. Pathway-specific microarray analysis interrogating 83 genes involved in apoptosis, cytokine production, and angiogenesis revealed up-regulation of expression of apoptosis-related genes in the thymus and spleen and to a lesser extent the liver 1 and 3 days after TCDD administration. Previous observations showing that the thymus is the most sensitive target of TCDD and that TCDD can induce apoptosis in a range of cell types were thus confirmed. In addition, differential regulation of cytokine gene expression was detected in the thymus, also confirming previous observations. Cytokine expression was affected in spleen as well, but the profile differed from that in the thymus. In this study, most genes were regulated at both time points, although the expression levels varied. When additional doses of TCDD were tested, a low dose was found to induce gene expression changes and a small subset of genes was regulated dose dependently (Zeytun *et al.*, 2002).

In another study, gene expression changes that underlie suppression of antibody production by TCDD were investigated in lymphocytes of mice immunized with ovalbumin and adjuvant (Nagai *et al.*, 2005). Gene expression was mainly up-regulated by immunization alone, particularly in T cells, while exposure to TCDD predominantly resulted in down-regulation of gene expression in the T cells 3 h and more so 24 h after immunization. Of the genes that were down-regulated by TCDD, three and seven genes were up-regulated by immunization alone after 3 h and 24 h respectively. Inhibition of antibody production by TCDD may thus be mediated by suppression of immunization-induced gene transcription in T cells. In B cells, TCDD mainly up-regulated gene expression, indicating that TCDD causes cell-type-specific effects (Nagai *et al.*, 2005).

Similar to the HCB study, these two experiments on TCDD-induced immunotoxicity demonstrate that microarray analysis is able both to substantiate known effects (even with a limited number of specific genes) and to expand the knowledge on specific mechanisms of action.

7.2.3   Bis(tri-*n*-butyltin)oxide

Bis(tri-*n*-butyltin)oxide (TBTO) is an organotin compound that has been used extensively as a biocide (for instance, in anti-fouling paints) and occurs as an environmental pollutant (EFSA, 2004). Immunotoxicity is the most sensitive toxicological endpoint of TBTO in rodents. Most apparent is the direct toxicity towards cortical thymocytes resulting in thymus involution and peripheral lymphocyte depletion and thereby a diminished cellular immune response. TBTO-induced thymus atrophy is most likely explained by inhibition of cell proliferation or, alternatively, by induction of apoptosis in thymocytes (for references, see Baken *et al.* (2006, 2007b)). In spite of many clues to the mechanisms of action of TBTO, the critical molecular target of TBTO remains unknown. A series of microarray studies in mice and rats has been performed to gain further insight into its cellular effects.

In C57BL/6 mice, the maximum tolerated dose of TBTO (300 ppm in food) was administered for 3, 7, or 14 days, and microarray analysis was performed on RNA from the primary target organ, the thymus. TBTO exposure resulted in thymic atrophy, and gene expression profiling revealed reduced expression of cell-surface determinants and receptors, as well as inhibition of cell-cycle-related processes, pointing to interference with T cell activation and inhibition of proliferation as a primary mechanism of action of TBTO. Hierarchical clustering and the main function of the differentially expressed genes are shown in Figure 7.2. Figure 7.3 shows an example of a cell-cycle-related pathway that was affected by TBTO. Some processes found to be affected by TBTO in this study (e.g. protein biosynthesis) have previously been reported to be interrupted by TBTO, while others had not been reported before. An example of the latter was stimulation of mitochondrial functioning and substrate metabolism, which may be a compensatory response to mitochondrial damage or to stress. Several nuclear receptors that may be involved in the effects on lipid metabolism were identified (Baken *et al.*, 2006).

A similar experiment was performed in rats, where gene expression profiles were assessed after 3, 7, 14, and 28 days of exposure to 80 ppm TBTO in food. Next to the thymus,

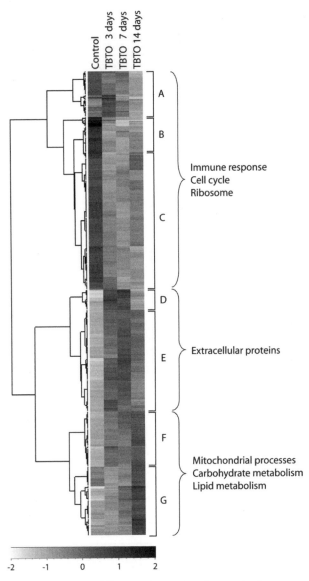

Figure 7.2 *Clustering of the 703 differentially regulated genes in thymuses of C57BL/6 mice after exposure to 300 ppm TBTO in the diet for 3, 7, or 14 days (n = 4 pools of two RNA samples). Gene expression levels, as assessed by microarray analysis, are indicated by the bar and relative ln values: white represents down-regulation and black up-regulation. Roughly, seven categories can be discriminated based on gene expression levels. A: late down-regulation; B: constant down-regulation; C: early down-regulation; D: temporary up-regulation; E: early up-regulation; F: gradual up-regulation; G: late up-regulation. Gene ontology term enrichment was determined for all regulated genes and the groups of down-regulated (A–C), early up-regulated (D, E), and gradually up-regulated (F, G) genes. The biological pathways and cellular components affected are shown beside the corresponding clusters (Baken KA., Pennings JLA., de Vries A., Breit TM., van Steeg H. and van Loveren H. J. Immunotoxicology 2006, 3(4), 227–244. Reprinted with permission of Taylor & Francis Ltd.)*

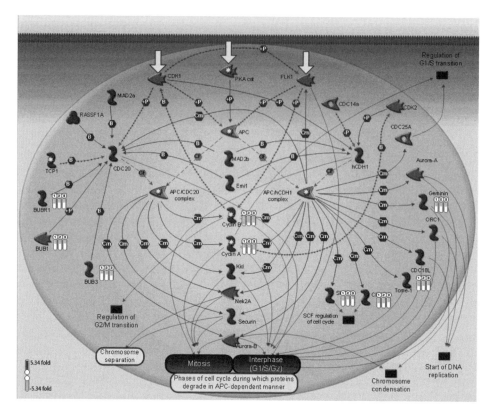

Figure 7.3 *Effect of TBTO on genes involved in cell cycle regulation by the anaphase-promoting complex (APC). C57BL/6 mice were exposed to 300 ppm TBTO in the diet for 3, 7, or 14 days (n = 4 pools of two RNA samples). Involvement of differentially expressed genes in specific cellular pathways and the extent of regulation of available pathways by TBTO were identified and visualized using MetaCore by GeneGo Inc. The pathway depicted involved a relatively high number of differentially expressed genes. Down-regulation of hits compared with controls is indicated by the bars beside the genes; (1) refers to the first time point of 3 days, (2) to the second time point, and (3) to the third time point. Arrows comprising a sign of the mechanism involved indicate a negative effect and dotted arrows a positive effect. For legend, see http://portal.genego.com/legends/legend_6.png (Baken KA., Pennings JLA., de Vries A., Breit TM., van Steeg H. and van Loveren H. J. Immunotoxicology 2006, 3(4), 227–244. Reprinted with permission of Taylor & Francis Ltd.)*

which involuted as a result of TBTO exposure, spleen and liver were examined. Hardly any gene expression changes were detected in the thymus and spleen. In the liver, however, down-regulation of lipid synthesis was detected. The differences between the results of the two *in vivo* experiments were ascribed to dose and/or interspecies differences (Baken *et al.*, 2006).

Since gene regulation by TBTO was relatively absent in rat thymuses *in vivo*, primary thymocytes were exposed to (non)cytotoxic doses of TBTO in a follow-up study and appropriate dose levels and exposure periods were determined by measuring viability and apoptosis. Microarray analysis showed that regulation of lipid metabolism by TBTO occurred at a dose that did not yet result in disruption of cell functioning. Higher doses caused regulation of genes involved in apoptotic processes even before apoptosis could be observed phenotypically. The highest dose level tested also repressed mitochondrial functioning and immune cell activation. Several results corresponded to previous findings; but, in contrast to previous *in vivo* and *in vitro* studies showing that inhibition of cell proliferation was the primary mechanism of action of TBTO, apoptosis in this *in vitro* study appeared to mediate the toxic effects of TBTO. This discrepancy was most likely a result of *in vitro* versus *in vivo* differences and the relatively short duration of exposure to TBTO *in vitro*. Induction of apoptosis may, therefore, precede inhibition of cell proliferation by TBTO in time (Baken *et al.*, 2007b).

7.2.4   Gene Expression Changes Shared by Immunotoxicants

For evaluation of the applicability of gene expression profiling as a tool to screen compounds for immunotoxic properties, the transcriptional effects of several immunotoxicants were compared by Baken *et al.* (2008) in order to identify overlap in gene expression profiles. Gene expression changes after exposure to TBTO, cyclosporine A (CsA), benzo[*a*]pyrene (B[*a*]P), and acetaminophen (APAP) were investigated in murine spleens. CsA is used as an immunosuppressive drug to reduce the risk of organ rejection after clinical transplantation and in the treatment of autoimmune and inflammatory diseases. Immunotoxic effects of CsA include histopathological changes in the thymus, spleen, and lymph nodes, altered lymphocyte subpopulations in the spleen, and interrupted T cell function and antibody responses (de Waal *et al.*, 1995; Lebrec *et al.*, 1995). B[*a*]P is formed in incomplete combustion and belongs to the group of carcinogenic polycyclic aromatic hydrocarbons (PAHs). Consequences of B[*a*]P exposure for the immune system include atrophy of the bone marrow, thymus, and spleen, and a reduction of circulating red and white blood cells and immunoglobulins (de Jong *et al.*, 1999). APAP is a common analgesic and antipyretic drug which performs its actions by inhibiting cyclooxygenase enzymes, thereby blocking prostaglandin production, and by indirect activation of cannabinoid CB(1) receptors (Bertolini *et al.*, 2006). At high dose levels, APAP induces hepatotoxicity (Bessems and Vermeulen, 2001; Kaplowitz, 2004); additionally, indications of immunomodulating properties of APAP have appeared in the literature (Yamaura *et al.*, 2002; Masson *et al.*, 2007; and others: see Baken *et al.* (2008)). The immunosuppressive effects observed were suggested to prevent allergic or autoimmune responses to protein adducts formed by APAP's reactive metabolite.

Several compound-specific and overlapping effects were detected by microarray analysis. TBTO, for instance, interrupted cellular respiration, and CsA and B[*a*]P inhibited immunological processes. All compounds induced xenobiotic metabolism. The process that was most significantly affected by all toxicants was cell division (down-regulated); therefore it was concluded that the immunosuppressive properties of the model compounds appear to be mediated by cell cycle arrest. Since strongly proliferating immune cells will be particularly sensitive to effects on cell division, evaluation of cell proliferation thus remains a valuable tool to assess immunosuppression (Baken *et al.*, 2008).

Patterson and Germolec (2006) examined gene expression changes induced by the prototype immunosuppressive agents TCDD, cyclophosphamide, diethylstilbestrol (DES), and dexamethason in murine thymus and spleen. Preliminary data showed that, although most transcriptional effects were compound specific, some genes were regulated by all compounds. These genes were mainly involved in apoptosis, immune cell activation, antigen presentation and processing, and again cell proliferation. Both studies show that microarray analysis offers opportunities to discover gene expression changes that may be indicative of immunosuppressive effects of compounds.

7.2.5   Sensitizing Compounds

Skin sensitization can be induced after dermal contact with certain chemicals, e.g. dinitrochlorobenzene or nickel. Sensitization is initiated when the chemical binds covalently to a protein and forms a hapten. This hapten is then presented by dendritic cells (DCs) to T cells, causing activation and proliferation of hapten-specific T cells. Upon a second encounter with the same chemical, an inflammatory response is induced that leads to allergic manifestations. In the skin, the process of sensitization is mediated by keratinocytes (cytokine production) and DCs (recognition, processing, and presentation of antigens). Also, in the respiratory tract, certain chemicals can induce allergic responses. Currently, the local lymph node assay (LLNA) is used to identify chemicals with sensitizing properties, but legislation and public opinions have stimulated the search for reliable alternative *in vitro* screening methods that can detect respiratory or skin sensitizers and discriminate them from chemicals that have only irritant properties (Vandebriel *et al.*, 2005). Gene expression profiling may be used as a read-out system in the development of alternative (*in vitro*) methods. Furthermore, microarray analysis may extend insight into mechanisms involved in chemical-induced sensitization.

Betts *et al.* (2003) measured effects on global gene expression in the draining lymph node (LN) 18 and 48 h after single topical exposure to the contact allergen dinitrofluorobenzene (DNFB) on the ears of mice. This technique appeared to be both sensitive (since thresholds of detection were similar to the LLNA) and robust (since the kinetics and dose-responses could be confirmed by northern blot and/or reverse transcriptase polymerase chain reaction (RT-PCR)). The authors described three genes that were significantly affected: GlyCAM-1 (down-regulated); guanylate-binding protein-2, an IFN-γ inducible GTPase present in macrophages; and onzin (both up-regulated). GlyCAM-1 regulates migration of circulating lymphocytes into the periphery, and an inverse relation exists between its expression and LN weight. The authors thus hypothesized that reduced GlyCAM-1 expression is involved in recruitment of lymphocytes into the LN. Guanylate-binding protein-2 and onzin are involved in Th1-type immune responses (Cella *et al.*, 2000; Penna *et al.*, 2002; Rissoan *et al.*, 2002). The up-regulation of these genes may thus mediate the induction of Th1 responses, which is generally accepted to occur after contact allergen exposure.

DCs are frequently used in *in vitro* alternatives to assess the sensitizing properties of chemicals, because these cells play an important role in the initiation of sensitization. Ryan *et al.* (2004) have demonstrated dose-dependent changes in gene expression in blood-derived DCs induced by the contact allergen dinitrobenzene sulfonic acid (DNBS) that are associated with DC maturation, a process that has been proposed to occur during DC migration to LNs after activation by encountering a chemical allergen (Pennie and Kimber, 2002). Microarray analysis revealed up-regulation of 60 genes and down-regulation of 58 genes after exposure to both doses used, representing various cellular processes. A number of gene expression changes were consistent with known features of DC maturation, and some of those were recently also reported to occur after exposure of human DCs to the contact allergen nickel sulfate (Schoeters *et al.*, 2006). In this study, reproducibility (among microarray platforms and DC donors) of the changes observed was also demonstrated, as well as specificity: expression of many genes was altered only by DNBS and not by a structurally similar nonallergenic chemical. From the results of this study, a list of target genes was derived that may serve as biomarkers to predict skin sensitization by chemicals (Gildea *et al.*, 2006). These genes were evaluated and further prioritized using real-time RT-PCR analysis of peripheral blood mononuclear cells (PBMCs) exposed to an extended set of chemicals, including irritants, contact allergens, and nonsensitizers. The 10 genes that were brought forth by this approach showed selectivity, specificity, and a proper dynamic range, and may, therefore, be used in screening for skin-sensitizing chemicals (Gildea *et al.*, 2006).

Schoeters *et al.* (2007) used CD34+ progenitor-derived DCs, which were exposed to four contact allergens and two irritants. Microarray analysis revealed 25 genes with altered expression patterns after exposure to contact allergens and not irritants. Five of these genes were confirmed by real-time RT-PCR. Gene ontology linked 17 of the genes to immune responses and/or DC maturation. This set of genes can be evaluated further for their capacity to predict sensitizing potential of different classes of chemicals.

In conclusion, clues to mechanisms of sensitization can be inferred from the data obtained, and global gene expression profiles after *in vitro* exposure of specific cell types to various model compounds may provide novel (sets of) markers that can predict *in vivo* adverse effects, such as sensitization.

7.3 Lessons from Immunotoxicogenomics Studies

The examples described in the previous paragraphs show the ability of microarray analysis to detect immunotoxic effects of a wide range of agents *in vivo* and *in vitro* and to expand the understanding of their mechanisms of action. Immunotoxicogenomics studies have also generated directions for the integration of gene expression profiling in immunotoxicological research.

7.3.1 Experimental setup

Dose- and time-dependency of gene expression changes have been found in all studies that included more than a single dose and time point, indicating that this is an important aspect to consider when designing microarray experiments. The use of a low and a high dose, such as in the study on HCB (Ezendam *et al.*, 2004a), may provide valuable insight into gene expression changes in the presence and absence of pathological or cellular change. When overt toxicity is present, it is possible that microarray analysis detects mainly secondary effects. However, the studies performed with TBTO described above have demonstrated that a high dose of TBTO administered to mice resulted in significant regulation of gene expression in the thymus, whereas absence of overt gene expression changes was found in rat thymus after exposure to a lower dose at which thymus atrophy was already observed (Baken *et al.*, 2006). Although these dissimilarities might reflect differences in species instead of (or in addition to) dose, this suggests that microarray analysis is not always able to demonstrate toxic effects at lower dose levels than traditional methods of studying toxicity. The impression that high dose levels are needed to induce differential gene expression was supported by the experiments in murine spleen, where even maximum tolerated dose levels of CsA and B[*a*]P did not induce significant regulation of individual genes (Baken *et al*., 2008). Therefore, an important conclusion from immunotoxicogenomics studies performed so far seems to be that microarray experiments should include dose ranges similar to those applied in methods used traditionally in toxicity testing.

It has been assumed that changes in gene expression can be detected at an earlier time point than other methods to study toxicity. This was demonstrated for TBTO in the *in vitro* study (Baken *et al*., 2007b). Induction of apoptosis appeared to precede inhibition of cell proliferation, and was detected before being observed phenotypically. In the study on TBTO, CsA, B[*a*]P, and APAP, some effects on gene expression appeared prior to effects on organ weight or histology (Baken *et al*., 2008). This sensitivity of microarray analysis in terms of duration of exposure might offer the advantage of assessing immunotoxicity after short exposure periods. A second important conclusion from the studies presented is thus that, for microarray experiments, the exposure periods can be shorter than for the methods used traditionally in toxicity testing.

Analysis of multiple organs may provide additional information on effects of toxicants. In the HCB exposure study, six organs were analyzed, including organs that are not considered to be a (primary) part of the immune system (Ezendam *et al*., 2004a). Since mediators produced outside of the immune system may still function in the immune response (Abbas *et al*., 2005), this approach can be useful for compounds such as HCB that are known to exert toxic effects on systems other than the immune system.

7.3.2   Interpretation of Results

Changes in expression of genes mediating a certain process may not necessarily all point in the same direction. An example is the induction of pro- and anti-apoptotic processes by TBTO in the *in vitro* study, whereas phenotypical apoptosis was observed (Baken *et al*., 2007b). In addition, often, not all genes taking part in a certain pathway are regulated. Furthermore, it is possible that induction of an immune response is required for immunomodulators to exert their effects, which may more easily be detected after stimulation by antigens or mitogens, such as in some immune function assays. Gene expression profiling may thus not always provide decisive answers on the occurrence of (immuno)toxic events.

The interpretation of microarray results may also be complicated by the effect of changes in cell populations on gene expression profiles. TBTO is, for instance, known to be particularly toxic for rapidly dividing cortical thymocytes. At a certain time point after the start of the exposure, these cells (and their transcriptomes) may have been removed from the cell population in the thymus *in vivo*, whereas they might be cleared less efficiently *in vitro*. When assessing effects in spleen, influx of cells via the blood (possibly as a result of xenobiotic exposure) may cause altered abundance of certain mRNAs and, thus, altered gene expression profiles, as was seen, for instance, after exposure to the high dose of HCB (Ezendam *et al*., 2004a). Furthermore, effects of xenobiotics may differ per cell type, as was found for TCDD (Nagai *et al*., 2005); and when effects of several xenobiotics are compared in the same organ, different compounds may affect different cell types.

In conclusion, when investigating mechanisms of action, comparison of gene expression changes with effects detected by other test methods and measurement of the same endpoints in different assays are thus important for correct interpretation of microarray results. Examples of confirmation of effects detected by gene expression profiling by other analyses are the studies on TBTO (Baken *et al.*, 2007b) and DNFB (Betts *et al.*, 2003). Tsangaris *et al.* (2002) investigated the effects of a low and a high concentration of the toxic metal cadmium in an immature T-cell line. Time- and dose-dependent effects on the gene expression profile were associated with cell function, cell differentiation, malignant transformation, and apoptosis, and correlated with other markers of viability and apoptosis. Another approach is to compare gene expression profiles with those of compounds with the same or different mechanisms of action. Royaee *et al.* (2006), for example, studied transcriptional responses to cholera toxin (CT; a powerful mucosal adjuvant) in cultured human lymphocytes and monocytes. The expression of genes associated with immunomodulation, inflammation, and oxidative stress changed time- and dose-dependently. Expression of Th1 markers was down-regulated, whereas Th2 markers were up-regulated. The gene expression profiles were compared with those induced by an activator and an inhibitor of adenylate cyclase, since this enzyme is activated by CT, resulting in intracellular cAMP accumulation. Overlap and differences of gene expression alterations induced by these three compounds yielded insight into the involvement of cAMP in CT toxicity (Royaee *et al.*, 2006). Transcriptome changes after DNBS exposure were related to those of a structurally similar nonallergen (Ryan *et al.*, 2004).

It is equally important to establish correlation of absence of changes in gene expression with functional effects, since effects may only be observable in specific experimental settings or at other levels than the transcriptome, such as post-transcriptional or post-translational. Results of *in vitro* approaches should most ideally be confirmed with *in vivo* effects, since functional differences may exist between cells in culture or *in vivo*, and *in vitro* designs lack interaction of various different cell types (de Longueville *et al.*, 2004). This is, for instance, illustrated by the Dere *et al.* (2006) study comparing the effects of TCDD on gene expression *in vivo* and *in vitro*: immunotoxic effects were not detected *in vitro*, whereas these have been found *in vivo* (Zeytun *et al.*, 2002; Nagai *et al.*, 2005).

7.3.3 Development of Screening Assays

Toxicogenomics is expected to contribute to the development of screening assays by identifying signature gene expression profiles for specific classes of compounds. It is envisaged that, ultimately, small sets of biomarkers are sufficient to identify immunotoxicity, enabling high-throughput screening (Pennie, 2000; Fielden and Kolaja, 2006). Two studies tested this assumption by comparing gene expression profiles of several model immunotoxicants. In both these studies it was shown that immunosuppressive compounds affect cell proliferation (Patterson and Germolec, 2006; Baken *et al.*, 2008). Obviously, the specificity and predictivity of inhibition of cell division for immunotoxicity in general should be confirmed by testing a larger range of compounds. Since, in both studies, immunosuppressive compounds were included, the value of (genes involved in) cell proliferation as a biomarker in other areas of immunotoxicology, such as sensitization and autoimmunity, should still be investigated as well.

Several microarray studies described above showed that the spleen is a suitable organ for detection of immunosuppression by gene expression profiling. This is a promising finding with respect to development of screening assays, since effects in this organ are presumably reflected in peripheral lymphocytes that can easily be obtained from blood.

The examples on gene expression profiling after exposure to chemicals with sensitizing properties showed that, besides yielding clues to mechanisms of sensitization, microarray analysis may also provide opportunities to screen compounds for sensitizing properties. The sensitivity, specificity, and robustness demonstrated in the different studies support the viewpoint that *in vitro* methods relying on microarray analysis have the potential to (at least in part) replace existing methods to uncover sensitizing effects of compounds. Evaluation of gene expression profiles in human cells, as was done by Ryan *et al.* (2004) and Schoeters *et al.* (2007), will elucidate the relevance of the findings for the human situation. Since various cell types of human origin, most notably DCs, can now be routinely obtained (by culture) from human peripheral blood (a readily accessible source), interspecies comparison may become superfluous. It is well imaginable, therefore, that transcript changes identified by microarray analyses, such as performed by Gildea *et al.* (2006) and the Schoeters group (Schoeters *et al.*, 2005, 2006, 2007; Verheyen *et al.*, 2005), can serve as new markers for allergenicity.

7.4   Conclusion

The application of toxicogenomics in evaluation of immunotoxicity is not yet without challenges. Several of the issues mentioned above were also discussed in a workshop addressing the application of toxicogenomics in immunotoxicity screening and in recent reviews on this topic (Burns-Naas *et al.*, 2006; Luebke *et al.*, 2006; Baken *et al.*, 2007a). The overall opinion is that, at present, toxicogenomics is not yet able to replace the current methods for assessment of immunotoxicity. It does, however, already contribute to the understanding of immunotoxic processes and to the development of *in vitro* screening assays and, therefore, is expected to be of value for mechanistic insight into immunotoxicity and for hazard identification of existing and novel compounds.

References

Abbas AR, Baldwin D, Ma Y, Ouyang W, Gurney A, Martin F, Fong S, van Lookeren CM, Godowski P, Williams PM, Chan AC, Clark HF (2005) Immune response *in silico* (IRIS): immune-specific genes identified from a compendium of microarray expression data, *Genes Immun* **6**, 319–331.

Baken KA, Pennings JLA, de Vries A, Breit TM, van Steeg H, van Loveren H (2006) Gene expression profiling of bis(tri-*n*-butyltin)oxide (TBTO) induced immunotoxicity in mice and rats. *J Immunotoxicol* **3**, 227–244.

Baken KA, Vandebriel RJ, Pennings JLA, Kleinjans JC, van Loveren H (2007a) Toxicogenomics in the assessment of immunotoxicity. *Methods* **41**, 132–141.

Baken KA, Arkusz J, Pennings JL, Vandebriel RJ, van Loveren H (2007b) *In vitro* immunotoxicity of bis(tri-*n*-butyltin)oxide (TBTO) studied by toxicogenomics. *Toxicology* **237**, 35–48.

Baken KA, Pennings JLA, Jonker MJ, Schaap MM, de Vries A, van Steeg H, Breit TM, van Loveren H (2008) Overlapping gene expression profiles of model compounds provide opportunities for immunotoxicity screening. *Toxicol Appl Pharmacol* **226**, 46–59.

Bertolini A, Ferrari A, Ottani A, Guerzoni S, Tacchi R, Leone S (2006) Paracetamol: new vistas of an old drug. *CNS Drug Rev* **12**, 250–275.

Bessems JG, Vermeulen NP (2001) Paracetamol (acetaminophen)-induced toxicity: molecular and biochemical mechanisms, analogues and protective approaches. *Crit Rev Toxicol* **31**, 55–138.

Betts CJ, Moggs JG, Caddick HT, Cumberbatch M, Orphanides G, Dearman RJ, Ryan CA, Hulette BC, Frank GG, Kimber I (2003) Assessment of glycosylation-dependent cell adhesion molecule 1 as a correlate of allergen-stimulated lymph node activation. *Toxicology* **185**, 103–117.

Burns-Naas LA, Dearman RJ, Germolec DR, Kaminski NE, Kimber I, Ladics GS, Luebke RW, Pfau JC, Pruett SB (2006) 'Omics' technologies and the immune system. *Tox Mech Methods*, **16**, 101–119.

Cella M, Facchetti F, Lanzavecchia A, Colonna M (2000) Plasmacytoid dendritic cells activated by influenza virus and CD40L drive a potent TH1 polarization. *Nat Immunol* **1**, 305–310.

De Jong WH, Kroese ED, Vos JG, van Loveren H (1999) Detection of immunotoxicity of benzo[*a*]pyrene in a subacute toxicity study after oral exposure in rats. *Toxicol Sci* **50**, 214–220.

De Jong WH, van Loveren H (2007) Screening of xenobiotics for direct immunotoxicity in an animal study. *Methods* **41**, 3–8.

De Longueville F, Bertholet V, Remacle J (2004) DNA microarrays as a tool in toxicogenomics. *Comb Chem High Throughput Screen* **7**, 207–211.

De Waal EJ, Timmerman HH, Dortant PM, Kranjc MA, van Loveren H (1995) Investigation of a screening battery for immunotoxicity of pharmaceuticals within a 28-day oral toxicity study using azathioprine and cyclosporine A as model compounds. *Regul Toxicol Pharmacol* **21**, 327–338.

Dere E, Boverhof DR, Burgoon LD, Zacharewski TR (2006) *In vivo–in vitro* toxicogenomic comparison of TCDD-elicited gene expression in Hepa1c1c7 mouse hepatoma cells and C57BL/6 hepatic tissue. *BMGenomics C* **7**, 80–97.

EFSA (2004) Opinion of the Scientific Panel on Contaminants in the Food Chain on a request from the Commission to assess the health risks to consumers associated with exposure to organotins in foodstuffs. *EFSA J* **102**, 1–119.

Ezendam J, Staedtler F, Pennings J, Vandebriel RJ, Pieters R, Harleman JH, Vos JG (2004a) Toxicogenomics of subchronic hexachlorobenzene exposure in brown Norway rats. *Environ Health Perspect* **112**, 782–791.

Ezendam J, Hassing I, Bleumink R, Vos JG, Pieters R (2004b) Hexachlorobenzene-induced immunopathology in brown Norway rats is partly mediated by T cells. *Toxicol Sci* **78**, 88–95.

Ezendam J, Kosterman K, Spijkerboer H, Bleumink R, Hassing I, van Rooijen N, Vos JG, Pieters R (2005) Macrophages are involved in hexachlorobenzene-induced adverse immune effects. *Toxicol Appl Pharmacol* **209**, 19–27.

Fielden MR, Kolaja KL (2006) The state-of-the-art in predictive toxicogenomics. *Curr Opin Drug Discov Dev* **9**, 84–91.

Gildea LA, Ryan CA, Foertsch LM, Kennedy JM, Dearman RJ, Kimber I, Gerberick GF (2006) Identification of gene expression changes induced by chemical allergens in dendritic cells: opportunities for skin sensitization testing. *J Invest Dermatol* **126**, 1813–1822.

Inadera H (2006) The immune system as a target for environmental chemicals: xenoestrogens and other compounds. *Toxicol Lett* **164**, 191–206.

Kaplowitz N (2004) Acetaminophen hepatoxicity: what do we know, what don't we know, and what do we do next? *Hepatology* **40**, 23–26.

Lebrec H, Roger R, Blot C, Burleson GR, Bohuon C, Pallardy M (1995) Immunotoxicological investigation using pharmaceutical drugs. *In vitro* evaluation of immune effects using rodent or human immune cells. *Toxicology* **96**, 147–156.

Luebke RW, Holsapple MP, Ladics GS, Luster MI, Selgrade M, Smialowicz RJ, Woolhiser MR, Germolec DR (2006) Immunotoxicogenomics: the potential of genomics technology in the immunotoxicity risk assessment process. *Toxicol Sci* **94**, 22–27.

Masson MJ, Peterson RA, Chung CJ, Graf ML, Carpenter LD, Ambroso JL, Krull DL, Sciarrotta J, Pohl LR (2007) Lymphocyte loss and immunosuppression following acetaminophen-induced hepatotoxicity in mice as a potential mechanism of tolerance, *Chem Res Toxicol* **20**, 20–26.

Nagai H, Takei T, Tohyama C, Kubo M, Abe R, Nohara K (2005) Search for the target genes involved in the suppression of antibody production by TCDD in C57BL/6 mice. *Int Immunopharmacol* **5**, 331–343.

Patterson RM, Germolec DR (2006) Gene expression alterations in immune system pathways following exposure to immunosuppressive chemicals. *Ann. N Y Acad Sci* **1076**, 718–727.

Penna G, Vulcano M, Roncari A, Facchetti F, Sozzani S, Adorini L (2002) Cutting edge: differential chemokine production by myeloid and plasmacytoid dendritic cells. *J Immunol* **169**, 6673–6676.

Pennie WD (2000) Use of cDNA microarrays to probe and understand the toxicological consequences of altered gene expression. *Toxicol Lett* **112–113**, 473–477.

Pennie WD, Kimber I (2002) Toxicogenomics; transcript profiling and potential application to chemical allergy. *Toxicol In Vitro* **16**, 319–326.

Rissoan MC, Duhen T, Bridon JM, Bendriss-Vermare N, Peronne C, Blandine de Saint Vis FB, Bates EE (2002) Subtractive hybridization reveals the expression of immunoglobulin-like transcript 7, Eph-B1, granzyme B, and 3 novel transcripts in human plasmacytoid dendritic cells. *Blood* **100**, 3295–3303.

Royaee AR, Mendis C, Das R, Jett M, Yang DC (2006) Cholera toxin induced gene expression alterations. *Mol Immunol* **43**, 702–709.

Ryan CA, Gildea LA, Hulette BC, Dearman RJ, Kimber I, Gerberick GF (2004) Gene expression changes in peripheral blood-derived dendritic cells following exposure to a contact allergen. *Toxicol Lett* **150**, 301–316.

Schoeters E, Nuijten JM, van den Heuvel RL, Nelissen I, Witters H, Schoeters GE, van Tendeloo VF, Berneman ZN, Verheyen GR (2006) Gene expression signatures in CD34+-progenitor-derived dendritic cells exposed to the chemical contact allergen nickel sulphate. *Toxicol Appl Pharmacol* **216**, 131–149.

Schoeters E, Verheyen GR, Nelissen I, van Rompay AR, Hooyberghs J, van den Heuvel RL, Witters H, Schoeters GE, van Tendeloo V, Berneman ZN (2007) Microarray analyses in dendritic cells reveal potential biomarkers for chemical-induced skin sensitization. *Mol Immunol* **44**, 3222–3233.

Schoeters E, Verheyen GR, van den Heuvel R, Nelissen I, Witters H, van Tendeloo V, Schoeters GE, Berneman ZN (2005) Expression analysis of immune-related genes in CD34(+) progenitor-derived dendritic cells after exposure to the chemical contact allergen DNCB. *Toxicol In Vitro* **19**, 909–913.

Tsangaris GTh, Botsonis A, Politis I, Tzortzatou-Stathopoulou F (2002) Evaluation of cadmium-induced transcriptome alterations by three color cDNA labeling microarray analysis on a T-cell line. *Toxicology* **178**, 135–160.

Vandebriel RJ, van Och FM, van Loveren H (2005) *In vitro* assessment of sensitizing activity of low molecular weight compounds. *Toxicol Appl Pharmacol* **207**, 142–148.

Verheyen GR, Schoeters E, Nuijten JM, van den Heuvel RL, Nelissen I, Witters H, van Tendeloo V, Berneman ZN, Schoeters GE (2005) Cytokine transcript profiling in CD34+-progenitor derived dendritic cells exposed to contact allergens and irritants. *Toxicol Lett* **155**, 187–194.

Yamaura K, Ogawa K, Yonekawa T, Nakamura T, Yano S, Ueno K (2002) Inhibition of the antibody production by acetaminophen independent of liver injury in mice. *Biol Pharm Bull* **25**, 201–205.

Zeytun A, McKallip RJ, Fisher M, Camacho I, Nagarkatti M, Nagarkatti PS (2002) Analysis of 2,3,7,8-tetrachlorodibenzo-*p*-dioxin-induced gene expression profile *in vivo* using pathway-specific cDNA arrays. *Toxicology* **178**, 241–260.

8

Toxicogenomics and Ecotoxicogenomics: Studying Chemical Effects and Basic Biology in Vertebrates and Invertebrates

Taisen Iguchi, Hajime Watanabe and Yoshinao Katsu

8.1 Introduction

Persistent toxic chemicals such as polychlorinated organic pollutants, including poly-chlorinated biphenyls (PCBs), dioxins, and various pesticides, have been controlled by governments of developing countries. We are still using vast amount of chemicals which are not too toxic evaluated by various toxicity tests. In this decade, a new concept of chemical issues has been emerged, the so-called endocrine disruptors (Colborn and Clement, 1992). Environmental endocrine-disrupting chemicals (EDCs), such as organochlorines and pesticides, as well as plasticizers, pharmaceutics, and natural hormones, can interact with various receptors, such as the estrogen receptor (ER), androgen receptor (AR), and arylhydrocarbon receptor (McLachlan, 2001; Iguchi *et al.*, 2002a). Clarifying the molecular basis of EDCs and endogenous estrogens on developing organisms is essential if we are to understand the linkages between exposure levels, timing of exposure, genes responsive to these chemicals, and adverse effects. Various modes of action of chemicals and nontraditional targets of EDCs have been summarized (Fox, 2005; Tarrant, 2005). Understanding the effects of EDCs on various species, from invertebrates to mammals, at the level of molecular biology is greatly needed to help explain observations at the cellular and organismal levels of organization that typically are obtained with traditional toxicological approaches. The subdiscipline combining the fields of genomics and mammalian

Toxicogenomics: A Powerful Tool for Toxicity Assessment   Edited by S. C. Sahu
© 2008 John Wiley & Sons, Ltd

toxicology is defined as *toxicogenomics* (Nuwaysir *et al.*, 1999; Inoue and Pennie, 2002), and has become an important field in toxicology. Originally, toxicogenomics was intended to be used to evaluate the risks of chemicals to humans, but the recent increase in genetic information has allowed the field to be extended to other organisms. *Ecotoxicogenomics* is the application of toxicogenomics to organisms that are representative of ecosystems and is used to study the hazardous effects of chemicals on ecosystems as well as individuals (Snape *et al.*, 2004; Miracle and Ankley, 2005). Although the availability of genomic information about nonmodel organisms is still limited, the application of toxicogenomics to a variety of organisms could be a powerful tool for evaluating the effects of chemicals on ecosystems (Iguchi *et al.*, 2006, 2007b; Watanabe and Iguchi, 2006).

DNA microarrays provide a high-throughput diagnostic tool to screen the many variables required to examine gene-expression patterns effectively. Profiling of transcripts, proteins, and metabolites can help discriminate classes of EDCs and toxicants and be helpful in understanding modes of action. Through systematic efforts to generate mechanistic information, diagnostic and predictive assessments of the risk of EDCs and toxic chemicals will be established in model species for ecological risk assessment. Toxicogenomics and ecotoxicogenomics also provide important data on the basic biology of animal species. In this review, we provide an overview of our current approach to understanding EDCs using toxicogenomics and ecotoxicogenomic approaches.

8.2   Application of Transcriptomics for the Analysis of Estrogen-responsive Genes and Estrogen Response Element

DNA microarray methodology has been applied to obtain genome-wide analyses of gene expression stimulated by hormones and/or chemicals (Inoue and Pennie, 2002; Watanabe and Iguchi, 2003). An understanding of the patterns of expression of estrogen-responsive genes is essential if we are to begin to understand the mechanisms of action of estrogenic chemicals on target organs. A large number of genes affected by estrogen have been selected from the mouse (Watanabe *et al.*, 2002a,b, 2003a; Moggs *et al.*, 2004) or rat uterus (Daston and Naciff, 2005). For most of the selected genes, expression is induced in a dose-dependent manner. Moreover, the expression of these 'estrogen-dependent' genes is not altered following E_2 treatment in ERα knockout mice, thus confirming the dependency of these genes on ERα for the induction of expression. Activation of these genes provides a basis for the marked uterotrophic effect observed several days following estrogen administration. Characteristic gene expression patterns are observed for each estrogenic chemical, and these are distinct from that of E_2, suggesting specific mechanisms of action for endocrine disruption that could be different from that induced by endogenous estrogen (Watanabe *et al.*, 2004a; Daston and Naciff, 2005). Physiological (estradiol-17β, E_2) and nonphysiological (diethylstilbestrol, DES) estrogens, nonylphenol (NP) and dioxin have distinct patterns of gene expression in the uterus (Watanabe *et al.*, 2003b, 2004b). In the liver, NP and dioxin activate a set of genes that are distinct from estrogen-responsive genes (Watanabe *et al.*, 2004a,b). Thus, these results suggest that only a small number of genes are directly involved in the uterotrophic effects of estrogen (Figure 8.1), and NP has very similar effects to E_2 on gene expression in the uterus but not in hepatic tissue. In the liver, gene expression was more markedly affected by NP than by E2. Organ-specific effects,

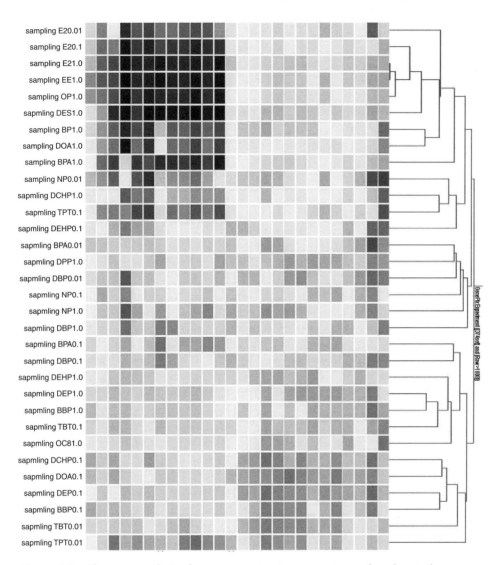

Figure 8.1 *Clustering analysis of gene expression in mouse uterus after chemical exposure. Ovariectomized adult mice were given a single injection of various chemicals. Numbers indicate the dose (mg kg −1) administered. Red and green bars indicate induced and repressed genes respectively. Dendrogram represents correlation of the gene expression profile.*

therefore, should be considered in order to elucidate the distinct effects of various EDCs. Fong *et al.* (2007) showed that anti-estrogen, 100 μg/kg body weight (bw) tamoxifen, and a synthetic estrogen, 100 μg/kg bw ethynylestradiol (EE_2) given orally, share similar gene expression profiles in the uterus from immature, ovariectomized mice. However, tamoxifen responses exhibit lower efficacy, while responses unique to EE_2 are consistent with the physiological differences elicited between compounds.

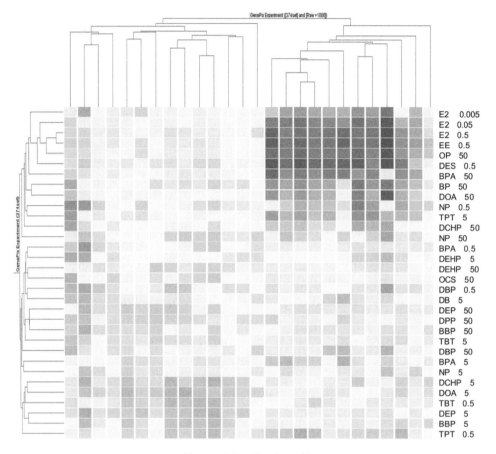

Figure 8.1   *(Continued)*

Estrogens play a central role in the reproduction and affect a variety of biological processes in vertebrates. The major target molecules of estrogens are ERs found in or near the nucleus that act as transcription factors; these receptors have been studied extensively at the molecular level. In contrast, analysis of estrogen response elements (EREs) in the promoter regions of genes remains limited. To identify genes regulated directly by ERs *in vivo*, chromatin-immunoprecipitation (ChIP)-mediated target cloning and microarray approaches have been applied to the mouse uterus. Aquaporin 5 (*AQP5*) (Kobayashi *et al.*, 2006) and adrenomedullin (Watanabe *et al.*, 2006) genes were precipitated with an antibody against ERα. Quantitative polymerase chain reaction (PCR) and DNA microarray analyses confirm that these genes are activated soon after estrogen administration, and the promoter region of these genes contains a functional ERE that was activated directly by estrogen (Kobayashi *et al.*, 2006).

The perinatal mouse model has been used to demonstrate the long-term effects of early sex hormone exposure on the female reproductive tract (Iguchi, 1992; Iguchi *et al.*, 2001; McLachlan, 2001; Newbold, 2004). Neonatal treatment of female mice with E₂ and

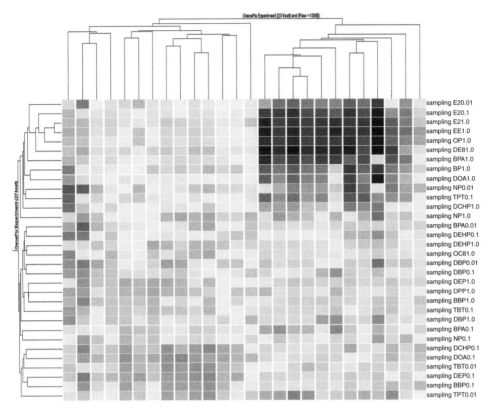

Figure 8.1 *(Continued)*

DES induces various abnormalities in the reproductive tracts, hypothalamo-hypophyseal-ovarian axis, immune function, and skeletal and muscular tissues. The growth response of perinatally DES-exposed reproductive organs to estrogen is altered, as are levels of ER, epidermal growth factor receptor (EGFR), prolactin receptor, and oncogenes and *Hox* genes (Bern, 1992; Iguchi *et al.*, 1993, 2002b; Newbold *et al.*, 2004; Yin and Ma, 2005). We found that activation cascade, such as persistent phosphorylation of erbB2 and ERα (Miyagawa *et al.*, 2004a), and also found sustained expression of JNK1, IGF-I receptor, and Akt in the vagina of neonatally DES-exposed mice (Miyagawa *et al.*, 2004b). Several genes have been cloned in the vagina after neonatal DES exposure (Katsu *et al.*, 2002, 2003; Katsu and Iguchi, 2006). The critical windows for induction of various abnormalities by postnatal exposure to exogenous estrogens varied from 3 to 30 days, thereby indicating the presence of tissue- and organ-specific critical windows in mice (Iguchi *et al.*, 2002a).

Gene expression pattern induced by DES was examined before and after the critical windows in mouse vagina. Vaginal epithelial cells and stromal cells showed proliferation after a single injection of DES at 20 and 70 days and at 0 and 5 days respectively. The number of genes induced 6 h after DES exposure in mouse vaginae at day 0 was lower than those induced at days 5–70. The DES-induced gene expression pattern in vaginae at day 5 was closer to the adult type. Several cell-cycle regulators and keratinocyte differentiation factors were induced by DES in vaginae from day 5 to adult. Thus, microarray analysis revealed that the gene expression pattern in vaginae during the critical period was different from that after the critical period (Suzuki *et al.*, 2006).

In utero exposure to DES induces various abnormalities in the Müllerian duct of the mouse (Suzuki *et al.*, 2002). To understand the underlying molecular mechanisms associated with DES-induced abnormalities of the Müllerian duct, gene expression was examined on gestation day 19 in mouse fetuses exposed to DES from gestation days 10 to 18. Microarray analysis revealed the presence of organ-specific changes in gene expression profiles in the oviduct, uterus, and vagina following DES exposure. About 400 genes were up-regulated and 200 genes were down-regulated in the oviduct and uterus by *in utero* DES exposure. The vagina showed changes in fewer than half the number of DES-regulated genes than those found in the oviduct and uterus. DES induced expression changes in genes such as *Dkk2*, *Nkd2* and *sFRP1*, as well as *Hox*, *Wnt* and *Eph* families in the female mouse fetal reproductive tract (Suzuki *et al.*, 2007a). We also compared estrogen responsive genes in the mature uterus, vagina and mammary gland of ovariectomized adult mice 6 h after an E_2 injection using a microarray method (Suzuki *et al.*, 2007b). Half of the E_2 up-regulated genes in the uterus were similar to those in the vagina. E_2 up-regulated the expression of *insulin-like growth factor 1 (Igf-1)* genes in the uterus and vagina. In the vagina, E_2 up-regulated the expression of IGF-binding proteins (*Igfbp2* and *Igfbp5*). In the mammary gland, unlike the uterus and vagina, no gene showed altered expression 6 h after the E_2 exposure. These results suggest that expression of *Igf-1* and morphogenesis genes is regulated by E_2 in an organ-specific manner, and it is supported by the results of bromodeoxyuridine labeling showing E_2-induced mitosis in the uterus and vagina but not in the mammary gland. The differences in organ specificity in response to E_2 may be attributed by differences in gene expression regulated by E_2 in female reproductive organs.

8.3   Mode of Action of Tributyltin Chloride in Vertebrates

Organotins are a diverse group of widely distributed environmental pollutants. Tributyltin chloride (TBT) and bis(triphenyltin) oxide (TPT) have pleiotropic adverse effects on both invertebrate and vertebrate endocrine systems. Organotins were first used in the mid 1960s as antifouling agents in marine shipping paints, although such use has been strictly restricted in recent years. Organotins persist as prevalent contaminants in dietary sources, such as fish and shellfish, and through pesticide use on high-value food crops (Appel, 2004; Golub and Doherty, 2004). Additional human exposure to organotins may occur through their use as antifungal agents in wood treatments, industrial water systems and textiles. Mono- and di-organotins are widely used as stabilizers in the manufacture of polyolefin plastics (polyvinyl chloride), which introduces the potential for transfer by contact with drinking water and foods.

Exposure to organotins such as TBT and TPT results in imposex, the superimposition of male sex characteristics in (or 'on') female gastropod molluscs (Blaber, 1970; Gibbs and Bryan, 1986; Horiguchi *et al.*, 2006). Imposex results in impaired reproductive fitness or sterility in severely affected animals, and TBT exposure represents one of the most clear-cut examples of environmental endocrine disruption. TBT exposure also leads to masculinization of at least two fish species (McAllister and Kime, 2003; Shimazaki *et al.*, 2003). In contrast, TBT exposure results in slight effects on the mammalian reproductive tracts and has not been reported to alter sex ratios (Ogata *et al.*, 2001; Omura *et al.*, 2001). Hepatic-, neuro- and immuno-toxicity are reported to be the major effects of organotin exposure in mammals (Boyer, 1989). Our current understanding of how organotins disrupt the endocrine system is based on how organotins affect the expression or activity of steroid regulatory enzymes such as P450 aromatase together with less specific toxic effects resulting from damage to mitochondria and immune cells (Powers and Beavis, 1991; Gennari *et al.*, 2000; Philbert *et al.*, 2000; Heidrich *et al.*, 2001; Cooke, 2002). The currently available data do not permit one to draw firm conclusions regarding whether organotins function primarily as protein and enzyme inhibitors *in vivo*, or instead regulate gene expression in a more direct manner.

It is widely believed that high-calorie diets coupled with reduced physical activity are the major, if not the only, cause of the rising worldwide incidence of obesity. The role of genetic factors is uncertain, but it is difficult to imagine a scenario in which individual genetic variations could be responsible for the rapid worldwide increase in obesity. A more reasonable idea is that interaction with the modern environment exposes underlying genetic differences that affect obesity. The Barker hypothesis postulates that *in utero* fetal nutritional status is a potential risk factor for obesity and related diseases (reviewed in Grün *et al.* (2006)). Developmental programming would change the setpoint for individual responses to diet, exercise, and environment. One such developmental programming event could be exposure to xenobiotic chemicals which are becoming increasingly prevalent in the environment. Our environmental 'obesogen' model predicted the existence of xenobiotic chemicals that inappropriately regulate lipid metabolism and adipogenesis to promote obesity (Grün *et al.*, 2006; Grün and Blumberg, 2006).

We demonstrated that the persistent and ubiquitous environmental contaminant TBT induces the differentiation of adipocytes *in vitro* and increases adipose mass *in vivo* (Grün *et al.*, 2006). TBT is a dual, nanomolar-affinity ligand for both the retinoid X receptor (RXR) and the peroxisome proliferator activated receptor γ (PPARγ). TBT promotes adipogenesis in the murine 3T3-L1 cell model and perturbs key regulators of adipogenesis and lipogenic pathways *in vivo*. Moreover, *in utero* exposure to TBT leads to strikingly elevated lipid accumulation in adipose depots, liver, and testis of neonatal mice and results in increased epididymal adipose mass in adults. In *Xenopus laevis*, ectopic adipocytes form in and around gonadal tissues following exposure to organotin, RXRα or PPARγ ligands; thus, organotins, including TBT, are potent and efficacious agonistic ligands of the vertebrate RXRs and PPARγ. The physiological consequences of receptor activation predict that permissive RXR heterodimer target genes and downstream signaling cascades are sensitive to organotin misregulation. We confirmed that TBT up-regulates adipose-related genes in the mouse liver by microarray. Therefore, TBT and related organotin compounds are the first of a potentially new class of EDCs that target adipogenesis by modulating the activity of key regulatory transcription factors in the adipogenic pathway, RXRα and PPARγ. Our results suggest that developmental exposure to TBT and its congeners, or other environmental contaminants, with the potential to activate RXR/PPARγ, could be expected to increase the incidence of obesity in exposed individuals and that chronic lifetime exposure could act as a potential chemical stressor for obesity and obesity-related disorders.

8.4   Induction Mechanism of Imposex in Snails

Detailed information concerning the effects and mechanisms of action of industrial chemicals on invertebrates has only been obtained from a few species, even though invertebrates represent more than 95% of the known species in the animal kingdom (deFur *et al.*, 1999). The masculinizing effects of organotin compounds such as TBT on snails (termed imposex and characterized by the development of a vas deferens and a penis in females) have been found in about 150 species of molluscs (Gibbs and Bryan, 1986; Horiguchi *et al.*, 1997). The mechanism by which TBT induces imposex in marine snails is still unknown, although TBT has been shown to inhibit the activity of the enzyme aromatase (Bettin *et al.*, 1996). Experimental exposure to TBT ($1\,pg\,l^{-1}$ for 4 weeks) induced imposex in rockshells (*Thais clavigera*). However, injection of the aromatase inhibitor fadrozole, alone or in combination with testosterone (T), did not induce imposex (Iguchi *et al.*, 2007a). Aromatization of [^{3}H]-T to [^{3}H]-E$_2$ was observed in the rockshell gonadal extract *in vitro* (Iguchi *et al.*, 2007a). These results suggest that neither inhibition of aromatase nor androgen action by TBT alone, or together, is the principal cause of imposex in rockshells.

To understand further the possible molecular mechanisms underlying imposex and estrogen function in snails, we have examined the function of an ER-like gene transcript. Cells transfected with mouse ERα showed estrogen-dependent reporter gene activation, whereas cells transfected with rockshell ER-like construct exhibited ligand-independent reporter gene activation (Iguchi *et al.*, 2007a,b), suggesting that the ER-like sequence found in rockshell snails has an as yet unknown function in this species, but it is unlikely that it acts like a vertebrate ER. In the freshwater snail *Marisa cornuarietis*, bisphenol-A (BPA) and octylphenol at concentrations of $1\,\mu g\,l^{-1}$ induced the development of an additional vagina, an enlargement of the accessory pallial sex glands, and an enhancement of oocyte production, whereas they reduced the length of the penis and size of the prostate gland in the marine prosobranch *Nucella lapillus* (Oehlmann *et al.*, 2000). These results suggest that these snails have a functional ER; thus, estrogenic chemicals could have a negative impact on these snails.

As reported above, human or amphibian (*Xenopus*) RXRα was activated by TBT in reporter gene assay systems (Grün *et al.*, 2006). TBT and 9-*cis*-retinoic acid have been shown to activate the rockshell RXR and induce imposex in this species (Nishikawa *et al.*, 2004). These results suggest a future avenue of research, as the mechanisms of action of various EDCs in invertebrates need to focus on other nuclear hormone receptors or mechanisms, distinct from the actions of the vertebrate-like ER and AR.

8.5   Male Production in Daphnids by Juvenile Hormone Agonists

Reproductive, acute or chronic, toxicity tests on daphnids have been widely used by aquatic toxicologists. Conflicting results on the molting frequency of *Daphnia magna* following exposure to estrogenic chemicals have been reported (Zou and Fingerman, 1997; Caspers, 1998; Niederlehner *et al.*, 1998; Tatarazako *et al.*, 2002).

For example, styrene dimers and trimers leached from disposable polystyrene cups reduced the number of offspring in *Ceriodaphnia dubia*. Styrenes (0.04–$1.7\mu g\,l^{-1}$), ecdysones (0.1–$1.08\mu g\,l^{-1}$) and juvenile hormone (JH) agonists ($1.05\,\mu g\,l^{-1}$) reduced fertility, whereas E_2 and BPA had no effect on the reproduction of *C. dubia*.NP (280 μg l^{-1}) influenced daphnid reproduction via membrane damage (Tatarazako *et al.*, 2002). We and others have recently cloned a full-length sequence of the ultraspiracle (a homologue of RXR; Kato *et al.*, 2007, 2008; Wang *et al.*, 2007), and ecdysone receptor from *D. magna* (Kato *et al.*, 2007, 2008), and established an ecdysone reporter gene assay (Kato *et al.*, 2007, 2008). This assay will be used to screen chemicals having ecdysone-like activity. We and others have observed that exposure of adult daphnids to JHs and their analogues, induces a pattern of parthenogenetic reproduction in *D. magna* so that females produce male neonates only (Olmstead and LeBlanc, 2002, 2003; Tatarazako *et al.*, 2003). Ten juvenoids (e.g. pyriproxyfen, fenoxycarb, methylfarnesoate (MF), JH I, JH II, JH III, methoprene, kinoprene, hydroprene, and epofenonane) have been shown to induce male neonate production (Tatarazako *et al.*, 2003; Oda *et al.*, 2005b). In addition, daphnids are susceptible to the male-sex-determining effect of juvenoids during early oogenesis, and the effect of juvenoids is reversible (Tatarazako *et al.*, 2003). The production of male neonates by fenoxycarb (0.6–9.3μg l^{-1}) in all five species examined (*D. magna*, *Moina macrocopa*, *M. micrura*, *C. dubia*, and *C. reticulata*) is a common response to juvenoids (Oda *et al.*, 2005a). In addition, acute and reproductive toxicity tests including male neonate production were conducted on seven strains of *D. magna* from six laboratories in five countries using 3,4-dichloroaniline and fenoxycarb (Oda *et al.*, 2007). Fenoxycarb exposure induced the male neonate production in all the strains used. Though estimated EC_{50} values for the induction of male offspring were highly valuable among strains – sensitivity of fenoxycarb differed by a factor of approximately 23 overall (0.45–10 μg l^{-1}) – the results suggest that induction of male sex in neonates by JH analogues is universal among genetically different strains. These findings suggest that JH agonists affect the chemical signaling responsible for the induction of male offspring. In order to understand the molecular functional mechanisms of JH agonists in the induction of male offspring, we have constructed a cDNA library of *D. magna* and characterized the ESTs of over 7000 clones (Watanabe *et al.*, 2005). In addition, the Daphnia Genomic Consortium has finished the genome reading of *D. pulex* (Colbourne *et al.*, 2007). We are currently analyzing the molecular mechanism of sex determination of *D. magna* (Kato *et al.*, 2008) and JH binding protein using a newly established microarray system for *D. magna*.

8.6   Evaluation of Toxicity of Environmental Chemicals on Daphnids using Microarray

Toxic chemical contaminants have a variety of detrimental effects on various species, and the impact of pollutants on ecosystems has become an urgent issue. However, the majority of studies regarding the effects of chemical contaminants have focused on vertebrates. Among aquatic organisms, *D. magna* has been used extensively to evaluate organism- and population-level responses of invertebrates to pollutants in acute toxicity or reproductive toxicity tests. Although these types of test can provide information concerning hazardous concentrations of chemicals, they provide no information about their mode of action. Toxicogenomics/ecotoxicogenomics have at least three major goals: an understanding of the relationship between environmental exposure and adverse effects; the identification of useful biomarkers of exposure to toxic substances; and the elucidation of the

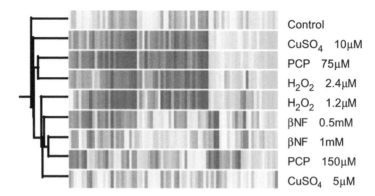

Figure 8.2   *Clustering analysis of gene expression in Daphnia magna after chemical exposure. Changes in gene expression estimated by DNA array analysis are indicated as the fold change in comparison with control samples. Chemicals and doses used for the exposure are indicated on the right. The numbers indicate the concentration of the chemicals in micrograms per liter. Red and green bars indicate induced and repressed genes respectively. Dendrogram represents correlation of the gene expression profile. PCP: pentachlorophenol; βNF: β-naphtoflavone.*

molecular mechanisms of toxicity – that is, an understanding of the biological responses and consequences of exposure. Changes in gene expression in response to chemical exposure can be detected using a DNA microarray, and multiple end points can be analyzed. We constructed an oligonucleotide-based DNA microarray and explored the acute toxicogenomic response of neonatal *D. magna* to several different types of chemical stressor: copper sulfate ($CuSO_4$), hydrogen peroxide (H_2O_2), pentachlorophenol (PCP), and β-naphthoflavone (βNF) (Watanabe *et al.*, 2007). Exposure of these chemicals resulted in characteristic patterns of gene expression that were chemical-specific (Figure 8.2), indicating that the *Daphnia* DNA microarray can be used for classification of toxic chemicals and for development of a mechanistic understanding of chemical toxicity on a common freshwater organism.

Microarray has been applied to the understanding of pathways of JH (MF; Eads *et al.*, 2007), since this hormone induced male broods in *D. magna* (Olmstead and LeBlanc, 2002). Twenty individuals were randomly selected for a 3-day exposure to either 400 nM MF or to methanol carrier. As a result, higher steady-state levels of 39 unique transcripts and lower levels of 16 transcripts have been identified. Hemoglobin and neuronal acetylcholine receptor transcripts were elevated under MF treatment, in addition to arginine kinase, amylase, cytochrome *c* oxidase and sytochrome *b*, several cuticle proteins, actin and a putative ribosomal biogenesis regulatory protein. Several transcripts were decreased in expression, such as glucosamine-6-phosphate deaminase, glucose-6-phosphatase, several proteases, an actin depolymerizing enzyme, a putative receptor of unknown function and a homologue of the Pheb G-coupled receptor (Eads *et al.*, 2007).

Heckmann *et al.* (2006) evaluated 10 candidate reference genes for real-time quantitative PCR in *D. magna* following a 24 h exposure to the nonsteroidal anti-inflammatory drug ibuprofen at 0, 29, 40 and 80 mg l^{-1}. Three of the ten candidates appeared suitable for use as reference genes, such as *glyceradldehyde-3-phosphate dehydrogenase*, *ubiquitin conjugating enzyme* and *actin*. Soetaert *et al.* (2006, 2007a) used a custom cDNA microarray consisting of in total 2455 gene fragments of *D. magna*, derived from three suppression subtractive hybridization libraries associated with energy metabolism, molting and life-stage processes. This custom cDNA array was successfully used to detect gene expression changes induced by the fungicides propioconazole (0.32, 1, 3.2, 10 and 31 mg l^{-1} for 48 h) and fenarimol (0.5, 0.75, 1 mg l^{-1} for 48 and 96 h). The gene expression data indicated a main effect on molting-specific pathways. At the highest concentration, a set of proteolytic enzymes (including different serine proteases and carboxypeptidases) was induced, whereas different cuticula proteins were down-regulated. The embryo-development-related gene vitellogenin was differentially expressed after 96 h of exposure together with a significant increase in embryo abnormalities in the offspring. Microarray analysis also revealed cadmium (from neonate) affected molecular pathways associated with processes such as digestion, oxygen transport, cuticula metabolism and embryo development, immune response, stress response, cell adhesion, visual perception, and signal transduction (Soetaert *et al.*, 2007b).

Poynton *et al.* (2007) used 16–18-day-old *D. magna* and demonstrated that distinct expression profiles in response to 24 h sublethal (1/10th of LC$_{50}$) copper (6 µg l^{-1}), cadmium (18 µg l^{-1}) and zinc (500 µg l^{-1}) exposures and found specific biomarkers of exposure, including two probable metallothioneins and a ferritin mRNA with a functional ion-response element. The gene expression supported known mechanisms of metal toxicity and revealed novel modes of action, including zinc inhibition of chitinase activity.

Classification of chemicals according to gene expression profiles or pathways may contribute to estimation of the relative risks of various chemicals. By increasing the number of genes on a DNA microarray, detailed gene expression profiles in response to chemicals can be obtained and new biomarkers and/or new pathways characteristic of environmental chemicals identified. In order to use these microarray data for risk assessment of chemicals, standardization of exposure regime, data analysis, and database establishment are critically needed.

8.7 Future Research Needs

The developing embryo is 'fragile' and sensitive to estrogenic agents (Bern, 1992). Therefore, information on gene and protein expression mediated by ERs is essential if we are to understand the effects of estrogenic chemicals. Species differences in the interaction of various chemicals with ERs and the metabolism of chemicals have been observed. Therefore, we are currently cloning steroid hormone receptors (ER, AR, PR, GR) and steroid and xenobiotic receptors (SXR) from various animal species (Milnes *et al.*, 2008), including amphibians, reptiles, birds, and fish, to examine whether large–scale species-specific differences in sensitivity to EDCs exist. We are also focusing on orphan nuclear receptors, which may provide new insights into the mechanisms of chemical action, as shown in the TBT activation of RXR in gastropods, *Xenopus*, and mice.

In order to clarify the adverse effect of chemicals, we need to understand the timing of gene expression (critical developmental window), degree of gene expression (versus exposure dose of chemicals) in specific organs, degradation of chemicals, and normal range of various biomarkers in each species. By the application of 'omic' technologies in the study of EDCs, we will begin to understand the detailed mechanisms of action of chemicals. In this review, we focused primarily on receptor-mediated gene expression; however, it is critical to broaden the spectrum of hormonal disruption beyond this approach to include hypothalamic-pituitary gland axes other than those associated with sex steroids and thyroid hormones, and the ability of animals to cope with stress or chemical communication (Propper, 2005; Guillette, 2006). An advance in our basic biological understanding of evolutionary physiology, comparative molecular endocrinology, genomics, and toxicology in animal key model species is essential in order to apply 'omic' technologies in wildlife species.

Acknowledgments

This study was supported in part by a Grant-in-Aid for Scientific Research from the Ministry of Education, Culture, Sports, Science and Technology of Japan, a Health Sciences research grant from the Ministry of Health, Labour and Welfare, Japan, and a research grant from the Ministry of Environment, Japan.

References

Appel KE (2004) Organotin compounds: toxicokinetics aspects. *Drug Metab Rev* **36**, 763–786.

Bern HA (1992) Fragile fetus. In *Chemically-Induced Alterations in Sexual and Functional Development: The Wildlife/Human Connection*, Colborn T, Clement C (eds). Princeton Scientific Publishing: Princeton, NJ.

Bettin C, Oehlmann J, Stroben E (1996) TBT-induced imposex in marine neogastropods is mediated by an increasing androgen level. *Helgoländer Meeresunters* **50**, 299–317.

Blaber SJM (1970) The occurrence of a penis-like outgrowth behind the right tentacle in spent females of *Nucella lapillus*. *Proc Malacolog Soc London* **39**, 231–233.

Boyer IJ (1989) Toxicity of dibutyltin, tributyltin and other organotin compounds to humans and to experimental animals. *Toxicology* **55**, 253–298.

Caspers N (1998) No estrogenic effects of bisphenol A in *Daphnia magna* Straus. *Bull Environ Contam Toxicol*, **61**, 143–148.

Colborn T, Clement C (eds) (1992) *Chemically-Induced Alterations in Sexual and Functional Development: The Wildlife/Human Connection*. Princeton Scientific Publishing: Princeton, NJ.

Colbourne JK, Eads BD, Shaw J, Bohuski E, Bauer DJ, Andrews J (2007) Sampling *Daphnia's* expressed genes: preservation, expansion and invention of crustacean genes with reference to insect genomes. *BMC Genomics* **8**, 217.

Cooke GM (2002) Effect of organotins on human aromatase activity *in vitro*. *Toxicol Lett* **126**, 121–130.

Daston GP, Naciff JM (2005) Gene expression changes related to growth and differentiation in the fetal and juvenile reproductive system of the female rat: evaluation of microarray results. *Reprod Toxicol* **19**, 381–394.

DeFur PL, Crane M, Ingersoll C, Tattersfield L (eds) (1999) *Endocrine Disruption in Invertebrates: Endocrinology, Testing, and Assessment*. SETAC Press: Pensacola, FL.

Eads BD, Andrews J, Colbourne JK (2007) Ecological genomics in *Daphnia*: stress responses and environmental sex determination. *Heredity* **100**, 184–190.

Fong CJ, Burgoon LD, Williams KJ, Forgacs AL, Zacharewski TR (2007) Comparative temporal and dose-dependent morphological and transcriptional uterine effects elicited by tamoxifen and ethynylestradiol in immature, ovariectomized mice. *BMC Genomics* **8**, 151.

Fox JE (2005) Non-traditional targets of endocrine disrupting chemicals: the roots of hormone signaling. *Integr Comp Biol* **45**, 179–188.

Gennari A, Viviani B, Galli CL, Marinovich M, Pieters R, Corsini E (2000) Organotins induce apoptosis by disturbance of [Ca^{2+}] and mitochondrial activity, causing oxidative stress and activation of caspases in rat thymocytes. *Toxicol Appl Pharmacol* **169**, 185–190.

Gibbs PE, Bryan GW (1986) Reproductive failure in populations of the dog-whelk, *Nucella lapillus*, caused by imposex induced by tributyltin from antifouling paints. *J Mar Biol Assoc UK* **66**, 767–777.

Golub M, Doherty J (2004) Triphenyltin as a potential human endocrine disruptor. *J Toxicol Environ Health B Crit Rev* **7**, 282–295.

Grün F, Blumberg B (2006) Environmental obesogens: organotins and endocrine disruption via nuclear receptor signaling. *Endocrinology* **147**, S50–S55.

Grün F, Watanabe H, Zamanian Z, Maeda L, Arima K, Chubacha R, Gardiner DM, Kanno J, Iguchi T, Blumberg B (2006) Endocrine disrupting organotin compounds are potent inducers of adipogenesis in vertebrates. *Mol Endocrinol* **20**, 2141–2155.

Guillette LJ Jr. 2006. Endocrine disrupting contaminants – beyond the dogma. *Environ Health Perspect* **114** (Suppl 1), 9–12.

Heckmann L-H, Connon R, Hutchinson TH, Maund SJ, Sibly RM, Callaghan A (2006) Expression of target and reference genes in *Daphnia magna* exposed to ibuprofen. *BMC Genomics* **7**, 175.

Heidrich DD, Steckelbroeck S, Klingmuller D (2001) Inhibition of human cytochrome P450 aromatase activity by butyltins. *Steroids* **66**, 763–769.

Horiguchi T, Shiraishi H, Shimizu M, Morita M (1997) Imposex in sea snails, caused by organotin (tributyltin and triphenyltin) pollution in Japan: a survey. *Appl Organomental Chem* **11**, 451–455.

Horiguchi T, Kojima M, Hamada F, Kajikawa A, Shiraishi H, Morita M, Shimizu M (2006) Impact of tributyltin and triphenyltin on ivory schell (*Babylonia japonica*) populations. *Environ Health Perspect* **114** (Suppl. 1), 13–19.

Iguchi T (1992) Cellular effects of early exposure to sex hormones and antihormones. *Int Rev Cytol* **139**, 1–57.

Iguchi T, Edery M, Tsai P-S, Ozawa S, Sato T, Bern HA (1993) Epidermal growth factor receptor levels in reproductive organs of female mice exposed neonatally to diethylstilbestrol. *Proc Soc Exp Biol Med* **204**, 110–116.

Iguchi T, Watanabe H, Katsu Y (2001) Developmental effects of estrogenic agents on mice, fish and frogs: a mini review. *Horm Behav* **40**, 248–251.

Iguchi T, Sumi M, Tanabe S (2002a) Endocrine disruptor issues in Japan. *Congen Anorm* **42**, 106–119.

Iguchi T, Watanabe H, Katsu Y, Mizutani T, Miyagawa S, Suzuki A, Sone K, Kato H (2002b) Developmental toxicity of estrogenic chemicals on rodents and other species. *Congen Anorm* **42**, 94–105.

Iguchi T, Watanabe H, Katsu Y (2006) Application of ecotoxicogenomics for studying endocrine disruption in vertebrates and invertebrates. *Environ Health Perspect* **114** (Suppl 1), 101–105.

Iguchi T, Katsu Y, Horiguchi T, Watanabe H, Blumberg B, Ohta Y (2007a) Endocrine disrupting organotin compounds are potent inducers of imposex in gastropods and adipogenesis in vertebrates. *Mol Cell Toxicol* **3**, 1–10.

Iguchi T, Watanabe H, Katsu Y (2007b) Toxicogenomics and ecotoxicogenomics for studying endocrine disruption and basic biology in vertebrates and invertebrates. *Gen Comp Endocr* **153**, 25–29.

Inoue T, Pennie WD (eds) (2002) *Toxicogenomics*. Springer-Verlag: Tokyo.

Kato Y, Kobayashi K, Oda S, Tatarazako N, Watanabe H, Iguchi T (2007) Cloning and characterization of the ecdysone receptor and ultraspiracle protein from the water flea *Daphnia magna*. *J Endocrinol* **193**, 183–194.

Kato Y, Kobayashi K, Oda S, Colbourn JK, Tatarazako N, Watanabe H, Iguchi, T. (2008). Molecular cloning and sexually dimorphic expression of DM-domain genes in *Daphnia magna*. Genomics **91**, 94–101.

Katsu Y, Iguchi T (2006) Tissue specific expression of Clec2g in mice. *Eur J Cell Biol* **85**, 345–354.

Katsu Y, Takasu E, Iguchi T (2002) Estrogen-independent expression of neuropsin, a serin protease in the vagina of mice exposed neonatally to diethylstilbestrol. *Mol Cell Endocrinol* **195**, 99–107.

Katsu Y, Lubahn D, Iguchi T (2003) Expression of novel C-type lectin in the mouse vagina. *Endocrinology* **144**, 2597–2605.

Kobayashi M, Takahashi E, Miyagawa S, Watanabe H, Iguchi T (2006) Chromatin immunoprecipitation-mediated identification of aquaporin 5 as a regulatory target of estrogen in the uterus. *Genes Cells* **11**, 1133–1143.

McAllister BG, Kime DE (2003) Early life exposure to environmental levels of the aromatase inhibitor tributyltin causes masculinization and irreversible sperm damage in zebrafish (*Danio rerio*). >*Aquat Toxicol* **65**, 309–316.

McLachlan JA (2001) Environmental signaling: what embryos and evolution teach us about endocrine disrupting chemicals. *Endocr Rev* **22**, 319–341.

Milnes MR, Garcia A, Grossman E, Grün F, Shiotsugu J, Tabb MM, Kawashima Y, Katsu Y, Watanabe H, Iguchi T, Blumberg B (2008) Activation of steroid and xenobiotic receptor (SXR, NR1I2) and its orthologs in laboratory, toxicological, and genome model species. *Eviron Health Perspect* (in press).

Miracle AL, Ankley GT (2005) Ecotoxicogenomics: linkages between exposure and effects in assessing risks of aquatic contaminants to fish. *Reprod Toxicol* **19**, 321–326.

Miyagawa S, Katsu Y, Watanabe H, Iguchi T (2004a) Estrogen-independent activation of ErbBs signaling and estrogen receptor a in the mouse vagina exposed neonatally to diethylstilbestrol. *Oncogene* **23**, 340–349.

Miyagawa S, Suzuki A, Katsu Y, Kobayashi M, Goto M, Handa H, Watanabe H, Iguchi T (2004b) Persistent gene expression in mouse vagina exposed neonatally to diethylstilbestrol. *J Mol Endocr* **32**, 663–677.

Moggs JG, Tinwell H, Spurway T, Chang H-S, Pate I, Lim FL, Moore DJ, Soames A, Stuckey R, Currie R, Zhu T, Kimber I, Ashby J, Orphanides G (2004) Phenotypic anchoring of gene expression changes during estrogen-induced uterine growth. *Environ Health Perspect* **112**, 1589–1606.

Newbold RR (2004) Lessons learned from perinatal exposure to diethylstilbestrol. *Toxicol Appl Pharmacol* **199**, 142–150.

Newbold RR, Jefferson WN, Padilla-Banks E, Haseman J (2004) Developmental exposure to diethylstilbestrol (DES) alters uterine response to estrogens in prepubescent mice: low versus high dose effects. *Reprod Toxicol* **18**, 399–406.

Niederlehner BR, Cairns J Jr, Smith EP (1998) Modeling acute and chronic toxicity of nonpolar narcotic chemicals and mixtures to *Ceriodaphnia dubia*. *Ecotoxicol Environ Saf* **39**, 136–146.

Nishikawa J, Mamiya S, Kanayama T, Nishikawa T, Shiraishi F, Horiguchi T (2004) Involvement of the retinoid X receptor in the development of imposex caused by organotins in gastropods. *Environ Sci Technol* **38**, 6271–6276.

Nuwaysir EF, Bittner M, Trent J, Barrett JC, Afshari CA (1999) Microarrays and toxicology: the advent of toxicogenomics. *Mol Carcinogenesis* **24**, 153–159.

Oda S, Tatarazako N, Watanabe H, Morita M, Iguchi T (2005a) Production of male neonates in 4 cladoceran species exposed to a juvenile hormone analog, fenoxycarb. *Chemosphere* **60**, 74–78.

Oda S, Tatarazako N, Watanabe H, Morita M, Iguchi T (2005b) Production of male neonates in *Daphnia magna* (Cladocera, Crustacea) exposed to juvenile hormones and their analogs. *Chemosphere* **61**, 1168–1174.

Oda S, Tatarazako N, Dorgerloh M, Johnson R, Kusk O, Leverett D, Marchini S, Nakari T, Williams T, Iguchi T (2007) Strain difference in sensitivity to 3,4-dichloroaniline and insect growth regulator, fenoxycarb, in *Daphnia magna*. *Ecotoxicol Environ Saf* **67**, 399–405.

Oehlmann J, Schulte-Oehlmann U, Tillmann M, Markert B (2000) Effects of endocrine disruptors on prosobranch snails (Mollusca: Gastropoda) in the laboratory. Part I: bisphenol A and octylphenol as xeno-estrogens. *Ecotoxicology* **9**, 383–397.

Ogata R, Omura M, Shimazaki Y, Kubo K, Oahima Y, Aou S, Inoue N (2001) Two-generation reproductive toxicity study of tributyltin chloride in female rats. *J Toxicol Environ Health A* **63**, 127–144.

Olmstead AW, LeBlanc GA (2002) Juvenoid hormone methyl farnesoate is a sex determinant in the crustacean *Daphnia magna*. *J Exp Zool* **293**, 736–739.

Olmstead AW, LeBlanc GA (2003) Insecticidal juvenile hormone analogs stimulate the production of male offspring in the crustacean *Daphnia magna*. *Environ Health Perspect* **111**, 919–924.

Omura M, Ogata R, Kubo K, Shimazaki Y, Aou S, Oshima Y, Tanaka A, Hirata M, Makita Y, Inoue N (2001) Two-generation reproductive toxicity study of tributyltin chloride in male rats. *Toxicol Sci* **64**, 224–232.

Philbert MA, Billingsley ML, Reuhl KR (2000) Mechanisms of injury in the central nervous system. *Toxicol Pathol* **28**, 43–53.

Powers MF, Beavis AD (1991) Triorganotins inhibit the mitochondrial inner membrane anion channel. *J Biol Chem* **266**, 17250–17256.

Poynton HC, Varshavsky JR, Chang B, Cavigiolio G, Chan S, Holman PS, Loguinov AV, Bauer DJ, Komachi K, Theil EC, Perkins EJ, Hughes O, Vupe CD (2007) *Daphnia magna* ecotoxicogenomics provides mechanistic insights into metal toxicity. *Environ Sci Technol* **41**, 1044–1050.

Propper CR (2005) The study of endocrine-disrupting compounds: past approaches and new directions. *Integr Comp Biol* **45**, 194–200.

Shimazaki Y, Kitano T, Oshima Y, Inoue S, Imada N, Honjo T (2003) Tributyltin causes masculization in fish. *Environ Toxicol Chem* **22**, 141–144.

Snape JR, Maund SJ, Pickford DB, Hutchinson TH (2004) Ecotoxicogenomics: the challenge of integrating genomics into aquatic and terrestrial ecotoxicology. *Aquat Toxicol* **67**, 143–154.

Soetaert A, Moens LN, van der Ven K, Van Leemput K, Naudts B, Blust R, De Coen WM (2006) Molecular impact of propiconazole on *Daphnia magna* using a reproduction-related cDNA array. *Comp Biochem Physiol Part C* **142**, 66–76.

Soetaert A, van der Ven K, Moens LN, Vandenbrouck T, van Remortel P, De Coen WM (2007a) *Daphnia magna* and ecotoxicogenomics: gene expression profiles of anti-ecdysteroidal fungicide fenarimol using energy-, molting- and life stage-related cDNA libraries. *Chemosphere* **67**, 60–71.

Soetaert A, Vandenbrouck T, van der Ven K, Maras M, van Remortel P, Blust R, De Coen WM (2007b) Molecular responses during cadmium-induced stress in *Daphnia magna*: integration of differential gene expression with higher-level effects. *Aquat Toxicol* **83**, 212–222.

Suzuki A, Sugihara A, Uchida K, Sato T, Ohta Y, Katsu Y, Watanabe H, Iguchi T (2002) Developmental effects of perinatal exposure to bisphenol A and diethylstilbestrol on reproductive organs in female mice. *Reprod Toxicol* **16**, 107–116.

Suzuki A, Watanabe H, Mizutani T, Sato T, Ohta Y, Iguchi T (2006) Global gene expression in mouse vaginae exposed to diethylstilbestrol at different ages. *Exp Biol Med* **231**, 632–640.

Suzuki A, Utushitani H, Sato T, Kobayashi T, Watanabe H, Ohta Y, Iguchi T (2007a) Gene expression change in the Müllerian duct of the mouse fetus exposed to diethylstilbestrol *in utero*. *Exp Biol Med* **232**, 503–514.

Suzuki A, Urushitani H, Watanabe H, Sato T, Iguchi T, Kobayashi T, Ohta Y (2007b) Comparison of estrogen responsive genes in the mouse uterus, vagina and mammary gland. *J Vet Med Sci* **69**, 725–731.

Tarrant AM (2005) Endocrine-like signaling in Cnidarians: current understanding and implications for ecophysiology. *Integr Comp Biol* **45**, 201–214.

Tatarazako N, Tkao Y, Kishi K, Onikura N, Arizono K, Iguchi T (2002) Styrene dimmers and trimers affect reproduction of daphnid (*Ceriodaphnia dubia*). *Chemosphere* **48**, 597–601.

Tatarazako N, Oda S, Watanabe H, Morita M, Iguchi T (2003) Juvenile hormone agonists affect the occurrence of male *Daphnia*. *Chemosphere* **53**, 827–833.

Wang YH, Wang G, LeBlanc GA. (2007) Cloning and characterization of the retinoid X receptor from a primitive crustacean *Daphnia magna*. *Gen Comp Endocrinol* **150**, 309–318.

Watanabe H, Iguchi T (2003) Evaluation of endocrine disruptors based on gene expression using a micorarray. *Environ Sci* **10**(Suppl), 61–67.

Watanabe H, Iguchi T (2006) Using ecotoxicogenomics to evaluate the impact of chemicals on aquatic organisms. *Mar Biol* **149**, 107–115.

Watanabe H, Suzuki A, Mizutani T, Handa H, Iguchi T (2002a) Large-scale gene expression analysis for evaluation of endocrine disruptors. In *Toxicogenomics*, Inoue T, Pennie WD (eds). Springer-Verlag: New York; 149–155.

Watanabe H, Suzuki A, Mizutani T, Kohno S, Lubahn DB, Handa H, Iguchi T (2002b) Genome-wide analysis of changes in early gene expression induced by estrogen. *Genes Cells* **7**, 497–507.

Watanabe H, Suzuki A, Kobayashi M, Lubahn D, Handa H, Iguchi T (2003a) Analysis of temporal changes in the expression of estrogen regulated genes in the uterus. *J Mol Endocr* **30**, 347–358.

Watanabe H, Suzuki A, Kobayashi M, Lubahn DB, Handa H, Iguchi T (2003b) Similarities and differences in uterine gene expression patterns caused by treatment with physiological and non-physiological estrogen. *J Mol Endocr* **31**, 487–497.

Watanabe H, Suzuki A, Goto M, Lubahn DB, Handa H, Iguchi T (2004a) Tissue-specific estrogenic and non-estrogenic effects of a xenoestrogen, nonylphenol. *J Mol Endocr* **33**, 243–252.

Watanabe H, Suzuki A, Goto M, Ohsako S, Tohyama C, Handa H, Iguchi T (2004b) Comparative uterine gene expression analysis after dioxin and estradiol administration. *J Mol Endocr* **33**, 763–771.

Watanabe H, Tatarazako N, Oda S, Nishide, H, Uchiyama I, Morita M, Iguchi T (2005) Analysis of expressed sequence tags of the waterflea *Daphnia magna*. *Genome* **48**, 606–609.

Watanabe H, Takahashi E, Kobayashi M, Goto M, Krust A, Chambon P, Iguchi T (2006) The estrogen-responsive adrenomedullin and receptor-modifying protein 3 gene identified by DNA microarray analysis are directly regulated by estrogen receptor. *J Mol Endocr* **36**, 81–89.

Watanabe H, Takahashi E, Nakamura Y, Oda S, Tatarazako N, Iguchi T (2007) Development of a *Daphnia magna* DNA microarray for evaluating the toxicity of environmental chemicals. *Environ Tox Chem* **26**, 669–676.

Yin Y. Ma L. (2005) Development of the mammalian female reproductive tract. *J. Biochem.*, **137**, 677–683.

Zou E, Fingerman M (1997) Synthetic estrogenic agents do not interfere with sex differentiation but do inhibit molting of the Cladoceran *Daphnia magna*. *Bull Environ Contam Toxicol* **58**, 596–602.

9

Gene Expression Profiling of Transplacental Arsenic Carcinogenesis in Mice

Jie Liu, Bhalchandra A. Diwan, Raymond Tennant and Michael P. Waalkes

9.1 Introduction

Inorganic arsenic is a naturally occurring environmental contaminant often found at excessive levels in drinking water (NRC, 2001). Exposure to inorganic arsenic can also occur from burning of coal containing high levels of inorganic arsenic (Liu *et al*., 2002). Epidemiologic studies have demonstrated that chronic arsenic exposure causes cancer of the skin, lung, liver, urinary bladder, prostate, and possibly other sites (NRC, 2001; IARC, 2004; NTP, 2004). However, the evidence of inorganic arsenic as an animal carcinogen has been considered limited (NRC, 2001; IARC, 2004). Indeed, various studies had yielded negative results for inorganic arsenic as a carcinogen in adult mice, rats, hamsters, rabbits, dogs, and monkeys (NRC, 2001; IARC, 2004), including our own prior studies in adult mice via repeated arsenic injections (Waalkes *et al*., 2000), or with the drinking water containing inorganic arsenicals (Liu *et al*., 2000). Skin cancer models have been developed in the last decade where inorganic arsenic is active in production of tumors only when combined with the skin tumor promoter 12-*O*-tetradecanoyl phorbol-13-acetate (TPA; Germolec *et al*., 1998), or with ultraviolet irradiation (Rossman *et al*., 2001). However, inorganic arsenic alone showed no carcinogenic potential in these models.

Thus, to establish relevant arsenic carcinogenesis animal models was a key research priority (Waalkes *et al*., 2007). In this regard, gestation is a period of high sensitivity to chemical carcinogenesis, including inorganics such as cisplatin (Diwan *et al*., 1993), lead (Waalkes *et al*., 1995) and nickel (Diwan *et al*., 1992), due to factors like

Toxicogenomics: A Powerful Tool for Toxicity Assessment Edited by S. C. Sahu
© 2008 John Wiley & Sons, Ltd

organogenesis-related cell differentiation, rapid and global proliferative tissue growth, and genetic programming and imprinting (Anderson *et al.*, 2000; Birnbaum and Fenton, 2003). Inorganic arsenic can readily cross the placenta and enter the fetus in humans (Concha *et al.*, 1998, NRC, 2001; IARC, 2004) and in rodents (Devesa *et al.*, 2006). Because the available data indicated adult rodents were insensitive to inorganic arsenic carcinogenesis, we have performed a series of transplacental arsenic carcinogenesis studies in mice (Waalkes *et al.*, 2007). Initially, a mouse strain of known generalized sensitivity to chemical carcinogenesis was used (C3H; Waalkes *et al.*, 2003), resulting in remarkable level of tumor formation in multiple tissues that include potential human targets of arsenic carcinogenesis (Waalkes *et al.*, 2003). This led us to perform additional studies using prenatal arsenic exposure combined with postnatal exposure to agents that can enhance tumor formation, including topical applications of the tumor promoter TPA in C3H mice (Waalkes *et al.*, 2004a), and subcutaneous injections of diethylstilbestrol (DES) or tamoxifen (TAM) in newborn CD1 mice (Waalkes *et al.*, 2006a,b).

Gene expression profiling was performed using tissue samples collected at various stages during these transplacental arsenic carcinogenesis studies. Like all the genomic studies, a lot of gene alterations were observed. Three major strategies were taken to digest the vast data information:

1. To confirm key gene expressions by various means including real-time reverse transcriptase polymerase chain reaction (RT-PCR), western-blot, immunohistochemistry and other toxicology-related endpoints.
2. To confirm key gene expression changes by biological duplication. In this regard, we have performed four transplacental arsenic carcinogenesis studies with largely reproducible results, and three of these studies are discussed in this chapter. We have also compared the gene expressions in various tissues and found some of gene expressions are tissue specific. Liver will be the focus of discussion in this chapter.
3. To make a connection or to make a story out of these gene expression data. In this regard, genomic profiling revealed multiple molecular events associated with transplacental arsenic carcinogenesis, but only the potential involvement of aberrant estrogen signaling in arsenic hepatocarcinogenesis will be discussed in this chapter.

9.2 Materials and Methods

9.2.1 Animals and Treatments

Mice were housed in the Association for Assessment and Accreditation of Laboratory Animal Care (AAALAC)-accredited animal facility at NCI-Frederick and the Institutional Animal Care and Use Committee approved the study protocol. C3H/HeNCr (C3H) were obtained from the Animal Production Area, NCI-Frederick, and CD1 mice were obtained from the Charles River Laboratory (Rowley, NC). The primigravid females were randomly divided into groups of 10 or more each and given drinking water containing sodium arsenite (NaAsO$_2$) at 0 (control), 42.5 or 85 ppm arsenite *ad libitum* from days 8 to 18 of gestation. Dams were allowed to give birth, and litters were culled to no more than eight at 4 days *post partum*. Offspring were weaned at 4 weeks, randomly put into separate groups ($n = 25$ to 35) of males and females according to maternal exposure level and subsequent postnatal treatments, and observed for at least 90 weeks. In all this work, neither maternal drinking water consumption nor body weight of the pregnant mice was altered by these levels of arsenic in the drinking water. Postnatal treatments included the subcutaneous injections of DES (2 µg/pup/day), TAM (10 µg/pup/day) or vehicle (corn oil) on postpartum days 1–5 in CD1 mice (Waalkes *et al.*, 2006a,b), or the topical application of TPA (2 µg/0.1 ml acetone) twice per week to a shaved area of dorsal skin after weaning for 21 weeks in C3H mice (Waalkes *et al.*, 2004a).

9.2.2   Pathology and Sample Collection

A complete necropsy was performed on all moribund animals, animals found dead, or on mice at terminal sacrifice. In most cases, the portion of the liver, urinary bladder, kidneys, lung, adrenal, gonads (ovaries or testes), skin and grossly abnormal tissues were fixed in 10 % neutral buffered formalin for histopathological analysis; other aliquots of tissue samples were also snap frozen in liquid nitrogen and stored at $-70\,^{\circ}\mathrm{C}$ prior to gene expression analysis.

9.2.3   Microarray Analysis

Total RNA was isolated from tissue samples with TRIzol reagent (Invitrogen, Carlsbad, CA), followed by purification with RNeasy mini kit (Qiagen, Valencia, CA). The quality of RNA was confirmed by an Agilent 2100 Bioanalyzer (Agilent Technologies, Palo Alto, CA). The 22K mouse oligo chip arrays (Liu *et al.*, 2006a, 2007) or custom-designed filter arrays (600 genes) (Liu *et al.*, 2004, 2006b; Xie *et al.*, 2007) were used for microarray analysis. For the oligo chip array, total RNA was amplified using the Agilent Low RNA Input Fluorescent Linear Amplification Kit protocol. Starting with 500 ng of total RNA, Cy3- or Cy5-labeled cRNA was produced according to the manufacturer's protocol. For each two-color comparison, 750 ng of each Cy3- and Cy5-labeled cRNAs was mixed and fragmented using the Agilent *In Situ* Hybridization Kit protocol. Hybridizations were performed overnight in a rotating hybridization oven. Slides were washed and scanned with an Agilent Scanner. Data were obtained using the Agilent Feature Extraction software (v7.5). Two hybridizations with fluor reversals were performed. For the filter array, approximately 5 µg of total RNA was converted to $[\alpha\text{-}^{32}\mathrm{P}]$-dATP-labeled cDNA probe using MMLV reverse transcriptase and the Atlas customer-array-specific cDNA synthesis primer mix, followed by purification with a NucleoSpin column (Clontech, Palo Alto, CA). The membranes were prehybridized with Expresshyb from Clontech for 2 h at 68 °C, followed by hybridization with the cDNA probe overnight. The membranes were then washed and exposed to a Molecular Dynamics Phosphoimage Screen. The images were analyzed densitometrically using Atlas Image software (version 2.01). The gene expression intensities were first corrected with the external background and then globally normalized.

9.2.4   Real-time RT-PCR Analysis

The expression levels of the selected genes were quantified using real-time RT-PCR analysis. The forward and reverse primers for selected genes were designed using ABI Primer Express software (Applied Biosystems, Foster City, CA). Purified total RNA was reverse transcribed with MuLV reverse transcriptase and oligo-dT primers and subjected to real-time PCR analysis using SYBR green master mix (Applied Biosystems, Cheshire, UK).

9.2.5   Western Blot Analysis

Western blot analysis of selected proteins was performed as described (Liu *et al.*, 2006a,b; Shen *et al.*, 2007). Briefly, tissues were homogenized in PER-Tissue Protein Extraction buffer (Pierce, Rockford, IL) containing protease inhibitors. Cytosolic protein (30 µg) was separated on NuPAGE Bis–Tris gels (Invitrogen, San Diego, CA), and transferred to nitrocellulose membranes. Membranes were blocked in 5 % dried milk in TBST, followed by incubation with the primary antibody. After washes, the membranes were incubated in HRP-conjugated secondary antibody and washed. Immunoblots were visualized using SuperSignal chemiluminescent substrate (Pierce, Rockford, IL).

9.2.6   Immunohistochemical Analysis

Tissues were studied immunohistochemically for localization and intensity of estrogen receptor-α (ER-α) The sections were treated and visualized as described (Waalkes *et al.*, 2004b, 2006a,b; Shen *et al.*, 2007), using a polyclonal rabbit anti-ER-α antibody (1:1000 dilution, Santa Cruz Biotechnology, Inc., Santa Cruz, CA) and Vector Elite kits (Vector Labs, Burlingame, CA). To define specificity, the primary antibodies were omitted from each staining series as a control.

9.2.7   Statistical Analysis

Images and gene expression markup language (GEML) files, including error and *p*-values, were exported from the Agilent Feature Extraction software and deposited into the Rosetta Revolver system (version 4.0, Rosetta Biosoftware, Kirkland, WA). The resultant ratio profiles were combined and genes were considered 'signature genes' if $p < 0.001$. The signature genes were also analyzed with Ingenuity Pathways software. For real-time RT-PCR analysis, the cycle time (Ct) values of genes of interest were first normalized with β-actin from the same sample, and then the relative differences between control and treatment groups were calculated and expressed as percentage of controls. Data are given as mean plus/minus standard error of the mean (SEM), and the significant levels were set at $p \leq 0.05$.

9.3   Results and Discussion

9.3.1   Transplacental Arsenic Carcinogenesis in Mice

Liver is a likely target organ of arsenic carcinogenesis in humans (Chen *et al.*, 1997; Morales *et al.*, 2000; Lu *et al.*, 2001; Zhou *et al.*, 2002; Centeno *et al.*, 2002; IARC, 2004). In transplacental arsenic carcinogenesis studies, a dose-dependent increase in hepatocellular carcinoma incidence was observed in male C3H (Waalkes *et al.*, 2003, 2004a) and in male CD1 (Waalkes *et al.*, 2006b). Thus, the gene expression analysis of liver samples from transplacental arsenic carcinogenesis studies is the research priority (see below).

Urogenital system cancers (such as urinary bladder tumors) are seen in humans exposed to inorganic arsenic (NRC, 2001; IARC, 2004). In transplacental arsenic carcinogenesis studies, increased urinary bladder proliferative lesions, including tumors were observed in CD1 mice (Waalkes *et al.*, 2006a,b), and increased ovary and adrenal tumors were seen in both C3H (Waalkes *et al.*, 2003,b) and CD1 mice (Waalkes *et al.*, 2006a,b). Expression analyses of urogenital tissues are very important and another research priority (Waalkes *et al.*, 2007).

Lung is also a target organ of arsenic carcinogenesis in humans (NRC, 2001; IARC, 2004). Transplacental arsenic exposure increased pulmonary tumor incidence in female mice (Waalkes *et al.*, 2003), and arsenic-induced lung tumors were promoted by TPA in both sexes (Waalkes *et al.*, 2004a). The lung tumor tissues were analyzed by immunohistochemistry and the fetal lung tissue samples were analyzed for gene expression (Shen *et al.*, 2007).

9.3.2 Initial Gene Expression Profiling in Liver Tumors of Adult Offspring Exposed to Arsenic *In Utero*

The gene expression of arsenic-induced hepatocellular carcinoma was initially analyzed using custom-made Clontech filter arrays (Liu *et al.*, 2004), with liver tumors and nontumorous surrounding liver tissue samples taken at necropsy from adult male mice exposed *in utero* to arsenic via the maternal drinking water (Waalkes *et al.*, 2003). Total RNA was purified and subjected to microarray analysis. Among 600 genes, arsenic-induced hepatocellular carcinoma showed a significantly higher rate of aberrant gene expression (more than twofold and $p < 0.05$, 14 %) than spontaneous tumors (7.8 %). In nontumorous liver samples of arsenic-exposed animals, 60 genes (10 %) were differentially expressed. To confirm and to extend these findings, liver samples from the second arsenic transplacental carcinogenesis study (Waalkes *et al.*, 2004a) were used for genome-wide microarray analysis using 22K mouse oligo chips. Arsenic exposure *in utero* resulted in significant alterations in the expression of 2010 genes (9 %) in arsenic-exposed liver samples and in the expression of 2540 genes (11.8 %) in arsenic-induced hepatocellular carcinoma in male mice. Ingenuity Pathway Analysis revealed that significant alterations in gene expression occurred in a number of biological networks; for instance, Myc plays a critical role in one of the primary networks (Liu *et al.*, 2006a). Real-time RT-PCR and western blot analysis of selected genes/proteins showed >90 % concordance. Table 9.1 exemplifies some of the consistent changes in these two gene expression analyses.

Consistent overexpression of α-fetoprotein, c-myc, c-met, cytokeratin-8 and -18 and plasminogen activator inhibitor-1 (PAI-1), all biomarkers of hepatocllular carcinoma, were evident after arsenic exposure in these two gene expression studies. Genes involved with proliferation, such as cyclin D1, cdk4, cdkn2b and PCNA, were consistently increased. The expression of IGF-I was decreased, while the expression of IGF-II and IGFBP1 was increased. For stress-related genes, glutathione *S*-transferases and EGR-1 were consistently increased. Heme oxygenase-1 was slightly increased in the first study, but was not evident in the second study, and down-regulation of metallothionein-1 was also noted (Liu *et al.*, 2006a). Betaine-homocystein methyltransferase (BHMT) was consistently decreased by arsenic in these two studies. Interestingly, a metabolic feminization was consistently observed in these two studies, with increases in female dominant CYP2A4 and CYP2B9, and decreases in male-dominant CYP2F2 and CYP7B1 in male mouse livers. Using the

Table 9.1 *Example of consistent changes in gene expression in two transplacental arsenic carcinogenesis studies*[a]

Gene name	Effects	Expression ratio: arsenic/control	
		1st study	2nd study
Oncogenes and liver tumor-related genes			
α-Fetoprotein	up	18.9	19.1
c-myc	up	1.5	1.8
c-met	up	2.0	3.4
Cytokeratin-8	up	2.9	2.9
Cytokeratin-18	up	4.9	11.1
Plasminogen activator inhibitor 1	up	12.0	9.2
Cell proliferation and growth factors			
Cyclin D1	up	3.8	5.1
Cdk4	up	3.9	4.4
Cdkn2b	up	3.7	6.3
IGF-1	down	0.3	0.6
IGFBP1	up	22.2	9.1
IGF-II	up	3.5	4.4
PCNA	up	1.8	2.6
Stress-related genes			
Early growth response protein	up	2.1	3.1
GST-theta	up	2.7	3.1
GST-mu	up	3.1	4.8
Heme oxygenase-1	up	2.0	0.9
Metabolic enzyme genes			
CYP2A4	up	18.4	25.3
CYP2B9	up	5.4	2.6
CYP2F2	down	0.2	0.5
CYP7B1	down	0.5	0.5
BHMT	down	0.3	0.3

[a] Data are expression ratios of arsenic to control (folds) with $p < 0.05$. The gene expression was based on real-time RT-PCR analysis in arsenic-induced liver tumors from the first study (Waalkes et al., 2003; Liu et al., 2004) and the second study (Waalkes et al., 2004a; Liu et al., 2006a).

22K mouse oligo chips, more aberrantly expressed genes were detected after arsenic exposure and their significances remain to be determined. This includes genes involved in cell–cell communication, signal transduction, and lipid metabolism (Liu et al., 2006a). Where should we go from this initial gene profiling?

9.3.3 Aberrant Estrogen Signaling as a Mechanism of Transplacental Arsenic Carcinogenesis

The spectrum of transplacental arsenic-induced tumors, such as liver, ovary, adrenal and uterus, resembles the targets of broad range or tissue-selective carcinogenic estrogens (Birnbaum and Fenton, 2003; Newbold, 2004). Indeed, aberrant expression of ER-α is associated with a variety of tumors in rodents and in humans (Fishman et al., 1995; Dickson and Stancel, 2000). Based on the observed arsenic-induced liver spectrum, a hypothesis that aberrant estrogen signaling is involved in arsenic hepatocarcinogenesis was proposed (Waalkes *et al.*, 2004b).

9.3.3.1 ER-α and ER-α-related Gene Expression

To test this hypothesis, efforts were initially directed to examine the ER-α and ER-α-related gene expression in adult mouse tissues. Consistent with microarray data, ER-α and ER-α-related gene expression were increased in adult livers of mice exposed to arsenic *in utero*. These findings are also consistent with our prior observation of overexpression of ER-α and cyclin D1 (an estrogen-responsive oncogene) in adult mouse livers (Chen *et al.*, 2004). The hepatic ER-α overexpression is associated with hypomethylation of the promoter region of the ER-α gene in livers of adult offspring exposed to arsenic *in utero* (Waalkes *et al.*, 2004b), or in chronic arsenic-exposed adult mouse livers (Chen *et al.*, 2004). Most interestingly, the liver metabolic feminization (increases in female dominant CYP2A4 and CYP2B9 and decreases in male-dominant CYP2F2 and CYP7B1) was associated with the increased expression of estrogen-linked/regulated genes, such as HSD17β7, Akr1c18, TFF3, CTGF and nidogen 1. Furthermore, immunostaining of ER-α and ER-α-linked cyclin D1 was widespread, and most intense in hepatic nucleus, indicative of the active form (Waalkes *et al.*, 2004b; Chen *et al.*, 2004). These data suggest that the disruption of estrogen signaling could be a potential mechanism in arsenic hepatocarcinogenesis.

9.3.3.2 Gene Expression Analysis in Female Mouse Livers following TPA Promotion

When *in utero* arsenic exposure was combined with postnatal treatments with TPA, liver tumor incidence in female mice was also increased over control or TPA alone (Waalkes *et al.*, 2004a). To help define any gender-related gene expression differences associated with arsenic hepatocarcinogenesis in mice, gene expression analysis was performed in liver from male and female mice exposed to arsenic *in utero* followed by postnatal TPA treatments (Liu *et al.*, 2006b). Some examples of expression differences are given in Table 9.2. The activation of oncogenes related to liver tumorigenesis was evident in arsenic-induced tumors, regardless of the gender. The basal expression of cell-proliferation genes, such as cyclin D1 and IGFBP1, was much higher in female control liver than in males. Arsenic exposure dramatically increased cyclin D1 and IGFBP1 expression in male livers and further increased their expression in females. For genes encoding metabolic enzymes, the basal expression of CYP2A4 is 54-fold higher in female liver, while arsenic increased male mouse expression towards female levels, and further increased female expression. On the other hand, the male-predominant CYP2F2 and CYP7B1 are lower in females, and arsenic exposure decreased their expression in males; and a further decrease was also observed in females. The gene expression in female mouse liver data further support the concept of aberrant estrogen signaling as a contributing factor to arsenic hepatocarcinogenesis. However, all these data are from adult animals: what would happen in the early life stages?

9.3.4 Gene Expression Profiling at the Early-life Stages following *In Utero* Arsenic Exposure

To define early expression changes possibly associated with transplacental arsenic carcinogenesis, global genomic analysis with 22K mouse oligo chips was performed using mouse liver tissues right after *in utero* arsenic exposure (fetal tissue), at birth, at

Table 9.2   *Real-time RT-PCR analysis of genes in male and female mice exposed to arsenic in utero plus postnatal TPA[a]*

Gene name	Basal level Female/male	As-tumor/control Male	As-tumor/control Female
Oncogenes and liver tumor related genes			
α-Fetoprotein	1.0	240	350
c-myc	1.5	1.8	2.3
Cytokeratin-8	1.3	3.2	3.2
Cytokeratin-18	1.2	3.7	3.6
Plasminogen activator inhibitor 1	3.2	15.9	14.2
Cell proliferation and growth factors			
Cyclin D1	11.4	20	3.5
Cdkn2b	2.9	29.8	29.1
IGF-1	1.5	0.5	0.2
IGFBP1	21.0	26.2	3.6
IGFBP3	2.5	4.0	5.3
Stress-related genes			
Early growth response protein	1.5	7.2	4.0
GST-theta	1.0	3.3	2.9
GST-alpha	1.3	4.6	2.1
Heme oxygenase-1	1.5	1.3	0.5
Metabolic enzyme genes			
CYP2A4	54.0	60.5	1.4
CYP2B9	50.0	8.1	0.2
CYP2F2	0.2	0.4	0.2
CYP7B1	0.1	0.5	0.7
BHMT	1.2	0.2	0.2

[a] Data are expressed as ratios (folds) with $p < 0.05$. All animals received TPA exposure with or without arsenic exposure *in utero*. The data were modified from the second transplacental arsenic carcinogenesis study (Waalkes *et al.*, 2004a; Liu et al., 2006b).

weaning (21 days old), at puberty (8 weeks old), and at adult (100 weeks old). However, no consistent gene expression pattern was evident (data not shown), and these vast data are currently being analyzed. In this chapter, only the gene expression data from fetal and neonatal tissues will be discussed.

9.3.4.1   Arsenic Speciation in Fetal Tissues

Inorganic arsenic exposure crosses the placental barrier and reaches the fetal tissues at levels similar to that seen in the dam (Devesa *et al.*, 2006). All the various forms of arsenic are seen in the mouse fetus, such as inorganic arsenicals and mono- and di-methylated forms. In the mouse fetal blood, levels of inorganic arsenicals ($12.1\,\mu g\,l^{-1}$) and monomethylated arsenicals (MMAs, $20.2\,\mu g\,l^{-1}$) after 42.5 ppm arsenic exposure are very similar to plasma levels (inorganic arsenic $8.2\,\mu g\,l^{-1}$; MMA $20.7\,\mu g\,l^{-1}$) in a human population from Inner Mongolia chronically exposed to drinking water containing 0.41 ppm arsenic (Pi *et al.*, 2002). In the fetal liver, the concentrations of total arsenicals were $293\,ng\,g^{-1}$ and $504\,ng\,g^{-1}$ after 42.5 ppm and 85 ppm arsenic exposure respectively (Devesa *et al.*, 2006). Human liver samples from drinking 0.22–2 ppm arsenic-contaminated water contained 500 to 6000 ng arsenic per gram dry weight (Mazumder, 2005), which corresponds to approximately 100–1200 ng per gram wet weight (Paul *et al.*, 2007). Thus, although the arsenic concentrations in the drinking water used in transplacental exposure studies are about 100 times greater than typical human exposure, no overt toxicity occurred to maternal and fetus, and the blood and tissue arsenic concentrations in fetal tissues are comparable to those of human tissues, indicating that arsenic clearance in rodent is faster than that in humans (Paul *et al.*, 2007; Waalkes *et al.*, 2007). Indeed, 3 days after termination of *in utero* arsenic exposure, the arsenic concentration in newborn mouse liver (Xie *et al.*, 2007) was only about 15 % of that occurring in fetal livers at gestation day 18 (Devesa *et al.*, 2006). Clearly, *in utero* arsenic exposure resulted in significant amount of arsenic in the fetal tissues, which could impact gene expression at the early life-stages.

9.3.4.2   Gene Expression in Fetal and Newborn Livers

To help define the early molecular events associated with arsenic hepatocarcinogenesis, pregnant C3H mice were given drinking water containing 0 (control) or 85 ppm arsenic from days 8 to 18 of gestation. RNA was isolated from fetal livers for microarray analysis using 22K oligo chips. Arsenic exposure *in utero* produced significant alterations in expression of 190 genes, with approximately 25 % of them related to estrogen signaling and steroid metabolism. These included X-inactive-specific transcript, anterior gradient-2, trefoil factor-1, CRP-ductin, ghrelin, small proline-rich protein-2A, cytokeratin 1–19 and CYP2A4. The increased expression of 17β-hydroxysteroid dehydrogenase-7 (HSD17β7; involved in estradiol production) and decreased expression of HSD17β5 (involved in testosterone production) indicated the disruption of steroid metabolism in the fetus. Thus, exposure of mouse fetus to inorganic arsenic during a critical period in development significantly alters the expression of various genes involved with estrogen signaling and steroid metabolism (Liu *et al.*, 2007).

In newborn C3H mouse livers, global DNA methylation and aberrant expression of genes relevant to the carcinogenic process were also examined. A significant reduction in methylation occurred globally in GC-rich regions of DNA after arsenic. Microarray and real-time RT-PCR analysis showed that arsenic exposure enhanced expression of genes encoding for glutathione production and genes related to insulin growth factor signaling and alterations in expression of metabolic enzyme genes (Xie *et al.*, 2007).

9.3.4.3   Gene Expression in Fetal Lung

Fetal lung was examined following *in utero* arsenic exposure for gene expression and compared with arsenic-induced lung tumors (Shen *et al.*, 2007). Increased expression of ER-α transcript and protein levels were observed in female fetal lung. An overexpression of various estrogen-regulated genes also occurred, including trefoil factor-3, anterior gradient-2, and the steroid metabolism gene HSD17β5 and genes in the insulin growth factor system. Similar to that seen in fetal liver, *in utero* arsenic exposure also induced overexpression of α-fetoprotein and several other genes potentially associated with oncogenesis in fetal lung. Most importantly, lung adenoma and adenocarcinoma from adult female mice exposed to arsenic *in utero* showed widespread, intense nuclear ER-α staining, consistent with positive ER-α staining in arsenic-induced tumors in adult liver. In contrast, diethylnitrosamine-induced lung adenocarcinoma showed little evidence of ER-α expression (Shen *et al.*, 2007). Thus, transplacental arsenic exposure at carcinogenic doses also produced aberrant estrogen-linked gene expression in the lung.

These alterations at the very early life-stages could disrupt genetic programming in early life, which could impact tumor formation much later in adulthood (Cook *et al.*, 2005). If arsenic really acts through aberrant estrogen signaling, then it would act synergistically with carcinogenic estrogens to increase tumor formation. To test this hypothesis, the third transplacental study was conducted (see below).

9.3.5   Synergistic Effects of *In Utero* Arsenic with Postnatal DES in Tumor Induction

Perhaps the most convincing evidence for aberrant estrogen signaling comes from the effects of *in utero* (from gestation days 8 to 18) arsenic exposure plus postnatal DES (a synthetic estrogen from postnatal days 1 to 5) in CD1 mice. Urogenital tumor incidence, including urinary bladder and adrenal tumors in both sexes, as well as ovary and uterus tumors in females, were dramatically increased by arsenic plus DES (Figure 9.1), an effect much more dramatic than theoretical additive (Waalkes *et al.*, 2006a,b). Liver tumor incidence, including hepatocellular adenoma and hepatocellular carcinoma, was also synergistically increased by arsenic plus DES in male mice (Waalkes *et al.*, 2006b). It should also be noted that no urinary transitional cell carcinoma in control CD1 mice was reported in many 2-year bioassays; thus, the increased incidence of transitional cell carcinoma in female offspring exposed to arsenic *in utero* followed by DES is quite remarkable (Waalkes *et al.*, 2007).

To further confirm aberrant estrogen signaling in arsenic carcinogenesis, the expression of ER-α and ER-linked pS2 were examined in adult offspring. Immunohistochemistry clearly demonstrated the intense and widespread immunostaining of ER-α and ER-linked

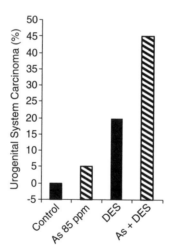

Figure 9.1   *Synergistic effect of in utero arsenic (85 ppm from gestation days 8 to 18) exposure plus postnatal treatment of DES (2 µg/pup/day from postpartum days 1 to 5) on urogenital system carcinoma (urinary bladder, ovary, adrenal) formation. Modified from Waalkes et al. (2006a).*

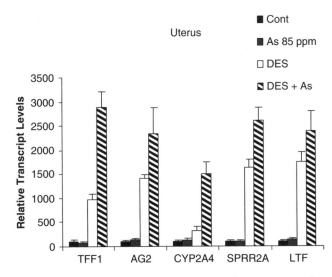

Figure 9.2 *Synergistic effect of in utero arsenic (85 ppm from gestation days 8 to 18) exposure plus postnatal treatment of DES (2 μg/pup/day from postpartum days 1 to 5) on expression of estrogen-related genes in the uterus. Data are mean plus SEM of five mice. *Significantly different from DES alone, p < 0.05.*

pS2 in arsenic-induced uterine adenocarcinoma and urinary bladder transitional cell carcinoma (Waalkes *et al.*, 2006a). The intense, wide-spreading staining for ER-α was also observed in arsenic-induced liver malignancies, while mouse hepatocellular carcinomas induced by diethylnitrosamine showed little or no staining for ER-α (Waalkes *et al.*, 2006b).

Neonatal tissues of CD1 mice exposed to arsenic *in utero* and postnatal DES were collected at postnatal day 12 for gene expression analysis. The expression of ER-α in neonatal uterus is low, but modest increases are observed after arsenic *in utero* (Waalkes *et al.*, 2006a). However, the expression of ER-linked genes was dramatically increased by arsenic plus DES. For example, arsenic plus DES increased the expression of TFF1, AG2, CYP2A4, SPRR2A, and lactoferrin (LTF) in the uterus (Figure 9.2); some of these gene expressions have been reported (Waalkes *et al.*, 2006a), and some are still under investigation.

These data provide evidence that *in utero* arsenic acts synergistically with estrogens in carcinogenesis, with aberrant ER signaling occurring in adult tissues, but also occurring in the neonatal tissues.

9.4 Summary

This series of studies provides genomic profiling during transplacental arsenic carcinogenesis in mice. The expression analysis was performed using cDNA microarray and bioinformatics, followed by confirmation with real-time RT-PCR, western blot analysis, immunohistochemistry, and toxicology-related endpoints. Genomic analysis reveals multiple molecular events associated arsenic carcinogenesis. This chapter gives an example of how we approach the hypothesis of aberrant estrogen signaling in arsenic hepato-carcinogenesis. However, this mechanism may be limited to certain tissues, such as liver and urogenital tissues, as the aberrant estrogen signaling does not apply to arsenic skin carcinogenesis in our fourth arsenic transplacental carcinogenesis (data not shown). It is likely that the mechanism of arsenic carcinogenesis is multifactorial and that the role of any signaling cascade may be limited to only certain tissues.

The disruption of critical gene expression during early life could lead to genetic reprogramming and errors in gene imprinting, both of which likely impact tumor formation later in life (Cook *et al.*, 2005). Recent work indicates increased lung cancer following *in utero* and early-life exposure to inorganic arsenic in humans (Smith *et al.*, 2006), and overexpression of ER-α and cyclin D1 was observed in liver biopsy samples of arsenicosis patients in Guizhou, China (Waalkes *et al.*, 2004b). Thus, data are emerging that support the human relevance of our mouse transplacental studies and potential involvement of aberrant estrogen signaling in arsenic hepatocarcinogenesis. Clearly, early-life exposure may be a key susceptibility factor in arsenic carcinogenesis. Identifying other susceptibility factors and sensitive populations for arsenic carcinogenesis should be a research priority.

Acknowledgments

We thank Dr Erik Tokar, Dr Hong Dang, Dr Yang Sun and Dr Larry Keefer for their critical review of this manuscript. Research was supported in part by the Intramural Research Program of the NIH, National Cancer Institute, Center for Cancer Research, the federal funds from the National Cancer Institute, National Institutes of Health, under contract No. N01-CO-12400, and the National Center for Toxicogenomics at NIEHS. The content of this publication does not necessarily reflect the views or policies of the Department of Health and Human Services, nor does mention of trade names, commercial products, or organizations imply endorsement by the US Government.

References

Anderson LM, Diwan BA, Fear NT, Roman E (2000) Critical windows of exposure for children's health: cancer in human epidemiological studies and neoplasms in experimental animal models. *Environ Health Perspect* **108**(Suppl 3), 573–594.

Birnbaum LS, Fenton SE (2003) Cancer and developmental exposure to endocrine disruptors. *Environ Health Perspect* **111**, 389–394.

Centeno JA, Mullick FG, Martinez L, Page NP, Gibb H, Longfellow D, Thompson C, Ladich ER (2002) Pathology related to chronic arsenic exposure. *Environ Health Perspect* **110**(Suppl 5), 883–886.

Chen CJ, Yu MW, Liaw YF (1997) Epidemiological characteristics and risk factors of hepatocellular carcinoma. *J Gastroenterol Hepatol* **12**, S294–S308.

Chen H, Li S, Liu J, Diwan BA, Barrett JC, Waalkes MP (2004) Chronic inorganic arsenic exposure induces hepatic global and individual gene hypomethylation: implications for arsenic hepatocarcinogenesis. *Carcinogenesis* **25**, 1779–1786.

Concha G, Vogler G, Lezcano D, Nermell B, Vahter M (1998) Exposure to inorganic arsenic metabolites during early human development. *Toxicol Sci* **44**, 185–190.

Cook JD, Davis BJ, Cai SL, Barrett JC, Conti CJ, Walker CL (2005) Interaction between genetic susceptibility and early life environmental exposure determines tumor suppressor gene penetrance. *Proc Natl Acad Sci U S A* **102**, 8644–8649.

Devesa V, Adair BM, Liu J, Waalkes MP, Diwan BA, Styblo M, Thomas DJ (2006) Speciation of arsenic in the maternal and fetal mouse tissues following gestational exposure to arsenite. *Toxicology* **224**, 147–155.

Dickson RB, Stancel GM (2000) Estrogen receptor-mediated processes in normal and cancer cells. *J Natl Cancer Inst Monogr* **27**, 135–145.

Diwan BA, Kasprzak KS, Rice JM (1992) Transplacental carcinogenic effects of nickel(II) acetate in the renal cortex, renal pelvis and adenohypophysis in F344/NCr rats. *Carcinogenesis* **13**, 1351–1357.

Diwan BA, Anderson LM, Rehm S, Rice JM (1993) Transplacental carcinogenicity of cisplatin: initiation of skin tumors and induction of other preneoplastic and neoplastic lesions in SENCAR mice. *Cancer Res* **53**, 3874–3876.

Fishman J, Osborne MP, Telang NT (1995) The role of estrogen in mammary carcinogenesis. *Ann N Y Acad Sci* **768**, 91–100.

Germolec DR, Spalding J, Yu HS, Chen GS, Simeonova PP, Humble MC, Bruccoleri A, Boorman GA, Foley JF, Yoshida T, Luster MI (1998) Arsenic enhancement of skin neoplasia by chronic stimulation of growth factors. *Am J Pathol* **153**, 1775–1785.

IARC (2004) Arsenic in drinking water. In *IARC Monographs on the Evaluation of Carcinogenic Risks to Humans. Vol. 84: Some Drinking Water Disinfectants and Contaminants, Including Arsenic*. World Health Organization, International Agency for Research on Cancer: Lyon, France; 269–477.

Liu J, Liu Y, Goyer RA, Achanzar W, Waalkes MP (2000) Metallothionein-I/II null mice are more sensitive than wild-type mice to the hepatotoxic and nephrotoxic effects of chronic oral or injected inorganic arsenicals. *Toxicol Sci* **55**, 460–467.

Liu J, Zheng B, Aposhian HV, Zhou Y, Cheng ML, Zhang A, Waalkes MP (2002) Chronic arsenic poisoning from burning high-arsenic-containing coal in Guizhou, China. *Environ Health Perspect* **110**, 119–122.

Liu J, Xie Y, Ward JM, Diwan BA, Waalkes MP (2004) Toxicogenomic analysis of aberrant gene expression in liver tumors and nontumorous livers of adult mice exposed *in utero* to inorganic arsenic. *Toxicol Sci* **77**, 249–257.

Liu J, Xie Y, Ducharme DMK, Shen J, Diwan BA, Merrick BA, Grissom SF, Tucker CJ, Paules PS, Tennant R, Waalkes MP (2006a) Global gene expression associated with hepatocarcinogenesis in adult male mice induced by *in utero* arsenic exposure. *Environ Health Perspect* **114**, 404–411.

Liu J, Xie Y, Merrick BA, Shen J, Ducharme DMK, Collins J, Diwan BA, Logsdon D, Waalkes MP (2006b) Transplacental arsenic plus postnatal 12-*O*-teradecanoyl phorbol-13-acetate exposures induced similar aberrant gene expressions in male and female mouse liver tumors. *Toxicol Appl Pharmacol* **213**, 216–223.

Liu J, Xie Y, Cooper R, Ducharme DM, Tennant R, Diwan BA, Waalkes MP (2007) Transplacental exposure to inorganic arsenic at a hepatocarcinogenic dose induces fetal gene expression changes in mice indicative of aberrant estrogen signaling and disrupted steroid metabolism. *Toxicol Appl Pharmacol* **220**, 284–291.

Lu T, Liu J, LeCluyse EL, Zhou YS, Cheng ML, Waalkes MP (2001) Application of cDNA microarray to the study of arsenic-induced liver diseases in the population of Guizhou, China. *Toxicol Sci* **59**, 185–192.

Mazumder DN (2005) Effect of chronic intake of arsenic-contaminated water on liver, *Toxicol Appl Pharmacol* **206**, 169–175.

Morales KH, Ryan L, Kuo T-L, Wu M-M, Chen C-J (2000) Risk of internal cancers from arsenic in the drinking water. *Environ Health Perspect* **108**, 655–666.

Newbold RR (2004) Lessons learned from perinatal exposure to diethylstilbestrol. *Toxicol Appl Pharmacol* **199**, 42–50.

NRC (2001) *Arsenic in the Drinking Water (Update)*. National Research Council, National Academy: Washington, DC; 1–225.

NTP (2004) Arsenic compounds, inorganic. In *Report on Carcinogens* (11th edition). US Department of Health and Human Services, Public Health Service, National Toxicology Program: Research Triangle Park, NC; III18–III20.

Paul DS, Hernandez-Zavala A, Walton FS, Adair BM, Dedina J, Matousek T, Styblo M (2007) Examination of the effects of arsenic on glucose homeostasis in cell culture and animal studies: development of a mouse model for arsenic-induced diabetes. *Toxicol Appl Pharmacol* **222** (3), 305–314.

Pi J, Yamauchi H, Kumagai Y, Sun G, Yoshida T, Aikawa H, Hopenhayn-Rich C, Shimojo N (2002) Evidence for induction of oxidative stress caused by chronic exposure of Chinese residents to arsenic contained in drinking water. *Environ Health Perspect* **110**, 331–336.

Rossman TG, Uddin AN, Burns FJ, Bosland MC (2001) Arsenite is a cocarcinogen with solar ultraviolet radiation for mouse skin: an animal model for arsenic carcinogenesis. *Toxicol Appl Pharmacol* **176**, 64–71.

Shen J, Liu J, Xie Y, Diwan BA, Waalkes MP (2007) Fetal onset of aberrant gene expression relevant to pulmonary carcinogenesis in lung adenocarcinoma development induced by *in utero* arsenic exposure. *Toxicol Sci* **95**, 313–320.

Smith AH, Marshall G, Yuan Y, Ferreccio C, Liaw L, von Ehrenstein O, Steinmaus C, Bates MN, Selvin S (2006) Increased mortality from lung cancer and bronchiectasis in young adults after exposure to arsenic *in utero* and in early childhood. *Environ Health Perspect* **114**, 1293–1296.

Waalkes MP, Diwan BA, Ward JM, Devor DE, Goyer RA (1995) Renal tubular tumors and atypical hyperplasias in B6C3F1 mice exposed to lead acetate during gestation and lactation occur with minimal chronic nephropathy. *Cancer Res* **55**, 5265–5271.

Waalkes MP, Keefer LK, Diwan BA (2000) Induction of proliferative lesions of the uterus, testes, and liver in Swiss mice given repeated injections of sodium arsenate: possible estrogenic mode of action. *Toxicol Appl Pharmacol* **166**, 24–35.

Waalkes MP, Ward JM, Liu J, Diwan BA (2003) Transplacental carcinogenicity of inorganic arsenic in the drinking water: induction of hepatic, ovarian, pulmonary, and adrenal tumors in mice. *Toxicol Appl Pharmacol* **186**, 7–17.

Waalkes MP, Ward JM, Diwan BA (2004a) Induction of tumors of the liver, lung, ovary and adrenal in adult mice after brief maternal gestational exposure to inorganic arsenic: promotional effects of postnatal phorbol ester exposure on hepatic and pulmonary, but not dermal cancers. *Carcinogenesis* **25**, 133–141.

Waalkes MP, Liu J, Chen H, Xie Y, Achanzar WE, Zhou YS, Cheng ML, Diwan BA (2004b) Estrogen signaling in livers of male mice with hepatocellular carcinoma induced by exposure to arsenic *in utero*. *J Natl Cancer Inst* **96**, 466–474.

Waalkes MP, Liu J, Ward JM, Powell DA, Diwan BA (2006a) Urogenitial system cancers in female CD1 mice induced by *in utero* arsenic exposure are exacerbated by postnatal diethylstilbestrol treatment. *Cancer Res* **66**, 1337–1445.

Waalkes MP, Liu J, Ward JM, Diwan BA (2006b) Enhanced urinary bladder and liver carcinogenesis in male CD1 mice exposed to transplacental inorganic arsenic and postnatal diethylstilbestrol or tamoxifen. *Toxicol Appl Pharmacol* **215**, 295–305.

Waalkes MP, Liu J, Diwan BA (2007) Transplacental arsenic carcinogenesis in mice. *Toxicol Appl Pharmacol* **222**(3), 271–280.

Xie Y, Liu J, Benbrahim-Tallaa L, Ward JM, Logsdon D, Diwan BA, Waalke MP (2007) Aberrant DNA methylation and gene expression in livers of newborn mice transplacentally exposed to a hepatocarcinogenic dose of inorganic arsenic. *Toxicology* **236**(1–2), 7–15.

Zhou YS, Du H, Cheng M-L, Liu J, Zhang XJ, Xu H (2002) The investigation of death from diseases caused by coal-burning type of arsenic poisoning. *Chin J Endemiol* **21**, 484–448.

10

Characterization of Estrogen-active Compounds and Estrogenic Signaling by Global Gene Expression Profiling *In Vitro*

Stephanie Simon, Kathleen Boehme, Susanne U. Schmidt and Stefan O. Mueller

10.1 Introduction

Estrogens are derivatives of cholesterol and play essential roles in the development, growth, differentiation and function of various tissues. They regulate the female sexual development, but their role is not restricted to female. It spans to male and female reproductive tissues, such as mammary gland, uterus, ovary, testes and prostate. Estrogens also have protective effects on the cardiovascular system, the bone, and lipid metabolism (Auchus and Fuqua, 1994). On the other hand, estrogens play a critical role in estrogen-dependent cancers of both the mammary gland and the uterus (Pike *et al*., 1993; Colditz *et al*., 1995; Grady *et al*., 1995).

The effects of estrogens are mediated via alternative pathways (Figure 10.1). The classical, well-described pathway is ligand binding to the intracellular estrogen receptors (ERs). ERs belong to the class of steroid hormone receptors (SHRs) within the superfamily of nuclear receptors (Evans, 1988; Mangelsdorf *et al*., 1995; Kliewer *et al*., 1999) and are ligand-inducible transcription factors. Upon ligand binding, the ER monomers dimerize, bind to specific regulatory DNA sequences, so-called estrogen response elements (EREs) in the promoter region of estrogen-dependent genes, and activate gene transcription (Truss *et al*., 1992). Therefore, this pathway is called the genomic or transcriptional pathway.

Toxicogenomics: A Powerful Tool for Toxicity Assessment   Edited by S. C. Sahu
© 2008 John Wiley & Sons, Ltd

Up to now, two subtypes, ERα and ERβ (Green *et al.*, 1986; Greene and Press, 1986; Kuiper *et al.*, 1996; Mosselman *et al.*, 1996), and several transcript variants of ER (Murphy *et al.*, 1997; Rubanyi *et al.*, 1997) were described. Since it was the accepted opinion for many years that estrogenic action is mediated mainly via ER signaling, an estrogen target tissue was defined as one expressing functional ER and showing a specific response to estrogen treatment. Classical target tissues of this sex hormone, therefore, are predominantly reproductive tissues, such as mammary gland, uterus, ovary, testes, and prostate. But nevertheless, the ERs are expressed in diverse tissues with a distinct distribution of ER subtypes ERα and ERβ and with different abundances (Couse *et al.*, 1997; Mueller *et al.*, 2002).

Besides this direct way to activate target genes upon ERE binding, the ER uses alternative mechanisms. The interaction of the liganded ER with other transcription factors, such as nuclear factor kappa B (NF-κB; McKay and Cidlowski, 1999), Sp1 (Safe *et al.*, 2001) or activating protein-1 (AP-1; Johnston *et al.*, 1999; Webb *et al.*, 1999; Kushner *et al.*, 2000) was reported. Additionally, indirect modulation of gene transcription by sequestration of general/common transcriptional components (Harnish *et al.*, 2000; Speir *et al.*, 2000) or epigenetic regulations, e.g. modulation of histone acetyltransferase (HAT) or histone deacetylase (HDAT) activity (reviewed in Leader *et al.*, 2006) was shown.

There is also strong evidence for various so-called nongenomic pathways. Membrane-associated ERs (Pietras and Szego, 1977; Levin, 1999; Razandi *et al.*, 1999; Coleman and Smith, 2001; Zivadinovic and Watson, 2005) are supposed to be involved in the activation of phospholipase C (PLC), G-protein-coupled receptors, membrane iron channels and adenylate cyclase production by estrogens (Nakajima *et al.*, 1995; Tesarik and Mendoza, 1995; Le Mellay *et al.*, 1997; Kelly and Wagner, 1999; Valverde *et al.*, 1999; Filardo and Thomas, 2000; Revankar *et al.*, 2005; Thomas *et al.*, 2005). Furthermore, ubiquitous regulatory cascades, such as mitogen-activated protein kinases (MAPKs; Migliaccio *et al.*, 1996; Watters *et al.*, 1997), the phosphatidylinositol 3-OH kinases (PIK), and tyrosine kinases are modulated through nontranscriptional mechanisms by estrogens (reviewed in Moggs and Orphanides (2001)) (Figure 10.1). These actions lead to very rapid (seconds to minutes from exposure to the hormone) cellular responses (reviewed in Moggs and Orphanides (2001) and Simoncini *et al.* (2003, 2006)). Another aspect of nongenomic estrogenic signaling is the crosstalk with growth factor signaling pathways. These actions take place in both nuclear and cytoplasmic compartments (Song, 2007; Song and Santen, 2006; Song *et al.*, 2006, 2007). The transcriptional activity of ERα can be regulated by phosphorylation through insulin-like growth factor receptor (IGFR) or epidermal growth factor receptor (EGFR) in the presence of estradiol (Vignon *et al.*, 1987; Ignar-Trowbridge *et al.*, 1995; Hewitt *et al.*, 2005).

For a long time it was believed that nonclassical target tissues (e.g. lung, brain, cardiovascular system, bone) are not steroid-regulated areas due to the lower concentrations of steroid hormone receptors (Ciocca and Vargas-Roig, 1995). But in the past few years, evidence is arising that nongenomic estrogenic signaling plays a crucial role in these tissues especially.

Many of these mechanisms are not very well understood to date, but understanding these complex processes is of outstanding importance to developing new approaches for the identification of pharmacological targets for drug development. Furthermore, endocrine activity could also be undesired, as it may lead to pathophysiological effects, i.e. so-called

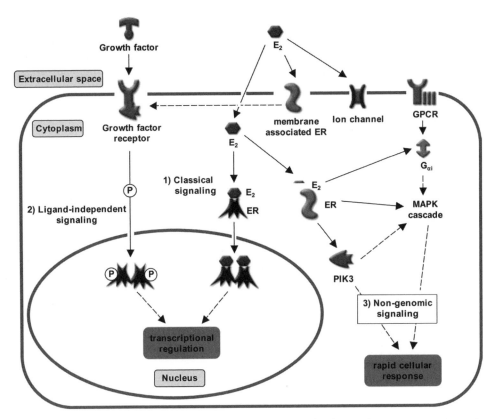

Figure 10.1 *Estrogen and estrogen receptor signaling. The figure demonstrates schematically three potential cellular signaling mechanisms of estrogen and estrogen receptors and the cellular distribution of the participants. (1) The classical signaling results in ligand-dependent transcriptional regulation by E_2–ER dimers. Additionally, nonclassical ER-mediated signaling triggered by interactions of liganded ER with other transcription factors as Activating Protein 1 (AP-1) may occur (Webb et al., 1999; Kushner et al., 2000). (2) Growth factors are able to phosphorylate and thus activate estrogen receptors without ligand binding. (3) Moreover, different transcriptional independent (nongenomic) estrogen actions have been described. These events possibly involve membrane-bound estrogen receptors, G proteins, ion channels and lead to rapid activation of intracellular signaling cascades like MAPK and phosphoinositide-3-kinase pathway. Furthermore, cross-talk between estrogen receptors and growth factor signaling has been described. (E_2: 17β-estradiol; ER: estrogen receptor; $G\alpha_i$: guanine nucleotide binding protein (G protein), alpha inhibiting activity; GPCR: G protein-coupled receptor; MAPK: mitogen-activated protein kinase; PIK3: phosphoinositide-3-kinase; figure generated using MetaCore^{TM}).*

endocrine disruption. Additionally, knowledge about toxic effects of chemicals is important during the drug development process. The understanding of both the pharmacology and toxicology of estrogen signaling has to be improved, and global expression profiling may help to elucidate these networks and to predict pharmacologic and toxic activity of chemicals.

10.2   Screening Tools for Estrogenic Activity

Owing to the crucial role of ER in cell proliferation and differentiation, disruption of ER signaling pathways may contribute to diverse pathophysiological effects (Mueller, 2002; Mueller *et al.*, 2004). Hence, in the past few years, considerable public attention has been drawn to the investigation of possible endocrine-disrupting effects, as well as mode of action of naturally occurring or anthropogenic toxicants, drugs or chemicals because of posing a potential hazard to human health and the environment.

10.2.1   *In Vivo* Screening Tools

Before *in vitro* assays had been developed to detect potential endocrine disrupters, rodents were used for screening for many decades. The uterotrophic bioassay measures the increase in the uterine weight after subacute administration of test compounds either to immature intact or adult ovariectomized female rats and was the first official screening assay for estrogenicity or antiestrogenicity validated and accepted by the Organisation for Economic Cooperation and Development (Gelbke *et al.*, 2004; OECD, 2006a,b 2007). The enhanced 28-day acute toxicity study in rats based on the OECD Guideline 407 includes the assessment of weights of endocrine-related organs and the histopathological analysis of endocrine target tissues and is commonly used for the detection of endocrine-related effects (OECD, 1995; Kennel *et al.*, 2003; Gelbke *et al.*, 2007).

10.2.2   *In Vitro* Screening Tools

Besides the large amount of animals necessary for *in vivo* tests, expenditure of time and costs urged the exploration and application of *in vitro* alternatives. The classical E-screen assay mimics cell proliferation *in vitro* by measuring growth induction to increasing doses of a test chemical in estrogen-responsive cell lines (Soto *et al.*, 1995).

Another possibility is to study binding affinities of radioactively or fluorescence-labeled ligands to estrogen receptors. Moreover, these assays provide only information on binding affinities, but are not capable of discriminating agonistic from antagonistic effects.

In contrast, with functional assays, activation or inhibition of estrogen receptors can be detected (Korach, 1979; Sun *et al.*, 1999; Mueller, 2002). In transactivation assays, ligand-induced estrogen receptor-mediated reporter gene transcription can be measured in cells, either with endogenous receptor expression or transient as well as stably transfected with ER. Furthermore, two-hybrid assays, fluorescent resonance energy transfer or glutathione-*S*-transferase pull-down assay, are suitable methods to investigate protein–protein interactions. Using these techniques, several studies have been performed to characterize different estrogen receptor ligands regarding their ability to regulate recruitment of coactivators, which is important for receptor action (Zhou *et al.*, 1998; Gee *et al.*, 1999; Nishikawa *et al.*, 1999; Routledge *et al.*, 2000; An *et al.*, 2001; Mueller, 2002; Mueller *et al.*, 2004).

10.3   Cellular Models for Estrogenicity Testing

To assess the estrogenic potential of chemicals, various cellular model systems exist, such as immortalized, primary, or established permanent cell lines. All *in vitro* models are *per se* artificial systems and are able to reflect only a small part of physiologically relevant effects. Therefore, it is important to use a model as close as possible to the real physiological conditions. The cellular environment has a crucial impact on the experimental results; hence, the choice of the adequate model system is a challenge, as it determines the quality and reliability of results.

A potentially appropriate model to analyze estrogenic effects is to immortalize tissue-specific cell lines with distinct endogenous ER expression. Two examples are immortalized testes (Mueller and Korach, 2001) and mammary (Mueller *et al.*, 2000) cell lines derived from ERα (αERKO) knock-out and wild-type mice (Lubahn *et al.*, 1993; Eddy *et al.*, 1996; Korach *et al.*, 1996; Couse and Korach, 1999; Couse *et al.*, 2000) with distinct expression of either ERα or ERβ. These cell lines allow one to analyze the roles of ER subtypes in testis and mammary function on the cellular level (Mueller and Korach, 2001) and may serve as a useful tissue-specific model to test xenoestrogens for their effects on male and female reproductive function (Cheek and McLachlan, 1998; de Kretser, 1998).

The most commonly used cell line to study estrogen-related effects is the human MCF-7 breast adenocarcinoma cell line. This cell line expresses both ERα and ERβ endogenously and was also found to express membrane-associated ERα species (Zivadinovic and Watson, 2005; Zivadinovic *et al.*, 2005). This cellular model was applied for various research projects, including screening for phytoestrogens (Matsumura *et al.*, 2005) and phytochemicals (van Meeuwen *et al.*, 2007a,b), analysis of non-genomic estrogen signaling pathways (e.g. Filardo *et al.*, 2000; Filardo and Thomas, 2005; Li *et al.*, 2006) and selective estrogen receptor modulators (SERMs; Diel *et al.*, 2002; Katzenellenbogen and Frasor, 2004).

The estrogen-responsive BG-1 ovarian cancer cell line expresses ER, IGF1R and EGFR and served as a model to study cross-talks of ER-mediated and peptide growth factor signaling pathways (Ignar-Trowbridge *et al.*, 1995; Baldwin *et al.*, 1998) and the estrogen-induced formation of micronuclei which may play a role in estrogen-related tumor progression (Stopper *et al.*, 2003).

Another model to study ER-independent signaling pathways is the ER-negative breast cancer cell lines MDA-MB-435S and MDA-MB-231. These cell lines are important to analyze hormone-related effects in ER-negative cancer cells and may help to elucidate new ways for breast cancer therapies (Platet *et al.*, 1998; Chernicky *et al.*, 2000; Nakagawa *et al.*, 2001). Jordan and coworkers used the ER-deficient MDA-MB-231 cell line stably transfected with human ERα to analyze estrogen-dependent regulation of TGF-α in breast cancer cells (Jiang and Jordan, 1992; Levenson *et al.*, 1998).

Mueller *et al.* (2003a) developed human Ishikawa endometrial adenocarcinoma cell lines stably transfected with either ERα or ERβ. These cell lines allowed for the first time a comprehensive analysis of ERα- or ERβ-specific effects that may be important to assess the potential hazard of phytoestrogens or xenoestrogens to human health (Mueller *et al.* 2003b, 2004). The parental Ishikawa cell line used for the development of stable transfectants was shown to express no functional ER and did not respond to estrogens (Mueller *et al.*, 2003a; Simon, 2007, unpublished data). Additionally, an estrogen-responsive Ishikawa cell line that expresses endogenous ERα but not ERβ is also used to assess estrogenic compounds (Holinka *et al.*, 1986; Simon, 2007, unpublished data).

Despite these cellular models, an appropriate model system to distinguish direct ER-mediated from indirect non-ER-mediated estrogenic effects in tissue-specific cell lines was still lacking. We have, therefore, used the above-described estrogen-responsive Ishikawa, termed here Ishikawa-plus, and the ER-deficient Ishikawa cells, termed Ishikawa-minus, as a model system to analyze estrogenic effects in a cell type and ER-dependent manner.

Several studies employed gene expression profiling in MCF-7 (e.g. Frasor *et al.*, 2003) or MDA-MB231 (Levenson *et al.*, 2002) to characterize estrogenic effects. To point out the crucial role of genomics methods in the future, we present here an example of gene expression profiling as a screening tool for ER-dependent and ER-independent estrogenic effects in endometrial Ishikawa cell lines.

10.4   Application of Gene Expression Profiling Methods

In this chapter we will focus on the global gene expression analysis and how it can be used to generate characteristic profiles of endocrine-active compounds *in vivo* and *in vitro* (Moggs and Orphanides, 2003; Orphanides and Kimber, 2003). These analyses may help to better understand the mode of action of various compounds and the signaling pathways involved and may provide a predictive model to characterize unknown compounds with regard to their estrogenic potential.

Besides the functional assays to assess the cellular response to endocrine-active compounds described above, the analysis of gene expression patterns gets more and more important. With the advent of microarray technology it is now possible to investigate the complex ER-mediated gene transcription with a global approach in contrast to single gene analysis.

The link of conventional toxicological research and functional genomics resulted in the emergence of toxicogenomics. Toxicogenomics studies the complex interactions between the structure and activity of the genome and adverse biological effects caused by exogenous agents (Aardema and MacGregor, 2002). Hence, using toxicogenomic technologies may support the characterization of various compounds and may serve as a tool to predict their properties by comparing gene expression profiles of unknown compounds with those of well-known compounds. Therefore, these analyses may open new ways of drug development. The broad application spectrum of gene expression microarray technology has led to the production of multiple platforms varying in the kind of probes (e.g. cDNA, short or long oligonucleotide), the hybridizing paradigm (competitive versus noncompetitive), the labeling method and the production method (e.g. *in situ* polymerization, spotting, microbeads; Barnes *et al.*, 2005).

10.4.1   Affymetrix GeneChips®

Affymetrix is the leading company in the field of microarray analysis, selling the first commercial GeneChip in 1994. Affymetrix GeneChips® are produced by *in situ* synthesis of 25-mer oligonucleotides in a photolithographic process directly on the surface of the glass slide (Chee *et al.*, 1996; Lockhart *et al.*, 1996). The oligonucleotide probes are arranged in a specific layout with each probe synthesized at a predefined location. A particular feature of Affymetrix GeneChips® is that multiple probes are used for each gene along with a one-base mismatch probe intended as a control for nonspecific and background hybridization, which is important for quantifying weakly expressed mRNAs. Another advantage of the Affymetrix platform is the large number of genes (approximately 38 000 on the human genome U133 Plus 2.0 array) that can be analyzed simultaneously on one chip. Affymetrix GeneChip HuGeneFL arrays, which enable the relative monitoring of mRNA transcripts of approximately 5600 full-length human genes, were used to identify novel estrogen receptor target genes in human ZR75-1 breast cancer cells (Soulez and Parker, 2001). Other researchers suggested correlations between the clinical outcome of breast cancer and gene expression changes elicited by the selective estrogen receptor modulator tamoxifen in ERα-positive MCF-7 cells applying Affymetrix human GeneChip U133A microarrays (Chang *et al.*, 2006; Frasor *et al.*, 2006).

10.4.2   Illumina Sentrix® BeadChip Technique

The Illumina Sentrix® BeadChip technique is one of the new microarray approaches that have emerged in the last few years. In contrast to Affymetrix, Illumina's longer oligonucleotides (50-mer probe oligonucelotide and 25-mer address sequence) are synthesized using standard oligonucleotide synthesis and are attached to 3 μm silica beads (Barnes *et al.*, 2005). Each bead type carries around 800 000 copies of a specific probe oligonucelotide sequence and is represented approximately 30 times on each array, providing an internal technical replication that Affymetrix lacks (Kuhn *et al.*, 2004). All beads are randomly arrayed into small wells of a patterned substrate on the chip surface (Steemers and Gunderson, 2005). The random assembly requires the identification of the localization of each bead type within the chip. Therefore, a highly efficient decoding algorithm based on a sequential hybridization procedure to the address sequence has been developed (Gunderson *et al.*, 2004). The high quality of Illumina microarrays is ensured by a variety of internal control beads (housekeeping genes, hybridization, signal generation, and background). The enormous advantages of Illumina BeadChips are the low per-sample array pricing compared with Affymetrix and the possibility of running 6–16 samples in parallel (Elvidge, 2006; Ragoussis and Elvidge, 2006). In spite of being completely different technologies, several studies have shown that Affymetrix and Illumina platforms deliver very comparable results (Barnes *et al.*, 2005; Shi *et al.*, 2006).

Illumina product offerings range from genotyping assays for SNP detection, customized microarrays, to whole genome chips (Steemers and Gunderson, 2005). The suitability of Illumina microarrays for toxicity screening was shown in our laboratory by Zidek *et al.* (2007), who developed a customized focused microarray containing 550 liver-specific genes for predictive screening of acute hepatotoxic compounds.

In our approach, we applied Illumina Sentrix HumanRef-8 Expression BeadChips, allowing profiling of eight samples per BeadChip with more than 23 000 transcript probes per sample. Sample preparation and analysis process comprises hybridization of biotin-labeled cRNA and detection with streptavidin-conjugated Cy3 by confocal laser scanning technology. Data analysis is a multistep process beginning with decoding the image spots and controlling the array quality. In the next step, data normalization is necessary to subtract the background noise (e.g. from unspecific dye binding or cross-hybridization) and make data from different arrays or chips comparable. Afterwards, statistical interpretation of the data and pathways analysis has to be done.

10.5   Genomics Methods as Tools for Estrogen Screening

10.5.1   Ishikawa Cell Lines Show Significant Differences in Global Gene Expression Patterns

We have chosen the Illumina Sentrix® human Ref-8 BeadArray chips to examine global gene expression profiles. In a first approach, Ishikawa-plus were distinguished from Ishikawa-minus by a principle components analysis (Yeung and Ruzzo, 2001) (Figure 10.2). Our observation was that Ishikawa-plus and Ishikawa-minus separate into two distinct, well-defined clusters, confirming their significant differences in gene expression patterns.

To identify cell-type-specific patterns of gene expression that are induced or inhibited by estradiol (E2), we compared gene expression profiles of Ishikawa-plus with those of the ER-deficient Ishikawa-minus cell line (Figure 10.3). The number of genes affected by E2 in Ishikawa-plus was strikingly higher than in Ishikawa-minus, confirming the major role of the ER in mediating the estrogenic response. Interestingly, in Ishikawa-plus, the majority (70 %) of regulated genes were repressed in response to estrogens (Figure 10.3B). Similar results were reported by Frasor *et al.* (2003) for gene expression profiles in MCF-7 cells. In contrast, in Ishikawa-minus, the distribution of induced and repressed genes was more balanced (Figure 10.3C and D). Nevertheless, the detection of gene expression changes in the ER-deficient cell line in response to estrogen supports alternative pathways (e.g. epidermal growth factor (EGF) or insulin-like growth factor (IGF)) involved in estrogenic signaling (Smith, 1998).

10.5.2   Identification of Marker Genes for ER-dependent and ER-independent Regulation

A major aim of our study was the identification of marker genes that are regulated either ER-dependently or ER-independently. Surprisingly, out of 23 000 queried target genes, only 17 were regulated in both cell types (Tables 10.1 and 10.2). Another interesting finding

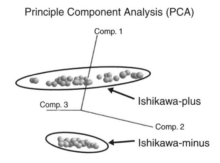

Figure 10.2   *Principle components analysis. This algorithm was applied to reduce the complexity of the huge data set and classifies the normalized fluorescence intensities based on the three outstanding features (principle components) of the data. Every sphere represents one replicate of a sample group. PCA revealed that samples of Ishikawa-plus (upper sphere cluster) and Ishikawa-minus (lower sphere cluster) separate in two well-defined, distinct clusters, indicating significant differences in gene expression profiles.*

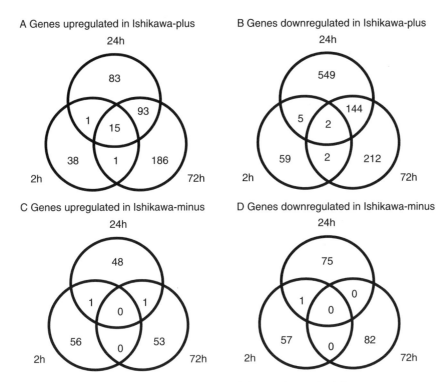

Figure 10.3 *Genes significantly regulated in Ishikawa-plus and Ishikawa-minus in response to E2 stimulation for 2, 24 and 72 h. Fold induction values were calculated relative to the time-matched vehicle control (set to unity). For all values, a Student's t-test was performed to determine significant differences between treated and control samples. Genes with a p < 0.05 and a regulation of at least 1.5-fold compared with the control were accepted to be significantly regulated and, therefore, were selected for further analysis. Based on these reductions, we were able to identify 1392 genes regulated by E_2 in Ishikawa-plus and 374 in Ishikawa-minus. Number of induced (A) and repressed (B) genes in Ishikawa-plus and number of induced (C) and repressed (D) genes in Ishikawa-minus are shown.*

was, that eight of them were regulated diametrically (Table 10.1). Therefore, the signal for these genes observed in Ishikawa-plus is likely ER dependent, and these genes might serve as putative marker genes for ER-dependent estrogen-induced signals. The regulation of RAB14, which belongs to the Ras oncogene family of small GTPases (Junutula *et al.*, 2004; Proikas-Cezanne *et al.*, 2006), and ALEX2, an armadillo repeat protein (Smith *et al.*, 2005) that may be involved in the Wnt/Wingless growth factor signaling pathway in both cell lines in opposite directions, may be evidence for the tissue-specific effects of estrogens in ER-positive or ER-negative tissues. The remaining nine genes were regulated unidirectionally in both cell lines (Table 10.2) and were not affected by cotreatment of the antiestrogen ICI, indicating estrogen-responsive but ER-independent targets (Simon, 2007, unpublished data).

Additionally, through comparison of genes whose regulation was abolished by cotreatment with the antiestrogen ICI with genes that were not affected by ICI cotreatment, we

Table 10.1 Genes diametrically regulated[a] in Ishikawa in response to E2

Gene	GeneBank Accession No.	Definition	Functions/biological processes	Regulation in Ishikawa-plus	Regulation in Ishikawa-minus	Time point of regulation (h)
RAB14	NM_016322.2	Member of RAS oncogene family; small GTP-binding protein 14	Small GTPase-mediated signal transduction; vesicle-mediated protein transport	−2.9	1.9	2
AK5	NM_174858.1	Adenylate kinase 5. Transcript variant 1	Adenylate kinase activity; ADP biosynthesis	−3.0	1.7	24
ALEX2	NM_177949.1	Armadillo repeat protein	Putative tumor suppressor	−2.8	3.9	24
FBX15	NM_152676.1	F-box only protein 15	Ubiquitin ligase	−4.0	7.1	24
LGALS2	NM_006498.1	Lectin galactoside-binding soluble 2 (galectin 2)	Heterophilic cell adhesion; binds to lymphotoxin-alpha	3.7	−8.4	24
BIRC7	NM_022161.2	Baculoviral IAP repeat-containing 7 (livin). Transcript variant 2	Apoptosis inhibitor; activation of JUNK, inhibition of caspase activation	2.3	−10.3	72
ELF3	NM_004433.2	E74-like factor 3 (ets domain transcription factor)	Epithelial-specific transcription factor	2.6	−1.6	72
HEY1	NM_012258.2	Hairy/enhancer-of-split related with YRPW motif 1	Transcription factor; DNA-dependent regulation of transcription	3.8	−1.5	72

[a]Genes were selected based on Student's t-test p-values ($p < 0.05$) and a more than 1.5-fold change (compared with the solvent control).

Table 10.2 Genes unidirectionally regulated[a] in Ishikawa in response to E2

Gene	GeneBank Accession No.	Definition	Functions/biological processes	Regulation in Ishikawa-plus	Regulation in Ishikawa-minus	Time point of regulation (h)
CCR3	NM.178329.1	Chemokine (C–C motif) receptor 3 transcript variant 2	Inflammation	7.6	7.7	24
MGC21621	NM.145015.2	G protein-coupled receptor MrgF	Function unknown	−2.0	−2.8	24
TEKT2	NM.014466.2	Tektin 2 (testicular)	Spermatogenesis	−1.8	−3.1	24
ASP	NM.031916.2	AKAP-associated sperm protein	Sperm protein	−3.7	−2.9	72
CCT6B	NM.006584.1	Chaperonin containing TCP1, subunit 6B (zeta 2)	Spermatogenesis; protein folding	−1.5	−1.7	72
DNASE1L3	NM.004944.1	Deoxyribonuclease I-like 3	Mediates the breakdown of DNA during apoptosis	15.0	34.0	72
INHA	NM.002191.2	Inhibin, alpha	Male gonadal differentiation and perineural development	1.7	1.6	72
KCNJ13	NM.002242.2	Potassium channel, subfamily J, member 13	Ion transport	−2.9	−3.0	72
PCYT1B	NM.004845.2	Phosphate cytidylyltransferase 1, choline, beta isoform	Defense response	−3.5	−7.5	72

[a] Genes were selected based on Student's t-test p-values ($p < 0.05$) and a more than 1.5-fold change (compared with the solvent control).

were able to distinguish ER-dependent from ER-independent signals in Ishikwawa-plus. We identified genes with a slight decrease in regulation in response to E2/ICI compared with E2 treatment, indicating a partially ER-mediated regulation. Interestingly, less than 5 % of all significantly E2-regulated genes showed no ICI-induced repression of E2-triggered regulation, indicating their ER-independency. Therefore, these genes may serve as suitable marker genes for E2-induced but ER-independent gene expression.

10.5.3   Crosstalk with Other Signaling Pathways

The important role of the peptide growth factors IGF-1 and EGF in oncogenic transformation by influencing the regulation of cell proliferation, apoptosis, and survival has long been established. EGF and IGF-1 are able to signal very potently, both through the ras/raf/mitogen-activated protein kinase (MAPK) pathway for proliferation and through the PIK3/Akt pathway mediating cell survival and inhibiting apoptosis (Normanno *et al.*, 2005; Kooijman, 2006). We examined genes of the IGF-1 and EGF pathways (Figures 10.4 and 10.5) and found 11 regulated genes that are involved in these pathways. Four of them were regulated ER-dependently in Ishikawa-plus (Table 10.3, bottom) and seven of them were regulated ER-independently in Ishikawa-plus and -minus (Table 10.4). The IGF-1 signaling pathway cross-talks with the E2 pathway, but the extent and precise mechanism of this interaction are not completely understood (Figure 10.5). In the uterus, IGF-1 signaling can be mediated by estradiol acting through its nuclear receptor to stimulate IGF-1 synthesis, which mediates the proliferation of uterine tissue. Conversely, *in vivo* studies using ER reporter mice have demonstrated that ER transcriptional activity can be induced by IGF-1 independently of E2 (Klotz *et al.*, 2002). Furthermore, gene expression analysis after IGF-1 treatment in wild-type (WT) and ERα knockout (αERKO) mice revealed a nearly identical gene expression pattern, suggesting a predominantly ERα-independent mechanism (Hewitt *et al.*, 2005). We observed similar ER-independent growth factor signaling in Ishikawa cells. Four genes involved in the IGF-1 or EGF-signaling pathway were identified in Ishikawa-plus, whose regulation was not abolished by ICI, indicating an ERα-independent mechanism (Table 10.4, Figures 10.4 and 10.5). Intriguingly, the PIK3C2B, which is member of both the EGF and IGF-1 pathways, was up-regulated by E2. PI3-kinases play key roles in signaling pathways involved in cell proliferation, oncogenic transformation, cell survival, and migration, as well as intracellular protein trafficking (Grimberg, 2003). This gene was previously described to be amplified in Glioblastomas (Knobbe and Reifenberger, 2003). PDPK1, another protein of the IGF-1 pathway, was regulated by E2. PDPK1 is a mediator of multiple signaling pathways coupled to growth factor receptor activation in human cancers. PDPK1-expressing cells exhibited a high degree of transformation that was associated with the activation of AKT1 and an elevation of PKC alpha expression (Lin *et al.*, 2005). PRKCG, known to be part of the EGF-pathway, belongs to the family of serine- and threonine-specific protein kinases that are known to be involved in diverse cellular signaling pathways. The gene product of RASA1 stimulates the GTPase activity of normal RAS p21 but not its oncogenic counterpart. Acting as a suppressor of RAS function, the protein enhances the weak intrinsic GTPase activity of RAS proteins, resulting in the inactive GDP-bound form of RAS, thereby allowing control of cellular proliferation and differentiation. Two additional genes, PIK3C2A and AKT2, involved in these pathways were identified in Ishikawa-minus. PIK3C2A has been found to be oncogenic and has been implicated in ovarian cancers (Campbell *et al.*, 2004). AKT2

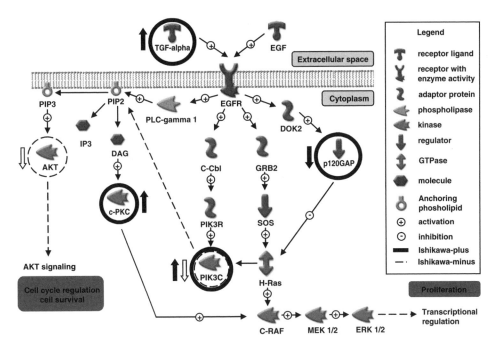

Figure 10.4   *Epidermal growth factor (EGF) signaling. Genes selected from gene expression studies of Ishikawa-plus (black circles) and -minus (dotted circles) cell lines were subjected to pathway analysis using GeneGo's MetaCore™ software. Four genes being part of EGF signaling were found to be significantly regulated in response to E2. TGF-α is a potent activator of EGF receptor and was found to be up-regulated in Ishikawa-plus in an ERα-dependent manner. Furthermore, regulation of c-PKC, p120GAP/ RASA1 and PIK3C genes acting downstream in EGF signaling could be identified. Up-regulated c-PKC gene is part of phospholipase C gamma signaling. In contrast, p120 GAP, an inhibitor of proliferative Ras/MAP kinase pathway, was down-regulated in Ishikawa-plus. PIK3C catalyzes the conversion of PIP2 to PIP3 and was diametrically regulated in Ishikawa-plus and -minus. Downstream of PIK3C, AKT2 was identified to be down-regulated in Ishikawa-minus. The AKT pathway negatively regulates apoptosis, allowing cell survival. (AKT: V-akt murine thymoma viral oncogene homologue; c-Cbl: E3 ubiquitin-protein ligase CBL; c-PKC: protein kinase C; cRAF: rapidly growing fibrosarcoma (serine/threonine-protein kinase); DAG: diacylglycerol; DOK2: docking protein 2; EGF: epidermal growth factor; EGFR: EGF receptor; ERK 1/2: extracellular-signal-regulated kinase; GRB2: growth factor receptor-bound protein 2; H-Ras: Harvey rat sarcoma viral (v-Ha-ras) oncogene homologue; IP3: inositol-trisphosphate; MAPK: mitogen-activated protein kinase; MEK 1/2: map-erk kinase; PIK3C: phosphoinositide-3-kinase C2B; PIP2: phosphatidylinositol-4,5-bisphosphate; PIP3: phosphatidylinositol-3,4,5-trisphosphate; PIK3R: PI3K reg class IA; PLC-gamma 1: phospholipase c gamma 1; p120GAP/RASA1: GTPase activating protein; SOS: son of sevenless homologue 1; TGF-α: tumor growth factor alpha; figure generated using MetaCore™.)*

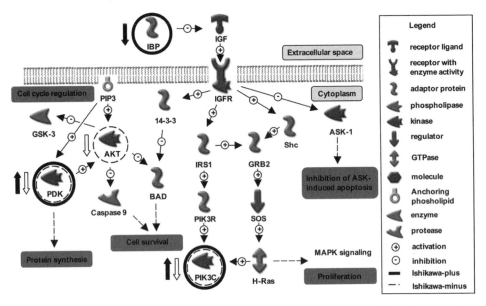

Figure 10.5 *Insulin growth factor (IGF) signaling. GeneGo's MetaCore^{TM} pathway analysis revealed four significantly regulated genes in Ishikawa cells involved in IGF signaling. The IGF-binding protein (IBP) is able to inhibit the activity of IGF and was found to be ERα-dependent down-regulated in Ishikawa-plus after E2 treatment. Another gene encoding phosphoinositide-dependent protein kinase (PDK) positively regulates the activity of AKT. PDK was up-regulated in Ishikawa-plus cells in response to E2. In contrast, PDK and AKT genes were both down-regulated in Ishikawa-minus, indicating pro-apoptotic signaling in this cell line. (AKT: V-akt murine thymoma viral oncogene homologue; ASK-1: mitogen-activated protein kinase kinase kinase 5; BAD: Bcl2 antagonist of cell death; IBP: IGF-binding protein; IGF: insulin-like growth factor I; IGFR: insulin-like growth factor 1 receptor; GRB2: growth factor receptor-bound protein 2; GSK-3: glycogen synthase kinase-3; IRS1: insulin receptor substrate 1; PDK: phosphoinositide-dependent protein kinase; PIP3: phosphatidylinositol-3,4,5-trisphosphate; PIK3R: PI3K reg class IA; PIK3C: phosphoinositide-3-kinase C2B; H-Ras: Harvey rat sarcoma viral (v-Ha-ras) oncogene homologue; Shc: src homology 2 domain-containing transforming protein C1; SOS: son of sevenless homologue 1; figure generated using MetaCore^{TM}.)*

is a putative oncogene encoding a protein belonging to a subfamily of serine/threonine kinases. Furthermore, AKT2 was amplified and overexpressed in ovarian carcinoma cell lines (Arboleda *et al.*, 2003).

Taken together, the regulation described of IGF-1/EGF target genes points to anti-apoptotic and pro-apoptotic signaling in Ishikawa-plus and Ishikawa-minus respectively. Intriguingly, the major downstream effectors of IGF-1/EGF signaling were not affected by cotreatment with the antiestrogen ICI, whereas the regulation of upstream modulators (i.e. TGF-α and IGFBPs) was abolished by ICI. These results give evidence for a dual (ER-dependent and -independent) modulation of proliferative and anti-apoptotic signaling by estrogens via growth factors. Furthermore, this ER-independent but E2-induced proliferative response exemplifies a possible mode of action of tumors resistant to antiestrogenic treatment.

Table 10.3 *Regulation[a] of selected ERα-dependent genes in Ishikawa-plus in response to E2*

Process	Gene	GeneBank Acc. No.	2 h	24 h	72 h
				Time point of regulation	
Proliferation markers	GREB1 a	NM_014668.2	n.d.[b]	2.4	4.0
	GREB1 b	NM_033090.1	n.d.	n.d.	11
Cell cycle control	Cyclin A1	NM_003914.2	n.d.	1.6	2.1
	Cyclin D1	NM_053056.1	n.d.	−1.6	−2.4
	Cyclin E2	NM_004702.2	n.d.	−1.6	n.d.
	Cyclin G2	NM_004354.1	n.d.	−2.0	n.d.
Cell cycle progression	CDKL 1	NM_004196.3	n.d.	−1.8	n.d.
	CDK 2	NM_001798.2	n.d.	−3.1	n.d.
	CDKN1A (p21)	NM_078467.1	−2.8	n.d.	n.d.
	CDKN2B (p15)	NM_004936.2	n.d.	−1.6	n.d.
	CDKN1C (p57)	NM_000076.1	n.d.	−1.6	n.d.
Receptors/ transcription factors	Progesterone receptor (PGR)	NM_000926.2	n.d.	24	120
IGF-1-signaling	IGFBP3	NM_000598.2	n.d.	−4.1	−2.6
	IGFBP4	NM_001552.1	n.d.	n.d.	1.9
	IGFBP7	NM_001553.1	n.d.	−1.9	−2.3
EGF-signaling	TGFA	NM_003236.1	n.d.	n.d.	2.6

[a] Genes were selected based on Student's *t*-test *p*-values ($p < 0.05$) and a more than 1.5-fold change (compared with the solvent control).
[b] n.d.: not significantly regulated.

Table 10.4 *Regulation[a] of ERα-independent genes in Ishikawa-plus and -minus in response to E2*

Cell line	Process	Gene	GeneBank Acc. No.	Regulation[b]
Ishikawa-plus	IGF-1-signaling	PDPK1	NM_002613.1	1.8 (24 h)
		PIK3C2B	NM_002646.2	1.9 (2 h)
	EGF-signaling	PIK3C2B	NM_002646.2	1.9 (2 h)
		PRKCG	NM_002739.2	1.9 (72 h)
		RASA1	NM_002890.1	−3.0 (72 h)
Ishikawa-minus	IGF-1 signaling	AKT2	NM_001626.2	−1.5 (2 h)
		PDPK1	NM_002613.1	−1.7 (72 h)
		PIK3C2A	NM_002645.1	−1.5 (72 h)
	EGF-signaling	AKT2	NM_001626.2	−1.5 (2 h)
		PIK3C2A	NM_002645.1	−1.5 (72 h)

[a] Genes were selected based on Student's *t*-test *p*-values ($p < 0.05$) and a more than 1.5-fold change (compared with the solvent control).
[b] Time point of regulation is given in parentheses.

10.5.4   Many Genes Involved in Cell Cycle Processes are Regulated ER Dependently

E2 is known to stimulate cell growth and proliferation. Hence, we analyzed the regulation of genes that are involved in cell cycle control and progression or were described as proliferation markers (Table 10.3). The pro-proliferative effects of E2 in Ishikawa-plus were supported by the ER-dependent up-regulation of typical proliferative markers such as cyclin A and variants of the gene regulated by estrogen in breast cancer 1 (GREB1a/GREB1b), along with down-regulation of anti-proliferative markers like cyclin G2 and p21. To date, the expression of GREB1 variants was reported in breast and prostate cancer cells only (Ghosh *et al.*, 2000, Rae *et al.*, 2005, 2006).

10.6   Conclusions

Estrogens are involved in many essential cellular processes and exert pleiotropic effects in various male and female tissues. There are many different pathways that mediate estrogenic effects, e.g. rapid nongenomic effects via membrane-bound ER or genomic effects via signaling through nuclear ER. The identification of estrogenic effects is of great importance for the subsequent toxicity assessment of chemicals (i.e. reproductive toxicities, hormone-dependent malignancies). For that, several screening tools *in vitro*, but also *in vivo*, are available that help to identify estrogen-active compounds. Endocrine disruption – the term that describes the adverse health effects related to endocrine action – has then to be determined in available regulatory toxicity testing for embryo–fetal, reproductive, and carcinogenic effects.

Similar screening systems can also be applied to identify potential drugs with desired estrogenic or antiestrogenic activity such as the SERMs to treat hormone-dependent diseases. Owing to the complex network of endocrine action, the sole identification of estrogen activity is usually not sufficient to define a physiologically relevant pharmacological or toxicological action. *In vivo* studies offer the possibilities to assess the complex physiological and phenotypical response of endocrine activity. We have described genomics tools here that enable a thorough analysis of various modes of action of endocrine-active compounds. Using these techniques and appropriate models, ER-dependent and -independent estrogen-mediated actions can be deciphered. In the example presented, we were able to characterize two different Ishikawa cell lines based on gene expression profiling and showed that these cell lines may provide a suitable model system to elucidate complex estrogen signaling. These models may, therefore, serve also as an advanced tool to analyze and characterize endocrine-active compounds more thoroughly.

In conclusion, this study identified marker genes for ER-mediated and non-ER-mediated estrogenic effects in human endometrial cells. The application of gene expression profiling revealed the complex network of estrogen signaling pathways, facilitated the analysis of estrogen-active compounds, and may help to identify new targets for drug development. The elucidation of these multiple regulatory pathways remains an interesting challenge for the future.

References

Aardema MJ, MacGregor JT (2002) Toxicology and genetic toxicology in the new era of 'toxicogenomics': impact of '-omics' technologies. *Mutat Res* **499**, 13–25.

An JP, Tzagarakis-Foster C, Scharschmidt TC, Lomri N, Leitman DC (2001) Estrogen receptor b-selective transcriptional activity and recruitment of coregulators by phytoestrogens. *J Biol Chem* **276**, 17808–17814.

Arboleda MJ, Lyons JF, Kabbinavar FF, Bray MR, Snow BE, Ayala R, Danino M, Karlan BY, Slamon DJ (2003) Overexpression of AKT2/protein kinase Bbeta leads to up-regulation of beta1 integrins, increased invasion, and metastasis of human breast and ovarian cancer cells. *Cancer Res* **63**, 196–206.

Auchus RJ, Fuqua SA (1994) Prognostic factors and variant estrogen receptor RNAs in clinical breast cancer. *Nucl Med Biol* **21**, 449–454.

Baldwin WS, Curtis SW, Cauthen CA, Risinger JI, Korach KS, Barrett JC (1998) BG-1 ovarian cell line: an alternative model for examining estrogen-dependent growth *in vitro*. *In Vitro Cell Dev Biol Anim* **34**, 649–654.

Barnes M, Freudenberg J, Thompson S, Aronow B, Pavlidis P (2005) Experimental comparison and cross-validation of the Affymetrix and Illumina gene expression analysis platforms. *Nucleic Acids Res* **33**, 5914–5923.

Campbell IG, Russell SE, Choong DY, Montgomery KG, Ciavarella ML, Hooi CS, Cristiano BE, Pearson RB, Phillips WA (2004) Mutation of the *PIK3CA* gene in ovarian and breast cancer. *Cancer Res* **64**, 7678–7681.

Chang EC, Frasor J, Komm B, Katzenellenbogen BS (2006) Impact of estrogen receptor beta on gene networks regulated by estrogen receptor alpha in breast cancer cells. *Endocrinology* **147**, 4831–4842.

Chee M, Yang R, Hubbell E, Berno A, Huang XC, Stern D, Winkler J, Lockhart DJ, Morris MS, Fodor SP (1996) Accessing genetic information with high-density DNA arrays. *Science* **274**, 610–614.

Cheek AO, McLachlan JA (1998) Environmental hormones and the male reproductive system. *J Androl* **19**, 5–10.

Chernicky CL, Yi LJ, Tan HQ, Gan SU, Ilan J (2000) Treatment of human breast cancer cells with antisense RNA to the type I insulin-like growth factor receptor inhibits cell growth, suppresses tumorigenesis, alters the metastatic potential, and prolongs survival *in vivo*. *Cancer Gene Ther* **7**, 384–395.

Ciocca DR, Vargas-Roig LM (1995) Estrogen receptors in human nontarget tissues: biological and clinical implications. *Endocrine Rev* **16**, 35–62.

Colditz GA, Hankinson SE, Hunter DJ, Willett WC, Manson JE, Stampfer MJ, Hennekens C, Rosner B, Speizer FE (1995) The use of estrogens and progestins and the risk of breast cancer in postmenopausal women. *N Engl J Med* **332**, 1589–1593.

Coleman KM, Smith CL (2001) Intracellular signaling pathways: nongenomic actions of estrogens and ligand-independent activation of estrogen receptors. *Front Biosci* **6**, D1379–D1391.

Couse JF, Korach KS (1999) Reproductive phenotypes in the estrogen receptor-alpha knockout mouse. *Ann Endocrinol (Paris)* **60**, 143–148.

Couse JF, Lindzey J, Grandien K, Gustafsson JA, Korach KS (1997) Tissue distribution and quantitative analysis of estrogen receptor-alpha (ERalpha) and estrogen receptor-beta (ERbeta) messenger ribonucleic acid in the wild-type and ERalpha-knockout mouse. *Endocrinology* **138**, 4613–4621.

Couse JF, Curtis Hewitt S, Korach KS (2000) Receptor null mice reveal contrasting roles for estrogen receptor alpha and beta in reproductive tissues. *J Steroid Biochem Mol Biol* **74**, 287–296.

De Kretser DM (1998) Are sperm counts really falling? *Reprod Fertil Dev* **10**, 93–95.

Diel P, Olff S, Schmidt S, Michna H (2002) Effects of the environmental estrogens bisphenol A, *o,p′*-DDT, *p-tert*-octylphenol and coumestrol on apoptosis induction, cell proliferation and the expression of estrogen sensitive molecular parameters in the human breast cancer cell line MCF-7. *J Steroid Biochem Mol Biol* **80**, 61–70.

Eddy EM, Wasburn TF, Bunch EH, Goulding EH, Gladen BC, Lubahn DB, Korach KS (1996) Targeted disruption of the estrogen receptor gene in male mice causes alteration of spermatogenesis and infertility. *Endocrinology* **137**, 4796–4805.

Elvidge G (2006) Microarray expression technology: from start to finish. *Pharmacogenomics* **7**, 123–134.

Evans RM (1988) The steroid and thyroid receptor superfamily. *Science* **240**, 889–895.

Filardo EJ, Thomas P (2005) GPR30, a seven-transmembrane-spanning estrogen receptor that triggers EGF release. *Trends Endocrinol Metab* **16**, 362–367.

Filardo EJ, Quinn JA, Bland KI, Frackelton Jr AR (2000) Estrogen-induced activation of Erk-1 and Erk-2 requires the G protein-coupled receptor homolog, GPR30, and occurs via trans-activation of the epidermal growth factor receptor through release of HB-EGF. *Mol Endocrinol* **14**, 1649–1660.

Frasor J, Danes JM, Komm B, Chang KC, Lyttle CR, Katzenellenbogen BS (2003) Profiling of estrogen up- and down-regulated gene expression in human breast cancer cells: insights into gene networks and pathways underlying estrogenic control of proliferation and cell phenotype. *Endocrinology* **144**, 4562–4574.

Frasor J, Chang EC, Komm B, Lin CY, Vega VB, Liu ET, Miller LD, Smeds J, Bergh J, Katzenellenbogen BS (2006) Gene expression preferentially regulated by tamoxifen in breast cancer cells and correlations with clinical outcome. *Cancer Res* **66**, 7334–7340.

Gee AC, Carlson KE, Martini PGV, Katzenellenbogen BS, Katzenellenbogen JA (1999) Coactivator peptides have a differential stabilizing effect on the binding of estrogens and antiestrogens with the estrogen receptor. *Mol Endocrinol* **13**, 1912–1923.

Gelbke HP, Kayser M, Poole A (2004) OECD test strategies and methods for endocrine disruptors. *Toxicology* **205**, 17–25.

Gelbke HP, Hofmann A, Owens JW, Freyberger A (2007) The enhancement of the subacute repeat dose toxicity test OECD TG 407 for the detection of endocrine active chemicals: comparison with toxicity tests of longer duration. *Arch Toxicol* **81**, 227–250.

Ghosh MG, Thompson DA, Weigel RJ (2000) *PDZK1* and *GREB1* are estrogen-regulated genes expressed in hormone-responsive breast cancer. *Cancer Res* **60**, 6367–6375.

Grady D, Gebretsadik T, Kerlikowske K, Ernster V, Petitti D (1995) Hormone replacement therapy and endometrial cancer risk: a meta-analysis. *Obstet Gynecol* **85**, 304–313.

Green S, Walter P, Greene G, Krust A, Goffin C, Jensen E, Scrace G, Waterfield M, Chambon P (1986) Cloning of the human oestrogen receptor cDNA. *J Steroid Biochem* **24**, 77–83.

Greene GL, Press MF (1986) Structure and dynamics of the estrogen receptor. *J Steroid Biochem* **24**, 1–7.

Grimberg A (2003) Mechanisms by which IGF-I may promote cancer. *Cancer Biol Ther* **2**, 630–635.

Gunderson KL, Kruglyak S, Graige MS, Garcia F, Kermani BG, Zhao C, Che D, Dickinson T, Wickham E, Bierle J, Doucet D, Milewski M, Yang R, Siegmund C, Haas J, Zhou L, Oliphant A, Fan JB, Barnard S, Chee MS (2004) Decoding randomly ordered DNA arrays. *Genome Res* **14**, 870–877.

Harnish DC, Scicchitano MS, Adelman SJ, Lyttle CR, Karathanasis SK (2000) The role of CBP in estrogen receptor cross-talk with nuclear factor-kappaB in HepG2 cells. *Endocrinology* **141**, 3403–3411.

Hewitt SC, Collins J, Grissom S, Deroo B, Korach KS (2005) Global uterine genomics in vivo: microarray evaluation of the estrogen receptor alpha-growth factor cross-talk mechanism. *Mol Endocrinol* **19**, 657–668.

Holinka CF, Hata H, Gravanis A, Kuramoto H, Gurpide E (1986) Effects of estradiol on proliferation of endometrial adenocarcinoma cells (Ishikawa line). *J Steroid Biochem* **25**, 781–786.

Ignar-Trowbridge DM, Pimentel M, Teng CT, Korach KS, McLachlan JA (1995) Cross talk between peptide growth factor and estrogen receptor signaling systems. *Environ Health Perspect* **103**, 35–38.

Jiang SY, Jordan VC (1992) Growth regulation of estrogen receptor-negative breast cancer cells transfected with complementary DNAs for estrogen receptor. *J Natl Cancer Inst* **84**, 580–591.

Johnston H, Kneer J, Chackalaparampil I, Yaciuk P, Chrivia J (1999) Identification of a novel SNF2/SWI2 protein family member, SRCAP, which interacts with CREB-binding protein. *J Biol Chem* **274**, 16370–16376.

Junutula JR, De Maziere AM, Peden AA, Ervin KE, Advani RJ, van Dijk SM, Klumperman J, Scheller RH (2004) Rab14 is involved in membrane trafficking between the Golgi complex and endosomes. *Mol Biol Cell* **15**, 2218–2229.

Katzenellenbogen BS, Frasor J (2004) Therapeutic targeting in the estrogen receptor hormonal pathway. *Semin Oncol* **31**, 28–38.

Kelly MJ, Wagner EJ (1999) Estrogen modulation of G-protein-coupled receptors. *Trends Endocrinol Metab* **10**, 369–374.

Kennel P, Pallen C, Barale-Thomas E, Espuna G, Bars R (2003) Tamoxifen: 28-day oral toxicity study in the rat based on the Enhanced OECD Test Guideline 407 to detect endocrine effects. *Arch Toxicol* **77**, 487–499.

Kliewer SA, Lehmann JM, Willson TM (1999) Orphan nuclear receptors: shifting endocrinology into reverse. *Science* **284**, 757–760.

Klotz DM, Hewitt SC, Ciana P, Raviscioni M, Lindzey JK, Foley J, Maggi A, DiAugustine RP, Korach KS (2002) Requirement of estrogen receptor-alpha in insulin-like growth factor-1 (IGF-1)-induced uterine responses and *in vivo* evidence for IGF-1/estrogen receptor cross-talk. *J Biol Chem* **277**, 8531–8537.

Knobbe CB, Reifenberger G (2003) Genetic alterations and aberrant expression of genes related to the phosphatidyl-inositol-3′-kinase/protein kinase B (Akt) signal transduction pathway in glioblastomas. *Brain Pathol* **13**, 507–518.

Kooijman R (2006) Regulation of apoptosis by insulin-like growth factor (IGF)-I. *Cytokine Growth Factor Rev* **17**, 305–323.

Korach KS (1979) Estrogen action in the mouse uterus: characterization of the cytosol and nuclear receptor systems. *Endocrinology* **104**, 1324–1332.

Korach KS, Couse JF, Curtis SW, Washburn TF, Lindzey J, Kimbro KS, Eddy EM, Migliaccio S, Snedeker SM, Lubahn DB, Schomberg DW, Smith EP (1996) Estrogen receptor gene disruption: molecular characterization and experimental and clinical phenotypes. *Recent Prog Horm Res* **51**, 159–186; discussion 186–158.

Kuhn K, Baker SC, Chudin E, Lieu M-H, Oeser S, Bennett H, Rigault P, Barker D, McDaniel T, Chee MS (2004) A novel, high-performance random array platform for quantitative gene expression profiling. *Genome Res* **14**, 2347–2356.

Kuiper GGJM, Enmark E, Pelto-Huikko M, Nilsson S, Gustafsson J (1996) Cloning of a novel estrogen receptor expressed in rat prostate and ovary. *Proc Natl Acad Sci U S A* **93**, 5925–5930.

Kushner PJ, Agard DA, Greene GL, Scanlan TS, Shiau AK, Uht RM, Webb P (2000) Estrogen receptor pathways to AP-1. *J Steroid Biochem Mol Biol* **74**, 311–317.

Le Mellay V, Grosse B, Lieberherr M (1997) Phospholipase C beta and membrane action of calcitriol and estradiol. *J Biol Chem* **272**, 11902–11907.

Leader JE, Wang C, Fu M, Pestell RG (2006) Epigenetic regulation of nuclear steroid receptors. *Biochem Pharmacol* **72** (11), 1589–1596.

Levenson AS, Kwaan HC, Svoboda KM, Weiss IM, Sakurai S, Jordan VC (1998) Oestradiol regulation of the components of the plasminogen–plasmin system in MDA-MB-231 human breast cancer cells stably expressing the oestrogen receptor. *Br J Cancer* **78**, 88–95.

Levenson AS, Svoboda KM, Pease KM, Kaiser SA, Chen B, Simons LA, Jovanovic BD, Dyck PA, Jordan VC (2002) Gene expression profiles with activation of the estrogen receptor alpha-selective estrogen receptor modulator complex in breast cancer cells expressing wild-type estrogen receptor. *Cancer Res* **62**, 4419–4426.

Levin ER (1999) Cellular functions of the plasma membrane estrogen receptor. *Trends Endocrinol Metab* **10**, 374–377.

Li X, Zhang S, Safe S (2006) Activation of kinase pathways in MCF-7 cells by 17beta-estradiol and structurally diverse estrogenic compounds. *J Steroid Biochem Mol Biol* **98**, 122–132.

Lin HJ, Hsieh FC, Song H, Lin J (2005) Elevated phosphorylation and activation of PDK-1/AKT pathway in human breast cancer. *Br J Cancer* **93**, 1372–1381.

Lockhart DJ, Dong H, Byrne MC, Follettie MT, Gallo MV, Chee MS, Mittmann M, Wang C, Kobayashi M, Horton H, Brown EL (1996) Expression monitoring by hybridization to high-density oligonucleotide arrays. *Nat Biotechnol* **14**, 1675–1680.

Lubahn DB, Moyer JS, Golding TS, Couse JF, Korach KS, Smithies O (1993) Alteration of reproductive function but not prenatal sexual development after insertional disruption of the mouse estrogen receptor gene. *Proc Natl Acad Sci U S A* **90**, 11162–11166.

Mangelsdorf DJ, Thummel C, Beato M, Herrlich P, Schuetz G, Umesono K, Blumberg B, Kastner P, Mark M, Chambon P, Evans RM (1995) The nuclear receptor superfamily: the second decade. *Cell* **83**, 835–839.

Matsumura A, Ghosh A, Pope GS, Darbre PD (2005) Comparative study of oestrogenic properties of eight phytoestrogens in MCF7 human breast cancer cells. *J Steroid Biochem Mol Biol* **94**, 431–443.

McKay LI, Cidlowski JA (1999) Molecular control of immune/inflammatory responses: interactions between nuclear factor-kappa B and steroid receptor-signaling pathways. *Endocr Rev* **20**, 435–459.

Migliaccio A, Di Domenico M, Castoria G, de Falco A, Bontempo P, Nola E, Auricchio F (1996) Tyrosine kinase/p21ras/MAP-kinase pathway activation by estradiol-receptor complex in MCF-7 cells. *EMBO J* **15**, 1292–1300.

Moggs JG, Orphanides G (2001) Estrogen receptors: orchestrators of pleiotropic cellular responses. *EMBO Rep* **2**, 775–781.

Moggs JG, Orphanides G (2003) Genomic analysis of stress response genes. *Toxicol Lett* **140–141**, 149–153.

Mosselman S, Polman J, Dijkema R (1996) ERβ: identification and characterization of a novel human estrogen receptor. *FEBS Lett* **392**, 49–53.

Mueller SO (2002) Overview of *in vitro* tools to assess the estrogenic and antiestrogenic activity of phytoestrogens. *J Chromatogr B Analyt Technol Biomed Life Sci* **777**, 155–165.

Mueller SO, Korach KS (2001) Immortalized testis cell lines from estrogen receptor (ER) alpha knock-out and wild-type mice expressing functional ERalpha or ERbeta. *J Androl* **22**, 652–664.

Mueller SO, Tahara H, Barrett JC, Korach KS (2000) Immortalization of mammary cells from estrogen receptor alpha knock-out and wild-type mice. *In Vitro Cell Dev Biol Anim* **36**, 620–624.

Mueller SO, Hall JM, Swope DL, Pedersen LC, Korach KS (2003a) Molecular determinants of the stereoselectivity of agonist activity of estrogen receptors (ER) alpha and beta. *J Biol Chem* **278**, 12255–12262.

Mueller SO, Kling M, Arifin Firzani P, Mecky A, Duranti E, Shields-Botella J, Delansorne R, Broschard T, Kramer PJ (2003b) Activation of estrogen receptor alpha and ERbeta by 4-methylbenzylidene-camphor in human and rat cells: comparison with phyto- and xenoestrogens. *Toxicol Lett* **142**, 89–101.

Mueller SO, Simon S, Chae K, Metzler M, Korach KS (2004) Phytoestrogens and their human metabolites show distinct agonistic and antagonistic properties on estrogen receptor alpha (ERalpha) and ERbeta in human cells. *Toxicol Sci* **80**, 14–25.

Murphy LC, Dotzlaw H, Leygue E, Douglas D, Coutts A, Watson PH (1997) Estrogen receptor variants and mutations. *J Steroid Biochem Mol Biol* **62**(5–6), 363–372.

Nakagawa H, Tsuta K, Kiuchi K, Senzaki H, Tanaka K, Hioki K, Tsubura A (2001) Growth inhibitory effects of diallyl disulfide on human breast cancer cell lines. *Carcinogenesis* **22**, 891–897.

Nakajima K, Matsuda T, Fujitani Y, Kojima H, Yamanaka Y, Nakae K, Takeda T, Hirano T (1995) Signal transduction through IL-6 receptor: involvement of multiple protein kinases, stat factors, and a novel H7-sensitive pathway. *Ann N Y Acad Sci* **762**, 55–70.

Nishikawa J, Saito K, Goto J, Dakeyama F, Matsuo M, Nishihara T (1999) New screening methods for chemicals with hormonal activities using interaction of nuclear hormone receptor with coactivator. *Toxicol Appl Pharmacol* **154**, 76–83.

Normanno N, Di Maio M, De Maio E, De Luca A, de Matteis A, Giordano A, Perrone F (2005) Mechanisms of endocrine resistance and novel therapeutic strategies in breast cancer. *Endocr Relat Cancer* **12**, 721–747.

OECD (1995) Repeated dose 28-day oral toxicity study in rodents. In *OECD Environment Health and Safety Publications Series on Testing and Assessment No. 407*. Environment Directorate OECD, Paris.

OECD (2006a) OECD report of the initial work towards the validation of the rodent uterotrophic assay – phase one. In *OECD Environment Health and Safety Publications Series on Testing and Assessment No. 65*. Environment Directorate OECD, Paris.

OECD (2006b) OECD report of the validation of the rodent uterotrophic bioassay – phase 2, testing of potent and weak oestrogen agonists by multiple laboratories. In *OECD Environment Health and Safety Publications Series on Testing and Assessment No. 66*. Environment Directorate OECD, Paris.

OECD (2007) Uterotrophic bioassay in rodents: a short-term screening test for oestrogenic properties. In *OECD Environment Health and Safety Publications Series on Testing and Assessment No. 440*. Environment Directorate OECD, Paris.

Orphanides G, Kimber I (2003) Toxicogenetics: applications and opportunities. *Toxicol Sci* **75**, 1–6.

Pietras RJ, Szego CM (1977) Specific binding sites for oestrogen at the outer surfaces of isolated endometrial cells. *Nature* **265**, 69–72.

Pike MC, Spicer DV, Dahmoush L, Press MF (1993) Estrogens, progestogens, normal breast cell proliferation, and breast cancer risk. *Epidemiol Rev* **15**, 17–35.

Platet N, Prevostel C, Derocq D, Joubert D, Rochefort H, Garcia M (1998) Breast cancer cell invasiveness: correlation with protein kinase C activity and differential regulation by phorbol ester in estrogen receptor-positive and -negative cells. *Int J Cancer* **75** (5), 750–756.

Proikas-Cezanne T, Gaugel A, Frickey T, Nordheim A (2006) Rab14 is part of the early endosomal clathrin-coated TGN microdomain. *FEBS Lett* **580**, 5241–5246.

Rae JM, Johnson MD, Scheys JO, Cordero KE, Larios JM, Lippman ME (2005) *GREB1* is a critical regulator of hormone dependent breast cancer growth. *Breast Cancer Res Treat* **92**, 141–149.

Rae JM, Johnson MD, Cordero KE, Scheys JO, Larios JM, Gottardis MM, Pienta KJ, Lippman ME (2006) *GREB1* is a novel androgen-regulated gene required for prostate cancer growth. *Prostate* **66**, 886–894.

Ragoussis J, Elvidge G (2006) Affymetrix GeneChip system: moving from research to the clinic. *Expert Rev Mol Diagn* **6**, 145–152.

Razandi M, Pedram A, Greene GL, Levin ER (1999) Cell membrane and nuclear estrogen receptors (ERs) originate from a single transcript: studies of ERalpha and ERbeta expressed in Chinese hamster ovary cells. *Mol Endocrinol* **13**, 307–319.

Revankar CM, Cimino DF, Sklar LA, Arterburn JB, Prossnitz ER (2005) A transmembrane intracellular estrogen receptor mediates rapid cell signaling. *Science* **307**, 1625–1630.

Routledge EJ, White R, Parker MG, Sumpter JP (2000) Differential effects of xenoestrogens on coactivator recruitment by estrogen receptor (ER) alpha and ERbeta. *J Biol Chem* **275**, 35986–35993.

Rubanyi GM, Freay AD, Kauser K, Sukovich D, Burton G, Lubahn DB, Couse JF, Curtis SW, Korach KS (1997) Vascular estrogen receptors and endothelium-derived nitric oxide production in the mouse aorta. Gender difference and effect of estrogen receptor gene disruption. *J Clin Invest* **99**, 2429–2437.

Safe SH, Pallaroni L, Yoon K, Gaido K, Ross S, Saville B, McDonnell D (2001) Toxicology of environmental estrogens. *Reprod Fertil Dev* **13**, 307–315.

Shi L, Reid LH, Jones WD, Shippy R, Warrington JA, Baker SC, Collins PJ, de Longueville F, Kawasaki ES, Lee KY, Luo Y, Sun YA, Willey JC, Setterquist RA, Fischer GM, Tong W, Dragan YP, Dix DJ, Frueh FW, Goodsaid FM, Herman D, Jensen RV, Johnson CD, Lobenhofer EK, Puri RK, Schrf U, Thierry-Mieg J, Wang C, Wilson M, Wolber PK, Zhang L, Amur S, Bao W, Barbacioru CC, Lucas AB, Bertholet V, Boysen C, Bromley B, Brown D, Brunner A, Canales R, Cao XM, Cebula TA, Chen JJ, Cheng J, Chu TM, Chudin E, Corson J, Corton JC, Croner LJ, Davies C, Davison TS, Delenstarr G, Deng X, Dorris D, Eklund AC, Fan XH, Fang H, Fulmer-Smentek S, Fuscoe JC, Gallagher K, Ge W, Guo L, Guo X, Hager J, Haje PK, Han J, Han T, Harbottle HC, Harris SC, Hatchwell E, Hauser CA, Hester S, Hong H, Hurban P, Jackson SA, Ji H, Knight CR, Kuo WP, LeClerc JE, Levy S, Li QZ, Liu C, Liu Y, Lombardi MJ, Ma Y, Magnuson SR, Maqsodi B, McDaniel T, Mei N, Myklebost O, Ning B, Novoradovskaya N, Orr MS, Osborn TW, Papallo A, Patterson TA, Perkins RG, Peters EH, Peterson R, Philips KL, Pine PS, Pusztai L, Qian F, Ren H, Rosen M, Rosenzweig BA, Samaha RR, Schena M, Schroth GP, Shchegrova S, Smith DD, Staedtler F, Su Z, Sun H, Szallasi Z, Tezak Z, Thierry-Mieg D, Thompson KL, Tikhonova I, Turpaz Y, Vallanat B, Van C, Walker SJ, Wang SJ, Wang Y, Wolfinger R, Wong A, Wu J, Xiao C, Xie Q, Xu J, Yang W, Zhong S, Zong Y, Slikker W, Jr (2006) The MicroArray Quality Control (MAQC) project shows inter- and intraplatform reproducibility of gene expression measurements. *Nat Biotechnol* **24**, 1151–1161.

Simoncini T, Rabkin E, Liao JK (2003) Molecular basis of cell membrane estrogen receptor interaction with phosphatidylinositol 3-kinase in endothelial cells. *Arterioscler Thromb Vasc Biol* **23**, 198–203.

Simoncini T, Mannella P, Genazzani AR (2006) Rapid estrogen actions in the cardiovascular system. *Ann N Y Acad Sci* **1089**, 424–430.

Smith CA, McClive PJ, Sinclair AH (2005) Temporal and spatial expression profile of the novel armadillo-related gene, *Alex2*, during testicular differentiation in the mouse embryo. *Dev Dyn* **233**, 188–193.

Smith CL (1998) Cross-talk between peptide growth factor and estrogen receptor signaling pathways. *Biol Reprod* **58** (3), 627–632.

Song RX (2007) Membrane-initiated steroid signaling action of estrogen and breast cancer. *Semin Reprod Med* **25**, 187–197.

Song RX, Santen RJ (2006) Membrane initiated estrogen signaling in breast cancer. *Biol Reprod* **75**, 9–16.

Song RX, Fan P, Yue W, Chen Y, Santen RJ (2006) Role of receptor complexes in the extranuclear actions of estrogen receptor alpha in breast cancer. *Endocr Relat Cancer* **13** (Suppl 1), S3–S13.

Song RX, Zhang Z, Chen Y, Bao Y, Santen RJ (2007) Estrogen signaling via a linear pathway involving insulin-like growth factor I receptor, matrix metalloproteinases, and epidermal growth factor receptor to activate mitogen-activated protein kinase in MCF-7 breast cancer cells. *Endocrinology* **148**, 4091–4101.

Soto AM, Sonnenschein C, Chung KL, Fernandez MF, Olea N, Serrano FO (1995) The E-SCREEN assay as a tool to identify estrogens: an update on estrogenic environmental pollutants. *Environ Health Perspect* **103**, 113–122.

Soulez M, Parker MG (2001) Identification of novel oestrogen receptor target genes in human ZR75-1 breast cancer cells by expression profiling. *J Mol Endocrinol* **27**, 259–274.

Speir E, Yu ZX, Takeda K, Ferrans VJ, Cannon III RO (2000) Competition for p300 regulates transcription by estrogen receptors and nuclear factor-kappaB in human coronary smooth muscle cells. *Circ Res* **87**, 1006–1011.

Steemers FJ, Gunderson KL (2005) Illumina, Inc. *Pharmacogenomics* **6**, 777–782.

Stopper H, Schmitt E, Gregor C, Mueller SO, Fischer WH (2003) Increased cell proliferation is associated with genomic instability: elevated micronuclei frequencies in estradiol-treated human ovarian cancer cells. *Mutagenesis* **18**, 243–247.

Sun J, Meyers MJ, Fink BE, Rajendran R, Katzenellenbogen JA, Katzenellenbogen BS (1999) Novel ligands that function as selective estrogens or antiestrogens for estrogen receptor-alpha or estrogen receptor-beta. *Endocrinology* **140**, 800–804.

Tesarik J, Mendoza C (1995) Nongenomic effects of 17 beta-estradiol on maturing human oocytes: relationship to oocyte developmental potential. *J Clin Endocrinol Metab* **80**, 1438–1443.

Thomas P, Pang Y, Filardo EJ, Dong J (2005) Identity of an estrogen membrane receptor coupled to a G protein in human breast cancer cells. *Endocrinology* **146**, 624–632.

Truss M, Chalepakis G, Pina B, Barettino D, Bruggemeier U, Kalff M, Slater EP, Beato M (1992) Transcriptional control by steroid hormones. *J Steroid Biochem Mol Biol* **41**, 241–248.

Valverde MA, Rojas P, Amigo J, Cosmelli D, Orio P, Bahamonde MI, Mann GE, Vergara C, Latorre R (1999) Acute activation of Maxi-K channels (hSlo) by estradiol binding to the beta subunit. *Science* **285**, 1929–1931.

Van Meeuwen JA, Korthagen N, de Jong PC, Piersma AH, Van Den Berg M (2007a) (Anti)estrogenic effects of phytochemicals on human primary mammary fibroblasts, MCF-7 cells and their co-culture. *Toxicol Appl Pharmacol* **221**, 372–383.

Van Meeuwen JA, Van Den Berg M, Sanderson JT, Verhoef A, Piersma AH (2007b) Estrogenic effects of mixtures of phyto- and synthetic chemicals on uterine growth of prepubertal rats. *Toxicol Lett* **170**, 165–176.

Vignon F, Bouton MM, Rochefort H (1987) Antiestrogens inhibit the mitogenic effect of growth factors on breast cancer cells in the total absence of estrogens. *Biochem Biophys Res Commun* **146**, 1502–1508.

Watters JJ, Campbell JS, Cunningham MJ, Krebs EG, Dorsa DM (1997) Rapid membrane effects of steroids in neuroblastoma cells: effects of estrogen on mitogen activated protein kinase signalling cascade and *c-fos* immediate early gene transcription. *Endocrinology* **138**, 4030–4033.

Webb P, Nguyen P, Valentine C, Lopez GN, Kwok GR, McInerney E, Katzenellenbogen BS, Enmark E, Gustafsson JA, Nilsson S, Kushner PJ (1999) The estrogen receptor enhances AP-1 activity by two distinct mechanisms with different requirements for receptor transactivation functions. *Mol Endocrinol* **13**, 1672–1685.

Yeung KY, Ruzzo WL (2001) Principle component analysis for clustering gene expression data. *Bioinformatics* **17**, 763–774.

Zhou G, Cummings R, Li Y, Mitra S, Wilkinson HA, Elbrecht A, Hermes JD, Schaeffer JM, Smith RG, Moller DE (1998) Nuclear receptors have distinct affinities for coactivators: characterization by fluorescence resonance energy transfer. *Mol Endocrinol* **12**, 1594–1604.

Zidek N, Hellmann J, Kramer PJ, Hewitt PG (2007) Acute hepatotoxicity: a predictive model based on focused illumina microarrays. *Toxicol Sci* **99**, 289–302.

Zivadinovic D, Watson CS (2005) Membrane estrogen receptor-alpha levels predict estrogen-induced ERK1/2 activation in MCF-7 cells. *Breast Cancer Res* **7**, R130–R144.

Zivadinovic D, Gametchu B, Watson CS (2005) Membrane estrogen receptor-alpha levels in MCF-7 breast cancer cells predict cAMP and proliferation responses. *Breast Cancer Res* **7**, R101–R112.

11

Escherichia coli Stress Response as a Tool for Detection of Toxicity

Arindam Mitra, Nabarun Chakraborti and Suman Mukhopadhyay

11.1 Introduction

The advent of the microarray has opened new avenues for toxicologists to collect and interpret data (Nuwaysir *et al.*, 1999; Storck *et al.*, 2002; de Longueville *et al.*, 2004; Shioda, 2004). It usually involves a comparison of global gene expression between normal and drug-treated cells under *in vitro* conditions. The incorporation of genomics, bioinformatics, and large-scale sequencing information has resulted in the construction of gene chips, which enable speedy screening of new targets for important cellular processes, including toxicity. The emerging branch of toxicogenomics integrates application of functional genomics technologies and offers several advantages over that of conventional toxicology in terms of cost and time effectiveness, sensitivity, and enhanced correlation between experimental models and human. Potential applications of this discipline are mechanistic insight of metabolic or biological pathways leading to toxicity, especially metabolic processes (at the level of transcription) affected by chemical, environmental, or xenobiotic treatments, screening of probable drug candidates, facilitating the prediction of toxicity of unknown compounds, and improving interspecies and *in vitro–in vivo* extrapolations (Hamadeh *et al.*, 2002; Aubrecht and Caba, 2005).

The evolution of toxicogenomics has matured over the years, with several series of developments in toxicological sciences. Previously, animal toxicity was assessed by traditional methods, such as tissue pathology, system-level toxicity, and overall mortality. However, animal bioassay was often lengthy, labor intensive, expensive and limited in information (Smith, 2001; Tennant, 2002; Gant, 2003). Screening of more than 50 000 known

Toxicogenomics: A Powerful Tool for Toxicity Assessment   Edited by S. C. Sahu
© 2008 John Wiley & Sons, Ltd

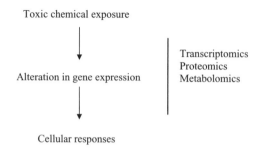

Figure 11.1　*Principle of evaluation of toxicity in toxicogenomics.*

chemicals for toxicity would be unfeasible using conventional methods; hence, newer, alternative strategies are needed.

In recent times, the focus has shifted towards understanding toxicity at the molecular level. In the last 30 years, evaluation of toxicity has undergone a remarkable transformation from assessing a single molecule change to the effect on the entire genome. Genomic information plays a key role in understanding of the molecular attributes of toxicity; for example, the genetic background of an individual could influence metabolism, absorption, excretion, or susceptibility of a metabolite or a chemical entity. The integration of genomics into the field of toxicological research will significantly advance our knowledge of molecular toxicity and key regulatory pathways that affect such processes (Figure 11.1). The potential usefulness of genomics could be immensely important and often involves approaches that utilize candidate targets which are affected by environmental stimulants. Conversely, a meticulous approach must be followed while analyzing genomic data, and experimentation for validation must be integrated within such studies (Pognan, 2007; Thybaud *et al.*, 2007).

11.2　Extraintestinal Pathogenic *Escherichia coli*

Escherichia coli, a bacteria-type species of the family Enterobacteriaceae, is naturally distributed in the intestinal microbial flora of homeothermic animals, including birds and humans (Janda and Abbott, 2005). Strains of *E. coli* are broadly categorized in three groups: commensals, intestinal pathogenic, and extraintestinal pathogenic. The recently added third group, termed extraintestinal pathogenic *E. coli* (ExPEC), has been formed based on the presence of specific virulence factors and the ability to cause organ infection outside the intestine (Johnson and Russo, 2002a; Kaper *et al.*, 2004; Smith *et al.*, 2007). Typically, ExPEC characteristic virulence factors aid in invasion and colonization of the microbe which lead to infection in extraintestinal sites. Some ExPEC-specific virulence factors include adhesins (e.g. Type 1 fimbriae or P fimbriae), factors that evade defense mechanisms (e.g. capsules, lipopolysaccharides), toxins (e.g. hemolysins), and factors to acquire nutrient availability (e.g. siderophores) (Johnson and Russo, 2005).

ExPECs are a growing concern, as evidenced by being causative agents of a plethora of diseases, including urinary tract infections (UTIs), neonatal meningitis, pneumonia, septicemia, osteomyelitis, and other extraintestinal infections (Johnson and Russo, 2002b;

Kim *et al.*, 2005; Lloyd *et al.*, 2007). Among the ExPECs, uropathogenic *E. coli* (UPEC) and avian pathogenic *E. coli* (APEC) cause significant morbidity and/or mortality in humans and poultry respectively. UPEC is the leading cause of UTIs in the USA. Every year in the USA, UPEC-associated UTI results in 6–8 billion cases of uncomplicated cystitis with a healthcare cost of $1 billion, 250 000 cases of uncomplicated pyelonephritis with a direct cost of $175 million, and 250 000–525 000 cases of catheter-associated UTI healthcare, the cost of which is $170–350 million (Johnson and Russo, 2003). APEC, on the other hand, is the leading cause of avian colibacillosis, characterized by air sacculitis, pericarditis, peritonitis, salpingitis, polyserositis, septicemia, synovitis, osteomyelitis, and yolk sac infection (Dho-Moulin and Fairbrother, 1999; Ewers *et al.*, 2004). In the USA, cellulitis caused by APEC is the second leading cause of condemnation of broiler chickens and results in an estimated loss of $40 million per year (Norton, 1997). The underlying mechanisms of pathogenesis and toxicity of *E. coli* have become more apparent with the application of genomics, bioinformatics, and molecular biology.

Unlike commensals, many pathogenic bacteria were demonstrated to switch between free-living and host-associated states. Apart from extraintestinal sites, ExPECs have been reported to asymptotically colonize in intestinal sites like commensals (Johnson and Russo, 2002b; Johnson *et al.*, 2003). In contrast, the intestinal pathogenic strains are not capable of asymptomatic colonization in the intestine. The environments in which ExPECs thrive vary and they must endure different stress conditions within the host. Often, the pathogenic bacteria have developed a complex signaling system that turns on specific sets of genes in a given environment and switches off those that are not required in that milieu. Multiple physiochemical cues, such as pH, osmolarity, temperature, and oxygen concentration, might effect such a change in gene expression. Interestingly, the gene expression pattern might be altered due to the presence of different environmental stimuli, including those of various toxic chemicals. Regulatory mechanisms which affect such changes are complex and take place at the levels of transcription and translation. The overall effect of such changes in the genome might be envisioned by the incorporation of genomics into this emerging field of toxicology.

11.3   Stress Responses

The effect of various stress responses on *E. coli* has been studied in greater detail (Pizarro, 1995; Semchyshyn and Lushchak, 2004; Gawande and Griffiths, 2005; Malone *et al.*, 2006; Nonaka *et al.*, 2006; Phadtare *et al.*, 2006; McMahon *et al.*, 2007). Molecular oxygen, for example, plays a crucial role in cellular metabolism; however, the effects of reactive oxygen species (ROS), such as superoxide radical, hydrogen peroxide, and hydroxyl radical, can be deleterious and may even cause apoptosis of aerobic cells. Various strategies, including enzymatic and nonenzymatic defenses have been employed to prevent such damage (Touati, 2000). Enzymatic defense systems, such as superoxide dismutase, catalase, and peroxidase, scavenge superoxide radicals and hydrogen peroxide and convert them into less reactive species. Nonenzymatic antioxidants include vitamins C and E, glutathione and β-carotene. Usually, a balance exists between ROS and antioxidants under normal conditions of the cell. A disruption in this critical balance could lead to oxidative stress either due to excess accumulation of ROS or depletion of antioxidants (Wick and

Egli, 2004). These, in turn, either damage cell components or trigger specific cell signaling pathways, leading to modulation of various cellular processes, improving the health of the cell, or leading to cell death (Imlay, 2003).

Release of ROS changes the oxidation reduction potential within the cell, leading to oxidative stress. The ROS molecules generated can carry out nucleophilic attacks on any electron-deficient group, including biomolecules such as DNA, protein, and lipids, leading to the formation of adduct, covalent binding of ROS to macromolecules, and disruption of cellular functions. The basic mechanisms to remove ROS involve chemical reactions that generate a nonreactive compound by altering gene expression to activate gene products that are designated to deal with toxic insults and turn off those that are not required. Cellular oxidative stresses are controlled either by direct or indirect alteration of gene expression. Chemicals or ROS may activate intracellular receptors that directly regulate transcription of target genes. Alternatively, ROS may interact with other molecules within the cell, which carries on the signal and elicits coordinated responses to cellular toxicity.

Bacteria have developed adaptive responses while shifting from anaerobic to aerobic growth conditions to counteract ROS (Partridge *et al.*, 2006). Usually, these responses are mediated in a coordinated manner by groups of genes termed regulons, each group under a common regulator. One key system is based on the *oxyR* system which acts in response to hydrogen peroxide and induces at least eight genes to counteract oxidative stress, including *ahpFC* encoding alkyl hydroperoxidase, glutathione reducatase encoded by *gor*, *katG* encoding catalase hydroperoxidase and *dps*, a DNA-binding protective protein. OxyR protein is thought to act by binding and stimulating transcription from various promoters upon receiving signals. Many of the OxyR regulon genes are also regulated by the stationery-phase starvation response system programmed by *rpoS*, a sigma 38 protein. The stationery-phase alternative sigma factor *rpoS* controls the expression of several genes involved in cell survival and is essential for expression of various stress resistances (Patten *et al.*, 2004; Vijayakumar *et al.*, 2004; Klauck *et al.*, 2007). Under laboratory conditions, *rpoS* mutants are sensitive to oxidative and osmotic stresses, as well as to temperature and acid shift. On the other hand, the SoxRS system induces many genes to combat the superoxide-generating agents and nitric oxide. The SoxRS response is initiated in two stages. Upon activation, the *soxR* sensor molecule induces *soxS* which, in turn, activates the transcription of the *soxRS* regulon. The stationery-phase alternative sigma factor σ^S is present in many bacterial species belonging to γ subdivision of proteobacteria. The regulation of sigmaS is complex and regulated at the level of transcription, post-transcription and protein stability (Lange and Hengge-Aronis, 1994). In *E. coli*, *rpoS* transcription is regulated by cAMP–CRP complex, as well as by several two-component signaling systems, including the BarA–UvrY system whose role is illustrated (Mukhopadhyay *et al.*, 2000; Hengge-Aronis, 2002; Sugiura *et al.*, 2003; Venturi V. 2003).

Genomics are increasingly more useful in exploring pathways and mechanisms underlying oxidative stress response. DNA microarrays have been used to characterize genes involved in oxidative stress responses. Interestingly, the patterns of gene expression altered in mammary cells in the presence of hydrogen peroxide, menedions, and *t*-butyl hydroperoxide were found to be quite similar regardless of the ROS source (Chuang *et al.*, 2002). Another study showed that the effect of 2,3-dimethoxy-1,4-naphthoquinone, an ROS-generating chemical, in HepG2 cells was comparable to that of heavy metal toxicity (Kawata *et al.*, 2007). Such studies have substantiated the notion that different stimuli can

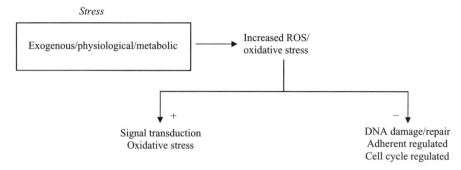

Figure 11.2   *Gene expression profile on increased reactive oxygen species levels.*

lead to generation of ROS and oxidative stress (Figure 11.2). Hence, production of ROS and oxidative stress might be considered as a general stress response.

11.4   Two-component Systems as Signal Transducers

A wide range of toxic insults often alters gene expression profiles in microbes specific to the nature of the chemicals tested. Adaptation to toxic compounds by bacterial species often enables that species to cope better in that environment. This response, appropriately called the adaptive response, refers to the ability of bacteria to withstand harmful, damaging effects of the given stress provided that it has previously been exposed to a similar stress environment at a lower dose. Several types of agent induce an adaptive response, including alkylating agents, heat stress, oxidative stress, and radiation. The adaptive response usually involves modulation of a plethora of genes in a coordinated manner. In bacteria, adaptation to a new environment largely relies on a signal transduction system called a two-component system (TCS). There is no common pathway for adaptation; however, there quite a few common themes exist. In *E. coli*, adaptation to a new environment often involves use of several TCSs that plays a crucial role for survival in an ever-changing environment. TCSs comprise a membrane-bound sensor histidine kinase (HPK) and a cognate response regulator (RR). The sensor kinase undergoes autophosphorylation at a conserved histidine residue upon reception of an appropriate environmental signal; this phosphate group is subsequently transferred to a conserved aspartate residue on the cognate RR. Upon phosphorylation, the response regulatory protein undergoes structural modification and acts as a gene transcription factor and often regulates gene expression or cellular responses, enabling the organism to adapt better in a new environment (Hotch and Silhavy, 1995; Kwon *et al.*, 2000; Robinson *et al.*, 2000). Approximately 60 such TCSs are present in *E. coli* and have been shown to be involved in adaptation, including intracellular metabolism, biofilm formation, global stress response, and virulence. One such system is the BarA-UvrY TCS involved in various physiological functions, including oxidative stress, sigmaS expression, biofilm formation and carbon metabolism.

The BarA (Bacterial Adaptive Response Gene A) sensor kinase was first identified for its ability to suppress a deletion *envZ* mutant by controlling expression of outer membrane proteins (Nagasawa *et al.*, 1992, 1993). BarA is a member of tripartite sensor kinase having three domains: an N-terminal transmitter domain with a conserved histidine residue (H1), a central receiver domain with a conserved aspartate residue (D1) and a C-terminal transmitter domain with a conserved histidine residue H2, also called an Hpt domain. Triggering of this system seems to be mediated in an ATP-dependent manner via a His–Asp–His–Asp phosphorelay cascade. UvrY is a member of the FixJ family and has been recently shown to be a cognate regulator of the sensor kinase BarA (Pernestig *et al.*, 2001). It has an N-terminal phosphoacceptor domain with a conserved aspartic acid residue at position 54, followed by a LuxR-type helix–turn–helix DNA-binding domain in the C-terminal region. It also has a close linkage with *uvrC*, a bicistronic mRNA, even though *uvrY* has no known role in the DNA repair system. Apparently, this system seems to be induced in response to a pH change.

The BarA–UvrY system plays a crucial role in carbon metabolism and biofilm formation. This TCS has also been implicated in hydrogen peroxide resistance. Both the *barA* and *uvrY* mutants were hypersensitive to hydrogen peroxide. It has been reported that the expression of the sensor kinase *barA* could be induced in the presence of weak acids, possibly indicating the significance of this TCS in survival of acid onslaught in stomach and inside macrophages. Additionally, this TCS could be induced in the presence of food preservatives, such as benzoate or bile salts, implying the importance of this TCS in adaptation to various stress responses and persistence.

11.5   Bacterial Biosensors for Detection of Toxicity

The presence of environmental stimulants or toxic chemicals often elicits a variety of stress responses in bacteria. Compounds demonstrating similar toxicities would ideally induce a specific pattern in gene expression. It is hypothesized that compounds that exhibit similar changes in gene expression might have similar mechanisms of action or act in similar biological processes or pathways. Thus, toxicity-induced alteration of gene expression might be used as a signature for classification and characterization of unknown chemicals. Genomic insults due to toxin-induced stimulation induce several stress responses, with alterations in gene expression that are often associated with diverse biological pathways. Once within the host, pathogenic bacteria often deal with diverse stress responses, such as pH, nutrient deprivation, high osmolarity, and oxidative stress. Inflammatory cells or phagocytes possess enzymes that are capable of generating ROS in response to invasion of pathogens. However, excess production of ROS also might affect the phagocytes and the surrounding tissue. Chronic renal scarring in pyelonephritis has been directly correlated with phagocytic oxidative damage. Hence, virulence genes involved in colonization or survival inside the host often have common genes that are affected by stress responses. Such genes have often been used as a sensor for detection and quantization of toxic chemicals in the environment. These sensors have the potential to be a warning system for toxicity detection and thereby reduce harmful effects on the environment.

Whole-cell bacterial biosensors detect gene products of reporter genes that are either naturally present or artificially introduced into the relevant bacterial strain. Commonly used reporter genes include *lacZ* encoding β-galactosidase (*E. coli*), *lux* encoding bacterial luciferase, *luc* encoding firefly luciferase and *gfp* encoding green fluorescent protein. In

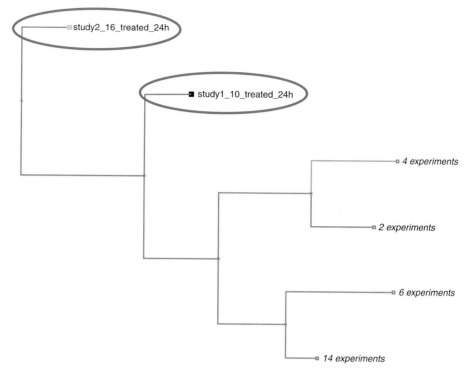

FIGURE 1.1 *Hierarchical clustering of all samples from study 1 and study 2 (positive correlation with average linkage).*

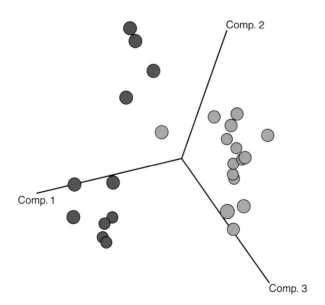

FIGURE 1.3 *Principal component analysis based on all 4327 genes (correlation matrix). (a) For all samples of study 1 (yellow) and all samples of study 2 (green). PC1 captures 16.7 % of variation in the data, PC2 displays 12.6 % and PC3 contains 9.8 %.*

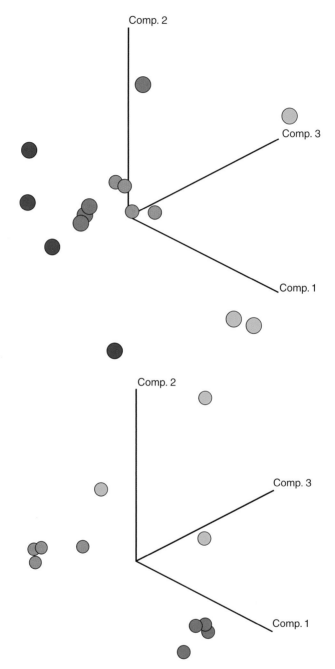

FIGURE 1.3 (b) For all samples of study 1 (brown: controls; yellow: 24 h; blue: 48 h; pink: 96 h). (c) For all samples of study 2 (brown: control; pink: 6 h; yellow: 24 h).

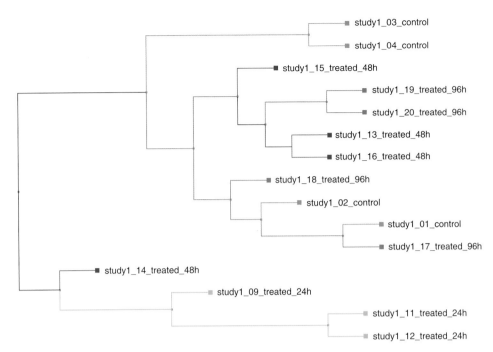

FIGURE 1.4 *Hierarchical clustering of all samples for study 1 (positive correlation with average linkage)*

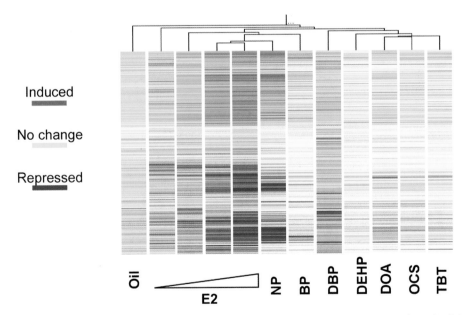

FIGURE 3.2 *Sample data for dendrographic analysis. Four columns, next to oil on the left, are from mice treated with a graded increase in dose of 17-beta-estradiol (E2). NP: nonylphenol; BP: benzophenone; DBP: dibutylphthalate; DEHP: diethylhexyl phthalate; DOA: dioctyl adipate; OCS: octachlorostyrene; TBT: tributyltin. (Unpublished data provided by Professor T. Iguchi, Okazaki National Biology Research Institute.)*

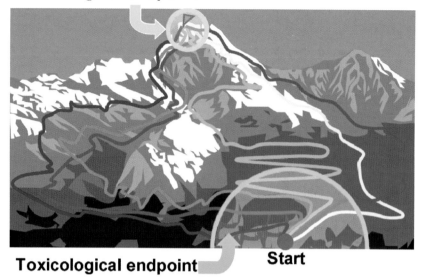

Pathological endpoint

Toxicological endpoint　　　　**Start**

FIGURE 3.4 *'Pathological endpoint' and 'toxicological endpoint'. Each route from the start to the summit represents an individual probabilistic variety of different gene expression profiles. Depending on the characteristics of toxicological impacts, responders may show different toxicological endpoints in different clusters, even if one uses an identical and homogeneous experimental protocol with a highly purified inbred strain (a probabilistic quantum effect based on the uncertainty principle; see text).*

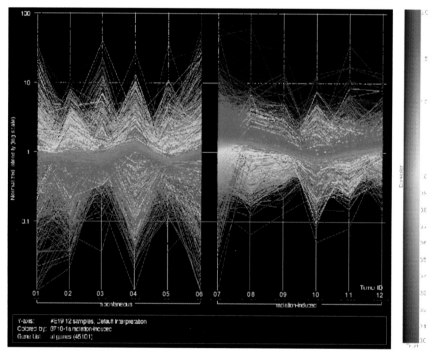

FIGURE 3.5 *Linear configurations of spontaneous and radiation-induced myelogenous leukemias. Six individual data on the left are from spontaneously developed myelogenous leukemias in C3H/He mice. The other six individual cases on the right are from radiation-induced myelogenous leukemias in C3H/He mice after 3 Gy X-ray exposure (Seki et al. 1991; Yoshida et al. 1997). Along the gene expression intensity from the highest (red) to the lowest (green) of group μ07, the same gene in the other groups was connected and designated with the same color. Accordingly, overexpressed genes in the radiation-induced myelogenous leukemia groups are largely repressed in the spontaneous myelogenous leukemias. See text.*

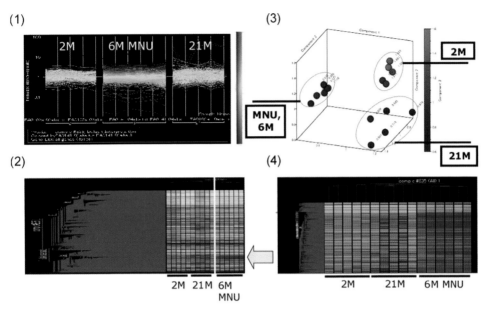

FIGURE 3.8 *(a) Age-related gene expression profiles determined from the bone marrow of 2-month-old and 21-month-old mice are shown in the line configuration. See text. (Six gene expression profiles of bone marrow obtained from mice 6 months after treatment with a single dose of methyl-nitroso-urea (MNU) at 50 mg kg^{-1} b.w.) (b) Two-dimensional dendrographic diagram of the same expressed genes from the bone marrow of 2-month-old and 21-month-old mice. (Six gene expression profiles of bone marrow obtained from mice 6 months after treatment with a single dose of MNU. Arrow indicates the gene cluster specifically up-regulated in the MNU-treated groups; an expanded view of the profiles is shown in (d).) (c) PCA of age control group, five mice each for the 2-month-old and 21-month-old groups, is shown in the three-dimensional contribution scores for components #1, #2 and #3, which discriminate between the clusters from the 2-month-old and 21-month-old groups. Six gene expression profiles of bone marrow obtained from mice 6 months after MNU treatment belong to another separate cluster. (d) Expanded view of profiles indicated by the arrow in (b) showing the gene cluster specifically up-regulated in the MNU-treated groups.*

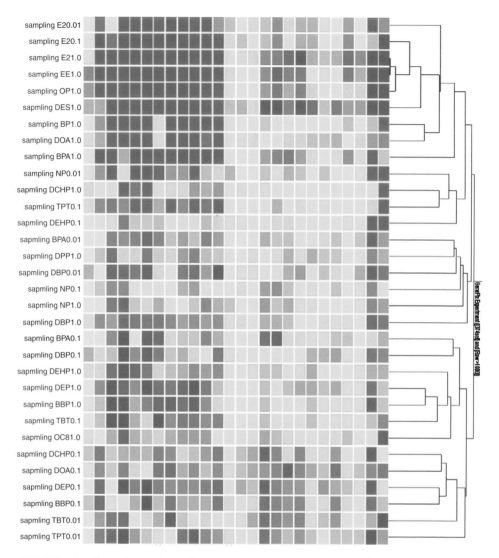

FIGURE 8.1 Clustering analysis of gene expression in mouse uterus after chemical exposure. Ovariectomized adult mice were given a single injection of various chemicals. Numbers indicate the dose (mg kg −1) administered. Red and green bars indicate induced and repressed genes respectively. Dendrogram represents correlation of the gene expression profile.

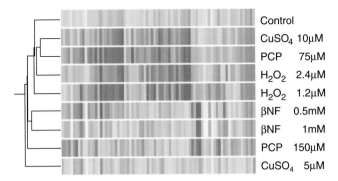

FIGURE 8.2 *Clustering analysis of gene expression in Daphnia magna after chemical exposure. Changes in gene expression estimated by DNA array analysis are indicated as the fold change in comparison with control samples. Chemicals and doses used for the exposure are indicated on the right. The numbers indicate the concentration of the chemicals in micrograms per liter. Red and green bars indicate induced and repressed genes respectively. Dendrogram represents correlation of the gene expression profile. PCP: pentachlorophenol; βNF: β-naphtoflavone.*

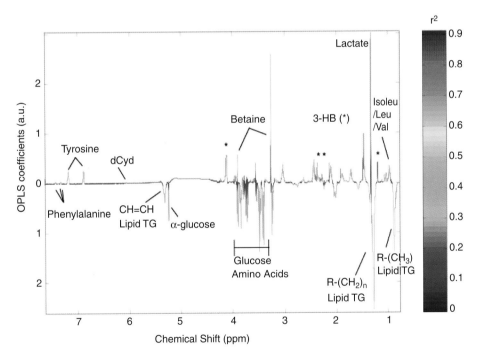

FIGURE 18.3 *O-PLS-DA coefficient loadings of full-resolution serum CPMG NMR data revealing metabolites responsible for discrimination between controls and galN-treated samples. The color scale represents correlation r2 to the discriminant variable. The upper section of the loadings plots represents metabolites increased in the treated class, whereas the lower part represents metabolites decreased in intensity. Key: Lipid TG, lipid triglyceride; 3-HB, d-3-hydroxybutyrate; Isoleu, isoleucine; Leu, leucine; Val, valine; dCyd, 2′-deoxycytidine. Author's unpublished data.*

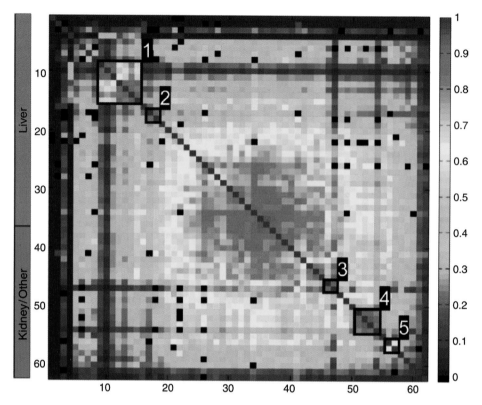

FIGURE 18.4 *COMET-2 expert system similarity matrix for 62 treatments sorted by hierarchical clustering; numbers 1–5 represent similarity blocks and are discussed in the text. Reproduced from Ebbels T, Keun H, Beckonert O, Bollard M, Lindon J, Holmes E, Nicholson J (2007) J Proteome Res 6(11), 4407–22. Reprinted with permission of the American Chemical Society.*

the case of general biosensors, the reporter gene is placed downstream to a constitutively expressed promoter, and a decrease in intensity of signal indicates a decrease in metabolic activity. On the other hand, semi-specific biosensors involve placing a reporter gene downstream to a stress-responsive promoter and an increase in reporter activity indicates an increase in stress (e.g. SOS or heat shock response). Furthermore, specific biosensors incorporate a reporter gene being placed downstream to a regulated promoter or regulatory protein, either activator or repressor. Even though general biosensors are the most popular due to their simplicity, they are nonspecific and could lead to false positives. In contrast, stress-responsive biosensors offer several advantages over general biosensors. As different stimulants often lead to a common stress response, such sensors can be good indicators of toxicity and stress-inducing conditions, such as DNA and protein damage, oxidative stress, and membrane damage. Their simplicity, selectivity, and sensitivity have made them extremely useful and popular. Specificity of such sensors might be increased by incorporating several different types of semi-specific biosensors to determine type and variety of toxicity. The stress promoter–reporter could be present in separate strains, or two reporters could be incorporated in the same strain. Identification of such stress-related genes for such sensors involves scanning through the transcription profile of the genome. Numerous stress gene promoters, including *sulA*, *katG*, *recA* and *uvrA*, have been fused with a reporter to construct biosensors for detection of compounds that cause DNA damage (Vollmer *et al.*, 1997; Rosen *et al.*, 2000). Panels of stress-responsive biosensors are also on the rise. Oxidative stress-sensitive cell array chips have been employed for identification of putative targets in the entire genome (Lee *et al.*, 2007). The sensitivity of such sensors could be significantly improved by fine-tuning the promoter and modification of host strains. Challenges for improvement of such sensors would encompass identification of strong promoters that are sensitive to a given stimulus, knowledge of the gene regulatory network, the design of instruments that are easy to use and inexpensive, the refinement of older reporters, and the creation of new reporter genes (Davidov *et al.*, 2000; Lee *et al.*, 2004, 2005; Mitchell and Gu, 2004).

11.6   Global Gene Expression Profiling of the BarA–UvrY TCS

To further identify downstream targets and pathways that are affected by the BarA–UvrY TCS, we have begun to study the effect of mutation of either *barA* or *uvrY* and compare it with a wild-type or a mutant expressing the UvrY protein from a low-copy plasmid-borne vector p-*uvrY* in UPEC CFT073. At first, the raw digitalized intensity of Affymetrix single-color slides was internally normalized using Microarray Suite version 5 (MAS 5.0, Affymetrix). The universally 'absent' genes from the normalized data were then eliminated. The noise generated due to chip–chip nonbiological variance was minimized through interchip-LOWESS normalization between the wild-type and individually treated samples using GeneSpring v6 (Agilent, Inc., CA). The resultant genomic regulation was determined as the ratio of the individual gene intensity of treated samples to that of the control samples. The normalized genes of the treated ensemble showing at least 1.15-fold difference (up- or down-regulation) from that of wild type were accepted for the remaining analysis. Approximately 1400 genes from the selected genome showed a similar regulatory trend between the *uvrY* and *barA* strains, of which around 570 genes were from the CFT073 genomic

Table 11.1 Microarray analysis of BarA–UvrY TCS in UPEC CFT073 background

Gene name/ID	Category/function	Fold induction	Description
rfaJ	Biosynthesis	2.2	Lipopolysaacharide 1,2-glycosyltransferase
serB		2.1	Phosphoserine phosphotase
hemA		2.2	Glutamyl tRNA reducatse
hisB		2.5	Histidine biosynthesis bifunctional protein
aroC		2.4	Chorismate synthase
dsdA	Metabolism	2.3	D-Serine dehydratase
bglA		2.5	6-Phospho-beta-glucosidase
ldcC		2.6	Lysine decarboxylase
c5039		2.4	Putative lactate dehydrogenase
ucpA		2.2	Oxidoreductase
kpsT	Transport	2.4	ATP binding transporter
kpsM		2.1	ATP binding
sitC		2.5	ABC transporters
iroN		2.4	Siderophore receptor
papC		2.5	Fimbrial usher protein
malK		2.2	Maltose transporter
focG	Adhesion	2.5	F1C minor fimbrial subunit protein
c4209		2.3	Putative minor fimbrial subunit precursor
c4214		2.7	Putative major fimbrial subunit precursor
csgA		2.1	Major curli subunit precursor
papH		2.5	Fimbrial protein
papI	Transcription	2.2	Fimbrial protein transcriptional regulators
flhC		2.6	Flagellar transcriptional activator
ymfL		2.4	Hypothetical protein
c2411		2.1	DNA-binding protein H-NS
pcnB		2.2	Poly (A) polymerase
yhiH		2.3	Hypothetical ABC transporter
fimB	Binding	2.7	Type 1 fimbriae regulatory protein
zntA		2.4	Lead, cadmium, zinc transporting ATPase
dppD		3.1	Dipeptide transport ATP binding protein
rseB		2.4	Sigma E factor regulatory protein
c0934		2.7	Hypothetical protein
dsdA	Catalysis	2.3	D-Serine dehydrates
nrdD		2.1	Ribonucleoside triphosphate reductase
trpB		2.5	Trytophan synthase beta chain
agp		2.4	Glucose 1-phosphatase precursor
ydjQ		2.7	Hypothetical protein
mtr	Membrane	2.4	Tryptophan specific transport protein
ompC		3.1	Outer membrane protein C precursor
ompA		2.4	Outer membrane protein A precursor
pitB		2.3	Low affinity inorganic phosphate transfer
yjaN	Unknown	2.6	Hypothetical protein
yfgJ		2.4	Hypothetical protein
ycjX		2.7	Hypothetical protein

segment and about 200 genes were from the intergenic region. Similar analysis identified roughly 900 genes, including ~270 entries from the CFT segment and ~100 entries from the intergenic segment that are expressed oppositely between *uvrY* and p-*uvrY* strains. Apparently, about 170 regulated genes according to the aforesaid null hypothesis showed similar regulation between *barA* and *uvrY* strains while simultaneously exhibiting reverse regulation between the *uvrY* and p-*uvrY* strains. This last genome contained ~50 CFT genes and ~20 intergenic entries. Unsupervised heretical clustering was performed for each of the three genomes independently using a standard correlation algorithm. To conclude, the biological, molecular, and cellular functions of each gene, part of the above-mentioned three genomes, were mined using the NetAffix GeneOntology (GO) analysis tool (Affymetrix, Inc., CA) and the genome was segmented according to its primary functions.

Several groups of genes have been annotated based on their function. Genes involved in metabolism, biosynthesis, cell adhesion, transcription and translation, catalysis, membrane, and many unknown functions were significantly affected by the mutation. Representative genes that are affected at least twofold by the mutation were reported (Table 11.1). This TCS, by virtue of its role in virulence, stress response, carbon regulation, and other key regulatory pathways in *E. coli*, could be a potential target for toxicity-detection studies in the future.

11.7 Conclusion

Toxicogenomics has now evolved into a multidisciplinary field by integrating several branches of biology, including toxicology, genetics, molecular biology, bioinformatics, functional genomics, transcript profiling, proteomics, metabolomics, and pharmacogenomics. With ongoing whole-genome sequencing efforts, the potential for identifying candidates for toxicity testing or pathways has been significantly accelerated using available high-throughput and inexpensive molecular genetics tools. An important strategy towards identification of novel toxic chemicals involves employing potential targets that are susceptible to various stresses in the presence of deleterious compounds. Genomics enables pinpointing such potential candidates by scanning through an entire genome in a high-throughput fashion. Identification, validation, and categorical classification of such targets will enhance future toxicity-detection studies.

References

Amin RP, Bennett L, Afshari CA, Hamadeh HK, Bushel PR, Paules RS (2002) Genomic interrogation of mechanism(s) underlying cellular responses to toxicants. *Toxicology* **181–182**, 555–563.

Aubrecht J, Caba E (2005) Gene expression profile analysis: an emerging approach to investigate mechanisms of genotoxicity. *Pharmacogenomics* **6**, 419–428.

Chancey ST, Wood DW, Pierson LS (1999) Two-component transcriptional regulation of *N*-acyl-homoserine lactone production in *Pseudomonas aureofaciens*. *Appl Environ Microbiol* **65**, 2294–2299.

Chuang Y-YE, Chandramouli JA, Tsai W, Cook D, Chen Y, Gadisetti VR, Yan S, Coffin M-H, DeGraff H (2002) Gene expression after treatment with hydrogen peroxide, menadione, or *t*-butyl hydroperoxide in breast cancer cells. *Cancer Res* **62**, 6246–6254.

Corbell N, Loper JE (1995) A global regulator of secondary metabolite production in *Pseudomonas fluorescens* Pf-5. *J Bacteriol* **177**, 6230–6236.

Davidov Y, Vollmer AC, Van Dyk TK, LaRossa RA, Rozen R, Smulski DR, Elsemore DA, Belkin S (2000) Improved bacterial SOS promoter::*lux* fusions for genotoxicity detection. *Mutat Res* **466**, 97–107.

De Longueville F, Bertholet V, Remacle J (2004) DNA microarrays as a tool in toxicogenomics. *Comb Chem High Throughput Screen* **7**, 207–211.

Dho-Moulin M, Fairbrother JM. (1999) Avian pathogenic *Escherichia coli* (APEC). *Vet Res* **30**, 299–316.

Eriksson AR, Palva ET, Andersson RA, Pirhonen M (1998) Two-component regulators involved in the global control of virulence in *Erwinia carotovora* subsp. *carotovora*. *Mol Plant Microbe Interact* **11**, 743–752.

Ewers C, Philipp H-C, Wieler LH, Janssen T, Kiessling S (2004) Molecular epidemiology of avian pathogenic *Escherichia coli* (APEC) isolated from colisepticemia in poultry. *Vet Microbiol* **104**, 91–101.

Gant TW (2003) Application of toxicogenomics in drug development. *Drug News Perspect* **16**, 217–221.

Gawande PV, Griffiths MW (2005) Effects of environmental stresses on the activities of the *uspA*, *grpE* and *rpoS* promoters of *Escherichia coli* O157:H7. *Int J Food Microbiol* **99**, 91–98.

Goodier RI, Ahmer BM (2001) SirA orthologs affect both motility and virulence. *J Bacteriol* **183**, 2249–2258.

Hamadeh HK, DiSorbo O, Stoll R, Martin K, Bennett L, Bushel PR, Jayadev S, Sieber S, Tennant R (2002) Gene expression analysis reveals chemical-specific profiles. *Toxicol Sci* **67**, 219–231.

Hengge-Aronis R (2002) Signal transduction and regulatory mechanisms involved in control of the sigma(S) (RpoS) subunit of RNA polymerase. *Microbiol Mol Biol Rev* **66**, 373–395.

Herren CD, Mitra A, Palaniyandi SK, Coleman A, Elankumaran S, Mukhopadhyay S (2006) The BarA–UvrY two-component system regulates virulence in avian pathogenic *Escherichia coli* O78:K80:H9. *Infect Immun* **74**, 4900–4909.

Hotch JA, Silhavy TJ (ed.) (1995) *Two-component Signal Transduction*. American Society for Microbiology Press: Washington, DC.

Imlay JA (2003) Pathways of oxidative damage. *Annu Rev Microbiol* **57**, 395–418.

Janda JM, Abbott SL (2005) *The Enterobacteria*. ASM Press: Washington, DC; 23–57.

Johnson JR, Russo TA, Gajewski A, Lesse AJ (2003) Extraintestinal pathogenic *Escherichia coli* as a cause of invasive nonurinary infections. *J Clin Microbiol* **41**, 5798–5802.

Johnson JR, Russo TA (2002a) Extraintestinal pathogenic *Escherichia coli*: 'the other bad *E coli*'. *J Lab Clin Med* **139**, 155–162.

Johnson JR, Russo TA (2002b) Uropathogenic *Escherichia coli* as agents of diverse non-urinary tract extraintestinal infections. *J Infect Dis* **186**, 859–864.

Johnson JR, Russo TA (2005) Molecular epidemiology of extraintestinal pathogenic (uropathogenic) *Escherichia coli*. *Int J Med Microbiol* **295**, 383–404.

Kaper JB, Nataro JP, Mobley HL (2004) Pathogenic *Escherichia coli*. *Nat Rev Microbiol* **2**, 123–140.

Kawata K, Okabe S, Yokoo H, Shimazaki R (2007) Classification of heavy-metal toxicity by human DNA microarray analysis. *Environ Sci Technol* **41**, 3769–3774.

Kay E, Riedel K, Valverde C, Spahr S, Humair B, Dénervaud V, Eberl L, Haas D (2006) Two GacA-dependent small RNAs modulate the quorum-sensing response in *Pseudomonas aeruginosa*. *J Bacteriol* **188**, 6026–6033.

Kim BY, Kang J, Kim KS (2005) Invasion processes of pathogenic *Escherichia coli*. *Int J Med Microbiol* **295**, 463–470.

Klauck E, Typas A, Hengge R (2007) The sigmaS subunit of RNA polymerase as a signal integrator and network master regulator in the general stress response in *Escherichia coli*. *Sci Prog* **90**, 103–127.

Kwon O, Georgellis D, Lin EC (2000) Phosphorelay as the sole physiological route of signal transmission by the arc two-component system of *Escherichia coli*. *J Bacteriol* **182**, 3858–3862.

Lange R, Hengge-Aronis R (1994) The cellular concentration of the sigma S subunit of RNA polymerase in *Escherichia coli* is controlled at the levels of transcription, translation, and protein stability. *Genes Dev* **8**, 1600–1612.

Lee JH, Cullen DC, Gu MB, Mitchell RJ, Kim BC (2005) A cell array biosensor for environmental toxicity analysis. *Biosens Bioelectron* **21**, 500–507.

Lee JH, Gu MB, Youn CH, Kim BC (2007) An oxidative stress-specific bacterial cell array chip for toxicity analysis. *Biosens Bioelectron* **22**, 2223–2229.

Lee JH, Mitchell RJ, Gu MB (2004) Enhancement of the multi-channel continuous monitoring system through the use of *Xenorhabdus luminescens lux* fusions. *Biosens Bioelectron* **20**, 475–481.

Lloyd AL, Mobley HLT, Rasko DA (2007) Defining genomic islands and uropathogen-specific genes in uropathogenic *Escherichia coli*. *J Bacteriol* **189** (9), 3532-3546.

Malone AS, Chung YK, Yousef AE (2006) Genes of *Escherichia coli* O157:H7 that are involved in high-pressure resistance. *Appl Environ Microbiol* **72**, 2661–2671.

McMahon MA, Xu J, Moore JE, Blair IS, McDowell DA (2007) Environmental stress and antibiotic resistance in food-related pathogens. *Appl Environ Microbiol* **73**, 211–217.

Mitchell RJ, Gu MB (2004) An *Escherichia coli* biosensor capable of detecting both genotoxic and oxidative damage. *Appl Microbiol Biotechnol* **64**, 46–52.

Mukhopadhyay S, Audia JP, Roy RN, Schellhorn HE (2000) Transcriptional induction of the conserved alternative sigma factor RpoS in *Escherichia coli* is dependent on BarA, a probable two-component regulator. *Mol Microbiol* **37**, 371–381.

Nagasawa S, Ishige K, Mizuno T (1993) Novel members of the two-component signal transduction genes in *Escherichia coli*. *J Biochem (Tokyo)* **114**, 350–357.

Nagasawa S, Tokishita S, Aiba H, Mizuno T (1992) A novel sensor-regulator protein that belongs to the homologous family of signal-transduction proteins involved in adaptive responses in *Escherichia coli*. *Mol Microbiol* **6**, 799–807.

Nonaka G, Blankschien M, Herman C, Gross CA, Rhodius VA (2006) Regulon and promoter analysis of the *E. coli* heat-shock factor, sigma32, reveals a multifaceted cellular response to heat stress. *Genes Dev* **20**, 1776–1789.

Norton RA (1997) Avian cellulitis. *World's Poult. Sci. J.* **53**, 337–349.

Nuwaysir EF, Afshari CA, Barrett JC, Bittner M, Trent J (1999) Microarrays and toxicology: the advent of toxicogenomics. *Mol Carcinog* **24**, 153–159.

Partridge JD, Poole RK, Green J, Scott C, Tang Y (2006) *Escherichia coli* transcriptome dynamics during the transition from anaerobic to aerobic conditions. *J Biol Chem* **281**, 27806–27815.

Patten CL, Kirchhof MG, Schertzberg MR, Morton RA, Schellhorn HE (2004) Microarray analysis of RpoS-mediated gene expression in *Escherichia coli* K-12. *Mol Genet Genomics* **272**, 580–591.

Pernestig AK, Melefors O, Georgellis D (2001) Identification of UvrY as the cognate response regulator for the BarA sensor kinase in *Escherichia coli*. *J Biol Chem* **276**, 225–231.

Phadtare S, Tadigotla V, Shin WH, Sengupta A, Severinov K (2006) Analysis of *Escherichia coli* global gene expression profiles in response to overexpression and deletion of CspC and CspE. *J Bacteriol* **188**, 2521–2527.

Pizarro RA (1995) UV-A oxidative damage modified by environmental conditions in *Escherichia coli*. *Int J Radiat Biol* **68**, 293–299.

Pognan F. (2007) Toxicogenomics applied to predictive and exploratory toxicology for the safety assessment of new chemical entities: a long road with deep potholes. In *Systems Biological Approaches in Infectious Diseases*, Boshoff HI, Barry III CE (eds). Progress in Drug Research, Vol. **64**. Birkhäuser: Basel; 217–238.

Rama G, Chhina DK, Chhina RS, Sharma S (2005) Urinary tract infections – microbial virulence determinants and reactive oxygen species. *Comp Immunol Microbiol Infect Dis* **28**, 339–349.

Robinson VL, Buckler DR, Stock AM (2000) A tale of two components: a novel kinase and a regulatory switch. *Nat Struct Biol* **7**, 626–633.

Rosen R, Belkin S, Davidov Y, LaRossa RA. (2000) Microbial sensors of ultraviolet radiation based on *recA′::lux* fusions. *Appl Biochem Biotechnol* **89**, 151–160.

Russo TA, Johnson JR (2003) Medical and economic impact of extraintestinal infections due to *Escherichia coli*: focus on an increasingly important endemic problem. *Microbes Infect* **5**, 449–456.

Sahu SN, Acharya S, Tuminaro H, Patel I, Dudley K, LeClerc JE, Cebula TA, Mukhopadhyay S (2003) The bacterial adaptive response gene, *barA*, encodes a novel conserved histidine kinase regulatory switch for adaptation and modulation of metabolism in *Escherichia coli*. *Mol Cell Biochem* **253**, 167–177.

Semchyshyn HM, Lushchak VI (2004) Oxidative stress and control of catalase activity in *Escherichia coli*. *Ukr Biokhim Zh* **76**, 31–42 (in Ukrainian).

Shioda T (2004) Application of DNA microarray to toxicological research. *J Environ Pathol Toxicol Oncol* **23**, 13–31.

Smith JL, Fratamico PM, Gunther NW (2007) Extraintestinal pathogenic *Escherichia coli*. *Foodborne Pathog Dis* **4**, 134–163.

Smith LL (2001) Key challenges for toxicologists in the 21st century. *Trends Pharmacol Sci* **22**, 281–285.

Storck T, Scheel J, Bach A, von Brevern M-C, Behrens CK (2002) Transcriptomics in predictive toxicology. *Curr Opin Drug Discov Devel* **5**, 90–97.

Sugiura M, Aiba H, Mizuno T (2003) Identification and classification of two-component systems that affect *rpoS* expression in *Escherichia coli*. *Biosci Biotechnol Biochem* **67**, 1612–1615.

Tennant RW (2002) The National Center for Toxicogenomics: using new technologies to inform mechanistic toxicology. *Environ Health Perspect* **110**, A8–A10.

Thybaud V, Le Fevre A-C, Boitier E (2007) Application of toxicogenomics to genetic toxicology risk assessment. *Environ Mol Mutagen* **48**, 369–379.

Tomenius H, Pernestig AK, Jonas K, Georgellis D, Möllby R, Normark S, Melefors O (2006) The *Escherichia coli* BarA–UvrY two-component system is a virulence determinant in the urinary tract. *BMC Microbiol* **6**, 27.

Touati D (2000) Sensing and protecting against superoxide stress in *Escherichia coli* – how many ways are there to trigger *soxRS* response? *Redox Rep* **5**, 287–293.

Venturi V (2003) Control of *rpoS* transcription in *Escherichia coli* and *Pseudomonas*: why so different? *Mol Microbiol* **49**, 1–9.

Vijayakumar SR, Kirchhof MG, Patten CL, Schellhorn HE (2004) RpoS-regulated genes of *Escherichia coli* identified by random lacZ fusion mutagenesis. *J Bacteriol* **186**, 8499–8507.

Vollmer AC, LaRossa RA, Van Dyk TK, Belkin S, Smulski DR (1997) Detection of DNA damage by use of *Escherichia coli* carrying *recA′::lux*, *uvrA′::lux*, or *alkA′::lux* reporter plasmids. *Appl Environ Microbiol* **63**, 2566–2571.

Wick LM, Egli T (2004) Molecular components of physiological stress responses in *Escherichia coli*. *Adv Biochem Eng Biotechnol* **89**, 1–45.

12

Toxicogenomics *In Vitro*: A Powerful Tool for Screening Hepatotoxic Potential of Food-related Products

Saura C. Sahu

12.1 Introduction

Liver is the primary organ involved in xenobiotic metabolism. It is the first filter of blood loaded with substances, including toxicants absorbed from the gut. Because of the metabolic activity and high levels of exposure to the toxicants, the liver is a major target organ for toxic reactions; thus, it is prone to injury in spite of its capacity for regenerative growth. Therefore, hepatotoxicity is an important parameter in safety evaluation.

Traditional preclinical hepatotoxicity testing *in vivo* involves the liver-specific enzyme levels in blood and histopathology of the liver. Animal studies are expensive, time consuming and involve differences in bioavailability and metabolism. Therefore, *in vitro* studies are important alternatives to complement and/or supplement the traditional animal studies *in vivo* for the toxicological risk assessment of regulated products (Green *et al.*, 2001; MacGregor *et al.*, 2001).

The *in vitro* studies minimize the use of animals, reduce variability between samples, require less test materials, allow more in-depth biochemical and molecular mechanisms of toxicity as well, and are adaptable to high-throughput screening tests. They provide well-defined and reproducible experimental conditions for toxicity testing. They are excellent systems to study mechanisms of toxicity and structure–activity relationships at the cellular and molecular levels. These attributes make them very useful tools for rapid screening of potential toxins. The use of *in vitro* cytotoxicity results can effectively establish the dose range for *in vivo* acute toxicity studies, thereby reducing the number of animals required for safety evaluation by the regulatory agencies (NTP, 2001).

Primary hepatocytes and many hepatocyte cell lines in culture retain many metabolic enzymes characteristic of the intact liver *in vivo* (Hengstler *et al.*, 2000, 2002; Runge *et al.*, 2001). They represent an excellent *in vitro* model for studying liver function, xenobiotic metabolism, pharmacology, and toxicology. Hepatocyte cultures are widely used for the evaluation of liver functions and toxicity of chemicals, drugs, and microbes (Barsig *et al.*, 1998; Castell *et al.*, 1997; Li *et al.*, 1997; Michalopoulos, 1999; Sahu *et al.*, 2001, 2006). Primary rat hepatocyte cultures are the most frequently used and best characterized *in vitro* model for testing liver toxicity of chemicals and drugs with a long history of use by many laboratories throughout the world (Sahu *et al.*, 2001, 2006; Beekman *et al.*, 2006).

12.2   Toxicogenomics

Toxicogenomics integrates toxicology and genomics. It is a relatively new discipline based on the principle that toxicity results from the structure–activity interactions of the genome in response to toxic exposure, leading to altered gene expression (Gatzidou *et al.*, 2007). Characteristic gene expression changes are associated with distinct classes of toxicants (Pennie, 2000; Pennie *et al.*, 2000; Bartosiewicz *et al.*, 2001). Microarray technology used in toxicogenomic studies has the ability to determine changes in the expression of thousands of genes simultaneously. Therefore, toxicogenomics is a powerful tool for large-scale screening of potential toxicants. Also, it enables to determine the molecular mechanisms of toxicity in greater detail (Pennie, 2000; Pennie *et al.*, 2000). Toxin-induced changes in gene expression are more sensitive and they are detected much earlier than the classical histopathology, clinical chemistry, and biochemical techniques. Therefore, gene expression analysis is a powerful tool for toxicity testing compared with the conventional endpoints. It is capable of predicting toxicity during early toxicity screening. In addition, gene expression changes can be verified by histopathological, clinical chemistry, and biochemical assays (Waring *et al.*, 2001; Ruepp *et al.*, 2005; Beyer *et al.*, 2007; Zidek *et al.*, 2007).

12.3   Toxicogenomic Techniques

Toxicogenomics technology uses molecular biological techniques, such as microarrays and real-time reverse transcription (RT) polymerase chain reaction (PCR) for gene expression analysis. In the microarray technique, RNA isolated from the exposed and control target cells is labeled with fluorescent dye probe and hybridized with oligos or c-DNA spotted on the array for gene expression analysis. The DNA microarray technique allows monitoring of the expression levels of a large number of genes simultaneously. In real-time RT-PCR, the amount of amplicon generated as the reaction progresses is measured by monitoring reporter dye fluorescence.

12.4   Toxicogenomics *In Vivo*

In vivo toxicogenomics uses an animal model for gene expression analysis that can be verified by histopathological, clinical chemistry, and biochemical parameters. Bartosiewicz *et al.* (2001) used the DNA microarray technique to examine gene expression patterns in mice in response to five classes of chemicals, such as benzo(*a*)pyrene, 3-methylcholanthrene, clofibrate, dimethylnitrosamine, cadmium chloride and carbon tetrachloride. Each class of chemicals yielded a distinctive gene expression profile. Chung *et al.* (2006) examined differential gene expression profiles by microarray analysis and histopathology on D-galactosamine-induced liver injury in mice. They identified genes associated with injury and regeneration stages of the mouse liver.

Schuppe-Koistinen *et al.* (2002) determined the hepatotoxicity of aminoguanidine carboxylate in rats *in vivo* and primary rat hepatocytes *in vitro* using oligonucleotide microarrays combined with cytotoxicity parameters. Their studies showed that the array-based gene expression combined with cytotoxicity data led to a better understanding of the molecular basis of liver injury. Waring *et al.* (2001, 2003) developed DNA microarrays for gene expression analysis in rats of several known hepatotoxins, such as allyl alcohol, amiodarone, aroclor 1254, arsenic, carbanazepine, carbon tetrachloride, dietlynitrosamine, dimethylformamide, diquat, etoposide, indomathacin, methylpyrilene, methotrexate, monocrotaline, and 3-methylcholanthrene. They correlated the gene expression findings with histopathology and clinical chemistry parameters. They observed a strong correlation between the gene expression profiles and the histopathological and clinical chemistry data. They demonstrated cluster formation by gene expression profiles of toxins with similar mechanisms of action. Huang *et al.* (2004) used gene expression profiling by microarray technology to study several known hepatotoxins, such as acetaminophen, methotrexate, methapyrilene, furan, and phenytoin, administered orally to Sprague-Dawley rats. They correlated gene expression data with classical endpoints of hepatotoxicity showing a good correlation. Hamadeh *et al.* (2004) exposed Sprague-Dawley rats to the known hepatotoxin furan and correlated gene expression profiling by microarray technology with liver histopathology and clinical chemistry parameters. They found that furan-induced gene expression analysis data correlated well with the traditional clinical endpoints and pathological changes in the rat liver. Minami *et al.* (2005) evaluated hepatic gene expression profiles of acetaminophen, bromobenzene, carbon tetrachloride, dimethyl amine and thioacetamide administered to male Sprague-Dawley rats by a single intraperitoneal injection. They used biochemical (liver enzymes in serum) and toxicogenomic (both microarray and real-time RT-PCR) assays for hepatotoxicity assessment. All the test chemicals produced direct correlation between the biochemical and toxicogenomic parameters in the rat liver.

Lu *et al.* (2001) applied the DNA microarray technique to study arsenic-induced hepatotoxicity in humans. The gene expression analysis was compared with the histopathology of liver showing liver lesions. They observed a direct correlation between the changes in gene expression and the liver degenerative lesions in arsenic-exposed patients.

The predictive accuracy for sensitive toxicogenomic analysis depends on the experimental variables that can influence the reproducibility of toxic effects. Therefore, simultaneous measurement of genomic, pathological, and biochemical parameters in the same treated animals is desirable for reliable predictions.

12.4.1   Inter-laboratory Comparison of Toxicogenomics *In Vivo*

Beyer *et al.* (2007) compared a multi-laboratory study of acetaminophen hepatotoxicity using genomic, chemical, enzymatic, and histological analysis. Data generated individually by seven laboratories using a mouse model and a standard protocol were compared. Their studies show that gene expression profiling accompanied by histopathology and clinical chemistry data provide accurate information on hepatotoxicity with greater confidence.

12.5   Toxicogenomics *In Vitro*

In vitro toxicogenomics uses an *in vitro* system for gene expression analysis. Within the last decade, *in vitro* toxicogenomics has been extensively used for testing hepatotoxicity of drugs in the early stage of drug development and for high-throughput screening of toxicity. The primary hepatocytes and hepatocyte cell lines are an excellent *in vitro* model system for hepatotoxicity studies (Castell *et al.*, 1997; Li *et al.*, 1997; Barsig *et al.*, 1998; Michalopoulos, 1999; Sahu *et al.*, 2001, 2006; Sahu, 2003; Beekman *et al.*, 2006). *In vitro* toxicogenomic analysis has been used successfully to distinguish genotoxic and nongenotoxic carcinogens (Van Delft *et al.*, 2004; Fielden *et al.*, 2007) using known genotoxic hepatocarcinogens (i.e. 2-acetylaminofluorene, aflatoxin B_1, hydrazine, N-nitrosodiethylamine) and nongenotoxic hepatotoxins (i.e. acetaminophen, carbon tetrachloride, chloroform, tetracycline).

Longueville *et al.* (2003) investigated the gene expression profiles of several different known hepatotoxins, such as acetaminophen, amiodarone, clofibrate, and Phenobarbital, based on their hepatocellular effects on primary rat hepatocytes. Their *in vitro* gene expression profiles correlated well with the *in vivo* hepatotoxic effects in rats reported in literature. They were able to cluster compounds of similar hepatotoxicity. Ruepp *et al.* (2005) have created a reference database using liver gene expression profiles in response to exposures of various known hepatotoxins both in primary rat hepatocytes *in vitro* and in rats *in vivo*. They observed good concordance of gene expression changes with histopatholgical findings and clinical chemistry data.

Boess *et al.* (2003) measured the mRNA expression profiles in several *in vitro* systems, such as rat liver cell line BRL 3A derived from Buffalo rats, rat clone-9 cell line derived from Sprague-Dawley rats, primary rat hepatocytes, and rat liver slices and compared them with that of the rat liver tissue *in vivo*. They were able to identify genes with pronounced changes in expression levels in the *in vitro* relative to the rat liver tissue *in vivo* liver slices exhibiting the strongest similarity. They observed a change in expression patterns soon after cell isolation and culture initiation, which stabilized with time in culture.

Jessen *et al.* (2003) compared gene expression changes induced in primary rat hepatocytes and liver slices *in vitro* and male Sprague-Dawley rat liver *in vivo* by known hepatotoxins with diverse mechanisms of toxicity. The test hepatototoxins were phenobarbital, carbon tetrachloride, napthylisothiocyanate, tacrine and Wy-14643. Comparison of gene expression analysis revealed a greater than 80 % concordance between the *in vivo* liver and both *in vitro* systems.

Boess *et al.* (2007) studied the hepatotoxicity of two serotonin receptor (5-HT6) antagonists in rat primary hepatocytes using gene expression analysis and traditional biochemical assays. They analyzed the CYP 2B2, CYP 2B Exon 9 and CYP 3A1 genes. The gene expression analysis was performed by both the microarray and RT-PCR techniques. This study demonstrated the power of *in vitro* gene expression analysis combined with traditional biochemical information for assessing heaptotoxicity of the test agents and the mechanisms of their action.

Kier *et al.* (2004) used microarray technology and traditional histopathology and clinical chemistry parameters to evaluate hepatotoxicity of a number of known hepatotoxins, such as acetaminophen, aflatoxin B_1, allyl alcohol, carbon tetrachloride, ethanol, and phenobarbital, in primary human hepatocytes *in vitro* and in rat liver tissue *in vivo*. Their data showed gene expression analysis combined with histopathology and clinical chemistry can predict hepatotoxicity with great accuracy.

Toxicogenomics *in vitro* with the microarray technique was used to investigate the gene expression changes in cultured human hepatoma cell line HepG2 (Burczynski *et al.*, 2000; Pennie, 2000; Harries *et al.*, 2001). These studies established a direct correlation between the cytotoxicity and gene expression changes induced by known hepatotoxins, such as ethanol and carbon tetrachloride (Pennie, 2000; Harries *et al.*, 2001) and cisplatin, duffunisal and flufenamic acid (Burczynski *et al.*, 2000).

Sawada *et al.* (2005, 2006) used ArrayPlate and real-time PCR assays to screen phospholipidosis (PLD) in human hepatoma cell line HepG2 induced by amiodarone and 80 other proprietary compounds. They observed direct correlation between both the assays. Atienzar *et al.* (2007) used gene expression analysis in HepG2 to screen PLD-inducing potential of drugs and chemicals. Nioi *et al.* (2007) characterized PLD in HepG2 cells induced by known inducers of PLD, such as amiodarone, amitriptyline, fluoxetine, tamoxifen, and loratadine, by gene expression profiles. By adding a fluorescent-labeled phospholipid (LipidTox) to the HepG2 growth media they correctly identified 100 % of the PLD-positive and PLD-negative compounds. They found that this assay was less time consuming, more sensitive and had a higher throughput than the gene expression analysis.

Harris *et al.* (2004) used DNA microarray and real-time RT-PCR techniques to determine gene expression profiles of three genotoxic hepatocarcinogens, (aflatoxin B_1, 2-acetylaminoifluorene and dimethylnitrosamine) and one nongenotoxic hepatotoxin (acetaminophen) in two *in vitro* systems. The two systems they used were human primary hepatocytes from different donors and the human hepatoma cell line HepG2. Both the systems expressed similar gene expression profiles. However, the primary human hepatocytes from different donors showed considerable variability. Out of total 2172 genes detectable in these hepatocytes from all donors, only 29 % were expressed in hepatocytes from a single donor, 31 % were expressed in hepatocytes from two donors and 40 % were expressed in hepatocytes from all donors (Harris *et al.*, 2004).

12.5.1 Interlaboratory Comparison of Toxicogenomics *In Vitro*

Beekman *et al.* (2006) undertook an interlaboratory study on gene expression analysis of the well-known nongenotoxic hepatotocarcinogen methapyrilene in primary rat hepatocyte cultures *in vitro*. They used a DNA microarray for gene expression analysis and compared it with cytotoxicity as determined by lactate dehydrogenase leakage from the hepatocytes. The comparison of genomic data from four different laboratories was complex because of the experimental and statistical variability. However, the analysis of data from all the participating laboratories concluded that *in vitro* toxicogenomics could be an excellent tool for predicting the hepatotoxic potential of test agents.

12.5.2 Limitations of *In Vitro* Toxicogenomics

In vitro systems have inherent limitations in their ability to model *in vivo* conditions. Intrinsic limitations of the *in vitro* model can affect the reproducibility of the toxicogenomic results. Gene expressions can be affected by the *in vivo* cellular environment and cell–cell interactions. Therefore, choice of an *in vitro* model comparable to the *in vivo* situation is a limiting factor for *in vitro* toxicogenomics. Also, the extrapolation of the data derived from *in vitro* systems to the animals and humans is a critical factor. Dere *et al.* (2006) compared the gene expression changes induced by 3,7,8-tetrachlorodibenzo-*p*-dioxin (TCDD) in Hepa1c1c7 mouse hepatoma cells *in vitro* and C57BL/6 mouse liver tissue *in vivo* using the microarray and real-time RT-PCR techniques. A direct comparison of their untreated *in vitro* and *in vivo* systems revealed a good correlation ($R = 0.75$) between the basal levels of the two systems, illustrating that many genes were basally expressed to similar levels. However, the TCDD treatment resulted in a number of divergent gene responses across both the systems. They observed inherent differences in TCDD-induced gene expression profiles between the two systems, highlighting the limitations of the *in vitro* test system.

12.6   Applications of Toxicogenomics

Pennie (2000; Pennie *et al.* 2000) grouped the applications of toxicogenomics into two broad classes: mechanistic toxicology and predictive toxicology. However, toxicogenomic applications can be many. A few examples are given below.

12.6.1   Mechanism of Hepatotoxicity

Toxicogenomics illustrates the mechanisms of toxicity leading to a better understanding of the mode of action of the toxic agent (Duggan *et al.*, 1999). This information is helpful for risk assessment of toxic agents.

Oxidative stress is a common mechanism of hepatotoxicity. Hepatocytes respond and adapt to oxidative stress. Etomoxir-induced oxidative stress in HepG2 cells has been detected by the gene expression analysis and confirmed by the biochemical and enzymatic assays (Merrill *et al.*, 2002). Different classes of hepatotoxins, such as reactive metabolites, peroxisome proliferators, and macrophage activators, all produce oxidative stress and hepatotoxicity, but each class has distinctive gene expression profiles (McMillian *et al.*, 2005). These distinctive gene sets provide useful biomarkers for hepatotoxicity screening.

Powell *et al.* (2006) examined the acetaminophen-induced oxidative stress in rat liver by a diverse panel of biomarkers, including gene expression analysis, immunohistochemistry, oxidative DNA damage, and peroxidation of lipids and proteins. They found gene expression profiling as a very sensitive tool capable of detecting cellular stress at lower doses and shorter time of exposure that could not be detected by classical toxicological endpoints.

Apoptosis, i.e. programmed cell death, plays an important role in toxicity. Apoptosis requires energy in order to accomplish a number of ATP-dependent steps, such as apoptotic volume decrease, bleb formation, caspase(s) activation, enzymatic hydrolysis of macromolecules, chromatin condensation, and movement of mitochondria toward the nucleus and apoptotic body formation. The intracellular ATP determines cell death by apoptosis or necrosis. *In vitro* toxicogenomics has been used successfully to study apoptosis in hepatocytes (Hilderbrand *et al.*, 1999).

Chemical-induced PLD is an adverse health concern. Testing for PLD potential is a part of new drug development. *In vitro* toxicogenomics has been used successfully for screening PLD-inducing agents (Sawada *et al.*, 2005, 2006; Atienzar *et al.*, 2007; Nioi *et al.*, 2007).

12.6.2 Predictive Screening Tool for Hepatotoxic Potential of Food-related Products

Toxicogenomic analysis is a useful tool for predictive toxicology (Maggioli *et al.*, 2006).

Gene expression analysis *in vitro* can be used for screening potential hepatotoxins. Hepatocyte cultures have been widely used for screening chemicals, drugs, carcinogens, mutagens, mycotoxins, microbial pathogens, and viruses for their hepatotoxic potential (Sahu, 2003; Dambach *et al.*, 2005). Hepatotoxicity is routinely assessed in the safety evaluation of food additives and is a common adverse health effect of food contaminants and dietary supplements. Many of the botanical herbal products labeled as dietary supplements show hepatotoxicity (Willett *et al.*, 2004; Stickel *et al.*, 2005). Dietary supplements such as Chaparral (Sheikh *et al.*, 1997), Ephedra (Bajaj *et al.*, 2003), Germander (Lekehal *et al.*, 1996), Kava (Teschke *et al.*, 2003), LipoKinetix (Favreu *et al.*, 2002; Novak and Lewis, 2003) and pyrrolizidine alkaloids (Stedman, 2002) are known hepatotoxins. Rat hepatocytes have been shown to be a good *in vitro* model for testing hepatotoxicity of herbal products (Lekehal *et al.*, 1996). Human hepatocellular carcinoma cell line HepG2 was found to be an excellent model for assessing cytotoxicity of medicinal plant extracts (Ruffa *et al.*, 2002).

Beekman *et al.* (2006) demonstrated *in vitro* genomics as a predictive tool for hepatotoxicity using gene expression patterns from cultured rat primary hepatocytes treated with the known hepatotoxin methapyrilene. Nishimura *et al.* (2006) investigated species difference in mRNA induction of various transporters in primary cultures of human and rat hepatocytes using real-time RT-PCR. They observed species differences and found both the *in vitro* and *in vivo* systems useful. Tuschl and Mueller (2006) used gene expression analysis by real-time RT-PCR to demonstrate that the rat primary hepatocyte culture in a sandwich between two layers of gelled collagen and in a serum-free medium was the most suitable model for long-term *in vitro* hepatotoxicity screening.

Zidek *et al.* (2007) have established a predictive screening system *in vivo* for acute hepatotoxicity by differential gene expression analysis. They used six known hepatotoxins, such as acetaminophen, carbon tetrachloride, and chloroform, and six nonhepatotoxic compounds, such as clofibrate, estradiol, and quinidine. A single dose of the test agent was administered to male Sprague-Dawley rats by intraperitoneal injections. They used the gene microarray successfully to predict hepatotoxicity of test compounds.

12.6.3 Safety Evaluation and Quantitative Risk Analysis

A toxicological profile of a toxic agent is required for its risk assessment. Toxicogenomics provides better understanding of the molecular mechanisms of toxicity of a substance. The greater understanding of the molecular basis of toxic exposure provides a more scientific basis for risk analysis. Oberemm *et al.* (2005) have discussed in greater detail the potential use of toxicogenomics in safety evaluation and risk assessment. An expensive 2-year rodent bioassay is used for risk assessment of nongenotoxic hepatocarcinogens. Recently, Fielden *et al.* (2007) have developed a multigene biomarker for nongenotoxic hepatocarcinogens using hepatic gene expression profiling by the microarray technique. They collected hepatic gene expression data from rats treated for 5 days with 47 test chemicals. Each chemical was evaluated in the short-term *in vivo* study using three rats. They observed that this short-term *in vivo* rodent model was more sensitive, more accurate, and provided quicker results than the traditional models for risk assessment of nongenotoxic hepatocarcinogens (Fielden *et al.*, 2007).

12.6.4   Interactive Hepatotoxicity of Mixed Exposures

The majority of real-life exposures are mixed exposures. Humans are exposed to many different potential toxins. Their interaction is a matter of concern. Synergistic enhancement of toxicity is not uncommon. For example, lipopolysaccharide (LPS) has been shown to influence toxicity of xenobiotics (Cebula *et al.*, 1984; Zhou *et al.*, 2000; Wiesenfeld *et al.*, 2007). Rat hepatocytes have been used as an experimental model to study the interactive toxicity of carbon tetrachloride and trichloroethylene (Kefalas and Stacey, 1993). Primary rat hepatocytes have been used to study the interactive hepatotoxicity of flunitrazepam and ethanol (Assaf and Abdel-Rahman, 1999) as well as LPS and allyl alcohol (Sneed *et al.*, 2000). Therefore, hepatocytes *in vitro* can be used as a model to predict interactive hepatotoxicity of test agents using toxicogenomic techniques.

12.6.5   Species Differences in Hepatotoxicity

In vitro toxicogenomics can be used to assess species differences in hepatotoxicity. Hepatocyte cultures are predictive models for metabolism and pharmacokinetics in different species. Good correlation between *in vitro* and *in vivo* drug metabolism has been established using animal and human hepatocyte cultures (Sandker *et al.*, 1994; Pathernik *et al.*, 1995). Human, dog, and rat hepatocyte cultures have been shown to be valuable *in vitro* tools for the study of species differences in drug metabolism and pharmacokinetics (Bayliss *et al.*, 1999). Differences in sensitivity towards chemical toxicity have been demonstrated in rat and human hepatocyte cultures (Merrill *et al.*, 1995). Similar sensitivity towards chloroform-induced cytolethality is reported in mouse and rat hepatocytes (Ammann, *et al.*, 1998). Species-dependent differences in glutathione conjugation of perchloroethylene have been studied in rat and mouse hepatocyte cultures (Lash *et al.*, 1998). A similar metabolism of thiabendazole in rat, rabbit, calf, pig, and sheep hepatocytes has been reported (Coulet *et al.*, 1998).

12.6.6   Gender Differences in Hepatotoxicity

Toxicogenomic techniques can be successfully applied to determine the role of gender in hepatotoxicity. Gender difference plays an important role in the toxic responses of the liver (Treinen-Moslen, 2001). Adverse reactions to therapeutic drugs are more common in women than men (Kando *et al.*, 1995). Epidemiological evidence shows that women have greater susceptibility to alcohol-induced liver damage than men (Jensen, 1996; Schenker, 1997). Drug-induced acute liver failure occurs more frequently in women than men (Miller, 2001). Females are predisposed to hepatotoxicity, and concomitant agents that induce cytochrome P450 enzymes also increase individual susceptibility (Stedman, 2002).

The female mouse is more susceptible to fumonisin B1-induced hepatocellular neoplasm than the male mouse (NTP, 1999). However, cocaine causes more liver damage in male mice than in female mice (Visalli *et al.*, 2004). There exists a distinct sexual dimorphism in rat liver cytochrome P450 (CYP) activity (Lewis *et al.*, 1998). Intrinsic sexual differences in hepatocytes of males and females appear to result in different levels of responses to CYP. This intrinsic sexual difference in hepatocytes is more pronounced in rats than in humans.

Successful use of the primary rat hepatocytes in culture as an *in vitro* model for evaluating the sex differences in hepatotoxicity has been demonstrated (Carfagna *et al.*, 1996). This model has been used to show the gender difference in hepatotoxicity of safingol (Carfagna *et al.*, 1996). Delongchamp *et al.* (2005) used gene expression profiles of donated human liver tissues from nine males and nine females *in vivo* and primary human hepatocytes *in vitro* to evaluate gender differences in hepatotoxicity. They tested a total of 31 100 genes and found that the gene expression of only 224 genes differed between sexes. The observed gender differences in gene expressions were small. False discovery rates exceeded 80 % for every set of genes selected, making it impossible to identify specific genes with gender differences. They observed high interindividual variability in phenotypes and genotypes of human liver, as well as large variations in individual consumption of various diets.

More research is needed for evaluation of gender effects in hepatotoxicity. Gender difference studies using gene expression analysis in rodent livers would be of interest.

12.6.7   Detection and Interaction of Food-related Pathogens

Toxicogenomic technology can be used successfully to investigate the detection and interaction of food-related pathogens (Cebula *et al.*, 2005). Toxicogenomic techniques have been used for detection of *Escherichia coli* in beef (O'Hanlon *et al.*, 2005). Toxicogenomics technology has been used to investigate host–microbe interactions (Cummings and Relman, 2000) and acute hepatitis C virus infection in liver (Bigger *et al.*, 2001). Plumet and Gerlier (2005) have developed a sensitive and specific assay based on real-time RT-PCR to quantify individual measles virus RNA. This can be useful for a reliable detection of measles virus in clinical samples. Cai *et al.* (2005) have assessed the efficacy of nitazoxanide and paromomycin against two strains of the *Cryptosporidium parvum* in HCT-8 cells *in vitro* using quantitative real-time RT-PCR assay. Both compounds displayed dose-dependent inhibitions.

12.7   Conclusions

The purpose of our efforts reported here is to review literature demonstrating toxicogenomics as a useful tool for predictive hepatotoxicity screening and safety assessment of food-related products that occur naturally or that are added deliberately to the foods and dietary supplements. The published reports clearly demonstrate the value of toxicogenomics as an end point of hepatotoxicity. Gene expression analysis in combination with histopathological and biochemical assays provides a powerful tool for accurate assessment of hepatotoxicity, as well as for studying mechanisms of toxicity. *In vitro* toxicogenomics offers a powerful cost-effective tool for faster high-throughput toxicity testing. It can be effectively used for high-throughput screening of food-related products for potential hepatotoxicity. The technology has already been used for this purpose. Stierum *et al.* (2005) have discussed the application of toxicogenomics to study hepatic effects of food additives such as butylated hydroxytoluene, curcumin, propyl gallate, and thiobendazole. Arbillaga *et al.* (2007) tested the toxicity of ochratoxin A (OTA – a mycotoxin found as a contaminant in cereals and agricultural products) in human renal proximal tubular epithelial cell line HK-2 using gene expression profiles. The comparison of their gene expression analysis with the cellular oxidative stress level and oxidative DNA damage indicated a DNA nonreactive mechanism of OTA toxicity.

References

Amman P, Laethem C, Kedderis G (1998) Choloroform-induced cytolethality in mouse and rat hepatocytes. *Toxicol Appl Pharmacol* **149**, 217–225.

Arbillaga L, Azqueta A, Delft J, Cerain A (2007) *In vitro* gene expression data supporting a DNA non-reactive genotoxic mechanism for ochratoxin A. *Toxicol Appl Pharmacol* **220**, 216–224.

Assaf MS, Abdel-Rahman M (1999) Hepatotoxicity of flunitrapezam and alcohol *in vitro. Toxicol In Vitro* **13**, 393–401.

Atienzar F, Gerets H, Dufrane S, Tilmant K, Cornet M, Dhalluin S, Ruty B, Rose G, Canning M (2007) determination of phospholipidosis potential based on gene expression analysis in HepG2 cells. *Toxicol Sci* **96**, 101–117.

Bajaj J, Knox J, Komorowski R, Saeian K (2003) The irony of herbal hepatitis: Ma–Huang-induced hepatotoxicity. *Dig Dis Sci* **48**, 1925–1928.

Barsig J, Flesch I, Kaufman S (1998) Hepatic cells as chemokine producers in murine listeriosis. *Immunobiology* **199**, 87–104.

Bartosiewicz MJ, Jenkins D, Penn S, Emery J, Buckpitt A (2001) Unique gene expression patterns in liver and kidney associated with exposure to chemical toxicants. *Pharmacol Exp Ther* **297**, 895–905.

Bayliss MK, Bell J, Jenner W, Park G, Wilson K (1999) Utility of hepatocytes to model species differences in the metabolism of loxtidine and to predict pharmacokinetic parameters in rat, dog and man. *Xenobiotica* **29**, 253–268.

Beekman JM, Boess F, Hildrebrand H, Kalkuhl A, Suter L (2006) Gene expression analysis of hepatotoxicant methapyrilene in primary rat hepatocytes: an interlaboratory study. *Environ Health Perspect* **114**, 92–99.

Beyer RP, Fry RC, Lasarev MR, McConnachie LA *et al.* (2007) Multicenter study of acetaminophen hepatotoxicity reveals the importance of biological endpoints in genomic analysis. *Toxicol Sci* **99**, 326–337.

Bigger CB, Brasky K, Lanford R (2001) DNA microarray analysis of chimpanzee liver during acute resolving hepatitis C virus infection. *J Virol* **75**, 7059–7066.

Boess F, Kamber M, Romer S, Gasser R, Muller D, Albertini S, Suter L (2003) Gene expression in two hepatic cell lines, cultured primary hepatocytes and liver slices compared to the *in vivo* liver gene expression in rats: possible implications for toxicogenomics use of *in vitro* systems. *Toxicol Sci* **73**, 386–402.

Boess F, Durr E, Halker M, Albertini S, Suter L (2007) An *in vitro* study on 5-HT$_6$ receptor antagonist induced hepatotoxicity based on biochemical assays and toxicogenomics. *Toxicol In Vitro* **21**, 1276–1286.

Burczynski ME, McMillan M, Ciervo J, Li L, Parker JB, Dunn RT, Hicken S, Farr S, Johnson MD (2000) Toxicogenomics-based discrimination of toxic mechanism in HepG2 human hepatoma cells. *Toxicol Sci* **58**, 399–415.

Cai X, Woods K, Upton SJ, Zhu G (2005) Application of quantitative real-time reverse-transcription-PCR in assessing drug efficacy against the intracellular pathogen *Cryptosporidium parvum in vitro*. *Antimicrob Agents Chemother* **49**, 4437–4442.

Carfagna MA, Young K, Susick R (1996) Sex differences in rat hepatic cytolethality of the protein kinase C inhibitor safingol: role of biotransformation. *Toxicol Appl Pharmacol* **137**, 173–181.

Castell J, Gomez-Leochon M, Ponsoda X, Bort R (1997) Use of cultured hepatocytes to investigate mechanisms of drug hepatotoxicity. *Cell Biol Toxicol* **13**, 331–338.

Cebula TA, El-Hage AN, Ferrans VJ (1984) Toxic interactions of benzyl alcohol with bacterial endotoxin. *Infect Immun* **44**, 91–96.

Cebula TA, Brown EW, Jackson SA, Mammal MK, Mukherjee A, LeClerc JE (2005) Molecular applications for identifying pathogens in the post-9/11 era. *Expert Rev Mol Diagn* **5**, 431–445.

Chung H, Kim H, Jang K, Kim M, Yang J, Kang K, Kim H, Yoon B, Lee B, Lee Y, Kong G (2006) Comprehensive analysis of differential gene expression profiles on D-galactosamine-induced acute mouse liver injury and regeneration. *Toxicology* **227**, 136–144.

Coulet M, Eeckhoutte C, Sutra J, Hoogenboom A, Kuiper H, Castell J, Alvinerie M, Galtier P (1998) Comparative metabolism of thiabendazole in cultured hepatocytes from rats, rabbits, calves, pigs and sheep. *J Agric Food Chem* **46**, 742–748.

Cummings CA, Relman DA (2000) Using DNA microarrays to study host–microbe interactions. *Emerg Infect Dis* **6**, 513–523.

Dambach DM, Andrews B, Moulin F (2005) New technologies and screening strategies for hepatotoxicity: use of *in vitro* models. *Toxicol Pathol* **33**, 17–26.

Delongchamp RR, Velasco C, Dial S, Harris AJ (2005) Genome-wide estimation of gender differences in the gene expression of human livers: statistical design and analysis. *BMC Bioinformatics* **6** (Suppl 2), S13–S26.

Dere E, Boverhof DR, Burgoon LD, Zacharewski TR (2006) In vivo–*in vitro* toxicogenomic comparison of TCDD-elicited gene expression in Hepa1c1c7 mouse hepatoma cells and C57BL/6 hepatic tissue. *BMC Genomics* **7**, 80.

Duggan DJ, Bittner M, Chen Y, Meltzer P, Trent JM (1999) Expression profiling using cDNA microarrays. *Nat Genet* **21**, 110–114.

Favreau JT, Ryu M, Braunstein G, Park S, Love L, Fong T (2002) Severe hepatotoxicity associated with the dietary supplement LipoKinetix. *Ann Intern Med* **136**, 590–595.

Fielden MR, Brennan R, Gollub J (2007) A gene expression biomarker provides early prediction and mechanistic assessment of hepatic tumor induction by nongenotoxic chemicals. *Toxicol Sci* **99**, 90–100.

Gatzidou ET, Zira AN, Theocharis SE (2007) Toxicogenomics: a pivotal piece in the puzzle of toxicological research. *J Appl Toxicol* **27**, 302–309.

Green S, Goldberg A, Zurlo J (2001) Test smart – high production volume chemicals: an approach to implementing alternatives into regulatory toxicology. *Tox Sci* **63**, 6–14.

Hamadeh HK, Jayadev S, Gaillard E, Huang Q, Stoll R, Blanchard K, Chou J, Tucker C, Collins J, Maronport R, Bushel P, Afshari CA (2004) Integration of clinical and gene expression endpoints to explore furan-mediated hepatoxicity. *Mutat Res Fundam Mol Mech Mutagen* **549**, 169–183.

Harries HM, Fletcher ST, Duggan CM, Baker VA (2001) The use of genomics technology to investigate gene expression changes in cultured human liver cells. *Toxicol In Vitro* **15**, 399–405.

Harris A, Dial SL, Casciano DA (2004) Comparison of basal gene expression profiles and effects of hepatocarcinogens on gene expression in cultured primary human hepatocytes and Hep G2 cells. *Mutat Res Fundam Mol Mech Mutagen* **549**, 79–99.

Hengstler J, Ringael M, Biefang K, *et al.* (2000) Culture with cryopreserved hepatocytes: applicability for studies of enzyme induction. *Chemico-Biol Interact* **125**, 51–73.

Hengstler J, Utesch D, Steinberg P, *et al.* (2002) Cryopreserved primary hepatocytes as a constantly available *in vitro* model for the evaluation of human and animal drug metabolism and enzyme induction. *Drug Metab Rev* **32**, 81–118.

Hilderbrand H, Kempka G, Mahnke A (1999) Determination of apoptosis in primary rat hepatocytes by real-time quantitative PCR (TaqMan PCR). *Toxicol In Vitro* **13**, 561–565.

Huang Q, Jin X, Gaillard E, Knight B, Pack B, Stoltz J, Jayadev S, Blanchard K.T (2004) Gene expression profiling reveals multiple toxicity endpoints induced by hepatotoxicants. *Mutat Res Fundam Mol Mech Mutagen* **549**, 147–167.

Jensen G (1996) Prediction of risk of liver diseases by alcohol intake, sex and age: a prospective population study. *Hepatology* **23**, 1025–1029.

Jessen BA, Mullins JS, Peyster AD Stevens GJ (2003) Assessment of hepatocytes and liver slices as *in vitro* test systems to predict *in vivo* gene expression. *Toxicol Sci* **75**, 208–222.

Kando JC, Yonkers K, Cole J (1995) Gender as risk factors for adverse events to medication. *Drugs* **50**, 1–6.

Kefalas V, Stacey NH (1993) Use of primary cultures of rat hepatocytes to study interactive toxicity. *Toxicol In Vitro* **7**, 235–240.

Kier LD, Neft R, Tang L, Suizu R, Cook T, Onsurez K, Tiegler K, Sakal Y, Ortiz M, Nolan T, Sankar U, Li AP (2004) Applications of microarrays with toxicologically relevant genes (*tox* genes) for the evaluation of chemical toxicants in Sprague-Dawley rats *in vivo* and human hepatocytes *in vitro*. *Mutat Res Fundam Mol Mech Mutagen* **549**, 101–113.

Lash LH, Qian D, Desai K, Elfarra A, Sicuri A, Parker J (1998) Glutathione conjugation of per-chloroethylene in rat and mice *in vitro*. *Toxicol Appl Pharamacol* **150**, 49–57.

Lekehal M, Pessayer D, Lereau J, Moulis C, Foureste I, Fau D (1996) Hepatotoxicity is the herbal medicine germander. *Hepatology* **24**, 212–218.

Lewis DFV, Ioannides C, Parke DV (1998) Cytochromes P450 and species differences in xenobiotic metabolism and activation of carcinogens. *Environ Health Perspect* **106**, 633–641.

Li AP, Mauriel P, Gomez-Lechon M, Leng L, Jurima-Romet M (1997) Preclinical evaluation of drug–drug interaction potential. *Chemico-Biol Interact* **107**, 5–16.

Longueville L, Atienzar FA, Marcq L, Evrard S, Leroux F, Gerin B, Arnould T, Remacle J, Canning M (2003) Use of low-density microarray for studying gene expression patterns induced by hepatotoxicants on primary cultures of rat hepatocytes. *Toxicol Sci* **75**, 378–392.

Lu T, Liu J, LeCluyse EL, Zhou Y, Cheng M, Waalkes MP (2001) Application of cDNA microarray to study arsenic-induced liver diseases in the population of Guizhou, China. *Toxicol Sci* **59**, 185–192.

MacGregor JT, Collins JM, Sugiyama Y, Tyson CA, Dean J (2001) *In vitro* human tissue models in risk assessment: report of a consensus-building workshop. *Toxicol Sci* **59**, 17–36.

Maggioli J, Hoover A, Weng L (2006) Toxicogenomic analysis methods for predictive toxicology. *J Pharmacol Toxicol Methods* **53**, 31–37.

McMillian M, Nie A, Parker B, Loone A, Bryant S, Bittner A, Wan J, Johnson MD, Lord P (2005) Drug-induced oxidative stress in rat liver from a toxicogenomic perspective. *Toxicol Appl Pharmacol* **207**, 171–178.

Merrill CL, Ni H, Yoon L, Easton MJ, Creech DR, Hann LM, Thomas HC, Morgan KT (2002) Etomoxir-induced oxidative stress in HepG2 cells detected by differential gene expression is confirmed biochemically. *Toxicol Sci* **68**, 93–101.

Merrill JC, Beck DJ, Kaminski DA, Li AP (1995) Polybrominated biphenyl induction of cytochrome P450 mixed function oxidase activity in primary rat and human hepatocytes. *Toxicology* **99**, 147–152.

Michalopoulos G (1999) Hepatocyte cultures: currently used systems and their applications for studies of hepatocyte growth and differentiation. *Tissue Eng Methods Protocols* **18**, 227–243.

Miller MA (2001) Gender-based differences in toxicity of pharmaceuticals – the Food and Drug Administration's perspective. *Int J Toxicol* **20**, 149–152.

Minami K, Saito T, Narahara M, Tomita H, Kato H, Katoh M, Nakajima M, Yokoi T (2005) Relationship between hepatic gene expression profiles and hepatotoxicity in five typical hepatotoxicant-administered rats. *Toxicol Sci* **87**, 296–305.

Nioi P, Perry BK, Wang E, Gu Y, Snyder RD (2007) *In vitro* detection of drug-induced phospholipidosis using gene expression and fluorescent phospholipid-based methodologies. *Toxicol Sci* **99**, 162–173.

Nishimura M, Koeda A, Suzuki E, Kawano Y, Satoh T, Narimatsu S, Naito S (2006) Regulation of mRNA expression of MDR1, MRP1, MRP2 and MRP3 by prototypical microsomal enzyme inducers in primary cultures of human and rat hepatocytes. *Drug Metab Pharmacokinet* **21**, 297–307.

Novak D, Lewis J (2003) Drug-induced liver disease. *Curr Opin Gastroenterol* **19**, 203–215.

NTP (1999) Toxicology and carcinogenesis studies of fumonisin B$_1$ in rats and mice. Summary Report TR-496, National Toxicology Program, National Institutes of Health; 7–9.

NTP (2001) *In vitro* methods for assessing acute systemic toxicity. Federal Register, September 28, Vol. 66, No. 189. National Toxicology Program, Department of Health and Human Services; 49686–49687.

O'Hanlon KA, Catarame TMG, Blair IS, McDowell DA, Duffy G (2005) Comparison of a real-time PCR and an IMS/culture method to detect *Escherichia coli* O26 and O111 in minced beef in the Republic of Ireland. *Food Microbiol* **22**, 553–560.

Oberemm A, Onyon L, Gundert-Remy U (2005) How can toxicogenomics inform risk assessment? *Toxicol Appl Pharmacol* **207**, 592–598.

Pathernik S, Smihmid J, Sauter T, Schidberg F, Koebe H (1995) Metabolism of pimobendan in long term human hepatocyte culture: *in vivo–in vitro* comparison. *Xenobiotica* **25**, 811–823.

Pennie WD (2000) Use of cDNA microarrays to probe and understand the toxicological consequences of altered gene expression. *Toxicol Lett* **112–113**, 434–477.

Pennie WD, Tugwood JD, Oliver G, Kimber I (2000) The principles and practice of toxicogenomics: applications and opportunities. *Toxicol Sci* **54**, 277–283.

Plumet S, Gerlier D (2005) Optimized SYBR green real-time PCR assay to quantify the absolute copy number of measles virus RNAs using gene specific primers. *J Virol Methods* **128**, 79–87.

Powell CL, Kosyk O, Ross P, Boysen G, Swwenberg JA, Boorman GA, Cunningham ML, Paules RS, Rusyn I (2006) Phenotypic anchoring of acetamoniphen-induced oxidative stress with gene expression profiles in rat liver. *Toxicol Sci* **93**, 213–222.

Ruepp S, Boess F, Suter L, Steiner G, Steele T, Weiser T, Alberini S (2005) Assessment of hepatotoxic liabilities by transcript profiling. *Toxicol Appl Pharmacol* **207**, 161–170.

Ruffa MJ, Ferraro G, Wagner M, Calcagno M, Campos R, Cavallaro L (2002) Cytotoxic effects of Argentine medicinal plant extracts on human hepatocellular carcinoma cell line. *J Ethnopharmacol* **79**, 335–339.

Runge D, Michalopoulos G, Strom S, Runge DM (2001) Recent advances in human hepatocyte culture systems. *Biochem Biophys Res Commun* **274**, 1–3.

Sahu SC (2003) Hepatocyte culture as an *in-vitro* model for evaluating the hepatotoxicity of food-borne toxicants and microbial pathogens: a review. *Toxicol Mech Methods* **13**, 111–119.

Sahu SC, Flynn TJ, Bradlaw JA, Roth WL, Barton CN, Yates JG (2001) Pro-oxidant effects of the flavonoid myricetin on rat hepatocytes in culture. *Toxicol Methods* **11**, 277–283.

Sahu SC, Ruggles DI, O'Donnell MW (2006) Prooxidant activity and toxicity of nordihydroguaiaretic acid in clone-9 rat hepatocyte cultures. *Food Chem Toxicol* **44**, 1751–1757.

Sandker G, Vos R, Delbressine L, Sloof M, Meijer D, Groothuis G (1994) Metabolism of three pharmacologically active drugs in isolated human and rat hepatocytes. *Xenobiotica* **24**, 143–155.

Sawada H, Takami K, Asahi S (2005) A toxicogenomic approach to drug-induced phospholipidosis: analysis of its induction mechanism and establishment of a novel *in vitro* screening system. *Toxicol Sci* **83**, 282–292.

Sawada H, Taniguchi T, Takami K (2006) Improved toxicogenomic screening for drug-induced phospholipidosis using a multiplexed quantitative gene expression Array Plate assay. *Toxicol In Vitro* **20**, 1506–1513.

Schenker S (1997) Medical consequences of alcohol abuse: is gender a factor? *Alcoholism Chem Exp Res* **21**, 179–181.

Schuppe-Koistinen I, Frisk A, Janzon L (2002) Molecular profiling of hepatotoxicity induced by aminoguanidine carboxylate in the rat: gene expression profiling. *Toxicology* **179**, 197–219.

Sheikh NM, Philen R, Love L (1997) Chaparral-associated hepatotoxicity. *Arch Intern Med* **157**, 913–919.

Sneed RA, Buchweitz J, Jean P, Geney P (2000) Pentoxifylline attenuates bacterial polysaccharide-induced enhancement of allyl alcohol hepatotoxicity. *Toxicol Sci* **56**, 203–210.

Stedman C (2002) Herbal hepatotoxicity. *Seminars Liver Dis* **22**, 195–206.

Stickel F, Patsenker E, Schuppan D (2005) Herbal hepatotoxicity. *J Hepatol* **43**, 901–910.

Stierum R, Heijne W, Kienhuis A, Ommen B, Groten J (2005) Toxicogenomics concepts and application to study hepatic effects of food additives and chemicals. *Toxicol Appl Pharmacol* **207**, 179–188.

Teschke R, Gaus W, Loew D (2003) Kava extracts: safety and risks including rare hepatotoxicity. *Phytomedicine* **10**, 440–446.

Treinen-Moslen M (2001) Toxic responses of the liver. In *Casarett and Doull's Toxicology: The Basic Sciences of Poison*, fourth edition, Klaassen CD (ed.). McGraw-Hill, New York; 471–490.

Tuschl G, Mueller SO (2006) Effects of cell culture conditions on primary rat hepatocytes – cell morphology and differential gene expression. *Toxicology* **218**, 205–215.

Van Delft JHM, Van Agen E, Van Breda S, Herwijnen M, Staal Y, Kleinjan J (2004) Discrimination of genotoxic and non-genotoxic carcinogens by gene expression profiling. *Carcinogenesis* **25**, 1265–1276.

Visalli T, Turkall R, Abdel-Rahman M (2004) Cocaine hepatotoxicity. *Int J Toxicol* **23**, 163–170.

Waring JF, Jolly RA, Ciurlionis R, Lum PY, Parestegaard JT, Morfitt D.C, Buratto B, Roberst C, Schadt E, Ulrich R.G (2001) Clustering of hepatotoxins based on mechanism of toxicity using gene expression profiles. *Toxicol Appl Pharmacol* **175**, 28–42.

Waring JF, Cavet G, Jolly RA, McDowell J, Dai H, Zhang C, Lum P, Ferguson A, Roberts CJ, Ulrich RG (2003) Development of a DNA microarray for toxicology based on hepatotoxin-regulated sequences. *Environ Health Perspect* **111**, 863–870.

Wiesenfeld PW, Garthoff LH, Sobotka TJ, Suagee J, Barton CN (2007) Acute oral toxicity of colchicine in rats: effects of gender, vehicle matrix and pre-exposure to lipopolysaccharide. *J Appl Tox* **27**, 421–433.

Willett KL, Roth RA, Walker L (2004) Workshop overview: hepatotoxicity assessment of botanical dietary supplements. *Toxicol Sci* **79**, 4–9.

Zhou H, Harkem J, Hotchkiss J, Yan D, Roth R, Pestka J (2000) Lipopolysacchardie and trichothecene vomitoxin (deoxynivalenol) synergistically induce apoptosis in murine lymphoid organs. *Toxicol Sci* **53**, 252–263.

Zidek N, Hellmann J, Kramer P, Hewitt PG (2007) Acute hepatotoxicity: a predictive model based on focused illumine microarrays. *Toxicol Sci* **99**, 289–302.

13

Toxicogenomics Approach to Drug-induced Phospholipidosis

Hiroshi Sawada

13.1 Introduction

Phospholipidosis is a lipid storage disorder in which lysosomes accumulate excess phospholipids. In animal studies, phospholipidosis has been shown to occur in a variety of organs, including the liver, kidney, heart, lymph node, and brain following the administration of cationic amphiphilic drugs (CADs), including antidepressants, antianginal, antimalarial, and cholesterol-lowering agents (Lullmann *et al.*, 1978; Halliwell, 1997; Reasor, 1989). In humans, amiodarone (Martin and Standing, 1988; Lewis *et al.*, 1990), chloroquine (Muller-Hocker *et al.*, 2003), 4,4′-diethylaminoethoxyhexestrol (Yamamoto *et al.*, 1971; Shikata *et al.*, 1972), fluoxetine (Gonzalez-Rothi *et al.*, 1995), gentamicin (De Broe *et al.*, 1984), perhexiline (Pressayre *et al.*, 1979; Lhermitte *et al.*, 1976) and tobramycin (De Broe *et al.*, 1984) have been reported to cause phospholipidosis. While CADs are thought to induce phospholipidosis by inhibiting lysosomal phospholipase activity, the mechanism by which this occurs has not been extensively studied and has not been well understood. As a matter of practice, then, if a candidate compound is found to have the potential to induce phospholipidosis, its development needs to proceed carefully through both its preclinical and clinical stages, and the cost and time for development can increase considerably.

13.2 DNA Microarray Analysis

The purpose of the microarray study was to examine the molecular mechanisms that contribute to the development of phospholipidosis and to identify specific markers that

Toxicogenomics: A Powerful Tool for Toxicity Assessment Edited by S. C. Sahu
© 2008 John Wiley & Sons, Ltd

might form the basis of an *in vitro* screening assay (Sawada *et al.*, 2005). Specifically, we performed microarray analysis on human hepatoma HepG2 cells after they were treated for 6 or 24 h with each of 12 compounds known to induce phospholipidosis. When examined by electron microscopy (EM), HepG2 cells developed lamellar myelin-like bodies in their lysosomes, the characteristic change of phospholipidosis, after treatment with these compounds for 72 hr (Figure 13.1).

(A)

(B)

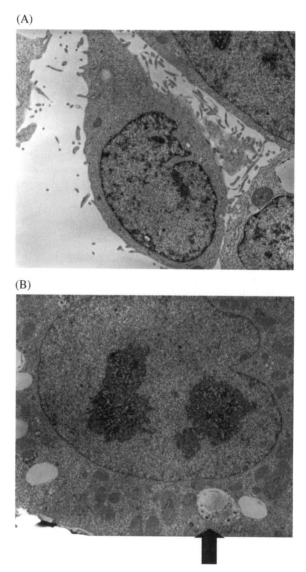

Figure 13.1 *Electron micrographs of HepG2 cells that were treated with vehicle (A) or 10 μmol l^{-1} amiodarone for 72 h. Arrows indicate abnormal lysosomes containing electron-dense deposits and membraneous structures arranged in whorled arrays (myelin figures), which are pathological changes that are characteristic of phospholipidosis (B).*

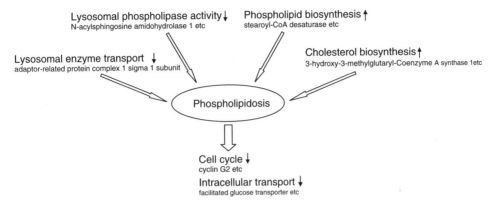

Figure 13.2   *Hypothetical mechanism of drug-induced phospholipidosis. Microarray data suggest that the inhibition of lysosomal phospholipase activity and enzyme transport and the promotion of phospholipid and cholesterol biosynthesis are involved in the induction of phospholipidosis. Cell proliferation and intracellular transport may be secondarily inhibited in this condition by increased intracellular phospholipid. Representative genes for each function are described in Table 13.1.*

Microarray analysis was carried out using the Affymetrix GeneChip System and Human Genome Array containing 22 283 human-specific probe sets (HG-U133A). In order to identify phospholipidosis-related genes and understand the molecular mechanisms, genes that were up- or down-regulated by more than 6 out of the 12 compounds at each time point were selected as putative biomarkers. Functional annotation and categorization of these genes revealed that increased synthesis and decreased degradation of phospholipids were associated with the induction of phospholipidosis, enabling us to propose mechanisms for drug-induced phospholipidosis. We hypothesize that the following four processes are involved in the induction of phospholipidosis (Figure 13.2) based on altered gene expression: (i) inhibition of lysosomal phospholipase activity, as suggested by the up-regulation of phospholipid degradation-related genes (*N*-acylsphingosine amidohydrolase 1, sphingomyelin phosphodiesterase, and hypothetical protein MGC4171); (ii) inhibition of lysosomal enzyme transport, as suggested by the down-regulation of genes involved in lysosomal enzyme transport (adaptor-related protein complex 1 sigma 1 subunit); (iii) enhanced phospholipid biosynthesis, as suggested by the up-regulation of fatty acid biosynthesis-related genes (ELOVL family member 6 and stearoyl-CoA desaturase); and (iv) enhanced cholesterol biosynthesis, as suggested by the induction of cholesterol biosynthesis-related genes (3-hydroxy-3-methylglutaryl-Coenzyme A synthase 1, 3-hydroxy-3-methylglutaryl-Coenzyme A reductase, squalene epoxidase, lanosterol synthase, and 7-dehydrocholesterol reductase). In addition, the up- or down-regulation of transporter genes (e.g. facilitated glucose transporter) and genes that control the cell cycle (e.g. cyclin G2) may also be involved, but modification of the expression of these genes likely reflects secondary changes that occur as an increase in cellular phospholipids.

From a set of 78 altered genes we selected 17 from various functional categories that showed a similar expression profile following treatment as candidate markers for phospholipidosis detection.

Table 13.1   *List of phospholipidosis markers for screening assay*

Gene symbol	Gene title
ASAH1	*N*-Acylsphingosine amidohydrolase (acid ceramidase) 1
MGC4171	Hypothetical protein MGC4171
LSS	Lanosterol synthase (2,3-oxidosqualene-lanosterol cyclase)
NR0B2	Nuclear receptor subfamily 0, group B, member 2
FABP1	Fatty acid binding protein 1, liver
HPN	Hepsin (transmembrane protease, serine 1)
SERPINA3	Serine (or cysteine) proteinase inhibitor, clade A (alpha-1 antiproteinase, antitrypsin), member 3
C10orf10	Chromosome 10 open reading frame 10
FLJ10055	Hypothetical protein FLJ10055
FRCP1	Likely ortholog of mouse fibronectin type III repeat containing protein 1
SLC2A3	Solute carrier family 2 (facilitated glucose transporter), member 3
TAGLN	Transgelin

13.3   Real-time Polymerase Chain Reaction Analysis

We performed real-time polymerase chain reaction (PCR) analysis at 24 h to validate the expression of the 17 candidate marker genes using HepG2 cells treated with 17 phospholipidosis-positive and 13 phospholipidosis-negative compounds and to establish an *in vitro* screening assay. The PCR data for five genes did not correspond well to the pathological phospholipidosis scores determined by EM (at 72 h). Thus, based on PCR validation, we identified 12 phospholipidosis-specific gene markers (Table 13.1) as the basis for an *in vitro* assay.

We attempted to establish a representative value reflecting the severity of phospholipidosis based on these phospholipidosis markers. We calculated the average fold change value of these gene markers, which we termed the 'phospholipidosis mRNA score.' The mRNA score correlated well with the cells' pathological scores determined by EM and was a good index of phospholipidosis induction potential for a given compound. Since these phospholipidosis markers included genes whose functions included phospholipid degradation, cholesterol biosynthesis, fatty acid transport, proteolysis and peptidolysis, and endopeptidase inhibition, among others, we were also able to monitor multiple intracellular events that were involved in the induction of phospholipidosis by monitoring these 12 marker genes. In conclusion, we established a novel *in vitro* PCR-based screening assay for drug-induced phospholipidosis. This assay requires a far lower amount of test compound and shorter periods of treatment (24 h) than do conventional *in vivo* toxicity studies.

13.4   ArrayPlate Assay

Subsequently, we transferred our PCR-based assay into a 96-well microplate-based assay (the ArrayPlate qNPA assay) measuring all 12 genes plus housekeeper control genes in each well of a microplate in order to improve the throughput of our *in vitro* screening system (Sawada *et al.*, 2006a,b). The ArrayPlate qNPA™ (quantitative nuclease protection assay, High Throughput Genomics) is a 96-well plate-based assay using an array of 16 distinct elements printed on the bottom of each well to provide a multiplexed quantitative measurement of gene expression. This assay can easily measure multiple samples at one time and only requires that the sample be lysed. This assay does not require RNA extraction, reverse transcription, or target amplification (Martel *et al.*, 2002). Detailed information about the qNPA assay is described in the next section.

In order to test the key performance criteria of sensitivity and repeatability, we measured the expression of the 12 phospholipidosis marker genes using real-time PCR and ArrayPlate qNPA on HepG2 cells after they were treated for 24 h with each of amiodarone and an additional set of 80 proprietary compounds. The sensitivity and linearity of the assay were determined using eight serial 1:2 dilutions of the pooled bulk cell lysate samples. The expression of all target genes was detected at quantifiable level in the ArrayPlate qNPA and showed linear responses of signal to sample size. The well-to-well repeatability of the ArrayPlate analysis was determined using aliquots from a pooled bulk cell lysate sample. The coefficient of variability (%CV – the standard deviation as a proportion of the average) for each gene measured in all wells from a single plate was within 13 %. To assess plate-to-plate and day-to-day repeatability, the ArrayPlate qNPA was carried out using four separate plates on different days, measuring vehicle-treated and amiodarone-treated cell lysates. The %CV of the fold change values between vehicle control and amiodarone-treated cell lysates for each gene, day-to-day, was ≤ 4 %. The correlation between the ArrayPlate qNPA and real-time PCR assays was determined using phospholipidosis mRNA scores in cells that were treated with 80 proprietary compounds, plotting the PCR score for each component versus the qNPA score. The results showed that the phospholipidosis mRNA scores correlated significantly ($R^2 = 0.95$) between the ArrayPlate qNPA and real-time PCR assays. Based on these data, it was concluded that the PCR-based assay had been successfully transferred to the ArrayPlate qNPA assay platform.

We proceeded to use this screening assay to test a set of 71 compounds which were structurally similar and included five compounds reported to cause phospholipidosis in rat (Figure 13.3), the development of which were consequently halted. From this screen, we found three chemical substituents that could reduce the phospholipidosis-inducing potential of compounds within the chemical series. Compound 13, with one such substituent, cleared the rat toxicity test and advanced into development. These results show that this novel gene expression biomarker *in vitro* screening assay is readily able to detect the phospholipidosis-inducing potential of compounds and to provide ranking scores that can be useful in structure–activity relationship studies.

Because the ArrayPlate qNPA is a 96-well microplate based assay, which does not require RNA extraction, reverse transcription, or gene amplification, the ArrayPlate qNPA provided significantly increased testing throughput and cost-efficiency. We have tested about 700 compounds per year using this ArrayPlate qNPA assay in our company in order to search for back-up compounds, to check the potential for inducing phospholipidosis before *in vivo* toxicity tests are initiated, and to compare our compounds with those of competitors.

13.5 Other Applications of ArrayPlate qNPA

The ArrayPlate qNPA is a versatile assay with many applications besides the one described in this chapter. It has been used for high-throughput screening to identify drug leads (Bakir *et al.*, 2007; Kris *et al.*, 2007), to provide precise EC_{50} data, and, perhaps most importantly, to validate signatures using formalin-fixed paraffin-embedded (FFPE) tissue (Roberts *et al.*,

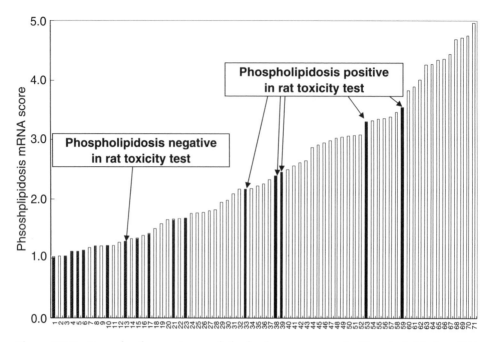

Figure 13.3   *Example of usage to search for back-up compounds. Phospholipidosis mRNA scores were examined on a set of 71 compounds which were structurally similar. On the X-axis, the 71 compounds are listed in ascending order of their phospholipidosis mRNA score. Compounds color-coded by black have been reported to cause phospholipidosis in rat toxicity tests, and dropped out of preclinical development because of phospholipidosis. As a result of screening these compounds, we found three chemical substituents (coded by diagonal lines) that reduced the phospholipidosis-inducing potential of analogs within the chemical series. Compound 13, with one of these substituents, cleared the rat toxicity test and advanced to the next stage of development.*

2007). In this latter application, it was shown that similar measurements of gene expression were obtained from fresh versus fixed or recently fixed versus 18-year archived FFPE. It was also shown that the level of gene expression correlated with the level of protein product measured by immunohistochemistry. Signatures that differentiated benign reactive lymph node from diffuse large B-cell lymphoma were identified retrospectively in this study using fixed-tissue biopsies archived for as long as 20 years. Thus, the opportunity exists to validate toxicity signatures from archived animal tissues collected from studies of the thousands of compounds tested by pharmaceutical companies over the past 30 or 40 years.

13.6   Toxicogenomics Approach for Blood Biomarkers

A suggested *in vivo* indicator for phospholipidosis is the presence of vacuolated lymphocytes (Rudmann *et al.*, 2004). Recent advancements in -omics methodology have allowed for the identification of candidate biomarkers of phospholipidosis in serum, plasma, and urine using metabonomics that employed liquid chromatography coupled to electrospray ionization mass spectrometry or nuclear magnetic resonance analysis (Nicholls *et al.*, 2000; Espina *et al.*, 2001; Idborg-Bjorkman *et al.*, 2003; Mortuza *et al.*, 2003). However, there have been no widely accepted *in vivo* biomarkers for phospholipidosis. Therefore, we also performed a large-scale gene expression analysis using the PAXgene Blood RNA System (Rainen *et al.*, 2002) and DNA microarrays on rat blood in order to identify a subset of genes in peripheral blood cells that could be used as potential clinical biomarkers for drug-induced phospholipidosis (Sawada *et al.*, 2006a,b). Male Crl:CD(SD) rats were treated daily for 1, 3, or 7 days with the following CADs that are known to induce phospholipidosis, and their blood was collected 24 h later: amiodarone, chloroquine, fluoxetine, imipramine (Lullmann-Rauch, 1974; Lullmann-Rauch and Scheid, 1975), perhexiline, quinacrine (Lullmann-Rauch *et al.*, 1996) or tamoxifen (Lullmann and Lullmann-Rauch, 1981; Drenckhahn *et al.*, 1983). Histopathological changes characteristic of phospholipidosis were observed in two or more organs of rats that received three or seven doses of the drugs. These animals also had high numbers of vacuolated lymphocytes in their peripheral blood. Microarray analysis was carried out using the Affymetrix GeneChip System and Rat Expression Array containing 15 924 rat-specific probe sets (RAE230A). We selected genes that were up- or down-regulated by all six compounds at each time point (number of probe sets 0, 36, and 23 after one, three, and seven daily doses respectively). Functional annotation and categorization of the selected genes suggested that these compounds enhanced lipid metabolism, repressed intracellular transport, interfered with the animals' response to oxidative stress, inhibited the cell cycle, and increased immune responsiveness in treated rats (Table 13.2). These genes require further validation before use in clinical research, applying the strategy depicted in Figure 13.4 that resulted in our successful ArrayPlate qNPA *in vitro* screen. However, our high-density array results suggest that a toxicogenomics

Table 13.2 *List of representative candidate blood biomarker genes for phospholipidosis*

Function	Gene title	Gene symbol	Up or down
Lipid metabolism	Annexin A1	Anxa1	Up
	Phosphatidylinositol binding clathrin assembly protein	Picalm	Up
	Arachidonate 12-lipoxygenase	Alox15	Down
	Diacylglycerol O-acyltransferase homolog 2 (mouse)	Dgat2	Up
Transport system	Calcium binding protein p22	Chp	Down
	Aquaporin 1	Aqp1	Down
	Similar to glycolipid transfer protein	—	Down
Antioxidant	Superoxide dismutase 2, mitochondrial	Sod2	Up
	Peroxiredoxin 1	Prdx1	Down
	Thioredoxin	Txn	Down
Cell cycle, proliferation, death	Similar to cell division cycle 34	—	Down
	Cyclin-dependent kinase inhibitor 3 (predicted)	—	Down
Immune response	Chemokine (C-C motif) ligand 6	Ccl6	Up
	Interleukin 1 beta	Il1b	Up
	Secretory leukocyte peptidase inhibitor	Slpi	Up

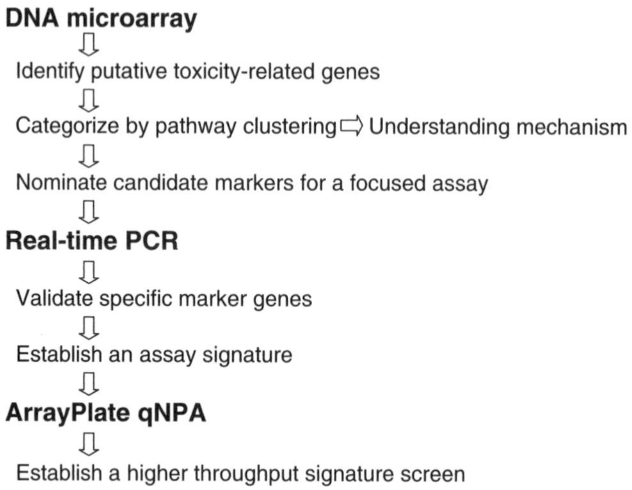

DNA microarray
⇩
Identify putative toxicity-related genes
⇩
Categorize by pathway clustering ⇨ Understanding mechanism
⇩
Nominate candidate markers for a focused assay
⇩
Real-time PCR
⇩
Validate specific marker genes
⇩
Establish an assay signature
⇩
ArrayPlate qNPA
⇩
Establish a higher throughput signature screen

Figure 13.4 *Strategy for toxicogenomics analysis. This figure depicts how we approached the problem of phospholipidosis to produce a screening assay useful for predicting the phospholipidosis potential of compound leads, for identifying backup compounds that did not induce phospholipidosis, and for providing structure–activity relationship (SAR) data to permit the modifications within a chemical series to be identified and subsets of analogs that do not induce phospholipidosis to be optimized. Having shown that PCR data correlate to qNPA data, the more efficient ArrayPlate qNPA can be used to validate the specific genes and to establish the assay signature, eliminating the use of real-time PCR from the process. In addition, future studies will not be limited to cells or fresh tissue, but can utilize FFPE from previous animal studies.*

approach targeting surrogate tissues, such as peripheral blood samples, should be useful in identifying candidate marker genes of phospholipidosis accessible from blood.

13.7 Discussion

We established a toxicogenomics screening system for drug-induced phospholipidosis. Our first paper represents an important advancement of toxicogenomics and a practical application of toxicogenomics to drug development by providing both an assessment of the mode of action and establishment of a novel risk-assessment tool (Cunningham and Lehman-McKeeman, 2005).

Drug-induced phospholipidosis is an area of great interest, both to the Food and Drug Administration and to the pharmaceutical industry. A Phospholipidosis Working Group was established in 2004 with the ultimate objective of developing 'guidance on phospholipidosis.' We discussed our results with this group both with regard to understanding the mechanism of drug-induced phospholipidosis and to developing clinically useful biomarkers (Berridge *et al.*, 2005).

Our research has used DNA microarray, real-time PCR, and ArrayPlate qNPA screening analysis to examine the mode of action of phospholipidosis-inducing compounds, identify and validate candidate gene markers, and establish a quantitative high sample throughput and cost-efficient screen to detect and rank the phospholipidosis-inducing potential of compounds and provide SAR data (Figure 13.4). In future studies we would go directly from the identification of putative markers using high-density arrays to use of the ArrayPlate qNPA to validate all those markers and then select the best markers for a screening and compound optimization assay. HTG is also actively developing a high-density qNPA array for gene marker identification, which will have the advantage of being able to test FFPE, enabling researchers to access the vast archives of animal and human tissues that are available from the past 50 years of safety studies.

References

Bakir F, Kher S, Pannala M, Wilson N, Nguyen T, Sircar I, Takedomi K, Fukushima C, Zapf J, Xu K, Zhang SH, Liu J, Morera L, Schneider L, Sakurai N, Jack R, Cheng JF (2007) Discovery and structure–activity relationship studies of indole derivatives as liver X receptor (LXR) agonists. *Bioorg Med Chem Lett* **17**, 3473–3479. DOI: 10.1016/j.bmcl.2007.03.076.

Berridge BR, Chatman LA, Odin M, Schultze AE, Losco PE, Meehan JT, Peters T, Vonderfecht SL (2007) Phospholipidosis in nonclinical toxicity studies. *Toxicol Pathol* **35**, 325.

Cunningham ML, Lehman-McKeeman L (2005) Applying toxicogenomics in mechanistic and predictive toxicology. *Toxicol Sci* **83**, 205–206.

De Broe ME, Paulus GJ, Verpooten GA, Roels F, Buyssens N, Wedeen R, Van Hoof F, Tulkens PM (1984) Early effects of gentamicin, tobramycin, and amikacin on the human kidney. *Kidney Int* **25**, 643–652.

Drenckhahn D, Jacobi B, Lullmann-Rauch R (1983) Corneal lipidosis in rats treated with amphiphilic cationic drugs. *Arzneimittelforschung* **33**, 827–831.

Espina JR, Shockcor JP, Herron WJ, Car BD, Contel NR, Ciaccio PJ, Lindon JC, Holmes E, Nicholson JK (2001) Detection of *in vivo* biomarkers of phospholipidosis using NMR-based metabonomic approaches. *Magn Reson Chem* **39**, 559–565.

Gonzalez-Rothi RJ, Zander DS, Ros PR (1995) Fluoxetine hydrochloride (Prozac)-induced pulmonary disease. *Chest* **107**, 1763–1765.

Halliwell WH (1997) Cationic amphiphilic drug-induced phospholipidosis. *Toxicol Pathol* **25**, 53–60.

Idborg-Bjorkman H, Edlund PO, Kvalheim OM, Schuppe-Koistinen I, Jacobsson SP (2003) Screening of biomarkers in rat urine using LC/electrospray ionization-MS and two-way data analysis. *Anal Chem* **75**, 4784–4792.

Kris RM, Felder S, Deyholos M, Lambert GM, Hinton J, Botros I, Martel R, Seligmann B, Galbraith DW (2007) High-throughput, high-sensitivity analysis of gene expression in *Arabidopsis*. *Plant Physiol* **144**, 1256–1266.

Lewis JH, Mullick F, Ishak KG, Ranard RC, Ragsdale B, Perse RM, Rusnock EJ, Wolke A, Benjamin SB, Seeff LB, Zimmerman HJ (1990) Histopathologic analysis of suspected amiodarone hepatotoxicity. *Hum Pathol* **21**, 59–67.

Lhermitte F, Fardeau M, Chedru F, Mallecourt J (1976) Polyneuropathy after perhexiline maleate therapy. *Br Med J* **1**, 1256.

Lullmann H, Lullmann-Rauch R (1981) Tamoxifen-induced generalized lipidosis in rats subchronically treated with high doses. *Toxicol Appl Pharmacol* **61**, 138–146.

Lullmann H, Lullmann-Rauch R, Wassermann O (1978) Lipidosis induced by amphiphilic cationic drugs. *Biochem Pharmacol* **27**, 1103–1108.

Lullmann-Rauch R (1974) Lipidosis-like ultrastructural alterations in rat lymph nodes after treatment with tricyclic antidepressants or neuroleptics. *Naunyn Schmiedebergs Arch Pharmacol* **286**, 165–179.

Lullmann-Rauch R, Scheid D (1975) Intraalveolar foam cells associated with lipidosis-like alterations in lung and liver of rats treated with tricyclic psychotropic drugs. *Virchows Arch B Cell Pathol* **19**, 255–268.

Lullmann-Rauch R, Pods R, von Witzendorff B (1996) The antimalarials quinacrine and chloroquine induce weak lysosomal storage of sulphated glycosaminoglycans in cell culture and *in vivo*. *Toxicology* **110**, 27–37.

Martel RR, Botros IW, Rounseville MP, Hinton JP, Staples RR, Morales DA, Farmer JB, Seligmann BE (2002) Multiplexed screening assay for mRNA combining nuclease protection with luminescent array detection. *Assay Drug Dev Technol* **1**, 61–71.

Martin II WJ, Standing JE (1988) Amiodarone pulmonary toxicity: biochemical evidence for a cellular phospholipidosis in the bronchoalveolar lavage fluid of human subjects. *J Pharmacol Exp Ther* **244**, 774–779.

Mortuza GB, Neville WA, Delany J, Waterfield CJ, Camilleri P (2003) Characterization of a potential biomarker of phospholipidosis from amiodarone-treated rats. *Biochim Biophys Acta* **1631**, 136–146.

Muller-Hocker J, Schmid H, Weiss M, Dendorfer U, Braun GS (2003) Chloroquine-induced phospholipidosis of the kidney mimicking Fabry's disease: case report and review of the literature. *Hum Pathol* **34**, 285–289.

Nicholls AW, Nicholson JK, Haselden JN, Waterfield CJ (2000) A metabonomic approach to the investigation of drug-induced phospholipidosis: an NMR spectroscopy and pattern recognition study. *Biomarkers* **5**, 410–423.

Pressayre D, Bichara M, Feldman G, Degott C, Potet F, Benhamou J-P (1979) Perhexiline maleate-induced cirrhosis. *Gastroenterology* **76**, 170–177.

Rainen L, Oelmueller U, Jurgensen S, Wyrich R, Ballas C, Schram J, Herdman C, Bankaitis-Davis D, Nicholls N, Trollinger D, Tryon V (2002) Stabilization of mRNA expression in whole blood samples. *Clin Chem* **48**, 1883–1890.

Reasor MJ (1989) A review of the biology and toxicologic implications of the induction of lysosomal lamellar bodies by drugs. *Toxicol Appl Pharmacol* **97**, 47–56.

Roberts RA, Sabalos CM, LeBlanc ML, Martel RR, Frutiger YM, Unger JM, Botros IW, Rounseville MP, Seligmann BE, Miller TP, Grogan TM, Rimsza LM (2007) Quantitative nuclease protection assay in paraffin-embedded tissue replicates prognostic microarray gene expression in diffuse large-B-cell lymphoma. *Lab Invest* **87**, 979–997. DOI: 10.1038/labinvest.3700665.

Rudmann DG, McNerney ME, VanderEide SL, Schemmer JK, Eversole RR, Vonderfecht SL (2004) Epididymal and systemic phospholipidosis in rats and dogs treated with the dopamine D3 selective antagonist PNU-177864. *Toxicol Pathol* **32**, 326–332.

Sawada H, Takami K, Asahi S (2005) A toxicogenomic approach to drug-induced phospholipidosis: analysis of its induction mechanism and establishment of a novel *in vitro* screening system. *Toxicol Sci* **83**, 282–292.

Sawada H, Taniguchi K, Takami K (2006a) Improved toxicogenomic screening for drug-induced phospholipidosis using a multiplexed quantitative gene expression ArrayPlate assay. *Toxicol In Vitro* **20**, 1506–1513.

Sawada H, Taniguchi, K. Mori I, Iwachido T, Nakashita Y, Takami K (2006b) Toxicogenomics approaches to drug-induced phospholipidosis: establishment of *in vitro* screening system and identification of candidate blood biomarkers. Itinerary Planner Abstract No. 1620. Society of Toxicology, San Diego CA.

Shikata T, Kanetaka T, Endo Y, Nagashima K (1972) Drug-induced generalized phospholipidosis. *Acta Pathol Jpn* **22**, 517–533.

Yamamoto A, Adachi S, Ishikawa K, Yokomura T, Kitani T (1971) Studies on drug-induced lipidosis. 3. Lipid composition of the liver and some other tissues in clinical cases of 'Niemann–Pick-like syndrome' induced by 4,4′diethylaminoethoxyhexestrol. *J Biochem* **70**, 775–784.

14

Use of Toxicogenomics as an Early Predictive Tool for Hepatotoxicity

Laura Suter

14.1   Introduction

There is a great and ever increasing need for better and earlier prediction of side effects, in particular hepatotoxicity. Several factors contribute to this phenomenon, including financial interest from the pharmaceutical industry, public health concern from the regulatory bodies, and ethical concerns regarding the use of animal testing for the assessment of the safety of drugs and chemicals.

Within the pharmaceutical industry, it is widely recognized that failure of compounds in development due to safety issues in late preclinical development or in the clinic represent an important economic burden. Accurate prediction of the potential toxicity of a new drug candidate at an early stage of development is highly desired in order to increase the quality of the selected compounds and decreasing attrition rates in the later development phases. Hence, the success of development of new efficacious and safe drugs depends in part on the rigorous weeding out of unsuitable compounds in the early stages, when screening and selection are relatively quick and cheap to perform. The application of 'omics' technologies, such as genomics, offers the possibility of moving the selection process upstream, bringing a new and rigorous filtering process into the early to very early low-cost phase.

The public sector, including regulatory bodies, has shown increasing concern in detecting and predicting adverse side effects of drugs more accurately and earlier than with the conventional toxicology methods. This is clearly reflected in the number of collaborations regarding new technologies and safety that have been entered or are partly funded by regulatory bodies or governments. Of major concern for industry, governments, and patient organizations alike is the likelihood of encountering adverse drug reactions while testing

Toxicogenomics: A Powerful Tool for Toxicity Assessment   Edited by S. C. Sahu
© 2008 John Wiley & Sons, Ltd

new drugs in healthy volunteers or patients, and the incidence of serious and fatal adverse drug reactions in the clinic. In US hospitals, the number of adverse drug reactions was found to be extremely high and represent an important clinical issue (Lazarou *et al.*, 1998). It is also of pivotal importance to characterize new drugs as accurately as possible in terms of safety (and pharmacology) in order to avoid unexpected adverse effects in patients, and hepatotoxicity is one of the most commonly observed adverse events.

It is also a major ethical concern that animal usage and animal suffering for testing of new drug candidates or chemicals should be minimized. More sophisticated endpoints such as toxicogenomics, *in vitro* systems, or a combination thereof are expected to be of use. This is of importance for the pharmaceutical industry, but also paramount for the chemical industry, which is facing thousands of chemicals that need to be tested in terms of safety in Europe following the regulation on Registration, Evaluation, Authorisation and Restriction of Chemicals (REACH) that came into force in June 2007.

All this leads to one major widespread need: new technologies which are able to provide better safety assessment are necessary: better in terms of sensitivity, detecting liabilities with shorter treatment periods and/or lower doses; better in terms of throughput and decreased animal usage or animal stress (for example, non- or less-invasive measurements or *in vitro* systems); better in providing additional knowledge on the molecular mechanisms underlying a given toxicity, addressing issues such as species specificity, exaggerated pharmacology, or target-related effects; better in identifying new specific biomarkers that could be used to monitor possible adverse effects in both the nonclinical and clinical settings. Holistic approaches, such as toxicogenomics, as well as metabonomics and proteomics, are believed to be ideally suited to fulfill some or all these expectations. These approaches provide the means to predict toxicity based on new endpoints and to increase the understanding of the molecular events underlying a given toxicity.

14.2   Managing Expectations

It can be concluded from available publications and presentations that experts in the field of safety assessment still do not fully agree on the expectations towards these new 'omics' technologies. Some scientists are convinced that understanding the intimate molecular mechanisms associated with a toxicity will greatly help the drug development process and advance our scientific knowledge. Others expect that these new technologies will be able to predict a toxic liability with high accuracy, within very short time-frames, and possibly with a minimal usage of compound and animal testing. Yet others rely on these technologies to identify novel biomarkers of toxicity. Ideally, these newly identified biomarkers could be subsequently validated, easily measured, and used to monitor progression and recovery of an injury in preclinical animal models and in patients. The technologies have the potential to address all these questions, but probably not simultaneously and certainly not following a one-size-fits-all solution. The matters that need to be addressed will impact the study design and the data analysis strategy. If the main question for which an answer is sought is the mechanism by which toxicity occurs, then the study design needs to be appropriate and include a meaningful number of sample collection time points, replicates, and doses. If biomarkers are the main goal, then the gold standard by which their performance is measured needs to be very clearly defined. Are we seeking a biomarker that parallels the histopathological findings, or are we willing to accept prodromal biomarkers that might give us a signal before other toxicological endpoints? Specifically, for the prediction of toxicity, acceptable performance of predictive models in terms of specificity and sensitivity need to be predefined for each experimental system. Is a prediction of hepatotoxicity with a 70 % sensitivity good enough? How many false positives, e.g. which specificity, are we willing to accept? Do we expect to predict specific human hepatotoxicants using a rodent *in vivo* or even *in vitro* model? Are we also trying to address idiosyncratic events?

Opinions on these matters are very varied, but experience so far can provide some guidance. In general, it can be said that the sensitivity and specificity of *in vitro* models is lower that that of the *in vivo* models. Also, we should assume that a correct prediction will always refer to the species in which the test is being performed. Extrapolations across species are necessary and possible; they provide indications regarding physiological or cellular functions that might or might not be affected in different organisms. But direct extrapolation across species and prediction of idiosyncrasy are extremely high hurdles for any experimental system and should not automatically be expected. Moreover, the selection of the appropriate statistical tools and the appropriate (cross-)validation for the construction and application of predictive models based on gene expression or other 'omics' data will also play a major role in the performance and ultimately on the outcome of any toxicogenomics experiment.

14.3 Technologies

There are a number of approaches that have been published and extensively reviewed in the fields of genomics, proteomics, and metabonomics for safety assessment in general and for hepatotoxicity in particular. It is completely outside the scope of this chapter to describe the technologies and platforms. However, it would not be appropriate to discuss the application of toxicogenomics without mentioning the other 'omics' commonly employed in toxicology for the prediction of hepatotoxicity.

Proteomics detects peptides and proteins using several platforms for the analysis of tissue or serum (or other body fluids such as urine). It is predominately used for the identification of protein-biomarkers that are associated with a given toxicity. Proteome analysis is very complex, and technologies are still being developed. They need to overcome the challenges posed by the large number and concentration range of proteins and peptides in a given biological sample (Elrick *et al.*, 2006). Indeed, proteomic approaches to identify treatment-related protein expression patterns have generally proven time consuming and insensitive, identifying generally few, abundant, stress-inducible proteins (Fountoulakis *et al.*, 2000, 2002; Yamamoto *et al.*, 2006). Despite this, proteomics has also been successfully used for the identification of protein markers amenable to better prediction of hepatotoxicity and carcinogenicity, with the aim of shortening of long-term studies (Kröger *et al.*, 2004).

Metabonomics analyzes endogenous metabolites using generally proton nuclear magnetic resonance (NMR) or gas or liquid chromatography coupled with mass spectroscopy of body fluids, such as plasma or urine. The data are utilized to identify metabolite biomarkers that can be monitored, to identify specific pathways that are affected by the test compound, and to identify fingerprints that are characteristic of a given toxicity. Matching these characteristic fingerprints to a database populated with metabolic profiles from known toxicants is a method often used to predict the toxicity of a given compound. For this purpose, the Consortium on Metabonomic Toxicology (COMET) constructed a database comprising urinary metabolic profiles and conventional toxiological endpoints for more than 100 test compounds. Their study constitutes the largest validation of the NMR-based metabonomic approach in toxicology. It focuses mainly on the liver and the kidney as target organs. Some of the results were recently published and showed that although the compounds elicited a relatively consistent metabolic response, the predictive power was limited. In particular, for hepatotoxicity, they achieved a sensitivity of 67 % and a specificity of 77 % (Ebbels *et al.*, 2007). In a similar study in our laboratories using urine metabolite profiles measured by NMR, we were able to differentiate two biphosphonates based on their metabonomic profiles. The results were in agreement with the different responses in terms of hepatotoxicity and nephrotoxicity caused by the compounds, and provided additional mechanistic information that could not be obtained by other conventional methods, including the identification of a putative biomarker (Dieterle *et al.*, 2007). In addition, several commercial providers have built proprietary databases using various technology platforms and generated statistical models that can be used for the prediction of several toxicities, including hepatotoxicity.

Genomics approaches involve a genome-wide measurement of transcripts expressed in a given tissue at a given time using commercially available microarrays. Genomics data can be used for mechanistic interpretation of the molecular and cellular events underlying the toxicity, so long as these events affect transcriptional regulation directly or indirectly. They can also serve to predict the toxicity of a compound by matching its gene expression profile to those stored in a database populated by transcriptomes from known toxicants. In the last 8–10 years, a large amount of genomic data was collected in the public and the private sectors. For safety assessment, the liver was, and remains, the most studied target organ using toxicogenomics technologies, and several examples will be discussed in more detail in this chapter.

14.4   Systems Toxicology

Technological progress will not fully replace the methods and knowledge available for safety assessment. Rather, these new technologies should not be applied in isolation. The combination of the three 'omics' technologies together with conventional endpoints, such as histopathology and clinical chemistry, is the holy grail of systems toxicology. There is an increasing interest in combining all or some of these technologies to improve our level of knowledge. This approach has been attempted by many, but little systematically collected and consolidated data are yet available (Craig *et al.*, 2006). The National Center for Toxicogenomics is developing the first public knowledge base that combines this kind of data. It is called the Chemical Effects in Biological Systems (CEBS; Waters *et al.*, 2003). Also, the PredTox Consortium in Europe is focusing its efforts on the liver using a systems toxicology approach. There are several reasons for the scarcity of large and complete data sets. One of them is the complexity of studies designed to collect this amount and variety of data. The technical challenges and costs involved in generating large-scale 'omics' data also play a major role. In addition, it is important to bear in mind that the task of systems toxicology is far from trivial. These new technologies are not wonders or magic solutions, but the source of very complex data sets that need careful analysis and interpretation. It is clear that the use of such multivariate, highly sensitive technologies carries the immense potential of providing a holistic view. However, technical issues in the detection of the analytes, statistical complexity due to few samples and many variables, and last but not least the use of an appropriate biological test system are factors that need to be carefully controlled to obtain biologically meaningful data (Gant and Zhang, 2005).

14.5   Timing Is Everything

As was stressed before, the earlier that (hepato)toxic liabilities are detected, the better. With the exception of *in silico* approaches that can deal with chemical structures without necessarily involving newly acquired experimental data, the earliest possible time a toxicogenomics approach can be applied is in the context of *in vitro* experiments, at the time when small amounts of chemical entities are first synthesized. A possible testing approach for hepatotoxicants using cell lines, hepatocyte and short-term *in vivo* studies combined with genomics technologies is depicted in Figure 14.1. The different stages of testing have different requirements linked with the strengths and caveats of each system. Each of these experimental systems has advantages and disadvantages that need to be considered. *In vitro* approaches involving cell lines or primary cell cultures have major advantages in terms of minimizing animal use and of allowing for high-throughput testing. However, and similar to any other endpoint, the reliability of an *in vitro* system in mimicking a living organism needs to be established first. As will be described in detail below, this is not always the case, and *in vitro* systems cannot fulfill all our needs for studying hepatotoxicity. Thus, for a more

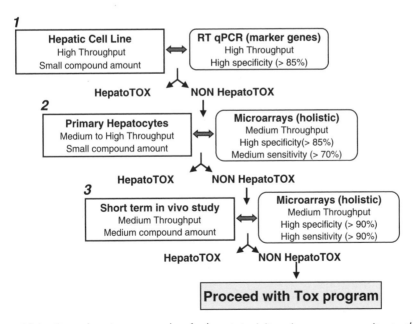

Figure 14.1   *Several testing approaches for hepatotoxicity using gene expression analysis.*

accurate safety assessment, animal testing is often necessary. But also in this context, early detection of hepatotoxicity would be very advantageous. Toxicogenomics should allow toxicologists to detect a possible liability after a single administration or a few administrations, e.g. earlier than after a 2- or 4-week toxicity study or a 2-year carcinogenicity study. These early *in vivo* assays have the major advantages of requiring less compound for dosing, implying less animal suffering and stress by decreasing the handling, and taking less time. In this context, genomics results can unveil (hepato)toxic liabilities before overt toxicity becomes evident by means of conventional endpoints – a truly predictive use of this technology. Later in the development phase, during exploratory or regulatory animal testing, genomics approaches are also very useful as additional endpoints collected during longer term animal studies. In these cases, the gene expression changes associated with a toxic phenotype, such as histopathological findings, can help in understanding the causes or mechanisms associated with the observed toxicity. Thus, although the earlier that hepatotoxicity can be detected the better, toxicogenomics has proven a useful tool at several different preclinical stages: from *in vitro* screening through to regulatory animal testing.

14.6   *In Vitro* Toxicogenomics

In vitro toxicogenomics has been used for a variety of toxicological endpoints, among them skin irritation, developmental toxicology, phospholipidosis, and genetic toxicology (Sawada *et al.*, 2006; Baken *et al.*, 2007; Ku *et al.*, 2007; Stigson *et al.*, 2007; Thybaud *et al.*, 2007). However, most of the efforts in the *in vitro* toxicogenomics field were focused on the liver and, in particular, the hepatocyte. Several authors have generated *in vitro* toxicogenomics databases with the aim of predicting and understanding hepatotoxicity. The choice of the cell system varies from primary cells obtained from rats or human, to a variety of more or less widespread cell lines.

Primary cells appear to be more representative of the physiological situation *in vivo*, whereas cell lines lack many of the metabolic capabilities characteristic of hepatocytes (Boess *et al.*, 2003). Primary cells, however, are more difficult to obtain, reflect the heterogeneity of the donor population, and can only be maintained in culture for a relatively short period of time. Nevertheless, all the large toxicogenomics projects have selected the rat primary hepatocyte in monolayer culture as the most suitable *in vitro* system for the study of hepatotoxicity. Among them are the Toxicogenomics Project in Japan (Kiyosawa *et al.*, 2006; Tamura *et al.*, 2006), the *in vitro* part of commercially available toxicogenomics databases such as DrugMatrix® by Iconix/Entelos (Fielden and Kolaja, 2006) and ToxExpress® by Genelogic, and several scientists throughout the pharmaceutical industry (Waring *et al.*, 2001, 2004; Hultin-Rosenberg *et al.*, 2006; Yang *et al.*, 2006; Boess *et al.*, 2007).

Cell lines, on the other hand, are usually easy to handle and store, stable in culture, and enable high-throughput approaches. The human hepatoma cell line HepG2 has been widely employed for many different kinds of investigations, including toxicogenomics (Burczynski *et al.*, 2000; Sawada *et al.*, 2006). This cell line, like many others, shows the disadvantage of lacking metabolic capability and being altogether very different from a hepatocyte. These differences are not restricted to the morphological and physiological characteristics, but they are also marked when studying the pattern of gene expression. Harris *et al.* (2004), for example, described that 31 % of the HepG2 transcriptome was unique to the cell line when compared with primary human hepatocytes. This major discrepancy greatly limits their use for accurate predictive toxicogenomics, since the ultimate goal of these studies is to reflect the situation in a patient population.

An interesting *in vitro* approach that has recently being undertaken is the use of cell lines derived from embryonic stem cells (ES) that can be differentiated in culture conditions into hepatocytes (Ahuja *et al.*, 2007; Ek *et al.*, 2007). In the future, this approach might provide a source of hepatocyte-like human cells that are relatively easy to obtain and that could be employed for the analysis of a variety of endpoints, including toxicogenomics. Currently, this approach is still in its infancy, and the use of human embryonic stem cells also poses some ethical issues. Thus, the time-frame for the application of this system for the investigation of hepatotoxicity using toxicogenomics or other endpoints cannot be foreseen.

In our hands, and in agreement with the majority of the reports published in the literature, toxicogenomics investigations performed with cell lines for the prediction of hepatotoxicty show a poor performance in terms of prediction. Thus, and despite being a relatively high-throughput biological system, there is not enough confidence in the results to make hepatotoxicity predictions based on cell lines a reliable tool for the classification and selection of compounds.

Primary rat hepatocytes, on the other hand, have been used by us and others with acceptable performance in terms of hepatotoxicity prediction. They also mimic the mechanisms of toxicity observed *in vivo*. Waring *et al.* (2001) showed that gene expression profiles for compounds with similar toxic mechanisms formed clusters, suggesting a similar effect on transcription. Using primary rat hepatocytes, we could also partially reproduce some of the treatment-related toxicological findings observed *in vivo* (Boess *et al.*, 2007). Also, preliminary, unpublished data generated in our laboratories indicate that models generated using primary rat hepatocyte cultures combined with toxicogenomics as an endpoint can achieve a sensitivity of approximately 70–80 % for the prediction of rat hepatotoxicity. One major point that needs to be taken into consideration when performing *in vitro* toxicogenomics experiments is to stay within a compound concentration range that does not cause cytotoxicity (Figure 14.2). Most drugs or chemicals will cause cytotoxicity at a high concentration; this cytotoxic effect will then trigger unspecific gene expression changes in the cells. Such unspecific transcriptional modulation adds little information to the standard cytotoxicity assays and might act as a confounding factor for the construction of predictive models and for the biological interpretation of the results. Hence, the design of the study in terms of the selection of appropriate concentrations and exposure times is vital.

A different approach to combine toxicogenomics results and *in vitro* test systems with primary hepatocytes for a predictive application could be to use the primary cell cultures to assay specific endpoints that have been derived from toxicogenomics results obtained *in vivo* or *in vitro*. In an unpublished case example, two drug candidates dosed during 2 weeks in an exploratory toxicological study caused hepatotoxicity in the rat. The effect was characterized by an increase in relative liver weights and an increase in serum bilirubin and liver enzymes. The concomitant histopathological findings included atrophic hepatocytes, hepatocellular hypertrophy, mixed cell infiltration, lipid deposits and vacuolation, multinucleated hepatocytes, and necrosis. Two other compounds addressing the same pharmacological target and tested following a similar study design did not elicit such

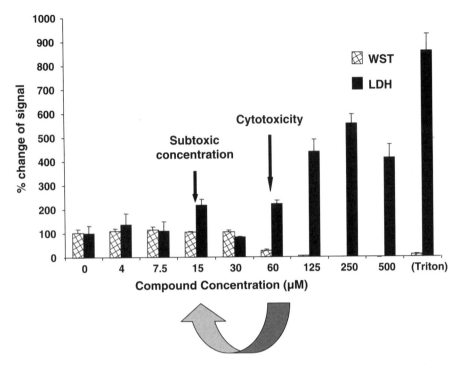

Figure 14.2   Cytotoxicity curve of a test compound. Cell viability was assessed by the WST-assay and the released LDH (lactate dehydrogenase) into the cell culture medium. Results are expressed as percentage of the control signal. Cytotoxic concentration and a subtoxic concentration (selected for transcriptomic analysis) are marked.

toxicity. For all these four compounds, toxicogenomics analysis was performed in the liver of the treated animals and gene expression analysis was carefully evaluated. As expected, there were clear differences in the liver gene expression profiles of the hepatototoxic and the nontoxic compounds. Among these differences, several genes coding for enzymes in the glutathione metabolic pathway were affected and their regulation seemed to be associated with the observed toxicity. Subsequent *in vitro* experiments performed in rat primary hepatocytes showed that the cellular glutathione content was significantly decreased by the two hepatotoxic compounds, while it was not affected by the other two. Thus, *in vitro* assays designed and performed based on the results of toxicogenomics investigations provided additional supporting evidence regarding a putative mechanism related to the observed toxicity. This evidence could also be used as a follow-up screening tool for other compounds that might be suspected of potentially eliciting a similar toxicity. Indeed, in this case study, additional compounds that were considered as possible drug candidates were tested in rat primary hepatocyte cultures to assess their potential to decrease cellular glutathione content, a biochemical response that in this case was probably associated with the observed toxicity. The results were very useful to rank the new drug candidates and select those with fewer liabilities.

Similarly, we investigated two compounds (R7199 and R0074) *in vitro* using rat hepatocytes in primary cultures to learn whether their differentiation in terms of hepatotoxic potential could be achieved *in vitro* as it had been *in vivo*. In the animal study, R7199, but not R0074, caused a dose-related increase in fatty change in the liver after single and repeated dosing, concomitant with hepatocellular vacuolation (microsteatosis) and with a slight increase in mitosis. These changes were accompanied by a decrease in serum cholesterol and triglycerides, as well as by a slight increase in some liver enzymes and by characteristic changes in the transcript profile (Suter *et al.*, 2003). In a subsequent *in vitro* experiment, several parameters were determined, including biochemical endpoints and gene expression analysis by microarray and qPCR. The exposure of primary hepatocytes in culture to R7199, but not R0074, caused accumulation of intracellular lipids and the induction of mRNA and activity of several cytochromes P450 (Cyp 2B, Cyp 3A and Cyp 1A families). Thus, the effects on primary hepatocytes were consistent with the effects in rat livers, and the *in vitro* results confirmed the steatotic potential of the compound (Boess *et al.*, 2007).

As an overall conclusion, it can be said that *in vitro* primary hepatocytes can be utilized as a suitable experimental model for specific applications, combining genomics with other cellular and biochemical parameters. They can provide supporting evidence and, therefore, help clarify the findings observed in the animal model. They can also be useful to rank compounds in terms of their potential to affect a certain pathway. At this point in time, the use of *in vitro* toxicogenomics as a pure predictive tool for the selection of the best molecules without any prior knowledge on the toxicological characteristics of the compound class might be overoptimistic. More publicly available data need to be available to support this latter claim.

14.7   *In Vivo* Toxicogenomics

A substantial amount of gene expression data has been generated in animal models (mostly rat) with known toxicants, mainly hepatotoxicants and nephrotoxicants, but also with genotoxic, cardiotoxic, and testicular toxic compounds. Several investigators in the pharmaceutical industry and in the public sector, as well as technology vendors, have published toxicogenomics results for the evaluation of hepatotoxicity in experiments performed in the rat. The results show that gene expression analysis can provide information to allow classification of compounds according to their mechanism of (hepato)toxicity, as well as identifying cellular pathways related to the toxic event. As a general rule, the use of toxicogenomics as a predictive tool requires prior knowledge of gene expression patterns related to toxicity. Consequently, this approach depends often on the availability of a reference gene expression database and robust software with appropriate algorithms for the comparison of complex fingerprints (Mattes *et al.*, 2004; Ganter *et al.*, 2005).

There are many published reports on compound characterization using gene expression profiles *in vivo* (Hamadeh *et al.*, 2002a,b; Ellinger-Ziegelbauer *et al.*, 2004; Steiner *et al.*, 2004; Ruepp *et al.*, 2005; Nie *et al.*, 2006; Zidek *et al.*, 2007; Nie *et al.*, 2006). Most of them present evidence supporting the idea that a manifestation of hepatotoxicity is reflected by associated characteristic gene expression changes. There are a number of technical differences, including the microarray platform and the rat strain, the types of compound, the time points investigated, the number of doses, and the algorithms and cross-validation models. However, the outcome consistently indicates that gene expression profiling in the liver reflects hepatotoxicity, can be predictive, and indicates the molecular mechanisms involved. A point that has not yet been elucidated is whether these toxicity-associated gene expression changes are the cause or the consequence of the toxic insult. In general, the reported gene expression changes are not limited to one or a few genes but involve profiles, or so-called fingerprints. These gene expression profiles parallel the histopathological findings, considered the gold-standard for the definition of toxicity. This information is useful in terms of understanding the underlying mechanisms of toxicity, but it is not always predictive in a strict sense. For predictive toxicogenomics, the common approach is the generation of a gene expression database with a large number of model compounds with known and well characterized (hepato)toxicity. The size of this database can vary, but it contains generally more than 100 compounds. The selection of molecules, doses, and time points of the compounds is accomplished by thorough literature research. The predictive models are usually generated using conventional endpoints, like histopathology of the liver, as phenotypic anchoring, and supervised algorithms to achieve a predictive outcome.

A very interesting example in terms of toxicogenomics in a strictly predictive sense is the publication of Ellinger-Ziegelbauer *et al.* (2004). Here, they showed that gene expression profiling of genotoxic carcinogens at doses known to produce liver tumors in 2-year carcinogenicity studies already revealed characteristic profiles after short-term treatment. These short-term *in vivo* studies combined with toxicogenomics as an endpoint allowed the scientists to identify genes and pathways commonly deregulated by genotoxic carcinogens. Among the modulated genes there were some which were biologically relevant, including a specific detoxification response, the activation of proliferative and survival signaling pathways, and some cell structural changes. These responses may be indicative of early events leading to tumorigenesis and, thus, may be truly predictive of later tumor development. Once validated, such fingerprints could provide immense value in the testing of nongenotoxic carcinogens in much shorter studies. Following a similar approach, Fielden *et al.* (2007) recently published the results of a large-scale experiment addressing nongenotoxic hepatic tumorigens by means of toxicogenomics. They measured hepatic gene expression data from rats treated for up to 7 days with 100 compounds (25 nongenotoxic hepato-carcinogens and 75 non-hepatocarcinogens). The analysis identified a characteristic gene expression signature of 37 genes. Among these statistically selected genes there were some that lacked annotation and others that regulated proliferation and apoptosis, as expected for carcinogens, thus adding biological plausibility to the results. Subsequent, independent, cross-validation studies including 47 additional compounds showed that this signature could predict the likelihood of a compound causing hepatic tumors with a sensitivity of 86 % and a specificity of 81 %, outperforming other putative biomarkers for carcinogens, such as clinical, histological, or genomic parameters. Moreover, and as expected, the signature appeared to be specific for chemicals inducing hepatic tumors through nongenotoxic mechanisms; it classified four genotoxic tumorigenic compounds as negative or only weakly positive. These two published examples are clear proof that the predictive toxicogenomics approach can yield faster, more accurate, and possibly more sensitive results than conventional toxicology assessment in the animal. In the field of nongenotoxic carcinogenicity testing, 1- to 2-week animal experiments might replace the long 2-year carcinogenicity study. This would be extremely useful as an exploratory tool at first, and then possibly as a regulatory test. However, validation is still ongoing and will require much more data. Apparently, the transcript profiles reveal the tumorigenic potential even after a single compound administration. Nie *et al.* (2006) also published data on nongenotoxic carcinogens that could be identified after a single administration by gene expression analysis of the livers.

Besides the ambitious goal of replacing 2-year nongenotoxic carcinogenicity studies with short-term studies combined with gene expression analysis, toxicogenomics also has the potential of predicting other forms of hepatotoxicity after very short drug administration periods. There are several publications analyzing the effects of peroxisome proliferators in the liver using gene expression analysis. Gene expression patterns revealed similarity in gene expression profiles between animals treated with different peroxisome proliferators: clofibrate (ethyl-*p*-chlorophenoxyisobutyrate), Wyeth 14,643 ([4-chloro-6(2,3-xylidino)-2-pyrimidinylthio]acetic acid), and gemfibrozil 5-(2,5-dimethylphenoxy)-2,2-dimethylpentanoic acid). The toxicogenomics results were specific enough to differentiate those peroxisome proliferators from phenobarbital, a typical enzyme inducer (Hamadeh *et al.*, 2002a). In addition, the knowledge gained on specific pathways whose regulation leads to, or is associated with, peroxisomal proliferation leading to tumorigenesis in the rodent has provided useful information for the assessment of the risk in non-rodents. It is a well-known fact that fibrates have been administered to patients for decades without causing liver tumors. However, when rodents are given PPAR α agonists, hepatic peroxisome proliferation, hypertrophy, hyperplasia, and eventually hepatocarcinogenesis become evident. Although we know that primates are relatively refractory to these effects, the mechanisms for the species differences remain unclear. Experimental evidence using gene expression profiling of monkeys treated with ciprofibrate, a PPAR α agonist, revealed clear species differences when compared with the reported effects in rodents. This might account for the differences in the toxicological profiles between rats and primates. On the one hand, the expected strong transcriptional modulation of specific pathways was observed in both species, due to the pharmacological properties of the compound. This included genes involved in fatty acid metabolism and mitochondrial oxidative phosphorylation, as well as genes related to ribosome and proteasome biosynthesis. Besides this similar pharmacologic response, several species differences became evident. In the primates, a number of key regulatory genes, including members of the JUN, MYC, and NFkappaB families were downregulated, while in rodents it has been reported that these genes are upregulated after PPAR α agonist treatment. Moreover, gene expression profiles in monkeys did not show any evidence for DNA damage or oxidative stress. In general, the data obtained in monkeys showed that the effects of the PPAR α agonist are qualitatively and quantitatively different from the effect reported in rodent livers. Thus, gene expression data provided evidence for the observed species specificity of the toxicological findings in the liver (Cariello *et al.*, 2005).

In our laboratories, we have generated an *in vivo* toxicogenomics database comprising well-known hepatotoxicants or compounds that showed hepatotoxicity during preclinical testing. For each animal, conventional endpoints were captured and served as phenotypic anchoring for the evaluation of the gene expression profiles. A supervised learning method (support vector machines (SVMs)) combined with recursive feature elimination was used to generate classification rules and a predictive model with high performance was obtained

(Steiner *et al.*, 2004). This model is now used with great success for the classification of gene expression profiles generated from livers of animals treated with drug candidates. In one selected example, we could demonstrate that gene expression profiles obtained after exposure of rats to two pharmacologically and chemically related compounds clearly indicated very different toxicological properties in the livers of the treated animals. This reflected and confirmed our prior knowledge on the toxicity of these compounds. Moreover, the hepatotoxicity could be detected earlier with toxicogenomics than with conventional endpoints, and a possible link between the affected pathways and the phenotypic manifestation of fat accumulation in the hepatocytes could also be established (Suter *et al.*, 2003). Additional unpublished applications of toxicogenomics for the study of hepatotoxicity further supported the usefulness of gene expression analysis for the study of liver toxicity. In one example, a prospective proprietary drug candidate caused liver injury characterized by increased liver weight and necrosis in rats dosed during 2 weeks. Gene expression analysis of the livers of the treated and control animals showed that they were classified correctly by the SVM model, in agreement with the histopathological findings. A retrospective, single-dose administration of the compound showed that the hepatotoxic liability could be detected by gene expression analysis before the lesions became detectable by histopathology or clinical chemistry. Other compounds of the same pharmacological class where thus assessed using toxicogenomics after a single administration to rule out a potential hepatotoxic liability. Table 14.1 summarizes the results of this experiment in which groups of rats treated once with vehicle or one of the four compounds in question underwent toxicogenomics analysis. The classification results assigned each individual liver gene expression profile to one of the toxicity classes in the SVM model (vehicle, cholestasis, direct reaction, steatosis, or hypertrophy). The classification of several animals in classes different from the 'vehicle' group is a clear and early indicator of their toxic potential. The toxicogenomics findings were subsequently confirmed by specially designed 2-week studies. Thus, this example further supported the utility of gene expression profiling combined with our SVM model as a predictive tool for the detection of hepatotoxicity.

Similarly, other proprietary examples show that hepatotoxicity can be identified by using gene expression patterns after a single administration, in the absence of clinical chemistry or histopathology signals. In these cases, subsequent confirmatory experiments in which rats were dosed subchronically elicited liver lesions that were concordant with the toxicogenomics prediction. Moreover, in our experience, gene expression profiles after single and subchronic administration are not identical, but lead to comparable conclusions regarding the type of hepatotoxicity observed. This means that, although the lists of modulated genes are not identical across doses and time points, the biological pathways that appear affected are similar. This observation is further supported by the results obtained by other scientists. Zidek and co-workers tested a number of hepatotoxicants in male rats and obtained accurate discrimination of all model compounds 24 h after a single administration of a high dose, if limiting the analysis to a number of specific genes related to specific pathways. Thus, similar compounds activate similar pathways which are associated with the underlying toxicity. This is very useful for the biological interpretation of the toxic effects, but also has some promising technical and practical aspects. The conclusion drawn from these experiments was that rather than whole-transcriptome approaches, focused microarrays could be sufficient to classify compounds with respect to toxicity prediction (Zidek *et al.*, 2007). This opinion is also supported by some data evaluating nongenotoxic hepatocarcinogens, where a selection algorithm yielded six genes which identified the

Table 14.1 *Classification results based on liver gene expression analysis of groups of rats (N = 4) treated either with vehicle or with one of four compounds (Cmp. A–D) at either a low dose (LD) or a high dose (HD). The genomics assessment was then summarized for each compound, as the outcome of toxicology studies*

Treatment	Genomics classification	Genomics assessment (after single dose)	Toxicity outcome (repeat dose studies)
Vehicle	Vehicle		
Vehicle	Vehicle	Similar to "control" in the	
Vehicle	Vehicle	database	
Vehicle	Vehicle		
Cmp A, LD	Vehicle		
Cmp A, LD	Vehicle		
Cmp A, LD	**Cholestasis**		
Cmp A, LD	Vehicle	Minor effects, unlikely to	No hepatotoxicity associated
Cmp A, HD	Vehicle	cause hepatotoxicity	with this compound
Cmp A, HD	Vehicle		
Cmp A, HD	Vehicle		
Cmp A, HD	Vehicle		
Cmp B, LD	Vehicle		
Cmp B, LD	**Cholestasis**		Compound caused
Cmp B, LD	Vehicle		hepatotoxicity at 2-weeks
Cmp B, LD	**Cholestasis**	Marked effects at LD and	as determined by
Cmp B, HD	Vehicle	HD, very likely to cause	histopathology and clinical
Cmp B, HD	**Direct_reaction**	hepatotoxicity	chemistry
Cmp B, HD	**Cholestasis**		
Cmp B, HD	**Direct_reaction**		
Cmp C, LD	**Direct_reaction**		
Cmp C, LD	**Cholestasis**		Compound caused
Cmp C, LD	**Cholestasis**		hepatotoxicity at 2-weeks
Cmp C, LD	Vehicle	Marked effects, very likely	as determined by
Cmp C, HD	Vehicle	to cause hepatotoxicity.	histopathology and clinical
Cmp C, HD	Vehicle		chemistry
Cmp C, HD	Vehicle		
Cmp C, HD	**Cholestasis**		
Cmp D, LD	Vehicle		
Cmp D, LD	Vehicle		
Cmp D, LD	Vehicle		
Cmp D, LD	Vehicle	No effects, unlikely to	No hepatotoxicity associated
Cmp D, HD	Vehicle	cause hepatotoxicity	with this compound
Cmp D, HD	Vehicle		
Cmp D, HD	Vehicle		
Cmp D, HD	Vehicle		

compounds with 88.5 % prediction accuracy (Nie *et al.*, 2006). The use of focused gene sets or signatures could greatly reduce the cost and increase the throughput of toxicogenomics analysis. The observed consistency of the gene expression results not only applies to groups of compounds and different time points, but is also valid across microarray platforms that have been shown to give comparable results: gene lists generated by specific statistical analysis showed consistent responses regarding the affected gene ontology terms and pathways, indicating that the biological impact of chemical exposure could be reliably deduced from all platforms analyzed (Guo *et al.*, 2006).

14.8   Consortia on the Use of 'Omics' for the Prediction of Hepatotoxicity

As toxicogenomics is a relatively new, technically sophisticated, and costly tool for the assessment of toxicity, several consortia are addressing this topic across the world. These consortia first focused on the technical aspects regarding technology platforms, robustness of the technology, identification of gene expression patterns (signatures), validation of identified signatures and/or biomarkers, and data analysis. Although many of these aspects apply to several fields in toxicology, liver gene expression has been very much in the limelight. Much work has been and is being performed, although few results have been yet published. The International Life Sciences Institute/Health and Environmental Sciences Institute Consortium provided one of the first forums for interdisciplinary, collegial study of the use of genomics in toxicology, with a main emphasis in hepatotoxicity (Pennie *et al.*, 2004; Ulrich *et al.*, 2004). Other currently prominent collaborations are the Predictive Safety Testing Consortium (PSTC) in the USA, the Predictive Toxicology (PredTox) consortium in Europe, and the Toxicogenomics Project in Japan (TGPJ). They all include public and private involvement. The PSTC brings pharmaceutical companies together to share and validate each other's safety testing methods under advice of the Food and Drug Administration. The members of the consortium share internally developed preclinical safety biomarkers, which have chiefly been identified using 'omics' technologies. They focus in four working groups: carcinogenicity, kidney, liver, and vascular injury (http://www.c-path.org/Portals/0/PSTC%20Overview.pdf). PredTox is part of Innovative Medicine (InnoMed) in Europe and is a joint endeavor between industry and the European Commission. The goal of PredTox is to assess the value of combining results from 'omics' technologies and more conventional toxicology methods. The aim is to achieve more informed decision making in preclinical safety evaluation (http://www.innomed-predtox.com/). The Toxicogenomics Project in Japan is also a cooperative research project joining both the nation and private companies. In this project, the liver and kidney were selected as the main target organs for toxicogenomics investigations *in vitro* and *in vivo* (http://wwwtgp.nibio.go.jp/index-e.html). The publication of studies with dissimilar and even contradictory results raised concerns with regard to the consistency of the data across laboratories and technology platforms. This major concern is being addressed by the MicroArray Quality Control (MAQC) project. Their work includes systematic testing of the technologies to enable the use of microarray data for clinical and regulatory purposes (MAQC Consortium, 2006). In addition, they have evaluated specific issues in the toxicogenomics field based on liver microarray data derived from animals treated with hepatotoxicants (Guo *et al.*, 2006).

14.9   Conclusions

The published results, some of which have been summarized in this chapter, clearly show the power of toxicogenomics as a tool for assessing hepatotoxicity. The data illustrate how toxicogenomics can provide truly predictive answers after short administration periods. Also, toxicogenomics is a powerful tool to elucidate the molecular mechanisms underlying hepatotoxicity. However, when applying global gene expression profiling to the study of hepatotoxicity, several points need to be taken into consideration. On the one hand, many xenobiotics will lead to the induction of genes which are unrelated to their mechanism of toxicity, such as early response genes. These gene expression changes are not necessarily characteristic of a particular type of toxicity, although they may be used as indicative that the biological system being tested is undergoing a stress response. On the other hand, technical issues, such as the selection of a suitable biological system (*in vitro* or *in vivo*) that best addresses the question posed, and careful statistical analysis of the data are very important for obtaining meaningful results that will help progress in this field. Also, it becomes clear from the results presented here that more work is necessary in the field of *in vitro* toxicogenomics.

References

Ahuja YR, Vijayalakshmi V, Polasa K (2007) Stem cell test: a practical tool in toxicogenomics. *Toxicology* **231**, 1–10.

Baken KA, Vandebriel RJ, Pennings JL, Kleinjans JC, van Loveren H (2007) Toxicogenomics in the assessment of immunotoxicity. *Methods* **41**, 132–141.

Boess F, Kamber M, Romer S, Gasser R, Muller D, Albertini S, Suter L (2003) Gene expression in two hepatic cell lines, cultured primary hepatocytes, and liver slices compared to the *in vivo* liver gene expression in rats: possible implications for toxicogenomics use of *in vitro* systems. *Toxicol Sci* **73**, 386–402.

Boess F, Durr E, Schaub N, Haiker M, Albertini S, Suter L (2007) An *in vitro* study on 5-HT6 receptor antagonist induced hepatotoxicity based on biochemical assays and toxicogenomics. *Toxicol In Vitro* **21**, 1276–1286.

Burczynski ME, McMillian M, Ciervo J, Li L, Parker JB, Dunn II RT, Hicken S, Farr S, Johnson MD (2000) Toxicogenomics-based discrimination of toxic mechanism in HepG2 human hepatoma cells. *Toxicol Sci* **58**, 399–415.

Cariello NF, Romach EH, Colton HM, Ni H, Yoon L, Falls JG, Casey W, Creech D, Anderson SP, Benavides GR, Hoivik DJ, Brown R, Miller RT (2005) Gene expression profiling of the PPAR-alpha agonist ciprofibrate in the cynomolgus monkey liver. *Toxicol Sci* **88**, 250–264.

Craig A, Sidaway J, Holmes E, Orton T, Jackson D, Rowlinson R, Nickson J, Tonge R, Wilson I, Nicholson J (2006) Systems toxicology: integrated genomic, proteomic and metabonomic analysis of methapyrilene induced hepatotoxicity in the rat. *J Proteome Res* **5**, 1586–1601.

Dieterle F, Schlotterbeck G, Binder M, Ross A, Suter L, Senn H (2007) Application of metabonomics in a comparative profiling study reveals *N*-acetylfelinine excretion as a biomarker for inhibition of the farnesyl pathway by bisphosphonates. *Chem Res Toxicol* **20**, 1291–1299.

Ebbels TM, Keun HC, Beckonert OP, Bollard ME, Lindon JC, Holmes E, Nicholson JK (2007) Prediction and classification of drug toxicity using probabilistic modeling of temporal metabolic data: the consortium on metabonomic toxicology screening approach. *J Proteome Res* **6**, 4407–4422.

Ek M, Söderdahl T, Küppers-Munther B, Edsbagge J, Andersson TB, Björquist P, Cotgreave I, Jernström B, Ingelman-Sundberg M, Johansson I (2007) Expression of drug metabolizing enzymes in hepatocyte-like cells derived from human embryonic stem cells. *Biochem Pharmacol* **74**, 496–503.

Ellinger-Ziegelbauer H, Stuart B, Wahle B, Bomann W, Ahr H-J (2004) Characteristic expression profiles induced by genotoxic carcinogens in rat liver. *Toxicol Sci* **77**, 19–34.

Elrick MM, Walgren JL, Mitchell MD, Thompson DC (2006) Proteomics: recent applications and new technologies. *Basic Clin Pharmacol Toxicol* **98**, 432–441.

Fielden MR, Kolaja KL (2006) The state-of-the-art in predictive toxicogenomics. *Curr Opin Drug Discov Dev* **9**, 84–91.

Fielden MR, Brennan R, Gollub J (2007) A gene expression biomarker provides early prediction and mechanistic assessment of hepatic tumor induction by nongenotoxic chemicals. *Toxicol Sci* **99**, 90–100.

Fountoulakis M, Berndt P, Boelsterli UA, Crameri F, Winter M, Albertini S, Suter L (2000) Two-dimensional database of mouse liver proteins: changes in hepatic protein levels following treatment with acetaminophen or its nontoxic regioisomer 3-acetamidophenol. *Electrophoresis* **21**, 2148–2161.

Fountoulakis M, de Vera MC, Crameri F, Boess F, Gasser R, Albertini S, Suter L (2002) Modulation of gene and protein expression by carbon tetrachloride in the rat liver. *Toxicol Appl Pharmacol* **183**, 71–80.

Gant TW, Zhang SD (2005) In pursuit of effective toxicogenomics. *Mutat Res* **575**, 4–16.

Ganter B, Tugendreich S, Pearson CI, Ayanoglu E, Baumhueter S, Bostian KA, Brady L, Browne LJ, Calvin JT, Day GJ, Breckenridge N, Dunlea S, Eynon BP, Furness LM, Ferng J, Fielden MR, Fujimoto SY, Gong L, Hu C, Idury R, Judo MS, Kolaja KL, Lee MD, McSorley C, Minor JM, Nair RV, Natsoulis G, Nguyen P, Nicholson SM, Pham H, Roter AH, Sun D, Tan S, Thode S, Tolley AM, Vladimirova A, Yang J, Zhou Z, Jarnagin K (2005) Development of a large-scale chemogenomics database to improve drug candidate selection and to understand mechanisms of chemical toxicity and action. *J Biotechnol* **119**, 219–244.

Guo L, Lobenhofer EK, Wang C, Shippy R, Harris SC, Zhang L, Mei N, Chen T, Herman D, Goodsaid FM, Hurban P, Phillips KL, Xu J, Deng X, Sun YA, Tong W, Dragan YP, Shi L (2006) Rat toxicogenomic study reveals analytical consistency across microarray platforms. *Nat Biotechnol* **24**, 1162–9116.

Hamadeh HK, Bushel PR, Jayadev S, Martin K, DiSorbo O, Sieber S, Bennett L, Tennant R, Stoll R, Barrett JC, Blanchard K, Paules RS, Afshari CA (2002a) Gene expression analysis reveals chemical-specific profiles. *Toxicol Sci.* **67**, 219–231.

Hamadeh HK, Bushel PR, Jayadev S, DiSorbo O, Bennett L, Li L, Tennant R, Stoll R, Barrett JC, Paules RS, Blanchard K, Afshari CA (2002b) Prediction of compound signature using high density gene expression profiling. *Toxicol Sci* **67**, 232–240.

Harris AJ, Dial SL, Casciano DA (2004) Comparison of basal gene expression profiles and effects of hepatocarcinogens on gene expression in cultured primary human hepatocytes and HepG2 cells. *Mutat Res* **549**, 79–99.

Hultin-Rosenberg L, Jagannathan S, Nilsson KC, Matis SA, Sjögren N, Huby RD, Salter AH, Tugwood JD (2006) Predictive models of hepatotoxicity using gene expression data from primary rat hepatocytes. *Xenobiotica* **36**, 1122–1139.

Kiyosawa N, Shiwaku K, Hirode M, Omura K, Uehara T, Shimizu T, Mizukawa Y, Miyagishima T, Ono A, Nagao T, Urushidani T (2006) Utilization of a one-dimensional score for surveying chemical-induced changes in expression levels of multiple biomarker gene sets using a large-scale toxicogenomics database. *J Toxicol Sci* **31**, 433–448.

Kröger M, Hellmann J, Toldo L, Glückmann M, von Eiff B, Fella K, Kramer PJ (2004) Toxico-proteomics: first experiences in a BMBF-study. *ALTEX: Altern Tierexp* **21**(Suppl 3), 28–40 (in German).

Ku WW, Aubrecht J, Mauthe RJ, Schiestl RH, Fornace Jr AJ (2007) Genetic toxicity assessment: employing the best science for human safety evaluation. Part VII. Why not start with a single test: a transformational alternative to genotoxicity hazard and risk assessment. *Toxicol Sci* **99**, 20–25.

Lazarou J, Pomeranz BH, Corey PN (1998) Incidence of adverse drug reactions in hospitalized patients: a meta-analysis of prospective studies. *J Am Med Assoc* **279**, 1200–1205.

MAQC Consortium (2006) The MicroArray Quality Control (MAQC) project shows inter- and intraplatform reproducibility of gene expression measurements. *Nat Biotechnol* **24**, 1151–1161.

Mattes WB, Pettit SD, Sansone S-A, Bushel PR, Waters MD (2004) Database development in toxicogenomics: issues and efforts. *Environ Health Perspect* **112**, 495–505.

Nie AY, McMillian M, Parker JB, Leone A, Bryant S, Yieh L, Bittner A, Nelson J, Carmen A, Wan J, Lord PG (2006) Predictive toxicogenomics approaches reveal underlying molecular mechanisms of nongenotoxic carcinogenicity. *Mol Carcinog* **45**, 914–933.

Pennie W, Pettit SD, Lord PG (2004) Toxicogenomics in risk assessment: an overview of an HESI collaborative research program. *Environ Health Perspect* **112**, 417–419.

Ruepp S, Boess F, Suter L, de Vera MC, Steiner G, Steele T, Weiser T, Albertini S (2005) Assessment of hepatotoxic liabilities by transcript profiling. *Toxicol Appl Pharmacol* **207**(2 Suppl), 161–170.

Sawada H, Taniguchi K, Takami K (2006) Improved toxicogenomic screening for drug-induced phospholipidosis using a multiplexed quantitative gene expression ArrayPlate assay. *Toxicol In Vitro* **20**, 1506–1513.

Steiner G, Suter L, Boess F, Gasser R, de Vera MC, Albertini S, Ruepp S (2004) Discriminating different classes of toxicants by transcript profiling. *Environ Health Perspect* **112**, 1236–1248.

Stigson M, Kultima K, Jergil M, Scholz B, Alm H, Gustafson AL, Dencker L (2007) Molecular targets and early response biomarkers for the prediction of developmental toxicity *in vitro*. *Altern Lab Anim* **35**, 335–342.

Suter L, Haiker M, De Vera MC, Albertini S (2003) Effect of two 5-HT6 receptor antagonists on the rat liver: a molecular approach. *Pharmacogenomics J* **3**, 320–334.

Tamura K, Ono A, Miyagishima T, Nagao T, Urushidani T (2006) Profiling of gene expression in rat liver and rat primary cultured hepatocytes treated with peroxisome proliferators. *J Toxicol Sci* **31**, 471–490.

Thybaud V, Le Fevre AC, Boitier E (2007) Application of toxicogenomics to genetic toxicology risk assessment. *Environ Mol Mutagen* **48**, 369–379.

Ulrich RG, Rockett JC, Gibson GG, Pettit SD (2004) Overview of an interlaboratory collaboration on evaluating the effects of model hepatotoxicants on hepatic gene expression. *Environ Health Perspect* **112**, 423–427.

Waring JF, Ciurlionis R, Jolly RA, Heindel M, Ulrich RG (2001) Microarray analysis of hepatotoxins *in vitro* reveals a correlation between gene expression profiles and mechanisms of toxicity. *Toxicol Lett* **120**, 359–368.

Waring JF, Ulrich RG, Flint N, Morfitt D, Kalkuhl A, Staedtler F, Lawton M, Beekman JM, Suter L (2004) Interlaboratory evaluation of rat hepatic gene expression changes induced by methapyrilene. *Environ Health Perspect* **112**, 439–448.

Waters M, Boorman G, Bushel P, Cunningham M, Irwin R, Merrick A, Olden K, Paules R, Selkirk J, Stasiewicz S, Weis B, Van Houten B, Walker N, Tennant R (2003) Systems toxicology and the Chemical Effects in Biological Systems (CEBS) knowledge base. *EHP Toxicogenomics* **111**(1T), 15–28.

Yamamoto T, Kikkawa R, Yamada H, Horii I (2006) Investigation of proteomic biomarkers in *in vivo* hepatotoxicity study of rat liver: toxicity differentiation in hepatotoxicants. *J Toxicol Sci* **31**, 49–60.

Yang Y, Abel SJ, Ciurlionis R, Waring JF (2006) Development of a toxicogenomics *in vitro* assay for the efficient characterization of compounds. *Pharmacogenomics* **7**, 177–186.

Zidek N, Hellmann J, Kramer PJ, Hewitt PG (2007) Acute hepatotoxicity: a predictive model based on focused illumina microarrays. *Toxicol Sci* **99**, 289–302.

15

Nutrigenomics: The Application of Genomics Signatures in Nutrition-related Research

Elisavet T. Gatzidou and Stamatios E. Theocharis

15.1 Introduction

The success of the Human Genome Project in association with the powerful and unprecedented advances in molecular biology promises to increase the depth of our understanding regarding the interactions between genes and nutrients.

Modern nutrition science, in order to enquire on the relations between genes and nutrition, ushers in the molecular biology field on nutritional research (Figure 15.1). This pioneering integration is revealed into two directions: nutrigenomics and nutrigenetics. Nutrigenomics explores the influence of dietary components on the expression regulation of genes mainly, but also proteins and metabolites (Ordovas and Mooser, 2004; Mutch *et al.*, 2005). Nutrigenetics investigates how the individual's genetic background coordinates the interaction between diet and the genome (Ordovas and Mooser, 2004).

Although both nutrigenomics and nutrigenetics focus on the evaluation of effects of dietary compounds using systems biology technologies, their goals remain different but complementary (Chadwick, 2004). Nutrigenomics expands in the direction of nutritional research to discover the best 'diet' recommendation from a given series of nutritional alternatives, while nutrigenetics is aimed at the clinical or public health direction in order to develop the best diet recommendation for an individual with a specific genetic background.

Toxicogenomics: A Powerful Tool for Toxicity Assessment   Edited by S. C. Sahu
© 2008 John Wiley & Sons, Ltd

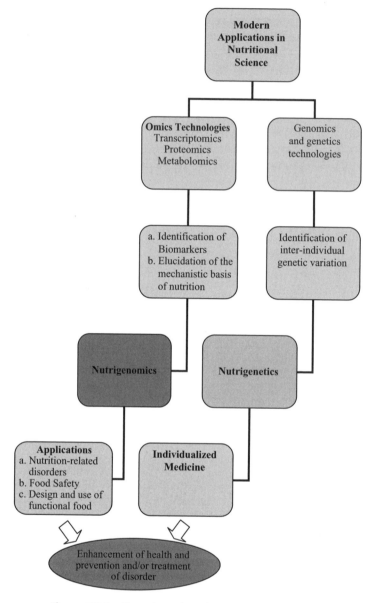

Figure 15.1   *Overview of modern nutritional science.*

However, both nutrigenomics and nutrigenetics possess as long-term goals the enhancement of health and prevention and/or treatment of diseases.

The present study outlines the two concepts of nutrigenomics and nutrigenetics. It particularly examines thoroughly the nutrigenomics concepts, its approaches, and strategies used in order to achieve its research target. Available technologies used by nutrigenomics with their advances and its applications are also presented.

15.2   Nutrigenomics: An Expectant Contrivance in Nutrition Research

Nutrigenomics is an integrative science that aims to determine the influence of dietary components on the genome, proteome, and metabolome in order to alter phenotype and balance between health and disease. It also attempts to relate the resulting different phenotypes with differences in cellular and/or genetic response of the biological system to nutritional stimulus (Ordovas and Mooser, 2004; Mutch *et al.*, 2005). This permits an improved understanding of how nutrition affects the regulation of gene expression and impacts metabolic pathways and homeostatic control, as well as how this regulation is disturbed during the early phases of diet-related diseases and the extent of an individual's genotype contribution to such disease (Gillies, 2003; Müller and Kersten, 2003). Nutrigenomics is also defined as the study of how genes and their products interact with dietary chemicals to alter phenotype and, conversely, how genes and their products metabolize nutrients (Kaput *et al.*, 2005). Nutrigenomics requires a systems biology approach that encompasses genetic and molecular understanding of the diverse tissue and organ effects of dietary chemicals on an individual's genetic background (nutrigenetics), DNA methylation and organization of chromatin (epigenomics), gene analysis (genomics), gene expression (transcriptomics), and protein expression (proteomics), as well as metabolite profiles (metabolomics), bioinformatics, nutritional epidemiology, pathology, and diverse clinical assessments in models ranging from cell culture to experimental animals and human populations (Kaput *et al.*, 2005) (Figure 15.1).

Two major approaches of nutrigenomics exist: the discovery of biomarkers and the investigation of mechanistic basis of nutrition (Afman and Müller, 2006). The identification of biomarkers appears to have an important aspect in human nutrition. It is based on the application of systems biology in order to develop measurable molecular markers for health promotion and disease prevention. Such biomarkers reveal early indicators for disease disposition and define early metabolic dysregulation and altered homeostasis that are modulated by diet and the efforts of the body to maintain this homeostasis (Kussmann *et al.*, 2006). They also help to differentiate responders from nonresponders in dietary terms and offer a better understanding of the health effects of certain diets or physiological factors and the beneficial or adverse effects of nutritional intervention in discovering bioactive, beneficial food. Last, but not least, biomarkers provide individuals with a critical feedback that is often the key to long-term compliance of nutritional supplementation (Kornman *et al.*, 2004).

Two biomarker discovery strategies exist that differ in type of starting material used for sample extraction and they produce complementary data (Kussmann *et al.*, 2006). One of them is focused on gene, protein, and metabolite expression profiling of body fluids, such as blood, urine, saliva, and tears, in order to identify 'molecular signatures' directly from easily accessible material (Kussmann *et al.*, 2006). This strategys allow direct assess to clinical biomarkers accurately reflecting the biological state of interest, but its use is restricted owing to the complexity of the started material and the possibility of diluted signals. The other strategy is based on gene, protein, and metabolite expression profiling of cell, tissue, or organ (Kussmann *et al.*, 2006). Studies using this type of strategy require animal or human biopsies derived from an established indicative biological model or clinical case. It excels because the type of model and specificity of samples used may trigger stronger signals in less complex analytical environments (Kussmann *et al.*, 2006).

The discovery of the mechanistic basis of nutrition provides detailed molecular information and increased understanding of the interaction between nutrition and the genome and offers new insights into mechanistic effects of dietary compounds on health and disease (Afman and Müller, 2006). In particular, it allows the identification of common nutrient response elements in promoter regions of genes which present modulations in their expression by diet. This favors the discovery of transcription factors and the elucidation of signaling pathways involved explaining the integrated response to a particular signal. In addition, it gives a measurement and validation of cell- and organ-specific gene expression profiles of the metabolic consequences of specific micronutrients and macronutrients. This approach also contributes to elucidation of interactions between nutrient-related regulatory pathways and proinflammatory stress pathways in an attempt to gain an improved understanding of the process of metabolic dysregulation resulting in diet-related disease. Last, but not least, it offers the identification of genotypes that are possibly risk factors for development and progression of diet-related chronic human diseases and the quantification of their impacts.

Nutritional research emanates largely from transgenic and knock-out mouse models and *in vitro* experiments using inducible expression systems, transdominant negative adenoviral vectors, and RNA interference. Besides these technologies, transcriptomics, proteomics, and metabolomics identify and elucidate responses as measurable indicators of dietary effect.

15.3 Mechanisms of Regulation of Gene Expression by Nutrients

Food provides dietary chemicals that are used as sources in energy metabolism and growth and for the development of structural components of the body. Furthermore, some nutrients are essential cofactors for the proper function of life-critical enzymes involved in various pathways of metabolism and tissue integrity.

Many nutrients also influence the expression of genes by these three mechanisms either directly or indirectly.

First, it has been shown that various nutrients bind to or directly activate or repress specific transcription factors, altering the transcription rate of genes (Kaput and Rodriguez, 2004). Nutrients act as ligands for members of the nuclear receptors superfamily of transcription factors (Jacobs and Lewis, 2002). Nuclear receptors bind to nutrients, and their metabolites enable cells to sense their nutrient environment and adjust cellular metabolism by altering gene expression. The nutrient–nuclear receptor complex binds to specific DNA sequences located within the promoter regions of target genes, activating or inhibiting their transcription. Upon ligand binding, nuclear receptors undergo a conformational change, providing coordinated dissociation of corepressors and recruitment of coactivators. Coactivators exist in multiprotein complexes that dock on transcription factors and modify chromatin, allowing the activation of transcription (Nettles and Greene *et al.*, 2005; Bain *et al.*, 2007). Thus, coregulators, coactivators and corepressors are substantial components of gene control.

For example, the peroxisome proliferator activator receptor (PPAR) family include three isotypes that act as nutrient sensors for dietary fatty acids and regulate the expression of specific genes (Kaput and Rodriguez, 2004; Bünger *et al.*, 2007). This complex of nuclear receptor–ligand binds as heterodimers together with Retinoid X Receptor (RXR), whose ligand is derived from another dietary chemical, vitamin A, to response elements in the promoter regions of genes (Kaput and Rodriguez, 2004; Bünger *et al.*, 2007). When a ligand activates the PPAR–RXR complex, this heterodimer complex stimulates the activation of transcription via binding to DNA response elements in and around the promoter region of a specific gene. In contrast, PPAR stimulates the repression of transcription with direct interaction with other transcription factors and interferes with their signaling pathways (Ricote and Glass, 2007).

Other nutrients, such as genistein and hyperforin, also bind directly to transcription factors and activate them. In particular, hyperforin, the active ingredient in St John's wort, binds to the ligand binding site of the pregnane X receptor (PXR) and induces transcription of reporter genes in cell-culture systems (Tirona *et al.*, 2004). Genistein, an isoflavone found in soya beans and other plants, binds to the active site of estrogen receptor β and induces specific gene expression in uteri of rats fed genistein-supplemented food (Naciff *et al.*, 2002). The sterol regulatory element binding proteins (SREBPs; Foufelle and Ferré, 2002) and carbohydrate-responsive element-binding protein (ChREBP) (Kawaguchi *et al.*, 2001; Yamashita *et al.*, 2001) are nutrients that alter the oxidation–reduction status of a cell in order to modulate transcription factor activity indirectly.

Second, dietary chemicals are converted by primary or secondary metabolic pathways altering the concentration of substrates or intermediates. This mechanism assists as a control for gene expression. For example, the levels of steroid hormones, originated from cholesterol, are regulated by enzymatic activities of multiple associated steps in the steroid biosynthetic pathway (Nobel *et al.*, 2001). Various intermediates are extended into other metabolic pathways such as degradative altering the concentrations of intermediates. Thus, specific combinations of alleles for these enzymatic steps affect the concentrations of ligands (Nobel *et al.*, 2001).

Third, dietary chemicals may bind to receptors at the cell surface and trigger a signal transduction cascade that ultimately facilitates the interaction between DNA and one or more transcription factors. Thus, the gene expression is either increased or decreased. For instance, dietary chemicals involved in ingredients of fruits and vegetables, such as 11-epigallocatechin galate (EGCG), theaflavins, resveratrol, inositol hexaphosphate, phenethyl isothiocyanate (PEITC), genistein and retinoids (vitamin A and its metabolites) directly affect, either positively or negatively, different signal transduction pathways (Dong, 2000).

As a consequence, nutrients are 'signaling molecules' that, through the above three mechanisms, result in transmission and translation of these dietary signals into alterations in genes, proteins, and metabolites expression. These characteristic expression profiles serve as dietary 'molecular signatures' or fingerprints that reflect the phenotype. The information that allows nutrients to activate specialized cellular mechanisms is contained within their molecular structure. Alterations in the structure of nutrients, such as saturated versus unsaturated fatty acids or cholesterol versus plant sterols, affect which of the sensor pathways is activated. For example, PPRAα binds and is activated by polyunsaturated fatty acids, as saturated fatty acids have a lower affinity for PPARα (Kliewer *et al.*, 2001).

15.4 The Other Side of the Same Coin, Nutrigenetics

In contrast to nutrigenomics, which focuses on differences among several dietary chemicals or conditions on quantitative measures of expression and their association with specific phenotypes, nutrigenetics focuses on differences among individuals in relation to the effects of dietary compounds (Ordovas and Mooser, 2004; Mutch *et al.*, 2005). Nutrigenetics examines the effects of genetic variations on the interaction between diet and disease. It is based on observations of clinical response to specific dietary components in individuals, testing also the hypothesis that differences in the observed response are associated with the presence or absence of specific genetic polymorphisms (Gillies, 2003).

Polymorphism can be qualitative, affecting the promoter region or coding/noncoding sequence of genes, or quantitative, affecting directly the level of expression. Qualitative polymorphism consists of single nucleotide polymorphism (SNP), which is base-pair alterations, such as small nucleotide deletions, duplications, or insertions. In contrast, quantitative polymorphism involves large duplications or deletion. The inherited genotypic variations in DNA sequence contribute to phenotypic variation and to differences in disease risk in response to the environment.

SNPs positively or negatively alter the response to dietary compounds by influencing genes coding for enzymes of transducers involved in absorption, metabolism, metabolic activation, detoxification, and/or site of action (Potter, 1999). The SNP itself also has an important functional impact on the protein product, directly affecting the disease.

Genomics through microarray technology is a powerful tool for the mapping, sequence identification, polymorphism detection, and analysis of all genes present in the genome of a given species (Hu *et al.*, 2005; Hunter, 2005). The identification and isolation of SNPs of human genomes among individuals that are unable to respond, or that respond abnormally, provide ways to disentangle the mechanisms by which a nutritional signal is transduced into a given response.

The aim of nutrigenetics is to identify individuals who are less efficient in specific metabolic pathways and to generate recommendations regarding the risks and benefits of specific diets or dietary components to the individual (Arab, 2004). This offers the promise of personalized nutrition or individualized nutrition depending on the genetic constitution of the consumer based on knowledge of variations in gene sequences of nutrient metabolism and genetic variations in nutrient targets (Subbiah, 2007).

15.5   Integration of Omics Technology

Transcriptomics is a relatively mature technology compared with other omics technologies and contributes to place proteomic and metabonomics into a larger biological perspective/prospect. Nowadays, it is possible to get a comprehensive analysis of the expression of all active genes in a microarray experiment suitable for a first 'round of discovery' in regulatory networks, but it is not yet possible to measure the whole proteome and metabolome (Kussmann *et al.*, 2006). Transcriptomics is the term for genome-wide gene expression analysis. It studies the mRNA in order to understand which genes are being transcribed. Transcriptomics through microarrays enables the simultaneous identification of tens of thousand of genes in a single experiment (Stears *et al.*, 2003). This suggests that genetic analysis can be done on a huge scale. Furthermore, it allows the determination of gene expression profiles, before and after the exposure, in a given cell type at a particular time and under various experimental conditions (Stears *et al.*, 2003). The examination of alterations in the gene expression profile at different stages in the cell cycle or during embryonic development can also be performed (Stears *et al.*, 2003). Transcriptomics is a powerful tool in nutritional research that can identify specific gene expression profiles in the pre-disease state that serve as diagnostic and prognostic biomarkers (Berger *et al.*, 2002), as well as potential targets (Dong *et al.*, 2002) for medical and nutritional intervention. It examines the physiological effects of diet, allowing for hypotheses about the potential site of action of dietary components and their interaction with molecular and cellular processes (Crott *et al.*, 2004; Ma *et al.*, 2005; Kussmann *et al.*, 2006).

However, transcriptomic studies require significant quantities of tissue material for extraction of the RNA needed, and the accessibility and collection of the mRNA from human tissues is restricted. In addition, transcription of genes does not ensure the activity of the product. The expression of the mRNA is not always associated with protein expression (Gygi *et al.*, 1999). Specifically, numerous alterations, like alternative splicing and post-translational modifications or simple shifts in the rate of synthesis and degradation, may occur in proteins that are not reflected in changes at the mRNA level (Panisko *et al.*, 2002).

Thus, proteomics examining the full complement of protein of a given tissue or cell (the proteome) using high throughput and automated methods of global protein separation, display, and identification (Patterson and Aebersold, 2003) comes a step closer to the activity of the gene. It is an essential component of systems biology and a valuable tool for the study of protein structure, levels of expression, subcellular localization, biochemical activity, modification states, protein to protein interaction, and interaction of protein with other types of biomolecule, like DNA and lipids. In a nutritional context, dietary compounds modify the translation of RNA to proteins, as well as post-translational modifications that influence protein activity (Dinkova-Kostova *et al.*, 2002; Knowles and Milner, 2003). Proteomics studies also facilitate the discovery of new and better biomarkers useful in the determination of nutrition status (Li *et al.*, 2002; Petricoin *et al.*, 2002; de Roos *et al.*, 2005).

Proteomics is more complex than the other omics technologies in terms of absolute number, chemical properties, and dynamic range of compounds present. Moreover, direct measurement of site of target tissue proteins in order to interpret the importance of protein differences is desirable but not feasible in human studies. Proteomic analysis is also restricted by small quantities of sample and detection of very small signals because of the lack of methods like polymerase chain reaction (PCR) in order to amplify proteins.

Metabolomics is the comprehensive analysis of endogenous and exogenous metabolites in a cell, tissue, or organ. It is a developing diagnostic tool for metabolic classification of individuals. Metabolomics is referred as the champion of omics because it provides the quantitative, noninvasive analysis of easily accessible human body fluids, such as urine, blood, saliva and tears (Whitfield *et al.*, 2004). Urine and blood samples are already tested for metabolites such as cholesterol and glucose, which are used as measurable indicators of health (Adams, 2003). However, metabolomics targets quantifying and classifying all small molecular weight compounds within a sample in order to find new biomarkers for diet-related diseases or metabolite expression profiles as indicators of nutritional status (German *et al.*, 2002).

Metabolomics needs to overcome a number of hurdles in order to gain a strong impact on nutritional research. First, it has to recover all metabolites from body fluids or tissue samples. This suggests access to the right tissue and measurement of metabolites at the right time. Second, metabolomics produces enormous amounts of data in association with measurements of variability within individuals, runs, and laboratories. Development of extensive databases is required to concentrate and store all this information about the nutritionally relevant metabolome. In addition, this large amount of metabolic data needs sophisticated technology and software in order to provide meaningful information and eliminate misinterpretation. For instance, some metabolic profiles, like the metabolic profile of butyrate, appear to be good for some diseases and bad for others.

Regarding transcriptomics profile gene expression at the mRNA level, the use of additional information from proteomic and metabolomic analyses is a major advantage in toxicological (Gatzidou *et al.*, 2007) and nutritional research (Clish *et al.*, 2004; Davis and Milner, 2004; Desiere, 2004; van Ommen, 2004). These various omics technologies are different but complementary to each other as the gene encodes RNA, which in turn encodes the enzymes that catalyze the conversion of metabolites. Although powerful different omics technologies are at different points in their development at this time, there is no overriding technology that provides answers to all research questions. A combination of the information generated from transcriptomics, proteomics, and metabolomics enables the study of relationships between diets, genetics, and metabolism in an entirely different depth (Corthésy-Theulaz *et al.*, 2005). This parallel use allows a complete description of the phenotype of a biological system, such as human being, in response to external stimuli, which is the major goal of nutritional research. However, a possible nutrigenomics strategy for successful utilization of the unprecedented abilities of the technologies included may be integration of emerging omics technologies with traditional molecular and biochemical approaches of classical nutrition research (Corthésy-Theulaz *et al.*, 2005). Moreover, the extensive amounts of information generated dictate a need for the development of software in order to store, integrate, and analyze these complex data. Bioinformatics uses advanced computing and clustering technologies in ways that facilitate 'data mining,' meaning the quick extraction of relevant parameters stored in a database (Desiere *et al.*, 2002).

15.6   Applications of Nutrigenomics in Nutrition-related Diseases

Nutrition is a key environmental factor with a permanent effect on the genome and is involved in the pathogenesis and progression of the typical diet-related diorders. Specifically, as mentioned previously, nutrients regulate the expression of genes that affect the cellular homeostasis directly or indirectly. Cells respond to this alteration in order to adapt to the new environment. Thus, diet can influence the proliferation and differentiation of cells (Müller and Kersten, 2003). An understanding of the biological impact of the interaction between genetic background and dietary exposure or nutritional therapy is the key to understanding the basis this diet-related disorder.

The gene–nutrient interactions first identified in monogenic disorders like galactosemia (Bosch, 2006) and phenylketonuria (PKU; Blau and Erlandsen, 2004). Galactosemia is a rare autosomal recessive defect of the metabolism of the sugar galactose. It is characterized by an inability of the enzyme galactose-1-phosphate uridyltransferase (GALT). This enzyme catalyzes the production of glucose-1-phosphate and uridyldiphosphate (UPD)-galactose from galactose-1-phosphate and UPD-glucose. The GALT defect results in the accumulation of lactose in blood, causing liver failure, neonatal death, sepsis, liver failure, and mental retardation. Another well-known autosomal recessive error of the metabolism is PKU, which is caused by a loss-of-function mutation on a gene that encodes the enzyme phenylalanine hydroxylase. This enzyme converts the amino acid phenylalanine to tyrosine. Tyrosine is essential for the production of certain hormones, neurotransmitters, and melanin. The lack of phenylalanine hydroxylase results in toxic levels of phenylalanine in the bloodstream. Phenylalanine is converted to a ketone that interferes with neural development and causes mental retardation. Both galactosemia and PKU are monogenic disorders, with a defined gene–nutrient interaction, and this fact facilitates their identification through genetic tests and prevention or treatment by diet alteration. Lactose-free and phenylalanine-restricted tyrosine-supplemented diets offer the nutritional treatment in galactosemia and PKU respectively.

In contrast to monogenic disorders, the major focus of nutrition research is on pathogenesis, progression, and prevention of common chronic diseases, such metabolic syndrome, cancer, type 2 diabetes mellitus (T2DM), cardiovascular disease, obesity, osteoporosis, atherosclerosis, birth defects, food allergies, and hypertension. Chronic diseases are also polygenic and are caused by a combination of genetic variations in multiple susceptibility genes. These multifactorial complex disorders are influenced by the integrated effects of genetic, lifestyle and environmental factors, including diet. This integration causes a high degree of heterogeneity in the pathophysiology, course, and secondary complications of disease, explaining the variance in clinical expression of a chronic disease among different individuals.

Metabolic syndrome is a very common diet-related polygenic disorder. It is characterized by insulin resistance, abdominal pain, dislipidemia, microalbuminuria, and hypertension. It often precedes T2DM and is associated with a greater risk of cardiovascular disease and atherosclerosis (Moller and Kaufman, 2005). The diverse clinical characteristics and multiple genetic targets involved in the pathogenesis and progression of this syndrome prove its complexity involving several dysregulated metabolic pathways, such insulin signaling, glucose homeostasis, lipoprotein metabolism, adipogenesis and inflammation, vascular functions, and coagulation (Roche *et al.*, 2005; Phillips *et al.*, 2006). An individual's genetic background can interact with dietary chemicals in order to affect the risk of metabolic syndrome. An excellent example of the relevance of gene–nutrient interaction in the development of metabolic syndrome and T2DM is the well-known polymorphism of the PPARγ gene resulting from a missense mutation (proline to alanine). This interaction between dietary fatty acids composition and the PPARγ Pro12Ala polymorphism is a fundamental part of the pathology of complex metabolic phenotypes (Roche *et al.*, 2005). PPARγ Pro12Ala polymorphism is a good candidate genetic marker and plays an important role in mediating the onset of insulin resistance (Roche *et al.*, 2005). Luan *et al.* (2001) showed that the PPARγ Pro12Ala polymorphism responds in different ways to alteration in the polysaturated/saturated fatty acids ratio (Luan *et al.*, 2001). When this ratio is low, Ala carriers have a greater body mass index than homozygote Pro allele carriers. Thus, homozygote carriers of the Pro allele are associated with an increased risk for metabolic syndrome. In contrast, some studies fail to account for previous findings that describe interactions between habitual fatty acids composition and the PPARγ polymorphism (Robitaille *et al.*, 2003).

The application of high-throughput systems biology technologies in nutritional research provides the opportunity to improve our molecular knowledge about the role of diet in prevention or cure of complex chronic disease. In particular, nutrigenomics allows the identification of molecular mechanisms underlying the pathogenesis of diet-associated disorders (Mathers, 2004; Davis and Hord, 2005). It identifies specific expression profiles of diet-regulated genes, protein, and metabolites that cause or contribute to the disease process and allows comparison between them. Alterations of specific expression profiles are then associated with the phenotype and can be explained by genetic variants in nuclear receptors, cis-acting elements in promoters, or differences in metabolism that produce altered concentrations of transcriptional ligands (Kaput and Rodriguez, 2004). Nutrigenomics studies focus on the pro-inflammatory stress and metabolic stress, the identification of transcription factors that act as nutrient sensors, and nutrient-induced homeostasis.

Cancer is an example of a multifactorial, complex, and polygenic disorder where diet is involved in the pathogenesis. It is well known that some types of cancer, such as of the breast, lung, liver, prostate, and colon, are related to dietary habits.

The application of nutrigenomics has contributed to the identification of mechanisms by which dietary compounds exert antitumor activity and chemopreventive properties. Polyunsaturated fatty acids and butyrate presented antitumor effects in human colorectal adenocarcinoma (Eitsuka *et al.*, 2004) and rat colon cancer (Williams *et al.*, 2003) respectively. Calcium (Lamprecht and Lipkin, 2003), vitamin D (Lamprecht and Lipkin, 2003), folate (Lamprecht and Lipkin, 2003), butyrate (Pool-Zobel *et al.*, 2005), and phenolic compounds and isothiocyanates (Hu and Kong, 2004) are some of the dietary cancer-chemopreventive compounds. Several studies performed comparisons of specific expression profiles elicited by nutrients and drugs (e.g. Mariadason *et al.*, 2000). Nutrigenomics also contributes to the discovery of specific expression patterns serving as molecular biomarkers that can be used for detecting or predicting a response elicited by a dietary compound (van der Meer-van Kraaij *et al.*, 2005). Moreover, it elucidates potential molecular and biochemical targets of dietary components, such as selenium (Dong *et al.*, 2003), quercetin (Wenzel *et al.*, 2004), butyrate (Tan *et al.*, 2002), and soy (Solanky *et al.*, 2003).

15.7   Food Safety

Another important application of nutrigenomics focuses on the improved procedures of food safety (Liu-Stratton *et al.*, 2004; Spielbauer and Stahl, 2005). Food safety involves analysis of food for authenticity, internal and external contamination occurring during harvesting, processing and storage, and genetically modified organisms (GMOs). Specifically, nutrigenomics through the integration of omics technologies offers an invaluable molecular approach for detection, typing, and identification of a pathogenic microorganism. It also provides an accurate labeling procedure, resulting in the avoidance of allergies from incorrect declarations of ingredients included in a product. The analysis of food for GMOs is restricted to monitoring specific gene expression profiles of GMO-derived food and genetically engineered ingredients in comparison with traditional food.

15.8   Functional Food

Fortified foods, functional foods, and nutraceuticals are intended to supplement human nutrition needs. Functional foods are dietary components or foods that provide health benefits or that promote or enhance desirable physiological effects beyond basic nutrition

(Clydesdale, 2004). Nutraceuticals are purified active ingredients that can be clinically tested and used for a specific metabolic problem (Subbiah, 2007). Nutrigenomics using omics technologies elucidates the effects of novel functional foods, dietary supplements, and nutraceuticals on global gene expression and cell function without making assumptions about what to look for in terms of risk.

Nutrigenomics can also be used to establish a link between the molecular mechanisms by which the genome perceives nutritional signals and mobilization of the organism to respond. This type of information will be a valuable tool to design and engineer food products to promote optimal health and reduce the risk of disease (Spielbauer and Stahl, 2005).

15.9   Conclusion

Genomes can respond to a nutritional stimulus by selective up- or down-regulation of the expression of specific genes. These responses are useful in order to elucidate the molecular events by which the genome perceives the nutritional signal and mobilizes the organism

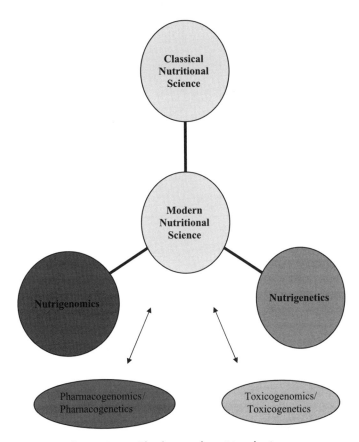

Figure 15.2   *The future of nutritional science.*

264   Toxicogenomics

to respond. Thus, nutrition can influence the expression regulation mainly of genes, but also of proteins and metabolites. As a consequence, nutrigenomics with the integrated use of different high-throughput omics technologies, such as transcriptomics, proteomics, and metabolomics, will lead to generation of information about pathways and provide the capacity to measure alterations of the pathway resulting from multifactorial nutritional influences on the human organism. Nutrigenomics using a systems biology approach will be the driving force of future nutritional research.

Furthermore, nutrigenetics with application of genomics and genetics technologies, screens for mutations and sequence variation, particularly SNPs, in genomic DNA and provides new insights into fundamental biological processes of human disease at the molecular level.

Nutrigenomics and nutrigenetics focus on the identification, validation, and incorporation of specific expression profiles and SNPs respectively into dietary strategy. Nutrigenomics offers an understanding of pathophysiology of disease by integration of the underlying molecular pathways and regulatory networks in cells, tissues, organs, and the whole organism. On the other hand, nutrigenetics offers the basis of personalized recommendation. They use different approaches, but their purpose is complementary. They target maintenance of health and prevention and/or treatment of nutrition-related diseases. For the completion of this purpose, it is essential that both fields of nutritional science combine in order to discover how dietary components in food alter the regulation of expression of genetic information in each individual. Association of nutrigenomics and nutrigenetics has tremendous potential to change the future of dietary guidelines and personal dietary recommendations.

Nowadays, it is possible to study the underlying mechanisms of human health in relation to diet and other environmental factors, such as drugs and toxins with pharmacogenomics/pharmacogenetics and toxicogenomics/toxicogenetics respectively. Findings from nutrigenomics research can be associated with results that arise from the application of other omics field (Figure 15.2). This integration can be used for optimization of health on an individual basis in relation to genotype and lifestyle.

References

Adams A (2003) Metabolomics: small molecules 'omics'. *The Scientist* **17**, 38–40.

Afman L, Müller M (2006) Nutrigenomics: from molecular nutrition to prevention of disease. *J Am Diet Assoc* **106**, 569–576.

Arab L (2004) Individualized nutritional recommendations: do we have the measurements needed to assess risk and make dietary recommendations? *Proc Nutr Soc* **63**, 167–172.

Bain DL, Heneghan AF, Connaghan-Jones KD, Miura MT (2007) Nuclear receptor structure: implications for function. *Annu Rev Physiol* **69**, 201–220.

Berger A, Mutch DM, German JB, Roberts MA (2002) Dietary effects of arachidonate-rich fungal oil and fish oil on murine hepatic and hippocampal gene expression. *Lipids Health Dis* **1**, 2.

Blau N, Erlandsen H (2004) The metabolic and molecular basis of tetrahydrobiopterin – responsive phenylalanine hydroxylase deficiency. *Mol Genet Metab* **82**, 101–111.

Bosch AM (2006) Classical galactosemia revisited. *J Inherit Metab Dis* **29**, 516–525.

Bünger M, Hooiveld GJEJ, Kersten S, Müller M (2007) Exploration of PPAR functions by microarray technology – a paradigm for nurtigenomics. *Biochim Biophys Acta* **1771**, 1046–1064.

Chadwick R (2004) Nutrigenomics, individualism and health. *Proc Nutr Soc* **63**, 161–166.

Clish CB, Davidov E, Oresic M, Plasterer TN, Lavine G, Londo T, Meys M, Snell P, Stochaj W, Adourian A, Zhang X, Morel N, Neumann E, Verheij E, Vogels JT, Havekes LM, Afeyan N, Regnier F, van der Greef J, Naylor S (2004) Integrative biological analysis of the APO*3-Leiden transgenic mouse. *Omics* **8**, 3–13.

Clydesdale F (2004) Functional foods: opportunities & challenges. *Food Technol* **58**, 35–40.

Corthésy-Theulaz I, den Dunnen JT, Ferré P, Geurts JM, Müller M, van Belzen N, van Ommen B (2005) Nutrigenomics: the impact of biomics technology on nutrition research. *Ann Nutr Metab* **49**, 355–365.

Crott JW, Choi SW, Ordovas JM, Ditelberg JS, Mason JB (2004) Effects of dietary folate and aging on gene expression in the colonic mucosa of rats: implications for carcinogenesis. *Carcinogenesis* **25**, 69–76.

Davis CD, Milner J (2004) Frontiers in nutrigenomics, proteomics, metabolomics and cancer prevention. *Mutat Res* **551**, 51–64.

Davis CD, Hord NG (2005) Nutritional 'omics' technologies for elucidating the role(s) of bioactive food components in colon cancer prevention. *J Nutr* **135**, 2694–2697.

De Roos B, Duivenvoorder I, Rucklidge G, Reid M, Ross K, Lamers RJ, Voshol PJ, Havekes LM, Teusink B (2005) Response of apolipoprotein E*3-Leiden transgenic mice to dietary fatty acids: combining liver proteomics with physiological data. *FASEB J* **19**, 813–815.

Desiere F, German B, Watzke H, Pfeifer A, Saguy S (2002) Bioinformatics and data knowledge: the new frontiers for nutrition and foods. *Trends Food Sci Technol* **12**, 215–229.

Desiere F (2004) Towards a systems biology understanding of human health: interplay between genotype, environment and nutrition. *Biotechnol Annu Rev* **10**, 51–84.

Dinkova-Kostova AT, Holtzclaw WD, Cole RN, Itoh K, Wakabayashi N, Katoh Y, Yamamota M, Talalay P (2002) Direct evidence that sulfhydryl groups of Keap 1 are the sensors regulating induction of phase 2 enzymes that protect against carcinogens and oxidants. *Proc Natl Acad Sci U S A* **99**, 11908–11913.

Dong Y, Ganther HE, Stewart C, Ip C (2002) Identification of molecular targets associated with selenium-induced growth inhibition in human breast cells using cDNA microarrays. *Cancer Res* **62**, 708–714.

Dong Y, Zhang H, Hawthorn L, Ganther HE, Ip C (2003) Delineation of the molecular basis for selenium-induced growth arrest in human prostate cancer cells by oligonucleotide array. *Cancer Res* **63**, 52–59.

Dong Z (2000) Effects of food factors on signal transduction pathways. *Biofactors* **12**, 17–28.

Eitsuka T, Nakagawa K, Miyazawa T (2004) Dual mechanisms for telomerase inhibition in DLD-1 human colorectal adenocarcinomas cells by polyunsaturated fatty acids. *Biofactors* **21**, 19–21.

Foufelle F, Ferré P (2002) New perspectives in the regulation of hepatic glucolytic and lipogenic genes by insulin and glucose: a role for the transcription factor sterol regulatory element binding protein-1c. *Biochem J* **366**, 377–391.

Gatzidou ET, Zira AN, Theocharis SE (2007) Toxicogenomics: a pivotal piece in the puzzle of toxicological research. *J Appl Toxicol* **27**, 302–309.

German JB, Roberts MA, Fay L, Watkins SM (2002) Metabolomics and individual metabolic assessment: the next great challenge for nutrition. *J Nutr* **132**, 2486–2487.

Gillies PJ (2003) Nutrigenomics: the Rubicon of molecular nutrition. *J Am Diet Assoc* **103** (Suppl), 50–55.

Gygi SP, Rochon Y, Franza BR, Aebersold R (1999) Correlation between protein and mRNA abundance in yeast. *Mol Cell Biol* **19**, 1720–1730.

Hu N, Wang C, Hu Y, Yang HH, Giffen C, Tang ZZ, Han XY, Goldstein AM, Emmert-Buck MR, Buetow KH, Taylor PR, Lee MP (2005) Genome-wide association study in esophageal cancer using GeneChip mapping 10K array. *Cancer Res* **65**, 2542–2546.

Hu R, Kong AN (2004) Activation of MAP kinases, apoptosis and nutrigenomics of gene expression elicited by dietary cancer-prevention compounds. *Nutrition* **20**, 83–88.

Hunter DJ (2005) Gene–environment interactions in human diseases. *Nat Rev Genet* **6**, 287–298.

Jacobs MN, Lewis DF (2002) Steroid hormone receptors and dietary ligands: a selected review. *Proc Nutr Soc* **61**, 105–122.

Kaput J, Rodriguez RL (2004) Nutritional genomics: the next frontier in the postgenomic era. *Physiol Genomics* **16**, 166–167.

Kaput J, Ordovas JM, Ferguson L, van Ommen B, Rodriguez RL, Allen L, Ames BN, Dawson K, German B, Krauss R, Malyj W, Archer MC, Barnes S, Bartholomew A, Birk R, van Bladeren P, Bradford KJ, Brown KH, Caetano R, Castle D, Chadwick R, Clarke S, Clément K, Cooney CA, Corella D, Manica da Cruz IB, Daniel H, Duster T, Ebbesson SO, Elliott R, Fairweather-Tait S, Felton J, Fenech M, Finley JW, Fogg-Johnson N, Gill-Garrison R, Gibney MJ, Gillies PJ, Gustafsson JA, Hartman IV JL, He L, Hwang JK, Jais JP, Jang Y, Joost H, Junien C, Kanter M, Kibbe WA, Koletzko B, Korf BR, Kornman K, Krempin DW, Langin D, Lauren DR, Ho Lee J, Leveille GA, Lin SJ, Mathers J, Mayne M, McNabb W, Milner JA, Morgan P, Muller M, Nikolsky Y, van der Ouderaa F, Park T, Pensel N, Perez-Jimenez F, Poutanen K, Roberts M, Saris WH, Schuster G, Shelling AN, Simopoulos AP, Southon S, Tai ES, Towne B, Trayhurn P, Uauy R, Visek WJ, Warden C, Weiss R, Wiencke J, Winkler J, Wolff GL, Zhao-Wilson X, Zucker JD (2005) The case for strategic international alliances to harness nutritional genomics for public and personal health. *Br J Nutr* **94**, 623–632.

Kawaguchi T, Takenoshita M, Kabashima T, Uyeda K (2001) Glucose and cAMP regulate the L-type pyruvate kinase gene by phosphorylation/dephosphorylation of the carbohydrate response element binding protein. *Proc Natl Acad Sci U S A* **98**, 13710–13715.

Kliewer SA, Xu HE, Lambert MH, Willson TM (2001) Peroxisome proliferator-activated receptors: from genes to physiology. *Recent Prog Horm Res* **56**, 239–263.

Knowles LM, Milner JA (2003) Diallyl disulfide induces ERK phosphorylation and alters gene expression profiles in human colon tumor cells. *J Nutr* **133**, 2901–2906.

Kornman KS, Martha PM, Duff GW (2004) Genetic variations and inflammation: a practical nutrigenomics opportunity. *Nutrition* **20**, 44–49.

Kussmann M, Raymond F, Affolter M (2006) OMICS-driven biomarker discovery in nutrition and health. *J Biotechnol* **124**, 758–787.

Lamprecht SA, Lipkin M (2003) Chemoprevention of colon cancer by calcium, vitamin D and folate: molecular mechanisms. *Nat Rev Cancer* **3**, 601–614.

Li J, Zhang Z, Rosenzweig J, Wang YY, Chan DW (2002) Proteomics and bioinformatics approaches for identification of serum biomarkers to detect breast cancer. *Clin Chem* **48**, 1296–1304.

Liu-Stratton Y, Roy S, Sen CK (2004) DNA microarray technology in nutraceutical and food safety. *Toxicol Lett* **150**, 29–42.

Luan J, Browne PO, Harding AH, Halsall DJ, O'Rahilly S, Chatterjee VK, Wareham NJ (2001) Evidence for gene–nutrient interaction at the PPARgamma locus. *Diabetes* **50**, 686–689.

Ma DW, Finnell RH, Davidson LA, Callaway ES, Spiegelstein O, Piedrahita JA, Salbaum JM, Kappen C, Weeks BR, James J, Bozinov D, Lupton JR, Chapkin RS (2005) Folate transport gene inactivation in mice increases sensitivity to colon carcinogenesis. *Cancer Res* **65**, 887–897.

Mariadason JM, Gorner GA, Augenlicht LH (2000) Genetic reprogramming in pathways of colonic cell maturation induced by short chain fatty acids: comparison with trichostatin A, sulindac and curcumin and implication for chemoprevention of colon cancer. *Cancer Res* **60**, 4561–4572.

Mathers JC (2004) The biological revolution – towards a mechanistic understanding of the impact of diet on cancer risk. *Mutat Res* **551**, 43–49.

Moller DE, Kaufman KD (2005) Metabolic syndrome: a clinical and molecular perspective. *Annu Rev Med* **56**, 45–62.

Müller M, Kersten S (2003) Nutrigenomics: goals and strategies. *Nat Rev Genet* **4**, 315–322.

Mutch DM, Wahli W, Williamson G (2005) Nutrigenomics and nutrigenetics: the emerging faces of nutrition. *FASEB J* **19**, 1602–1616.

Naciff JM, Jump ML, Torontali SM, Carr GJ, Tiesman JP, Overmann GJ, Daston GP (2002) Gene expression profile induced by 17-alpha-ethynyl estradiol, bisphenol A, and genistein in the developing female reproductive system of the rat. *Toxicol Sci* **68**, 184–199.

Nettles KW, Greene GL (2005) Ligand control of coregulator recruitment to nuclear receptors. *Annu Rev Physiol* **67**, 309–333.

Nobel S, Abrahmsen L, Oppermann U (2001) Metabolic conversion as a pre-receptor control mechanism for lipophilic hormones. *Eur J Biochem* **268**, 4113–4125.

Ordovas JM, Mooser V (2004) Nutrigenomics and nutrigenetics. *Curr Opin Lipidol* **15**, 101–108.

Panisko EA, Conrads TP, Goshe MB, Veenstra TD (2002) The postgenomic age: characterization of proteomes. *Exp Hematol* **30**, 97–107.

Patterson SD, Aebersold RH (2003) Proteomics: the first decade and beyond. *Nat Genet* **33** (Suppl), 311–323.

Petricoin EF, Ardekani AM, Hitt BA, Levine PJ, Fusaro VA, Steinberg SM, Mills GB, Simone C, Fishman DA, Kohn EC, Liotta LA (2002) Use of proteomic patterns in serum to identify ovarian cancer. *Lancet* **359**, 572–577.

Phillips C, Lopez-Miranda J, Perez-Jimenez F, McManus R, Roche HM (2006) Genetic and nutrient determinants of the metabolic syndrome. *Curr Opin Cardiol* **21**, 185–193.

Pool-Zobel BL, Selvaraju V, Sauer J, Kautenburger T, Kiefer J, Richter KK, Soom M, Wölfl S (2005) Butyrate may enhance toxicological defence in primary, adenoma and tumor human colon cells by favourably modulating expression of glutathione *S*-transferases genes, an approach in nutrigenomics. *Carcinogenesis* **26**, 1064–1076.

Potter JD (1999) Colorectal cancer: molecules and populations. *J Natl Cancer Inst* **91**, 916–932.

Ricote M, Glass CK (2007) PPARs and molecular mechanisms of transrepression. *Biochim Biophys Acta* **1771**, 926–935.

Robitaille J, Després JP, Pérusse L, Vohl MC (2003) The PPAR-gamma P12A polymorphism modulates the relationship between dietary fat intake and components of the metabolic syndrome: results from Québec Family Study. *Clin Genet* **63**, 109–116.

Roche HM, Phillips C, Gibney MJ (2005) The metabolic syndrome: the crossroads of diet and genetics. *Proc Nutr Soc* **64**, 371–377.

Solanky KS, Bailey NJ, Beckwith-Hall BM, Davis A, Bingham S, Holmes E, Nicholson JK, Cassidy A (2003) Application of biofluid ^{1}H nuclear magnetic resonance-based metabonomic techniques for the analysis of the biochemical effects of dietary isoflavones on human plasma profile. *Anal Biochem* **323**, 197–204.

Spielbauer B, Stahl F (2005) Impact of microarray technology in nutrition and food research. *Mol Nutr Food Res* **49**, 908–916.

Stears RL, Martinsky T, Schena M (2003) Trends in microarray analysis. *Nat Med* **9**, 140–145.

Subbiah MT (2007) Nutrigenetics and nutraceuticals: the next wave riding on personalized medicine. *Transl Res* **149**, 55–61.

Tan S, Seow TK, Liang RC, Koh S, Lee CP, Chung MC, Hooi SC (2002) Proteome analysis of butyrate-treated human colon cancer cells (HT-29). *Int J Cancer* **98**, 523–531.

Tirona RG, Leake BF, Podust LM, Kim RB (2004) Identification of amino acids in rat pregnane X receptor that determine species-specific activation. *Mol Pharmacol* **65**, 36–44.

van der Meer-van Kraaij C, Kramer E, Jonker-Termont D, Katan MB, van der Meer R, Keijer J (2005) Differential gene expression in rat colon by dietary heme and calcium. *Carcinogenesis* **26**, 73–79.

Van Ommen B (2004) Nutrigenomics: exploiting systems biology in the nutrition and health arenas. *Nutrition* **20**, 4–8.

Wenzel U, Herzog A, Kuntz S, Daniel H (2004) Protein expression profiling identifies molecular targets of quercetin as a major dietary flavonoid in human colon cancer cells. *Proteomics* **4**, 2160–2174.

Whitfield PD, German AJ, Noble PJ (2004) Metabolomics: an emerging post-genomic tool for nutrition. *Br J Nutr* **92**, 549–555.

Williams EA, Coxhead JM, Mathers JC (2003) Anti-cancer effects of butyrate: use of micro-array technology to investigate mechanisms. *Proc Nutr Soc* **62**, 107–115.

Yamashita H, Takenoshita M, Sakurai M, Bruick RK, Henzel WJ, Shillinglaw W, Arnot D, Uyeda K (2001) A glucose-responsive transcription factor that regulates carbohydrate metabolism in the liver. *Proc Natl Acad Sci U S A* **98**, 9116–9121.

16

Application of Toxicogenomics in Drug Discovery

Michael J. Liguori, Amy C. Ditewig and Jeffrey F. Waring

16.1 Introduction

As a field, toxicogenomics is relatively new as applied to drug discovery. Even with this caveat, numerous examples exist for the effective and advantageous application of the technology toward the drug discovery process. In this chapter, several documented examples will be presented to demonstrate this utility. First, the applicability of *in vivo* toxicogenomics, in particular its use as a predictive tool for early identification of potential adverse effects, will be discussed. Second, examples will be given discussing the use of toxicogenomics for understanding and characterizing mechanisms of toxicity. Many of these mechanistic studies have been performed for toxicity occurring in the liver; however, the technology is maturing to the point where other organ systems can now be confidently analyzed to yield valuable molecular information. In addition, while *in vivo* toxicogenomics has been the predominant implementation of this technology, much progress has been made toward successful and beneficial use of *in vitro* systems to predict and understand toxicities. Finally, while toxicogenomics has clearly been most beneficial to early lead selection and lead optimization, in some cases the technology can be instrumental to a better understanding of side effects or unanticipated toxicities of candidates later in development and all the way to post-marketing of the pharmaceutical.

16.2 Toxicogenomics Objectives and Study Design

A toxicogenomic analysis will only be of value if the in-life parameters are critically planned according to a study's objectives. In other words, just like any other scientific data,

Toxicogenomics: A Powerful Tool for Toxicity Assessment   Edited by S. C. Sahu
© 2008 John Wiley & Sons, Ltd

toxicogenomics itself cannot be expected to yield a valuable output if the initial study and objectives are not well planned. Sprague-Dawley rats are the small laboratory species of choice for first-pass evaluation of pharmaceutical new molecular entities (NMEs) due to their relatively low compound input requirements, although, in certain cases, other common species used in investigative toxicology and toxicogenomics can also be used (e.g. mouse or dog). Typically, 5-day or 14-day toxicology studies are designed once initial relevant pharmacokinetic properties are established. In most cases, at least three dosage groups are included, which could be more or less depending on the study's objectives. At the end of the dosing period, traditional toxicologic endpoints are typically performed, such as clinical chemistry, gross clinical observations, organ weights, and histopathological sectioning. In addition to these readouts, tissues of relevance to the majority of toxic reactions (e.g. liver, spleen, heart, kidney) are also collected by flash freezing or by storage in a reagent that stabilizes mRNAs (e.g. Qiagen's RNA Later™) for subsequent microarray analysis.

It is important to realize that toxicogenomics is not normally meant to replace traditional toxicology. Rather, it is most successfully employed as a supplementation to the other classical parameters to predict and understand better the molecular events that underlie adverse reactions. It is in this manner that toxicogenomics is most powerful as a technique. While traditional endpoints are usually most relevant in longer term studies (i.e. > 14 days), toxicogenomics yields the maximum benefit during shorter term studies (i.e. 3–7 days). It is in this interval that the majority of primary toxic events at the molecular level are most clearly defined. If toxicogenomics is used at later time points of an *in vivo* study (> 14 days), then there is a risk that tissues are already damaged and that toxicogenomics is detecting and characterizing secondary or even tertiary adverse events and, thus, lacking information on the actual cause of the toxicity. While this is the general trend, there are certainly instances where later time points are the most informative, and this is especially true if the molecule's pharmacokinetic profile results in gradual accumulation of the substance or if a metabolic intermediate is primarily responsible for the toxic event.

16.3   Toxicogenomics as a Predictive Tool

Clearly, one of the areas where toxicogenomics has been extensively applied in the drug discovery process is as a predictive tool. The theory behind this application is that gene expression changes will often occur before hispathologic or clinical pathologic variations (Waring *et al.*, 2002; Waring and Halbert, 2002). Thus, by monitoring gene expression changes that correlate strongly with toxicity it may be possible to estimate safety margins and eliminate compounds with undesirable toxicologic profiles from a 3–5-day rat study, which would use a minimal amount of compound, rather than the usual 2–4 weeks of treatment. These types of study are not intended to replace long-term toxicology studies; however, they can be useful for rank-ordering and prioritizing compounds for longer term studies.

A significant challenge in applying predictive toxicogenomics is to identify specific gene expression changes that are highly correlated and predictive of a toxicological reaction. These gene expression changes are often termed 'signatures'. In the vast majority of cases where predictive toxicogenomics has been successful, this has been accomplished by first establishing toxicogenomic databases of gene expression profiles (Yang *et al.*,

2004). Ideally, these databases consist of many known pharmaceutical agents, toxicants, and control compounds, at multiple doses and time points, with biological replicates for each condition (Guerreiro *et al.*, 2003; Waring *et al.*, 2003; Karpinets *et al.*, 2004; Tong *et al.*, 2004; Sawada *et al.*, 2005). The reference compounds profiled in the database usually consist of different structure–activity relationships and represent a variety of toxic mechanisms.

Often, pharmaceutical companies or laboratories will establish their own toxicogenomic databases, which allows for the flexibility of building the database using proprietary compounds within their libraries (Dai *et al.*, 2006; Yang *et al.*, 2006; Foster *et al.*, 2007; Zidek *et al.*, 2007). In other cases, a number of publicly available or commercial databases are available. For instance, Chemical Effects in Biological Systems (CEBS) is an integrated public repository for toxicogenomics data. It includes the study design, as well as the clinical chemistry and histopathology findings and microarray and proteomics data (Dix *et al.*, 2007). The Mount Desert Island Biological Laboratory created another publicly available database. This database, termed the Comparative Toxicogenomics Database, identifies interactions between chemicals and genes and facilitates cross-species comparative studies of these genes (Mattingly *et al.*, 2006). Likewise, companies such as GeneLogic (Gaithersburg, MD) or Iconix (Mountain View, CA) offer commercial toxicogenomics reference databases. These databases contain extensive gene expression profiles of a large number of prototypical reference compounds with corroborating toxicologic and pathologic endpoints (Ganter *et al.*, 2005; Mattes *et al.*, 2004). Most of the databases established to date have used rats, since rats are a main species used in traditional toxicology studies. However, toxicogenomics databases have also been established in mouse and in several *in vitro* cell lines, to name a few (Burczynski *et al.*, 2000; Hayes *et al.*, 2005; Yang *et al.*, 2006). With the use of these databases, it is possible in many cases to identify gene expression patterns, or signatures, that are highly associated with a known toxic mechanism or pathological outcome.

The identification of signatures that correlate with a particular endpoint can be aided by a number of different statistical methods. Because a typical profile can contain thousands of gene expression changes, the first step in developing signatures for predictive toxicogenomics is to identify the most robust gene expression changes that correlate with the endpoint. Initially, the number of gene expression changes can be greatly reduced by ranking genes with respect to differences in expression between experimental groups. The genes that are differentially expressed at a specified significance level can then be used for signature creation. For example, in an effort to distinguish phenobarbital, an enzyme inducer, from peroxisome proliferators, gene expression profiles from livers of rats treated for 24 h with phenobarbital were compared with the pooled expression data from samples treated with Wyeth 14,643, gemfibrozil and clofibrate (Bushel *et al.*, 2002). Application of a single-gene analysis of variance method was able to identify signature genes that had the most consistent and dramatic expression difference between the two classes of compounds.

Once the parameters have been reduced and signature genes have been selected, computational algorithms can be applied in order to establish a set of rules or formulas that can be applied to classify unknown samples. An array of computational algorithms has been used in gene expression data from the fields of yeast biology, oncology, and, certainly, toxicology. Examples of these methods include logistic regression, linear discriminant analysis, naïve Bayesian classifiers, artificial neural networks, and support vector machines (Yang *et al.*, 2004).

Over the past few years, as toxicogenomic screening has become more extensive in the pharmaceutical industry, there have been a number of examples where toxicogenomics has been effective for predicting a toxicological outcome that would not manifest for several more weeks. One example of this type of approach was demonstrated for identifying drugs with the potential to induce drug-induced renal tubular toxicity (Fielden *et al.*, 2005). In this study, rats were treated with 64 nephrotoxic or nonnephrotoxic compounds and the kidneys were harvested after 5 days and 4 weeks of treatment. There was no evidence of nephrotoxicity using traditional toxicology endpoints at day 5, although after 4 weeks all of the nephrotoxic compounds induced kidney toxicity. Gene expression analysis was conducted on the kidneys from day 5. A gene expression signature consisting of 35 genes was identified from these data that could be used to predict from 5-day studies whether a compound would induce renal tubular toxicity after 4 weeks of dosing. The signature was tested on new compounds and correctly classified the compounds as nephrotoxic or nonnephrotoxic 76% of the time.

A study by Thukrak *et al.* (2005) also demonstrated the feasibility of using gene expression profiling to predict gene expression patterns associated with proximal tubular-toxicity in Sprague-Dawley rats. In this study, rats were treated with the known kidney toxicants mercuric chloride, 2-bromoethylamine hydrobromide, hexachlorobutadiene, and puromycin aminonucleoside. Clinical chemistry analysis, histopathology, and microarray analysis was performed on the kidneys from treated rats. The results indicated that the gene expression analysis correlated with histopathology findings. Furthermore, using support vector machine analysis, it was possible to classify correctly the expression profiles used for forward validation into four pathological classes: no pathology; mild severity (tubular degeneration/regeneration); medium severity (degeneration/regeneration and glomerular damage); and high severity (tubular necrosis).

We have used a similar approach to identify whether a compound has the potential to cause hepatotoxicity after 3 days of dosing (Dai *et al.*, 2006). In this study, rats were treated with over 50 hepatotoxic or nonhepatotoxic compounds for 3 days. The majority of the rats treated with the hepatotoxic compounds did not display hepatotoxicity after 3 days of treatment, as determined by traditional toxicology measurements. However, a distinct difference in the gene expression patterns from the livers of rats treated with hepatotoxic versus nonhepatotoxic compounds could be observed.

By analyzing the gene expression changes in this database, and using a computational algorithm that classifies unknown samples (artificial neural network), we have identified a subset of 40 genes whose expression changes are highly correlated with drug-induced hepatotoxicity. This assay has proven to be highly reliable and predictive. For instance, a retrospective analysis of 52 compounds, some of which were associated with hepatotoxicity, while others were considered nonhepatotoxic, was conducted in our laboratories. Overall, the assay correctly identified eight compounds as hepatotoxic and 42 compounds as nonhepatotoxic. Only two compounds were incorrectly classified: one false positive and one false negative, indicating a sensitivity of 88.9% and a specificity of 97.7% with an overall predictive accuracy of 96.2%. Similar approaches have been used by other companies in order to identify gene expression changes that correlate with and predict the onset of hepatotoxicity (Bulera *et al.*, 2001; Jolly *et al.*, 2005; Minami *et al.*, 2005; Zidek *et al.*, 2007).

Another application where toxcicogenomics was applied in a predictive manner was illustrated in the study by Mutlib *et al*. (2006). In this study, toxicogenomics was applied in order to identify a set of gene expression changes that correlate with the generation of a reactive metabolite. Mice were dosed with a soft-electrophile-producing hepatotoxic compound, *N*-methylformamide, against corresponding deuterium-labeled analogues resistant to metabolic processing. Livers were harvested from the mice and a set of gene expression changes were identified that correlated with activation of the metabolic pathway leading to the production of reactive methyl isocyanate. These gene expression changes ultimately could serve as potential genomic biomarkers of hepatotoxicity induced by soft-electrophile-producing compounds.

Clearly, several studies have shown that, in some cases, toxicogenomics can be a powerful tool for predicting or identifying toxicity at an early time point. While there are limitations to this approach, applying toxicogenomics in this manner can, nonetheless, result in a tremendous savings in both time and resources for pharmaceutical companies, allowing them to eliminate or deprioritize drug candidates with unacceptable safety margins much earlier in the drug discovery process.

16.4 *In vivo* Toxicogenomics Using Liver as a Target Organ for Characterizing Mechanisms of Toxicity

By identifying gene expression changes that occur as a result of drug treatment, toxicogenomics can be a useful tool in order to understand and characterize the mechanisms of toxicity associated with new drug candidates. Since liver as an organ is exposed *a priori* to xenobiotic substances, it is often a major site of toxic events. As such, hepatic gene expression profiling is currently the most mature technique in the toxicogenomics arsenal (Dai *et al*., 2006; Chang and Schiano 2007; Lum *et al*., 2007). Alterations to hepatic homeostasis can be successfully monitored using expression profiling. In our laboratory, toxicogenomics has successfully been employed to identify NMEs that adversely affect hepatic functioning, including changes in fatty acid β-oxidation, lipid metabolism, mitochondrial homeostasis, peroxisomal proliferation, P450 induction, phase II metabolism, oncogenesis, ATP production, hypoxia signaling, oxidative stress, hepatocellular hypertrophy, apoptosis, necrosis, proteosomal changes, and many other functions. This type of information can assist in the characterization of an NME's impact on both off-target and on-target associated toxicities. Such knowledge can assist and increase productivity of lead optimization activities, target viability, rank ordering of lead compounds, identification of potential biomarkers of adverse reactions, and description of potential compound liabilities early in the characterization of new molecules. All of these activities ultimately improve the efficiency of the drug discovery process and rapidly provide relevant molecular information which previously was difficult, if not impossible, to unravel and understand effectively.

A recent example case study where hepatic gene expression analysis was used for mechanistic evaluation was using an inhibitor to Acetyl CoA carboxylase 2 (ACC2), which catalyzes the carboxylation of acetyl-CoA to form malonyl-CoA, and has been identified as a potential target for type 2 diabetes and obesity (Waring *et al*., 2008). A novel antagonist of ACC2, A-908292, resulted in, as expected, reduction in serum glucose levels.

However, evaluation of an inactive enantiomer of A-908292, unexpectedly, also reduced serum glucose and triglyceride levels. Therefore, it was postulated that A-908292 was not only acting through ACC2 inhibition, but also through a yet unidentified nonspecific mechanism.

An in-life study in rats was designed in an attempt to characterize the off-target interactions of A-908292 better at a molecular level. Rats were dosed for 3 days with 30 or 100 $mgkg^{-1}day^{-1}$ of A-908282 or its inactive enantiomer. Microarray analysis was conducted on the livers from the treated rats. Similar gene expression profiles were observed between the two compounds. Approximately 100 genes were identified that were regulated in common between animals and treatment. Comparison of these genes with the Drug-Matrix™ reference gene expression database showed that both experimental compounds showed a high degree of similarity to known peroxisome proliferator activated receptor (PPAR) agonists, such as bezafibrate, clofibrate, fenofibrate, and nafenopin.

PPAR agonists are known to have an impact on the major lipid and carbohydrate biosynthetic pathways, and classification of the genes significantly regulated by A-908282 and its inactive counterpart were also involved in these metabolic pathways. Toxicogenomic evaluation of these experimental compounds was also conducted *in vitro* in primary rat hepatocytes, which resulted in the activation of a gene expression signature for PPAR agonism. Further biochemical and physiological evaluations confirmed an involvement of PPAR agonism. Thus, it was concluded that A-908282 was affecting metabolic carbohydrate levels not only through ACC2 inhibition, but also through an off-target route mediated by PPAR stimulation. These data were useful in screening for new compounds that were pharmacologically active through only an ACC2 cascade and did not display any nonspecific activity.

Another interesting example of a pharmaceutically relevant toxicogenomic mechanistic characterization of hepatic toxicity concerns results from 5-hydroxytrypamine-6 (5-HT6) receptor antagonists with potential therapeutic use in nervous-system disorders (Suter *et al.*, 2003). Both Ro 65-7199 and Ro 66-0074 showed similar activities to their pharmacological target (5-HT6), but the former compound showed toxic hepatic changes in rats indicative of steatosis while the latter did not. Using microarray analysis of rat liver exposed to Ro 65-7199 or Ro 66-0074 for a period up to 7 days, gene expression profiles were generated that characterized the potential molecular changes that may underlie the pathogenesis. The two compounds clearly possessed distinct hepatic gene expression profiles, despite their similar pharmacology. Furthermore, many of the changes potentially relevant to the toxicity displayed a magnitude of expression change that correlated to the severity of the lesion.

Several of the modulated transcripts provided clues to potential mechanistic interactions behind the toxicity observed for Ro 65-7199 (Suter *et al.*, 2003). In particular, the dose-dependent induction of mRNA and protein of the drug metabolizing enzymes CYP2B2 and CYP3A1 were specific for the lesion-inducing compound Ro 65-7199, but not for its nonhepatotoxic counterpart. The changes in gene expression of the hepatotoxic 5-HT6 inhibitor were discussed as they relate to transcriptomic effects from phenobarbital in rodents, which also heightens the expression of CYP2B. All of these findings ultimately led to a possible involvement of inhibition of sterol biosynthetic machinery and a modulation of normal constitutive androstane receptor (CAR) signaling, which participates in the governing of the CYP2B expression. Of further relevance was the observation that Ro

65-7199 impacted the level of other gene products that are associated with maintenance of lipid homeostasis, such as fatty acid binding protein and epoxide hydrolase. This example serves to highlight that toxicogenomic techniques can generate plausible mechanistic interactions responsible for observed toxicities.

16.5 Monitoring Expression Changes of Drug Metabolizing Enzymes Using Genomics

Hepatic P450 induction or repression can have great consequences to the intended functioning of a xenobiotic (Tompkins and Wallace, 2007). For example, heightened expression of P450s that are specifically involved in metabolizing the drug can lead to sporadic changes in pharmacokinetic exposure, enhanced risk of drug–drug interactions, increased risk for hepatocellular hypertrophy or hyperplasia, and potential for increased production of deleterious reactive intermediates and reactive oxygen species. Microarray technologies coupled to quantitative polymerase chain reaction can rapidly identify potential induction liabilities for multiple P450 isoforms in one experiment. These types of observations can yield early information on the impact of erratic metabolism and exposure alterations for new drug candidates.

A documented example where the monitoring of P450 expression resulted in mechanistic information was for the evaluation of a preclinical pharmaceutical candidate A-972611, a novel HIV protease inhibitor (Healan-Greenberg *et al.*, 2008). Treatment of hepatocytes with many compounds of this class (e.g. ritonavir) typically results in overexpression of relevant human P450 enzymes, such as CYP3A4. Exposure to A-972611, however, resulted in a distinct reduction of the CYP3A4 gene product in human hepatocytes, but not in rat hepatocytes. Interestingly, co-treatment of human hepatocytes with A-972611 and with compounds typically associated with induction of CYP3A4 mRNA, e.g. rifampin and efavirenz, attenuated the expression of CYP3A4. Toxicogenomic analysis of human hepatocytes exposed to A-971611 revealed a reduction in expression of other P450s, including CYP2B6, CYP2C8, and CYP2C9. All of these observations suggested that A-972611 may inhibit the nuclear receptor and transcription factor pregnane X receptor (PXR). Confirmation of this notion was observed when treatment with A-972611 and rifampin or ritonavir resulted in reduced activation of PXR in luciferase reporter assays. Thus, toxicogenomics clearly has the potential to impact drug development via enhanced understanding of impact on drug metabolism.

In recent experiences, our laboratory has encountered several compounds whose pharmacokinetic properties were not only governed by hepatic phase I enzyme metabolism, but also intestinal P450 expression in the rat. This class of compounds was previously shown to be hepatic inducers and substrates of the CYP1A1 monooxygenase, which had great consequences for overall pharmacokinetic exposure. Using toxicogenomic analysis of sections from the jejunum and duodenum, it was possible to monitor CYP1A1 and other P450 expression activity in these tissues, which served as primary sites of exposure to the administered xenobiotic species. It was concluded that these intestinal tissues were equivalently or more relevant to the drug exposure level than hepatic-derived P450 counterparts. This example magnified the importance of surveying additional tissues which may also play a part in the metabolic profile of certain classes of pharmaceutical species.

16.6   Application of *In Vivo* Toxicogenomics to Nonhepatic Tissues

Hepatic toxicogenomics has made great strides in the last 10 years, and numerous examples have been presented that document the utility of the technique to the drug discovery process. However, it is obvious that liver is not the only target organ for toxic events, and the application of toxicogenomic technologies to other organ systems has the potential for great benefit. In many ways, the liver is an optimal organ for a toxicogenomic evaluation. It is relatively homogenous in cell composition, with the hepatocyte as the predominant cell type. This allows for maximum signal from the affected cell type and little interference from nonrelevant cells, which are not associated with pathogenesis. Furthermore, the liver is rich in the presence of stable RNA species, allowing for an optimal yield of the nucleic acid to perform the analysis comfortably. The liver also is an abundant tissue where large sections are available for storage.

Unfortunately, the optimal nature of the liver toward genomic analysis does not translate to a variety of other tissues. For example, if a compound has a specific pathogenesis to a small area of a target organ, such as the β-islet cells of the pancreas, then the analysis is complicated by dilution of the signal from the relevant cell type by other epithelial or endothelial layers. In our experience, it has been apparent that heterogeneous organs such as heart and kidney can, in many cases, be analyzed for a specific pathogenesis by planning and ensuring that the organ is sectioned in such a way that as many relevant cell types as possible are present for processing. Other heterogeneous tissues, such as brain or analogous nervous tissues, can only be successfully analyzed after careful dissection of the appropriate section. This process is often laborious and is typically not amenable to routine evaluations or screening. Furthermore, if, upon dissection, a contaminating tissue is sporadically present, then modulations in gene expression are often the result of differences in tissue type, rather than test-article-specific changes. These challenges may be overcome as advances in enrichment of relevant tissues are introduced for routine use and as lower levels of input RNAs are amenable to an accurate analysis.

Albeit more challenging, the successful application of toxicogenomics to nonhepatic organs has been successfully achieved as described for cardiotoxicity in the following example from our laboratory. Compound R was a novel molecule being investigated as a potential treatment for inflammation. In a 5-day dose range-finding toxicology study in rats, the compound was shown to induce cardiac necrosis with associated edema and inflammatory infiltrate starting at the lowest dose given ($150 \, \text{mg} \, \text{kg}^{-1} \, \text{day}^{-1}$) and above. The lesion manifested itself in such clinical signs as labored breathing, lethargy, ataxia, pericardial and pleural effusion, and eventually death.

Toxicogenomic analysis of the heart was planned in order to yield potential mechanistic insights. For mechanistic studies, it can be useful to monitor gene expression changes over a number of time points for maximum information gathering. Therefore, an *in vivo* study was performed with one dosage group of $150 \, \text{mg} \, \text{kg}^{-1} \, \text{day}^{-1}$ of compound R with heart collection for gene expression analysis at days 1, 3, and 5 of the study ($n = 3$ animals per timepoint). Treatment with compound R resulted in over 1300 cardiac gene expression changes relative to vehicle control rats. Critical evaluation of the gene expression changes showed that compound R was inducing changes in calmodulin signaling and in leukocyte cell adhesion factors. Furthermore, increased modulation of antigen presentation factors and NRF2-mediated oxidative stress protection systems was observed.

Comparing the gene expression profiles of compound R with a database (DrugMatrix™ from Iconix pharmaceuticals) of gene expression profiles of >100 cardiotoxicants, it was found that compound R had a high correlation to compounds that were known agonists of β-adrenergic receptors such as isoprenaline, norepinephrine, dobutamine, and racepinephrine. Subsequently, a set of genes modulated in common between these adrenergic agonists resulted in a high level of similarity to those from compound R. Further data mining revealed that compound R induced a multitude of changes in expression of genes involved with G-protein signal transduction. It was concluded, therefore, that compound R acts through nonspecific agonism of receptor(s) that are coupled downstream to adenylate cyclase and G-protein cascades, or perhaps to the G-protein itself. This, in turn, could result in disruption of calcium signaling, MAP kinase, and apoptotic cascades, ultimately leading to cell death and activation of inflammatory signals. The unraveling of this mechanism was useful for design of follow-on leads and for future compound decision-making.

Another example of a challenging tissue for successful toxicogenomic analysis is whole blood (Spijker *et al.*, 2004). This tissue has great appeal to drug discovery research as a surrogate due to its abundance, ease of accessibility, potential for biomarker identification, potential for enhanced mechanistic understanding, and enrichment of toxicological information gathered to more advanced preclinical species, and ultimately to the clinic for diagnostic and surveillance activities. The unique challenge for whole-blood analyses lies in the superabundance of globin transcripts (e.g. myoglobin or hemoglobins), which overwhelm the signal from other, less abundant transcripts, which are potentially critical for test-article-induced pathogenesis or related surrogate gene expression profiles. This greatly lowers the sensitivity of the analysis and makes data mining of the blood transcriptome difficult, if not impossible. Fortunately, advancements in globin-reducing sample-preparation techniques have given new insight into the successful implementation of transcriptomic profiling of this potentially informative tissue.

16.7 *In vitro* Toxicogenomics

Use of *in vitro* systems has great appeal to the discovery toxicologist. Use of these systems is more convenient and cost-effective than in-life studies. Furthermore, the amount of input test article compound is measured in milligrams rather than gram quantities typically required by traditional in-life experiments. This is especially crucial in the discovery phase of drug development, since compound availability is usually in short supply. Also, throughput is vastly increased *in vitro* because of the larger number of compounds and an increased number of concentrations capable of being evaluated; thus, *in vitro* methods become a substantial time-saver. All of these advantages make the concept of *in vitro* toxicology an optimal technique in the discovery toxicology arsenal.

Unfortunately, in current practice, *in vitro* systems have significant limitations. Foremost, the biological complexity for *in vitro* cell systems is greatly reduced compared with the *in vivo* situation. Paracrine interactions cannot be replicated *in vitro*, such as interaction between organ systems. For example, no changes due to hormonal signaling can be replicated, nor can interactions between the immune system and the hepatic system. This is especially problematic for more complex organs such as the kidney, where many different cell types are present that influence normal functioning. Furthermore, it has been demonstrated that isolation of cells from an intact organ to an *in vitro* environment can have major changes to the cell's native transcriptome (Richert *et al.*, 2006). There can also be problems with the chemical properties of the compound(s) under investigation. The substance can have limited solubility in an aqueous environment, thus disallowing proper exposure to the cell or tissue layer. If the substance forms a precipitate or crystals on the cell layer, then this can lead to artifactual results, especially for gene expression measurements. Some compounds have poor stability in aqueous systems, potentially resulting in misleading results; this is especially formidable at the discovery stage, since many chemical properties, such as stability, are not yet characterized.

Even with these challenges, many examples exist where *in vitro* systems have yielded valuable information about underlying toxicities, and sustained work in the area may lead to more and continuing improvements in the field. In the scope of toxicogenomics, advancements have been made with *in vitro* methodologies making them well suited and preferred for certain applications. For example, measurements of phase I and II enzyme induction can be accurately predicted *in vitro* using primary hepatocytes. Development and optimization of tissue slices allows for an environment that better replicates the *in vivo* situation (Boess *et al.*, 2003; Vickers and Fisher 2005). Several institutions have established or are working toward *in vitro* toxicogenomic databases with toxicologically relevant reference compounds in cells such as primary hepatocytes (Fielden and Kolaja, 2006). Examples of *in vitro* toxicogenomics to classify mechanisms of toxicity have been documented for compounds known to activate PPAR pathways and aryl hydrocarbon signaling (Abel *et al.*, 2006; Yang *et al.*, 2006). Gene expression studies with prototypical toxicants have been compared *in vivo* for liver and *in vitro* for hepatocytes and liver slices, which revealed a high concordance in gene expression (Jessen *et al.*, 2003). Clearly, with a sustained effort, *in vitro* toxicogenomics, while not optimal for all situations, can provide added value for many discovery toxicology applications.

16.8   Toxicogenomics of Human Cell Culture Systems

Prediction and molecular understanding of drug effects on the human situation has always been most challenging to molecular toxicologists. Complications arise when using human-sourced biological materials, including genetic and environmental variations between patients and the difficulty in procuring the affected human tissues themselves, especially in an undamaged state where useful information can still be derived. Whole-blood genomics, as discussed previously, is one potential avenue to circumvent some of the difficulties in accessing potential adverse reactions in humans. Idiosyncratic drug reactions, as discussed in detail elsewhere in this chapter, are an example of the tremendous difficulties and risks that the pharmaceutical industry encounters with novel molecular entities. There exist examples using human cell systems to attempt enhanced mechanistic understanding of idiosyncratic reactions, as examined later in this chapter.

Toxicogenomics for human systems almost exclusively relies on *in vitro* methodologies. Primary human hepatocytes (PHHs) and certain human cell lines such as HepG2 cells possess the most characterized transcriptomic changes (Harries *et al.*, 2001; Harris *et al.*, 2004). A major caveat lies in the fact that these cells possess transcriptomes considerably different from the *in vivo* situation (Richert *et al.*, 2006). However, advances in stem cell or similar technologies potentially could result in the creation of a cell type that more closely resembles that of the native organ (Hengstler *et al.*, 2005).

Current reports suggest that genomic technologies can successfully be employed to characterize *in vitro* models of human toxicities, such as steatohepatitis, using novel human cell systems. For example, a recent communication has examined SV40T antigen immortalized human hepatocytes as a model for hepatosteatosis (De Gottardi *et al.*, 2007). By exposing this novel cell line to oleic acid for a prolonged period, conditions for a steatotic state could be mimicked, which was confirmed and characterized using standard biochemical methods, including thin-layer chromatography and immunofluorescence. Genomic analysis of these cells revealed a dysregulation of several genes involved with critical lipid regulation processes and cholesterol maintenance. This included transcripts coding for sterol regulatory element binding protein 1 (SREBP1), a critical regulator of intracellular lipid homeostasis, and also fatty acid synthase and stearoyl CoA desaturase, additional key enzymes for normal fatty acid and lipid metabolism. The analysis also revealed a change in cellular factors that are linked to apoptotic activities, including caspases. These data show the utility of toxicogenomics in the characterization of cell systems that potentially could be of value in understanding the toxicities of NMEs.

Furthermore, new information continues to be gathered about nonhepatic human cell cultures and their relevance to toxic reactions. For example, a study involving *in vitro* human skin models to characterize irritation reactions using toxicogenomics has been reported (Borlon *et al.*, 2007). Several investigations have focused on gene expression changes and on stress response of renal cell culture systems (van de Water *et al.*, 2006). Toxicogenomics has also been employed to study models of gene therapy and to understand better the safety issues associated with this therapeutic approach (Omidi *et al.*, 2005). Also, searches for biomarkers of toxicity continue to rely on toxicogenomic technologies (Merrick and Bruno, 2004).

16.9   Use of Toxicogenomics for Detection of Idiosyncratic Drug Reactions

Adverse drug reactions can be categorized into two categories: type A reactions, which are predictable based on the drug's pharmacology; and type B reactions, which are unpredictable or idiosyncratic. Idiosyncratic drug reactions (IDRs) are unpredictable by traditional preclinical and clinical studies and occur in a small subset of the population (1 in 1000 to 1 in 100 000 patients (Lee, 2003). In many cases, these reactions are devastating and can result in significant organ damage, organ failure, or death.

The unpredictability of IDRs can cause a significant impact on public health, resulting in drugs that may be pulled from the market or receive a black box warning. A black box warning, also referred to as a black label warning, appears on the package insert of prescription drugs. This warning indicates that this pharmaceutical may cause serious adverse reactions and indicates that medical studies have been conducted and reveal that serious, or life-threatening, side effects may result from taking this drug. A black box warning is the strongest warning the Food and Drug Administration (FDA) can put on a drug. These black box warnings not only restrict the use of that particular pharmaceutical, but they may cause a loss of the public's confidence in the industry. In the USA, from 1975 until 2000, over 10% of newly approved drugs were either pulled from the market or received black box warnings, inevitably costing the pharmaceutical industry a significant amount of time and resources (Lasser *et al.*, 2002). It can take up to 10 years and cost an estimated 800 million dollars to market a pharmaceutical successfully (Suter *et al.*, 2004). The loss of time and resources associated with IDRs, combined with health risks for patients, can result in a major hurdle for the pharmaceutical industry.

To date, efforts to develop a successful animal model for IDR prediction, or to pinpoint biomarkers, have proven unsuccessful. However, new technologies such as toxicogenomics have provided a potential tool to better understand IDRs. The next section will review current research that has used toxicogenomics as a tool for characterization for IDRs.

16.10   Use of Toxicogenomics to Further Characterize Potential Animal Models

Development of an animal model for IDRs would provide a means by which mechanisms and biomarkers could be detected. Attempts at mimicking IDRs in animals by dosing with IDR-causing drugs, particularly in the rat, have met with limited success. However, some models have shown promise. Coupling these models with toxicogenomics could provide a potential means to identify and characterize compounds with the potential to induce IDRs.

One animal model for IDR that has utilized toxicogenomics is the rat inflammatory model. The theory behind this model is that some IDRs may occur due to mild levels of inflammation, which lowers the threshold for potential hepatic reactions (Buchweitz *et al.*, 2002; Luyendyk *et al.*, 2003). Several IDR compounds that do not result in hepatotoxicity in rats have been shown to cause adverse liver reactions in this species when co-dosed with low levels of lipopolysacharide (LPS; Liguori and Waring, 2006). One such example is trovafloxacin (TVX).

TVX is a novel quinolone antibiotic approved by the FDA in 1997 (Bertino and Fish, 2000). However, shortly after being launched on the market, it resulted in 150 cases of IDR, including 14 cases of liver failure and five deaths (Liguori and Waring, 2006). No significant findings during preclinical or clinical trials predicted this reaction (Waring *et al.*, 2006). However, rats co-dosed with the IDR-causing TVX and a nonhepatotoxic dose of bacterial LPS displayed significantly higher levels of alanine aminotransferase (ALT), a marker of liver damage, than rats dosed with TVX alone, or LPS and levofloxacin, a drug with low potential for hepatic IDR (Liguori *et al.*, 2005). Toxicogenomics analysis of the livers from treated rats revealed that animals treated with LPS and TVX had unique hepatic gene expression profiles when compared with other treatment groups. Use of toxicogenomics also revealed several potential biomarkers that could be used to predict TVX-induced IDRs, and also revealed mechanistic insights (Waring *et al.*, 2006). For example, polymorphonuclear neutrophils (PMNs) were shown to accumulate in the livers of animals co-dosed with TVX and LPS. PMNs have been shown to be involved in liver injury, specifically by inducing chemokines. If PMNs are depleted, then a hepatoprotective effect is observed. Rats treated with LPS and TVX had hepatic PMN accumulation, which could result in oxidative stress and the release of reactive oxygen species.

Another IDR-causing drug that has been studied in the inflammatory rat model and utilized toxicogenomics is ranitidine (RAN). Ranitidine is a histamine 2 receptor antagonist;

treatment with this compound has resulted in some adverse liver reactions (Luyendyk *et al.*, 2003). Similar to TVX, rats treated with a low dose of LPS and RAN displayed evidence of hepatotoxicity, based on clinical chemistry parameters and histopathology. Gene expression changes seen in the liver following co-treatment of LPS and RAN were associated with hypoxia and PMN activation. More specifically, livers treated with LPS and RAN had increased expression of several gene products, including tumor necrosis factor (TNF), suggesting its involvement in LPS/RAN-induced hepatotoxicity by enhancing hemostasis and inflammatory cytokine production (Tukov *et al.*, 2007).

A separate study further demonstrated how changes in gene expression could be applied in order to characterize the mechanism of IDR-causing drugs. This study assessed the gene expression profile of mouse livers following exposure to diclofenac, a nonsteroidal anti-inflammatory drug that has been associated with idiosyncractic hepatic toxicity (Skrebliukov, 1971; Manoukian and Carson, 1996; Chung *et al.*, 2006). Use of toxicogenomics revealed that multiple genes involved in eicosanoid synthesis, apoptosis, oxidative stress, and ATP synthesis showed variable transcript levels following acute diclofenac exposure (Chung *et al.*, 2006). The upregulated genes in the mouse could be involved in the mechanism of diclofenac hepatotoxicity, and could also serve as markers of hepatotoxicity. These studies indicate that gene expression analysis, coupled with animal models for IDR, can provide insights into identifying and characterizing idiosyncratic drug reactions.

16.11   Use of *In Vitro* Toxicogenomics for Understanding IDRs

Toxicogenomics has also proven to be a useful tool for assessing IDRs when applied to *in vitro* systems such as PHHs. Such an analysis was performed with isolated human hepatocytes treated with TVX. In a recent experiment, four separate human donors were tested to determine the gene expression changes caused by TVX relative to five other quinolones not associated with hepatotoxicity (levofloxacin, grepfloxacin, clinafloxacin, ciprofloxacin, gatifloxacin). Results indicate that the gene expression profile from TVX was unique from that of the other quinolones and that TVX caused the greatest number of gene expression changes. More importantly, gene expression changes induced by TVX fell into several critical functional categories, including oxidative stress, inflammation, mitochondrial function, and RNA transcript processing (Liguori *et al.*, 2005). Several of these TVX-induced changes in gene expression may play a role in hepatotoxic outcomes.

A similar study was performed with the antidepressant drug Nefazodone (NFD). NFD has been linked to IDR reactions causing human liver toxicity (Spigset *et al.*, 2003). Recent research combining the use of in-life rat studies with PHH culture determined that NFD inhibits the human bile salt export pump (ABCB11), which inhibits transport and elimination of bile acid resulting in its increase in hepatocytes. This is a potential mechanism for hepatotoxicity induced by NFD (Kostrubsky *et al.*, 2006). Therefore, screening for compounds that inhibit bile transport may be useful in predicting compounds likely to cause IDRs.

Finally, use of *in vitro* toxicogenomics has shown that the hepatotoxic compound troglitazone can be distinguished from its structurally similar analogue rosiglitazone by its large number of gene expression changes following treatment of human hepatocytes with both compounds (Kier *et al.*, 2004). Use of *in vitro* systems coupled with toxicogenomics can be a useful tool to predict human hepatotoxicity.

In summary, toxicogenomics is a promising tool for mechanistic analysis and detection of potential biomarkers underlying IDRs. Potential progress has been made toward development of animal models and *in vitro* systems for characterization of IDRs. The use of these systems and toxicogenomics are useful in furthering our understanding of IDR development and detection.

16.12   Summary

This chapter has provided an update on the use of toxicogenomic technologies in pharmaceutical discovery. Clearly, these technologies have greatly advanced in recent years and have a promising outlook for future applications. A substantial effort using toxicogenomics has revealed great benefit and impact on the manner in which discovery toxicology is used. As such, toxicogenomics use in discovery is clearly one of the most exciting and promising advances in modern toxicological research.

References

Abel S, Yang Y, Waring JF (2006) Development of an *in vitro* gene expression assay for predicting hepatotoxicity. *Altex* **23**(Suppl), 326–331.

Bertino J, Fish D (2000) The safety profile of the fluoroquinolones. *Clin Ther* **22**(7), 798–817.

Boess F, Kamber M, Romer S, Gasser R, Muller D, Albertini S, Suter L (2003) Gene expression in two hepatic cell lines, cultured primary hepatocytes, and liver slices compared to the *in vivo* liver gene expression in rats: possible implications for toxicogenomics use of *in vitro* systems. *Toxicol Sci* **73**(2), 386–402.

Borlon C, Godard P, Eskes C, Hartung T, Zuang V, Toussaint O (2007) The usefulness of toxicogenomics for predicting acute skin irritation on *in vitro* reconstructed human epidermis. *Toxicology* **241**(3), 157–166.

Buchweitz JP, Ganey PE, Bursian SJ, Roth RA (2002) Underlying endotoxemia augments toxic responses to chlorpromazine: is there a relationship to drug idiosyncrasy? *J Pharmacol Exp Ther* **300**(2), 460–467.

Bulera SJ, Eddy SM, Ferguson E, Jatkoe TA, Reidel JF, Bleavins MR, DeLaIglesia FA (2001) RNA expression in the early characterization of hepatotoxicants in Wistar rats by high-density DNA microarrays. *Hepatology* **33**(5), 1239–1258.

Burczynski ME, McMillian M, Cirvo J, Li L, Parker JB, Dunn RT, Hicken S, Farr S, Johnson MD (2000) Toxicogenomics-based discrimination of toxic mechanism in HepG2 human hepatoma cells. *Toxicol Sci* **58**, 399–415.

Bushel PR, Hamadeh HK, Bennett L, Green J, Ableson A, Misener S, Afshari CA, Paules RS (2002) Computational selection of distinct class- and subclass-specific gene expression signatures. *J Biomed Inform* **35**(3), 160–170.

Chang CY, Schiano TD (2007) Review article: drug hepatotoxicity. *Aliment Pharmacol Ther* **25**(10), 1135–1151.

Chung H, Kim HJ, Jang KS, Kim M, Yang J, Kim JH, Lee YS, Kong G (2006) Comprehensive analysis of differential gene expression profiles on diclofenac-induced acute mouse liver injury and recovery. *Toxicol Lett* **166**(1), 77–87.

Dai X, He YD, Dai H, Lum PY, Roberts CJ, Waring JF, Ulrich RG (2006) Development of an approach for *ab initio* estimation of compound-induced liver injury based on global gene transcriptional profiles. *Genome Inform* **17**(2), 77–88.

De Gottardi A, Vinciguerra M, Sgroi A, Moukil M, Ravier-Dall'Antonia F, Pazienza V, Pugnale P, Foti M, Hadengue A (2007) Microarray analyses and molecular profiling of steatosis induction in immortalized human hepatocytes. *Lab Invest* **87**(8), 792–806.

Dix DJ, Houck KA, Martin MT, Richard AM, Setzer RW, Kavlock RJ (2007) The ToxCast program for prioritizing toxicity testing of environmental chemicals. *Toxicol Sci* **95**(1), 5–12.

Fielden MR, Kolaja KL (2006) The state-of-the-art in predictive toxicogenomics. *Curr Opin Drug Discov Dev* **9**(1), 84–91.

Fielden MR, Eynon BP, Natsoulis G, Jarnagin K, Banas D, Kolaja KL (2005) A gene expression signature that predicts the future onset of drug-induced renal tubular toxicity. *Toxicol Pathol* **33**(6), 675–683.

Foster WR, Chen SJ, He A, Truong A, Bhaskaran V, Nelson DM, Dambach DM, Lehman-McKeeman LD, Car BD (2007) A retrospective analysis of toxicogenomics in the safety assessment of drug candidates. *Toxicol Pathol* **35**(5), 621–635.

Ganter B, Tugendreich S, Pearson CI, Ayanoglu E, Baumhueter S, Bostian KA, Brady L, Browne LJ, Calvin JT, Day GJ, Breckenridge N, Dunlea S, Eynon BP, Furness LM, Ferng J, Fielden MR, Fujimoto SY, Gong L, Hu C, Idury R, Judo MS, Kolaja KL, Lee MD, McSorley C, Minor JM, Nair RV, Natsoulis G, Nguyen P, Nicholson SM, Pham H, Roter AH, Sun D, Tan S, Thode S, Tolley AM, Vladimirova A, Yang J, Zhou Z, Jarnagin K (2005) Development of a large-scale chemogenomics database to improve drug candidate selection and to understand mechanisms of chemical toxicity and action. *J Biotechnol* **119**(3), 219–244.

Guerreiro N, Staedtler F, Grenet O, Kehren J, Chibout SD (2003) Toxicogenomics in drug development. *Toxicol Pathol* **31**(5), 471–479.

Harries HM, Fletcher ST, Duggan CM, Baker VA (2001) The use of genomics technology to investigate gene expression changes in cultured human cells. *Toxicol In Vitro* **15**, 399–405.

Harris AJ, Dial SL, Casciano DA (2004) Comparison of basal gene expression profiles and effects of hepatocarcinogens on gene expression in cultured primary human hepatocytes and HepG2 cells. *Mutat Res* **549**(1–2), 79–99.

Hayes KR, Vollrath AL, Zastrow GM, McMillan BJ, Craven M, Jovanovich S, Rank DR, Penn S, Walisser JA, Reddy JK, Thomas RS, Bradfield CA (2005) EDGE: a centralized resource for the comparison, analysis, and distribution of toxicogenomic information. *Mol Pharmacol* **67**(4), 1360–1368.

Healan-Greenberg C, Waring JF, Kempf DJ, Blomme EA, Tirona RG, Kim RB (2008) HIV protease inhibitor A-792611 is a novel functional inhibitor of human PXR. *Drug Metab Dispos* **36**(3), 500–507.

Hengstler JG, Brulport M, Schormann W, Bauer A, Hermes M, Nussler AK, Fandrich F, Ruhnke M, Ungefroren H, Griffin L, Bockamp E, Oesch F, von Mach MA (2005) Generation of human hepatocytes by stem cell technology: definition of the hepatocyte. *Expert Opin Drug Metab Toxicol* **1**(1), 61–74.

Jessen BA, Mullins JS, De Peyster A, Stevens GJ (2003) Assessment of hepatocytes and liver slices as *in vitro* test systems to predict *in vivo* gene expression. *Toxicol Sci* **75**(1), 208–222.

Jolly RA, Goldstein KM, Wei T, Gao H, Chen P, Huang S, Colet JM, Ryan TP, Thomas CE, Estrem ST (2005) Pooling samples within microarray studies: a comparative analysis of rat liver transcription response to prototypical toxicants. *Physiol Genomics* **22**(3), 346–355.

Karpinets TV, Foy BD, Frazier JM (2004) Tailored gene array databases: applications in mechanistic toxicology. *Bioinformatics* **20**(4), 507–517.

Kier LD, Neft R, Tang L, Suizu R, Cook T, Onsurez K, Tiegler K, Sakai Y, Ortiz M, Nolan T, Sankar U, Li AP (2004) Applications of microarrays with toxicologically relevant genes (*tox* genes) for the evaluation of chemical toxicants in Sprague Dawley rats *in vivo* and human hepatocytes *in vitro*. *Mutat Res* **549**(1–2), 101–113.

Kostrubsky SE, Strom SC, Kalgutkar AS, Kulkarni S, Atherton J, Mireles R, Feng B, Kubik R, Hanson J, Urda E, Mutlib AE (2006) Inhibition of hepatobiliary transport as a predictive method for clinical hepatotoxicity of nefazodone. *Toxicol Sci* **90**(2), 451–459.

Lasser KE, Allen PD, Woolhandler SJ, Himmelstein DU, Wolfe SM, Bor DH. (2002) Timing of new black box warnings and withdrawals for prescription medications. *JAMA* **287**, 2215–2220.

Lee WM (2003) Drug-induced hepatotoxicity. *N Engl J Med* **349**(5), 474–485.

Liguori MJ, Waring JF (2006) Investigations toward enhanced understanding of hepatic idiosyncratic drug reactions. *Expert Opin Drug Metab Toxicol* **2**(6), 835–846.

Liguori MJ, Anderson MG, Bukofzer S, McKim J, Pregenzer JF, Retief J, Spear BB, Waring JF (2005) Microarray analysis in human hepatocytes suggests a mechanism for hepatotoxicity induced by trovafloxacin. *Hepatology* **41**(1), 177–186.

Lum PY, He YD, Slatter JG, Waring JF, Zelinsky N, Cavet G, Dai X, Fong O, Gum R, Jin L, Adamson GE, Roberts CJ, Olsen DB, Hazuda DJ, Ulrich RG (2007) Gene expression profiling of rat liver reveals a mechanistic basis for ritonavir-induced hyperlipidemia. *Genomics* **90**(4), 464–473.

Luyendyk JP, Maddox JF, Cosma GN, Ganey PE, Cockerell GL, Roth RA (2003) Ranitidine treatment during a modest inflammatory response precipitates idiosyncrasy-like liver injury in rats. *J Pharmacol Exp Ther* **307**(1), 9–16.

Manoukian AV, Carson JL (1996) Nonsteroidal anti-inflammatory drug-induced hepatic disorders. Incidence and prevention. *Drug Saf* **15**(1), 64–71.

Mattes WB, Pettit SD, Sansone SA, Bushel PR, Waters MD (2004) Database development in toxicogenomics: issues and efforts. *Environ Health Perspect* **112**(4), 495–505.

Mattingly CJ, Rosenstein MC, Davis AP, Colby GT, Forrest Jr JN, Boyer JL (2006) The comparative toxicogenomics database: a cross-species resource for building chemical–gene interaction networks. *Toxicol Sci* **92**(2), 587–595.

Merrick BA, Bruno ME (2004) Genomic and proteomic profiling for biomarkers and signature profiles of toxicity. *Curr Opin Mol Ther* **6**(6), 600–607.

Minami K, Saito T, Narahara M, Tomita H, Kato H, Sugiyama H, Katoh M, Nakajima M, Yokoi T (2005) Relationship between hepatic gene expression profiles and hepatotoxicity in five typical hepatotoxicant-administered rats. *Toxicol Sci* **87**(1), 296–305.

Mutlib A, Jiang P, Atherton J, Obert L, Kostrubsky S, Madore S, Nelson S (2006) Identification of potential genomic biomarkers of hepatotoxicity caused by reactive metabolites of *N*-methylformamide: Application of stable isotope labeled compounds in toxicogenomic studies. *Chem Res Toxicol* **19**(10), 1270–1283.

Omidi Y, Barar J, Akhtar S (2005) Toxicogenomics of cationic lipid-based vectors for gene therapy: impact of microarray technology. *Curr Drug Deliv* **2**(4), 429–441.

Richert L, Liguori MJ, Abadie C, Heyd B, Mantion G, Halkic N, Waring JF (2006) Gene expression in human hepatocytes in suspension after isolation is similar to the liver of origin, is not affected by hepatocyte cold storage and cryopreservation, but is strongly changed after hepatocyte plating. *Drug Metab Dispos* **34**(5), 870–879.

Sawada H, Takami K, Asahi S (2005) A toxicogenomic approach to drug-induced phospholipidosis: analysis of its induction mechanism and establishment of a novel *in vitro* screening system. *Toxicol Sci* **83**(2), 282–292.

Skrebliukov IE (1971) Experience in a health nurse's function in a milk farm. *Feldsher Akush* **36**(5), 39–41 (in Russian).

Spigset O, Hagg S, Bate A (2003) Hepatic injury and pancreatitis during treatment with serotonin reuptake inhibitors: data from the World Health Organization (WHO) database of adverse drug reactions. *Int Clin Psychopharmacol* **18**(3), 157–161.

Spijker S, van de Leemput JC, Hoekstra C, Boomsma DI, Smit AB (2004) Profiling gene expression in whole blood samples following an *in-vitro* challenge. *Twin Res* **7**(6), 564–570.

Suter L, Haiker M, De Vera MC, Albertini S (2003) Effect of two 5-HT6 receptor antagonists on the rat liver: a molecular approach. *Pharmacogenomics J* **3**(6), 320–334.

Suter L, Babiss LE, Wheeldon EB (2004) Toxicogenomics in predictive toxicology in drug development. *Chem Biol* **11**(2), 161–171.

Thukral SK, Nordone PJ, Hu R, Sullivan L, Galambos E, Fitzpatrick VD, Healy L, Bass MB, Cosenza ME, Afshari CA (2005) Prediction of nephrotoxicant action and identification of candidate toxicity-related biomarkers. *Toxicol Pathol* **33**(3), 343–355.

Tompkins LM, Wallace AD (2007) Mechanisms of cytochrome P450 induction. *J Biochem Mol Toxicol* **21**(4), 176–181.

Tong W, Harris S, Cao X, Fang H, Shi L, Sun H, Fuscoe J, Harris A, Hong H, Xie Q, Perkins R, Casciano D (2004) Development of public toxicogenomics software for microarray data management and analysis. *Mutat Res* **549**(1–2), 241–253.

Tukov FF, Luyendyk JP, Ganey PE, Roth RA (2007) The role of tumor necrosis factor alpha in lipopolysaccharide/ranitidine-induced inflammatory liver injury. *Toxicol Sci* **100**(1), 267–280.

Van de Water B, de Graauw M, Le Devedec S, Alderliesten M (2006) Cellular stress responses and molecular mechanisms of nephrotoxicity. *Toxicol Lett* **162**(1), 83–93.

Vickers AE, Fisher RL (2005) Precision-cut organ slices to investigate target organ injury. *Expert Opin Drug Metab Toxicol* **1**(4), 687–699.

Waring JF, Halbert DN (2002) The promise of toxicogenomics. *Curr Opin Mol Ther* **4**(3), 229–235.

Waring JF, Gum R, Morfitt D, Jolly RA, Ciurlionis R, Heindel M, Gallenberg L, Buratto B, Ulrich RG (2002) Identifying toxic mechanisms using DNA microarrays: evidence that an experimental inhibitor of cell adhesion molecule expression signals through the aryl hydrocarbon nuclear receptor. *Toxicology* **181–182**, 537–550.

Waring JF, Cavet G, Jolly RA, McDowell J, Dai H, Ciurlionis R, Zhang C, Stoughton R, Lum P, Ferguson A, Roberts CJ, Ulrich RG (2003) Development of a DNA microarray for toxicology based on hepatotoxin-regulated sequences. *EHP Toxicogenomics* **111**(1T), 53–60.

Waring JF, Liguori MJ, Luyendyk JP, Maddox JF, Ganey PE, Stachlewitz RF, North C, Blomme EA, Roth RA (2006) Microarray analysis of lipopolysaccharide potentiation of trovafloxacin-induced liver injury in rats suggests a role for proinflammatory chemokines and neutrophils. *J Pharmacol Exp Ther* **316**(3), 1080–1087.

Waring JF, Yang Y, Healan-Greenberg CH, Adler AL, Dickinson R, McNally T, Wang X, Weitzberg M, Xu X, Lisowski AR, Warder SE, Gu YG, Zinker BA, Blomme EA, Camp HS (2008) Gene expression analysis in rats treated with experimental acetyl-CoA carboxylase inhibitors suggests interactions with the PPAR alpha pathway. *J Pharmacol Exp Ther* **324**(2), 507–516.

Yang Y, Blomme EA, Waring JF (2004) Toxicogenomics in drug discovery: from preclinical studies to clinical trials. *Chem Biol Interact* **150**(1), 71–85.

Yang Y, Abel SJ, Ciurlionis R, Waring JF (2006) Development of a toxicogenomics *in vitro* assay for the efficient characterization of compounds. *Pharmacogenomics* **7**(2), 177–186.

Zidek N, Hellmann J, Kramer PJ, Hewitt PG (2007) Acute hepatotoxicity: a predictive model based on focused illumina microarrays. *Toxicol Sci* **99**(1), 289–302.

Section 3

Toxicogenomics and Risk Assessment

17

Natural Products from Medicinal Plants and Risk Assessment

Leila Chekir-Ghedira

17.1 Introduction

Toxicogenomics is the study of the response to toxic agent exposure; it has been described as a tool of unprecedented power in toxicology (Nuwaysir *et al.*, 1999; Marchant, 2002). The term 'toxicgenomics' in its broadest meaning encompasses profiling of gene expressions, protein composition (proteomics), and the metabolic constituents of a cell.

A key toxicogenomic technique is to profile (using a DNA microarray or 'gene chip') the cell-wide changes in gene expression following exposure to toxins (Marchant, 2002). This approach creates the potential to provide a molecular 'fingerprint' of exposure to specific classes of substances (Nuwaysir *et al.*, 1999; Thomas *et al.*, 2001; Aardema and MacGregor, 2002).

Gene expression changes measured by DNA microarrays can identify genes that are co-regulated with drug targets. Both targets and co-regulated genes could be potential surrogate biomarkers for use in preclinical and clinical studies. High-throughput sequencing and transcript profiling have been applied to cell-based and animal models of disease or directly to human tissues to identify rapidly gene targets that initiate the drug discovery process.

The term DNA array refers to a solid surface to which is attached numerous small pieces of DNA representing short sequences of specific genes. Thousands of genes can be analyzed in a single experiment.

The application of this technique to the fields of cancer research, anti-HIV research, inflammatory diseases and others has led to significant advances in our understanding of the mechanisms underlying disease and its treatment (Calvano *et al.*, 2005; Hudson and

Toxicogenomics: A Powerful Tool for Toxicity Assessment Edited by S. C. Sahu
© 2008 John Wiley & Sons, Ltd

Altamirano 2006; Raetz *et al.*, 2006). Likewise, similar advances were made in understanding how herbal medicines work.

17.2   Advances in Risk Assessment

The process of risk assessment has undergone many refinements over the years (Oberemm *et al.*, 2005). Human exposure assessment and dose–concentration dependency form the basis for an expert judgment which concludes whether there exists a risk for human health or not. Hence, it encompasses not only qualitative descriptions of toxic properties, but also an approach to quantify both exposure and toxic responses (Oberemm *et al.*, 2005). Toxicological profiling of substances depends on the performance of time-consuming and costly animal studies. In consequence, new molecular techniques have been increasingly applied in the field of toxicology, which has led to the development of a new research field, namely toxicogenomics (Nuwaysir *et al.*, 1999; Schmidt, 2002).

Toxicogenomics combines genetics, genomic-scale mRNA expression (transcriptomics), cell and tissue-wide protein expression (proteomics), metabolite profiling (metabonomics), and bioinformatics with conventional toxicology in an effort to understand the role of gene–environment interactions in effect and disease (Oberemm *et al.*, 2005).

With the establishment of polymerase chain reaction (PCR) and DNA-microarrays, these tools are used to screen for individual genetic variabilities, which are well known to cause major differences of susceptibility against chemical exposure (Orphanides and Kimber, 2003).

Pattern changes caused by exposure to a chemical substance may be detected and classified in a descriptive manner, without being based on a deep understanding of underlying mechanisms. 'Fingerprints' of toxicogenomic responses can be used to classify compounds with a similar mode of action. If other methods are applied simultaneously (e.g. histopathologic antibody-arrays of selected biochemical markers), then there is information provided on specific endpoints which can be linked to pattern changes of toxic genomics to provide mechanistic information on cellular perturbations and pathways, and to identify biomarkers specific to particular classes of molecular damage. Toxicogenomics is expected to provide data on the molecular basis of exposure to chemicals, as well as toxicological effects and this improved prediction of human outcomes (Oberemm *et al.*, 2005).

Other aspects include the evaluation of dose–response relations at low doses and toxicity of mixtures. Toxicogenomics could be incorporated into routinely applied existing regulatory tests to retrieve mechanistic information in addition to conventional toxicity endpoints, which could be useful to predict chronic toxicity and to reduce long-term animal testing. By establishing 'fingerprints' of toxicological modes of action, adverse effects could be recognized and potentially hazardous substances could be identified and classified at an early stage of testing. This would be of particular interest in the field of the existing chemicals, since a more effective priority setting could be used to detect and evaluate problem substances earlier (Oberemm *et al.*, 2005).

Advanced knowledge of toxicogenetics would enhance the identification and characterization of sensitive life stages, subpopulations, and individuals, which is also a key challenge for risk assessment (Smith, 2001).

In order to link data from toxicogenomic studies, more efforts are needed to broaden access to available data and to develop publicly available databases. Finally, it must be mentioned that an effective use of toxicogenomics data in risk assessment requires some changes of the existing regulatory system, since not all endpoints evaluated by animal studies may be covered by toxicogenomic data and additional procedures have to be defined to combine traditional and new endpoints (Cunningham *et al.*, 2003).

17.3 Applications of DNA Microarrays in Herbal Drug Research and Development

Natural product research is often based on ethnobotanical information, and many of the drugs used today were employed in indigenous societies (Patwardhan, 2005). One of the aims of ethnopharmaceutical research is better understanding of the pharmacological effects of different medicinal plants traditionally used in healthcare (Heinrich, 2003). Plants are regarded as a promising source of novel therapeutic agents due to their higher structural diversity compared with standard synthetic chemistry. Plants have applications in development of therapeutic agents: as a source of bioactive compounds for possible use as drugs. There are three approaches to natural product-based drug discovery: screening of crude extracts; screening of prefractionated extracts; screening of pure compounds (Vuorela *et al.*, 2004).

The concept of multi-target therapy exists in traditional medical treatments that employ multicomponent extracts of natural products. With the exception of pure substances, botanical derivatives obtained from medicinal plants usually contain several classes of compounds, which simultaneously act on multiple targets. However, very few herbal medicinal products have documentation comparable to synthetic drugs. Preparation of standardized herbal products with respect to chemical composition and biological activity is complicated due to the myriad range of chemicals. This also makes it difficult to predict precisely the molecular targets, the mechanism of action, and the side effects of such products. However, many minor compounds with potential biological effects remain neglected from quantitative analysis of chemical markers. Advances in high-throughput experimentation have resulted in massive databases of genomic, proteomic, and chemical data which, in combination with efficient separation methods and powerful spectrometric methods for identification and structure elucidation, can be used for identification of active compounds. In addition, the past two decades has witnessed a number of citations that describe a diverse range of molecular mechanisms that govern cellular behavior. A systems biology approach that integrates such large and diverse sources of information together will serve to make useful biological predictions about the pharmacological effects of natural products (Wagner, 1999). DNA microarrays may provide a suitable high-throughput platform for research and development of drugs from natural products. In natural products, a broad repertoire of chemical entities acts together on multiple targets that make it necessary to study the changes in expression of multiple genes simultaneously. Novel technologies, such as serial analysis of gene expression (SAGE) and DNA microarrays, allow rapid and detailed analysis of thousands of transcripts, providing a revolutionary approach to the investigation of gene expression (Chavan *et al.*, 2006).

In this chapter, we discuss the application of DNA microarray technology in herbal drug research and development with suitable examples. Twenty-five studies (published in the period 2001–2007) in which some form of herbal medicine has been evaluated in the cell cultures or in animals, by means of DNA arrays, were chosen.

17.4   Olive Leaf Extract

Olive leaf extract (OLE) was reported to be effective in treating fever and malaria (Hanbury, 1854). OLE contains compounds with potent antimicrobial activities against bacteria, fungi, and mycoplasma (Juven and Henis, 1970; Fleming *et al.*, 1973; Aziz *et al.*, 1998; Bisignano *et al.*, 1999; Furneri *et al.*, 2002). In addition, it has antioxidant (Visioli and Galli, 1994, 2002; Caruso *et al.*, 1999; Coni *et al.*, 2000; Owen *et al.*, 2000) and anti-inflammatory (Petroni *et al.*, 1995; Visioli *et al.*, 1998; De la Puerta *et al.*, 2000) activities. Some of these effects have been proposed to contribute to the anti-atherogenic properties of a diet rich in olive oil (Visioli and Galli, 2001, Visioli *et al.*, 2002; Carluccio *et al.*, 2003). Recently, AIDS patients have begun to use OLE for a variety of indications, among them to strengthen the immune system, to relieve chronic fatigue, to boost the effects of anti-HIV medications, and to treat HIV-associated Kaposi's sarcoma and herpes simplex virus infections.

Lee-Huang *et al.* (2003) decided to examine the effect of OLE preparations standardized by liquid chromatography–coupled mass spectrometry on HIV-1 infection and replication. They observed inhibitory effect of OLE towards acute infection and cell-to-cell transmission of HIV-1 as assayed by syncytia formation using uninfected MT2 cells co-cultured with HIV-1-infected H9 T lymphocytes. They reported that OLE also inhibits HIV-1 replication as assayed by p24 expression in infected H9 cells.

These anti-HIV effects of OLE are dose dependent, with EC_{50} values of around $0.2\,\mu\mathrm{g\,ml^{-1}}$. In the effective dose range, no cytotoxicity on uninfected target cells was detected. To identify viral and host targets for OLE, Lee-Huang *et al.* (2003) characterized gene expression profiles associated with HIV-1 infection and OLE treatment using DNA microarrays. HIV-1 infection modulates the expression patterns of cellular genes involved in apoptosis, stress, cytokine, protein kinase C, and hedgehog signaling. HIV-1 infection up-regulates the expression of the heat shock proteins hsp27 and hsp90, the DNA damage inducible transcript 1 Gadd45, the p53-binding protein mdm2, and the hedgehog signal protein patched 1, while it down-regulates the expression of the anti-apoptotic Bcl2-associated X protein Bax. Treatment with OLE reverses many of these HIV-1 infection-associated changes. It also up-regulates the expression of the apoptosis inhibitor proteins IAP1 and 2 as well as the calcium and protein kinase C pathway signaling molecules IL-2, IL-2Rα and ornithine decarboxylase ODC1.

17.5   *Rhamnus alaternus*

Rhamnus alaternus extract is used in Tunisian traditional medicine as a laxative, purgative, diuretic, antihypertensive, and depurative. The study conducted by Ben Ammar *et al.* (2007) revealed the ability of enriched total oligomers flavonoids (TOF), ethyl acetate, and methanolic extracts as antigenotoxicant against mutagens such as AFB1 and nifuroxazide. The antigenotoxic activity of these extracts could be ascribed, as shown by Ben Ammar *et al.* (2007), at least in part to their antioxidant properties. However, other additional mechanisms, such as DNA repair enzyme induction, are not excluded.

Microarray analysis indicated that TOF extract exhibited the most important ability to increase antioxidant enzyme (SOD1) expression and the thioredoxin control systems (TXN, AOE 372). This is in accordance with its high superoxide scavenging activity when compared with the other extracts tested, whereas methanolic extract, which exhibited the lower superoxide scavenging activity, modulates in a less important manner the expression of transcripts involved in the antioxidant pathway. However, we cannot exclude that other genes are involved in cell oxidative defense as our array contained only 82 cDNA. On the other hand, methanolic extract showed a larger induction of the expression of genes coding for proteins involved in the DNA damage repairing pathway (POLD2, GADD45A, RPA3, hMSH2, XPA, TDG, RPA2, RAD23B, CCNH, PCNA, DDIT3) compared with TOF and ethyl acetate extracts, whereas it did not show more important antigenotoxic activity than TOF and/or ethyl acetate extract. This result supports the fact that antigenotoxic activity detected in the extracts tested may be ascribed to the antioxidant and radical scavenging effect of these extracts as well as to the induction of some transcripts involved in the DNA repair system (XPC, LIG4, XRCC5). This could be explained by the high flavonoids and polyphenols contents of these extracts, as such compounds were previously described to possess antioxidant activities (Urquiaga and Leighton, 2000; Shon *et al.*, 2004; Dasgupta and De, 2007; Kanatt *et al.*, 2007).

17.6 *Pistacia lentiscus*

In a recent investigation of flavonoids from medicinal plants, Abdelwehed *et al.* (2007) assessed the ability of two polyphenols isolated from the fruits of *Pistacia lentiscus* to scavenge free radicals and to induce antimutagenic effect assayed with SOS chromotest using *E. coli* PQ37 and Comet assay using the K562 cell line.

The results obtained demonstrated that gallic acid (GA) and pentagalloylglucose (PGA) exhibit an interesting antiradical potential, using free radical scavenging activity and lipid peroxidation inhibitory activity; indeed, they have the ability to scavenge the DPPH˙ radical by a hydrogen donating mechanism. They are more effective than vitamin E ($IC_{50} = 3\,\mu g\,ml^{-1}$) in scavenging DPPH˙ radicals. Usually, free-radical scavengers inhibit lipid peroxidation (Husain *et al.*, 1987; Ohinishi *et al.*, 1994; Liebert *et al.*, 1999; Lee *et al.*, 2003), although there are some exceptions. However, the potent scavenging activity of both GA and PGA was not accompanied by antioxidative properties *in vitro*. The two compounds exert strong inhibitory action against lipid peroxidation and rather no $O_2^{˙-}$ scavenging properties. These results were similar for some polyphenols tested by Hanasaki *et al.* (1994), such as morin, which have neither OH˙ nor $O_2^{˙-}$ scavenging effects but which were found to have a strong anti-lipid peroxidation effect. Furthermore, PGA shows maximum pro-oxidant effect using the X/XO assay system, followed by GA. It seems that the pro-oxidant effect increases by increasing the number of OH groups, which is more important in PGA than in GA (Maria *et al.*, 2003). This is in accordance with previous findings which suggest that galloyl derivatives are negligible inhibitors of xanthine oxidase (Chang *et al.*, 1993; Nagno *et al.*, 1999). Whereas antioxidants have attracted much interest with respect to their protective effect against free-radical damage as they may be the cause of many diseases, including cancer (Ohkawa *et al.*, 1979; Shon *et al.*, 2004), the antimutagenic activity of GA and PGA was investigated in the present study. Our results indicate that these polyphenols are able to interact and neutralize electrophiles such as nifuroxazide or may inhibit microsomal activation of AFB1 to electrophilic metabolite. PGA and GA may act, as described for others polyphenols such as flavonoids, by inhibiting the mutation or initiation caused by inhibition of activation of pro-mutagens and trap the electrophiles by chemical reaction or conjugation, antioxidant activity or scavenging of reactive oxygen species (De Flora, 1998; Shon *et al.*, 2004). Previous studies reported anti-carcinogenic activity of GA (Soleas *et al.*, 2002). The results obtained in this bacterial test should be confirmed by other tests or methods with eukaryotic cells, as eukaryotic cells, compared with prokaryotic cells, have more complicated morphological and biochemical structures. Therefore, in presenting this work, the Comet assay with human leukemia lymphoblast was used. This technique indicated the absence of potential genotoxic effect of GA and PGA. No significant DNA disturbance was observed with the concentrations of PGA and GA and the time of exposure tested. Furthermore, after 24 h exposure to GA, an apparent decrease in DNA damage induced by 2 h exposure to H_2O_2 was observed. This could be linked to the antioxidant activity of this compound, whereas PGA showed no antigenotoxic activity when using the Comet assay. We believe that the absence of antigenotoxicity of PGA is closely related to the number of galloyl moieties. The different sensitivity observed for the Comet assay and the SOS chromotest in the presence of PGA could be explained by considering the different types of cell used for these tests (eukaryotic cells with Comet assay and prokaryotic cells with SOS chromotest). This study showed that anti-lipid peroxidation activity of PGA and GA were within the same range (CI_{50} values were 200 $\mu g\,ml^{-1}$ and 220 $\mu g\,ml^{-1}$ respectively), but only GA was able to inhibit genotoxicity induced by H_2O_2 using the Comet assay. We deduce that the antigenotoxic activity of GA can be ascribed not only to the antioxidant effect of this molecule, but also to other additional mechanisms, such as DNA repair. That is why we undertook to compare gene expression related to cellular defense systems of stressed cells, treated by each of the tested compounds, using a specific array. For this purpose, we used a specific array to focus our attention on the analysis of gene expression related to cellular defense systems. From the analysis of hybridization signal in our array, we edited transcriptional profiles for the K562 stressed cells treated with GA or PGA and untreated K562 stressed cells. The majority of transcripts were related to antioxidant and DNA repair enzymes. The transcripts of thioredoxin reductase 1 (TXNRD1) tended to increase in cells treated by both H_2O_2 and GA. In contrast, TXNRD1 was absent in cells treated with only H_2O_2 or with both PGA and H_2O_2, whereas the transcription of selenoprotein W (SEPW1), involved in oxidation–reduction reactions, was repressed as well as in cells treated by H_2O_2 and GA or H_2O_2 and PGA. In contrast, it was induced in cells treated with only H_2O_2.

Moreover, glutathione peroxidase 1 (GPX1), which is involved in detoxification of H_2O_2, was more repressed in cells treated by both H_2O_2 and PGA than in cells treated by H_2O_2 and GA, but it was not detected in cells treated with only H_2O_2. Transcripts of glutathione synthetase (GSS) and heme oxygenase 2 (HMOX2) were expressed only in stressed cells treated by GA. However, thioredoxin peroxidase (AOE 372), which is related to antioxidant enzymes and involved in redox regulation of the cell, was not detected in cells treated with only H_2O_2, but it was expressed in cells treated with both H_2O_2 and GA and repressed in cells treated with H_2O_2 and PGA. The transcript of thioredoxin (TXN) involved in redox control and defense against oxidative stress is expressed at the same level as well in cells treated with H_2O_2 only or both H_2O_2 and GA, but it was repressed in cells treated with H_2O_2 and PGA. The repression of some transcripts related to antioxidant activity could be ascribed to the pro-oxidant effect detected using the X/XO assay system.

Besides the modulation of enzymes occurring in the antioxidizing system, we also observe a modulation of a certain number of DNA repair enzymes. AP endonuclease (APEX), a repair oxidative DNA damage enzyme, and Poly ADP-ribose polymerase (PARP), involved in the execution phase of apoptosis, were expressed in stressed cells. In contrast, they were repressed in cells treated with H_2O_2 and PGA and they were not detected in stressed cells treated with GA, whereas the level of transcripts involved in DNA repair as DNA ligase 4 (LIG4) and DNA-3-methyladenine glycosidase (MPG) was higher in cells treated by H_2O_2 and GA than in stressed cells. These transcripts were not detected in cells treated by H_2O_2 and PGA. Moreover, transcripts related to DNA repair by growth arrest-DNA damage-inducible gamma (GADD45A), xeroderma pigmentation complementation group A (XPA), thymine DNA glycosylase (TDG), replication protein 32 kDa (RPA2), excision repair cross-complementing rodent repair deficiency implicated in complementing group 1 (ERCC1), general transcription factor IIH peptide 1 (62 kDa) (GTF2H1), proliferation cell nuclear antigen (PCNA), and DNA damage-inducible transcript 3 (DDIT3) were induced only in stressed cells treated with GA. The transcript of DNA polymerase delta catalytic subunit 2 (POLD2) was expressed in cells treated with H_2O_2 and GA or with only H_2O_2; in contrast, it was repressed in stressed cells treated with H_2O_2 and PGA. However, the transcript of X-ray repair cross-complementing defective repair (XRCC5) was expressed only in stressed cells treated by PGA. Moreover, the transcript of Src homology 2 domain-containing (SHC1) induced in stressed cells was repressed in cells treated with H_2O_2 and PGA or GA.

The expression of genes related to cell defense in human leukemia cell line K562 treated by GA or PGA, using cDNA arrays, was essentially represented by antioxidant and DNA repair proteins. We can notice that the number of transcript modulated in the presence of GA was higher than transcripts modulated in the presence of PGA.

The results showed that the number and/or expression level of transcripts related to DNA repair were more important when cells were treated with GA than cells treated with PGA or only H_2O_2. This is in accordance with antigenotoxic activity exhibited by GA against H_2O_2-stressed cells detected by the Comet assay. These results suggest that this compound acts as an antimutagen by directly influencing the activity of DNA repair enzymes through modulating their gene expression. However, genes related to DNA repair were not expressed when PGA was added to the assay system; this result may explain the absence of antigenotoxic activity of this compound using the Comet assay. The PGA differs from the GA by the presence of a sugar moiety and the number of galloyls. Thus, these results reveal that the compounds tested isolated from *P. lentiscus* were nonmutagenic but exhibit antimutagenic and antioxidant activities. This suggests that these compounds may be phytopharmaceutical molecules of interest. Other biological properties should be studied to evaluate its pharmacological potential and to understand the mechanism by which these compounds act.

17.7   *Gingko biloba*

Leaf extracts of *Gingko biloba* have been advocated for many years as dietary supplements to ameliorate symptoms associated with various brain and circulatory disorders, including memory problems, and are currently among the most popular items sold in health food stores. Some specific bioactive components have been examined, such as the gingkolides and flavonoids, although it has often been assumed that the whole-leaf extract is more useful because of the synergistic actions of many bioactive constituents.

Watanabe *et al.* (2001) decided to examine the effect of *Gingko biloba* leaf extract (a standardized commercial extract, EGb 761, which has been used in clinical trials) for neuro-modulatory effects on gene expression in mouse cortex and hippocampus.

Mice were maintained for a month on a low-flavonoid diet, or one supplemented with *Gingko biloba* extract, 300 mg per kilogram of food pellets. They were then sacrificed and the cortex and hippocampi separately removed and frozen. RNA was extracted from pooled tissues by standard 'Trizol' methodology, amplified, and converted into biotinylated probes for hybridization to Affymetrix chips, containing oligonucleotides representing 6000 discrete mouse genes plus 6000 expressed sequence tags (ESTs).

They presumably used standard Affymetrix array analysis and interpretation equipment and software, and they recorded reproducible changes of a least twofold between control and treated tissues. Forty-three genes were significantly affected in the cortex and 13 in the hippocampus (<1 %), although only four were activated in both tissues (Gohil, 2002). Various functional groups were represented, including growth factors, transcription factors and signaling pathway components, and a few specific ion channels and cytoskeletal proteins. Some of the more conspicuous changes were verified by reverse transcription (RT)-PCR measurements on the same RNA preparations. The most impressive change was in hippocampal transthyretin (increased 16-fold), which is considered to play an important function in hormone transport in the brain and possibly in amyloid protein production.

Obviously, there are numerous limitations in the model used, such as the use of pooled heterogeneous tissues, and particularly the choice of a single time point during the 1 month administration of extract. However, a more detailed analysis of gene changes over time would require a lot more work, in addition to the substantial expense incurred in using numerous Affymetrix chips.

Gohil (2002) subsequently described results from a study of the effect of the same *Gingko biloba* extract on cultures of human bladder carcinoma cells, the line T24. Following a 72 h incubation with extract, treated and control cells were analyzed for changes in gene expression by means of a 2000 gene chip. Presumably the protocols were the same as in the previous report (described above). One hundred and fifty-five genes were affected by twofold or greater (approximately 8 % of total genes tested), mostly increases, and these represented a variety of functional groups, including some that could be important in antioxidant defenses. These could be significant to the long-term human consumption of the herb.

17.8   St John's Wort and Hypericin

St John's wort, *Hypericum perforatum* (SJW), is one of the most popular herbal medicines in Europe and North America. Its principal application is for people who suffer from seasonal mood disorders and other types of depression. Clinical studies have been conducted, with both positive and negative outcomes. As is the case for most herbal medicines, the nature of the SJW preparation has probably contributed significantly to its apparent success or lack thereof. Alcoholic extracts/tinctures, and water-based teas, have been promoted as anti-stress and relaxation drinks.

Among the many bioactive ingredients, hypericin and its related analogues, and hyperforin, have been incriminated in its beneficial actions (Agostinis *et al.*, 2002), although extracts usually contain many other phenolic compounds, and the composition varies between different *Hypericum* species (Rabanal *et al.*, 2005). Some of the bioactive compounds, such as hypericin and its analogues, are also photosensitizers (Towers *et al.*, 1997; Xie *et al.*, 2001), a fact which has not always been taken into account in laboratory studies. Hypericin itself, and some extracts, possess antiviral and antibiotic activities, which might have accounted for some of the uses of this plant in Ayurvedic medicine (Hudson and Towers, 1999).

A potential drawback in the use of SJW has been popularized recently, namely the fact that its interaction with cytochrome P450s (CYPs) could lead to interference with the metabolism of drugs that are consumed simultaneously (Wang *et al.*, 2000; Schultz, 2001). Such claims have led to a negative press on SJW, although one could turn the argument around and suggest that people taking SJW or hypericin should avoid certain synthetic drugs that require CYP metabolism.

Wong *et al.* (2004) used DNA arrays to study the effects of SJW extract, in comparison with the synthetic antidepressant drug imipramine, on gene expression in the hypothalamus of rats. Hypothalamus was chosen because of its known importance in mood-altering functions, and it was suggested that the SJW and imipramine could be expected to share some common pathways leading to their similar beneficial effects. Previous studies had indicated that anti-stress effects in rats required a period of several weeks for the test materials to show their effects.

Accordingly, following administration of a partly characterized SJW (by daily gavages) or imipramine (daily intraperitoneal injection), and appropriate controls, rats were sacrificed after 8 weeks and their hypothalami frozen. Subsequently, RNAs were extracted from these tissues (Trizol followed by 'RNeasy'), one hypothalamus for each array, and theRNAs reverse-transcribed and used to prepare biotinylated RNA for use with Affymetrix chips. The arrays contained 8799 rat genes plus ESTs. Following hybridization and processing, the chips were scanned and analyzed by means of Affymetrix software.

The results were disappointing: SJW differentially regulated a total of only 66 genes/ESTs, representing <1%, in comparison with imipramine, which affected 74 genes/ESTs. A variety of different pathways were represented by these genes, but the surprising finding was that only six genes were common to the treatments, in spite of the similar physiological effects on the animals. In addition to these small numbers, the actual magnitude of the changes was generally small, less than twofold, although they were claimed to be significant.

These relatively minor effects could have been due to the use of heterogeneous tissues (unavoidable in animal experiments) in which real significant effects in certain cell types could have been masked by nonresponding (or differently responding) cells. It is also conceivable that the animals may have adapted their metabolism during 8 weeks of daily exposure to the test materials. Another drawback was the use of a pure compound to compare with a crude extract, since the latter may have induced different responses from different ingredients. It would be useful to examine the effects of SJW/hypericin in appropriate cell lines, where one could anticipate a more significant response, although the choice of appropriate cell type is obviously open to discussion.

17.9   Cannabinoids

The subject of cannabis, and its well known major constituents the cannabinoids, needs no introduction. According to the myriad of anecdotal observations and studies, based on both traditional and modern experiences, various preparations derived from the plant *Cannabis sativa* (Fam. Cannabaceae) are used primarily for two purposes, the psychoactive one, due principally to delta-9-THC (tetrahydrocannabinol), and with which we personally cannot profess to be experts, and the applications of so-called 'medical marijuana,' which has been claimed to possess benefits to many patients suffering from different kinds of chronic disorders, such as cancer, AIDS, pain, multiple sclerosis, epilepsy, and others (Grinspoon and Balakar, 1997). In general, the psychoactive preparations are rich in delta-9-THC, whereas the medically applied preparations are low in THC but contain substantial quantities of cannabindiol and cannabinol, plus other cannabinoids. However, both types of preparation also contain numerous other phytochemicals, some of which probably contribute to the overall bioactivities (Turner *et al.*, 1980; Klein *et al.*, 1998; Zurier, 2003).

Studies performed to date with pure compounds have revealed the presence of receptors, CB1, CB2, and likely additional ones, that bind different cannabinoids to different degrees, and thereby activate various signaling pathways, with a variety of gene expression changes, depending upon the nature of the cannabinoid and the cell type (Klein *et al.*, 1998; Zurier, 2003). It has also become clear that there are several endogenous compounds, chemically unrelated to the cannabinoids, which share these receptors and which can mimic some of the bioactivities of the cannabinoids.

Since the cannabinoids, even individual compounds, clearly possess multifunctional attributes, then a DNA microarray analysis might prove fruitful. Kittler *et al.* (2000) and Parmentier-Batteur *et al.* (2002) attempted this in rats and mice respectively. At the outset there are potential problems associated with such an approach. First of all is the obvious one that mice and rats are not necessarily good models for humans, and we really cannot say if animals respond to THC in the same way that humans do; even individual humans differ widely in their responses (Grinspoon and Balakar, 1997). Second, there is the problem of tissue heterogeneity, as mentioned already; consequently, if there is a specific target site (e.g. cell type) in the brain for a compound such as THC, then its effects would be diluted by inclusion of unaffected tissues/cells. Furthermore, different cell types might respond in opposite ways to the compound, and analysis of the whole tissue could not reveal this. On the positive side, the study was at least *in vivo*, which should make it theoretically more relevant, although the method of administration of THC in animal studies is unlikely to reflect human usage; we are not aware of any reports of intraperitoneal inoculations of cannabinoids or extracts in humans.

Kittler *et al.* (2000) used rats, which were given daily intraperitoneal injections of delta-9-THC. Animals were sacrificed after 1, 7, or 21 days and hypothalami were removed and frozen at $-80\,^{\circ}\text{C}$ for subsequent RNA extractions. Commercial kits were used to extract and purify the mRNAs, which were then reverse transcribed to yield ^{32}P-labelled cDNAs.

They produced rat gene arrays in-house from a library of more than 24 000 cloned cDNAs and spotted these on two sets of nylon membranes. Following standard hybridization procedures, they analyzed the membranes by means of scanning and software programs. About a quarter of the cloned cDNAs hybridized to test RNA, but only 28 of them (<1 %) showed alterations in response to delta-9-THC. However, the significance of these changes is open to question, since most of them were less than twofold and the authors did not indicate if they made allowance for differential ^{32}P-labelling between preparations.

Parmentier-Batteur *et al.* (2002) made their analysis in the brain tissues of THC-treated mice. They administered delta-9-THC, or the dimethylsulfoxide vehicle alone, or in some cases a synthetic agonist, intraperitoneally in mice, and after 12 h killed the mice, removed brains, and extracted RNA. The cDNAs were labeled with Cy3 or Cy5, for controls and test compound respectively, and hybridized to slides containing 11 200 ESTs derived from a mouse brain library. The data were scanned and analyzed by commercial software to give ratios of Cy3:Cy5. Averages were taken from three separate experiments, which in retrospect might be considered dubious, although the ratios had to exceed 1.8 in all three experiments to qualify for acceptance.

A total of 86 genes (<1 %) were affected by one or both agents, mostly decreases by two- to three-fold, but only 20 genes in common. A variety of functionally different genes were affected, although they pointed out that their results were not comparable to those of Kittler *et al.* (2000) (see above). This might be explicable by the use of different animal species and experimental protocols, but it could just as easily reflect differences in the array systems used and their processing and analyses.

Parmentier-Batteur *et al.* (2002) also mentioned that their data were not comparable to those of Derocq *et al.* (2000), who had previously studied the effect of a cannabinoid CB2 receptor agonist in a human promyelocytic cell line, HL-60. This cell line has been the object of many investigations in connection with cell differentiation and signaling pathways in the immune system. It has the theoretical advantage of a 'homogeneous' cell culture, in which significant changes in gene expression should be more obvious than in animal tissue models. Nevertheless, this apparent 'homogeneity' could be less real in a cell line that undergoes continuous differentiation in culture. In other words, its attraction as a model of differentiation militates against its value in the analysis of gene expression changes induced by an agent.

Derocq *et al.* (2000) amplified the likelihood of seeing receptor-mediated changes by transfecting multiple copies of the receptor gene into these cells. They then treated the transfected cells with a synthetic cannabinoid agonist (CP 55,940), or synthetic antagonist (SR 144,528), for various times. RNA was extracted from these different cell populations, converted into biotinylated cRNAs and used with standard Affymetrix DNA arrays (huGene FL chips containing 6800 genes). These were processed and analyzed by means of the usual Affymetrix prescribed equipment and software.

Ten genes (=1 %) were concluded to be significantly altered (fold ratios, treated to control, in the range 2.2–4.7), mostly stimulated, following a 1 h exposure to the receptor agonist, and these represented genes involved in cytokine, transcription, and cell cycle processes. The changes were confirmed by combinations of northern blots and enzyme-linked immunosorbent assays. The authors were able to relate these changes to a demonstrable translocation of activated NFkB transcription factor, frequently implicated in immune activation and differentiation-related events. They suggested that this finding could reflect an involvement of the receptor in normal differentiation; but clearly, the system bears no relationship to the animal studies described above. However, considering the abundance of genes involved in differentiation processes and immune activation, it is surprising that so few genes in the HL-60 cells were altered.

17.10   Catalposide (*Catalpa ovata*), *Boswellia serata*, and Inflammatory Bowel Disease

Extracts of both catalposide (*Catalpa ovata*) and *Boswellia serata* have been used traditionally in Korea and India respectively to treat inflammatory conditions. People with inflammatory bowel disease (IBD) often have raised levels of certain pro-inflammatory cytokines, such as IL-6, IL-8, and TNFa, in colonic tissues. This has led to the prospect of therapy by means of anti-inflammatory compounds and herbals. Kim *et al.* (2004) used a combination of cell culture and mouse models to study the effect of catalposide, an iridoid isolated from *Catalpa ovata*, on gene expression. They had previously shown that this iridoid inhibited the activation of NFkB, and hence the secretion of several cytokines, in a line of mouse macrophage-like cells.

In the study by Hudson and Altamirano (2006), they incubated the HT-29 human intestinal cell line with TNFa (to stimulate pro-inflammatory cytokine secretion) with and without catalposide (500 ng ml^{-1}) for 16 h, extracted RNA, and converted the RNAs into Cy3- and Cy5-labeled cDNAs, by means of a commercial kit. These were hybridized to cDNAs on a commercial human 8000 cDNA chip and the results were analyzed by means of standard imaging and processing programs.

The data were not presented, although the authors stated that a number of cytokine- and chemokine-related genes were down-regulated, as expected in view of the anti-inflammatory effects. In some instances individual transcripts or proteins were measured, for confirmation, by means of various northern blots, western blots, and histochemical staining procedures.

The chemically induced mouse IBD model was also analyzed by a similar combination of techniques. Mice received two rectal doses of trinitrobenzene sulfonic acid (TNBS), with or without intra-colonic catalposide given before and after the TNBS. However, it appears that DNA microarray studies were not done on these tissues. The authors stressed the inhibition of activated NFkB as the focus of the anti-inflammatory effect of catalposide.

The *Boswellia serrata* tree gives rise to a resinous gum, which has been used traditionally as an anti-inflammatory agent in Ayurvedic medicine. Among the potential active ingredients is a class of terpenoids called boswellic acids, which have themselves been shown to possess anti-inflammatory activity in cell culture and animal models.

In the study reported by Kiela *et al.* (2005), colitis was induced in mice by either dextran sulfate in drinking water or TNBS administered rectally. Two different kinds of *Boswellia* extract were tested for protective effects: a hexane extract and a methanol extract. No protective effects were observed. Liver RNA was isolated from the various treated mice and converted into biotin-labeled cRNA for hybridization to Affymetrix mouse array chips, containing 13 672 genes. They searched for genes reproducibly affected by threefold or more. A total of 58 genes (<1 %) were modulated only by the hexane fraction and

20 genes only by the methanol fraction; in addition, 24 genes were common to both. The differences could be anticipated, since the fractions were chemically distinct and probably contained a variety of different bioactive compounds. Among the genes affected were several CYP genes and glutathione *S*-transferase, suggestive of hepatotoxic reactions, and this conclusion accorded with the histologic findings; i.e. rather than protection, the extracts were in fact hepatotoxic.

In the human intestinal cell line CaCo$_2$, the six individual boswellic acids either inhibited experimentally induced NFkB levels or had no effect on them. Perhaps the boswellic acids operate through the mediation of alternative transcription factors and pathways. It is unfortunate that investigators in general have become preoccupied with explaining inflammatory responses by means of effects solely on NFkB.

17.11   Herbal Glycosides

Various recipes of herbal glycosides, including baicalein and dioscin, have been used for a long time in Chinese medicine to enhance memory problems in stroke victims. Wang *et al.* (2004) used a mouse model of surgically induced cerebral ischemia to investigate the spatial memory abilities of mice with and without glycoside treatment, which was given regularly before and after surgery until the animals were sacrificed. Their hippocampi were frozen, and subsequently used for the extraction and purification of RNAs. ^{32}P-labelled cDNAs were produced and hybridized to arrays of 1176 mouse genes in the Clontech Atlas system.

About half of the array genes reacted with transcripts in the different RNA preparations; and of these, 33–46 (approximately 8 %) showed significant (>1.8-fold change) increases or decreases compared with sham or vehicle-treated animals. However, although these gene changes correlated with the performance of the animals in spatial memory tests, there was little overlap among the nature of the genes affected by three different doses of the glycoside formulations, which makes complete interpretation of the data difficult.

17.12   Genistein-treated Prostate Carcinoma Cells

Genistein, a major isoflavone isolated from soybeans, was already known to inhibit the growth of cultured PC3 prostate carcinoma cells (human origin) and to inhibit NFkB and AKT signaling pathways. Li and Sarkar (2002) wanted to determine which genes were involved in this process. Therefore, they treated the cells with 50 μmol l^{-1} of genistein for 6, 36, or 72 h (to examine early and delayed responding genes, an important parameter to consider), extracted RNA by means of standard commercial kits, and prepared the labeled probes according to the Affymetrix protocols, followed by hybridization to chips containing more than 12 500 human genes (cDNAs). Results were obtained by using the prescribed software and tools. Unfortunately, it is not clear if the relative cell numbers of treated and untreated cultures were taken into account; presumably they were.

They found what they considered to be significant alterations, i.e. a change by twofold or more, in 832 genes (about 6.7 %), of which 774 were decreased and 58 were increased. The majority of these changes were observed at the 6 h time point, with further decreases noticeable at the later times. Alterations in transcription levels were confirmed in 26 cases by means of RT-PCR measurements. A large number of the genes affected involved players in signaling pathways, transcription factors, protein kinases, apoptosis, and cell cycle functions, which is not surprising in view of the original rationale of the study.

This study is interesting because of the large number of genes affected, in contrast to some of the others described below, and also this represents the activities of a single component of soybean extracts. It would be interesting to learn whether these activities are modulated by other components of soybean. This is clearly relevant to prospective applications of such extracts, or genistein itself, in the treatment of prostate and other cancers.

17.13   PC-SPES and Prostate Carcinoma

PC-SPES, the crude mix of eight different herbs, has been advocated for some years in the prevention and treatment of cancer, especially for prostate carcinoma. The usual formulation comprises *Scutallaria baicalensis*, *Glycyrrhiza glabra*, *Ganoderma lucidum*, *Isatis indigotica*, *Panax pseudo-ginseng*, *Dendranthema morifolium*, *Rabdosia rebescens*, and *Serenoa repens*. Needless to say, many members of the medical and scientific establishments are suspicious of such a 'witches' brew,' although many of these individual extracts are known to contain bioactive ingredients and, in fact, some trials have shown decreases in blood prostate-specific antigen levels over time with daily consumption of the mixture.

It would seem to be an ambitious task to analyze gene responses to such a mixture of uncharacterized phytochemicals; nevertheless, Bonham *et al.* (2002a,b) have attempted to do this, and the data are spread over the two manuscripts cited. Of the three prostate carcinoma cell lines available to them, PC-3, DU-145 and LNCaP, the latter was selected for the gene array analysis. They used glass microarrays spotted with 3000 cDNAs obtained from their library of prostate cDNA clones. They compared each RNA preparation (PC-SPES-treated or solvent control-treated) with a reference prostate RNA, one labeled with Cy3 and the other with Cy5. They also did comparisons with the dye labels reversed, essential to compensate for possible differences in dye–dUTP incorporation rates.

They used their own combination of processing and analytical tools and searched for genes that were reproducibly increased or decreased by a factor of 1.5, which is lower than most investigators would want to use. The number and intensity of changes increased with time up to 48 h of treatment, resulting in 144 genes increased and 175 genes decreased, i.e. approximately 10 % of the genes represented. A variety of genes with different functions were affected, including numerous cytoskeletal functions, which they followed up with further studies. Other functions included apoptosis, stress, cell cycle, and proliferation, as well as androgen-regulated genes, all of which could be anticipated from the objectives of the study.

17.14   *Paeoniae radix* and Apoptosis

The root of *Paeoniae lactiflora pallas* has been used traditionally in China to treat liver diseases and various other disorders. This plant is often found as a constituent of multi-herb formulations, and extracts of PRE (*P. radix* extract) have been shown to induce apoptosis in two human hepatoma cell lines.

In this study Lee *et al.* (2002), prepared a hot water extract of *P. radix* and used the filtered and lyophilized preparation, in 5–10 mg ml^{-1} concentrations, to treat HepG2 cells. RNA was used to make Cy3- and Cy5-labeled cDNAs, which were hybridized to 374 oligomer spots on the Operon human apoptosis array, with the usual collection of control spots. The most remarkable thing about this study is that only four genes were affected (one increased and three decreased, a total of $\sim$1 %), even though the array was heavily biased toward apoptosis genes. On this basis, apart from the confirmation by other tests that apoptosis was induced by such treatment, it would be difficult to conclude anything from the array analysis.

17.15 *Coptidis rhizoma* and Berberine

Coptidis rhizoma, and one of its major constituents berberine, has been shown to possess anti-proliferative effects on pancreatic carcinoma cell lines. Iizuka *et al.* (2003) attempted to compare ID$_{50}$ values for both extract and berberine against the eight cell lines with corresponding effects on specific gene expression. The extracted RNAs (from cells treated with berberine or hot water extract) were purified and processed according to Affymetrix protocols, and hybridized to 11 000 oligomers on the huU95A chips, along with various controls.

Comparisons between ID$_{50}$ and a number of specific genes was made, although it was not really clear what exactly was being measured in terms of gene expression. Out of 33 genes apparently altered (ratios or levels of significant changes were not given), some correlated with both extract and berberine ID$_{50}$, but others only correlated with one or the other ID$_{50}$. Among the genes allegedly altered were the usual mixture of functionally diverse entities, such as signaling, transcription, DNA repair, and cell cycle genes.

In a more recent publication (Hara *et al.*, 2005), the same group refined their analyses, using the original data, by concentrating on 27 specific genes affected by several specific compounds, including berberine, isolated from the extract.

17.16 *Tripterygium* Alkaloids and Apoptosis

Tripterygium hypoglaucum root has been used traditionally in Chinese medicine to treat a number of inflammatory diseases. An alkaloid fraction, prepared by organic solvent extractions, has been shown to cause apoptosis in some cultured cells.

In the study by Zhuang *et al.* (2004), the human promyelocytic leukemia cell line HL-60 was treated for 8 h with 40 μg ml^{-1} of the alkaloid fraction and the RNA extracted and processed to produce Cy3- and Cy5-labeled (control and treated) probes, which were hybridized to 3000 spots derived from a commercial human leukocyte cDNA library. Unfortunately, it was not clear what controls were used, whether background values were subtracted, or whether the labels were reversed. However, 16 ($<$1 %) genes were reported to show ratios of more than twofold change, and these included apoptosis, signaling pathways, cell cycle, and differentiation functions. Thus, the array data really did not add to the results obtained by the other techniques described in the report.

17.17   *Anoectochilus formosanus* and Plumbagin

The anticancer extract from *Anoectochilus formosanus* and the pure phytochemical plumbagin (a naphthoquinone from *Plumbago rosea*) were compared with respect to their effects on gene expression in MCF-7 cells, a line of breast adenocarcinoma origin (Yang *et al.*, 2004). RNA extracted from the various treated and control cells was converted into biotinylated cDNA probes, which were hybridized to a collection of 9600 genes on nylon membranes. Various equipment and software programs were used in the analysis and interpretation of the results.

Data for individual genes were not shown, although Yang *et al.* (2004) concluded that 59 genes (only <1 %) were significantly affected (greater than threefold change) by the extract, whereas plumbagin modulated 80 genes. It was not clear to what extent the effects of the two agents had related effects, if at all.

17.18   Propolis and Differentiation

Propolis has been advocated for innumerable purposes, in health and disease. Its major drawback is the very nature of the product, which, because of its high variability in composition, is almost impossible to standardize in terms of phytochemistry and bioactivities. Consequently, a meaningful DNA array analysis of its effects on gene expression is, to say the least, ambitious.

In the study by Mishima *et al.* (2005), the scenario was made even more complex by evaluating two different extracts of propolis in a cell line undergoing differentiation, the promyelocytic leukemia HL-60 cells. The investigators were then faced with the challenge of deciding which genes are affected by the extracts themselves and which are affected secondarily as a consequence of the process of differentiation. The results were compared with the effects of *all-trans*-retinoic acid (ATRA), which is known to induce differentiation in these cells.

The RNAs extracted from the various treated cells were converted to Cy3- and Cy5-labeled cDNAs and hybridized to AceGene Human oligo chips, containing 10 000 genes (all in the form of 50-mers). Appropriate controls and dye-reversal reactions were carried out. One hundred and eighteen genes (~1 %) were affected grater than twofold by ATRA, 79 by aqueous extract of propolis, and only 6 by an ethanolic extract. In general, there were more decreases than increases, which might be expected in the face of a differentiation process. However, correlations were difficult because of the different degrees of differentiation (and possibly the type of differentiation?) shown by the different agents.

References

Aardema MJ, MacGregor JT (2002) Toxicology and genetic toxicology in the new area of 'toxicogenomics': impact of 'omics' technologies. *Mutat Res* **499**, 13–25.

Abdelwahed A, Bouhlel I, Skandrani I, Valenti K, Kadri M, Guiraud P, Steiman R, Mariotte AM, Ghedira K, Laporte F, Dijoux-Franca MG, Chekir-Ghedira L (2007) Study of antimutagenic and antioxidant activities of gallic acid and 1,2,3,4,6-pentagalloylglucose from *Pistacia lentiscus*. Confirmation by microarray expression profiling. *Chem Biol Interact* **165**, 1–13.

Agostinis P, Vantieghem A, Merlevede W, de Witte PAM (2002) Review: hypericin in cancer treatment: more light on the way. *Int J Biochem Cell Biol* **34**, 221–241.

Aziz NH, Farag SE, Moussa LA, Abo-zaid MA (1998) Comparative antibacterial and antifungal effects of some phenolic compounds. *Microbios* **93**, 43–54.

Ben Ammar R, Bouhlel I, Valenti K, Ben Sghaier M, Kilani S, Mariotte AM, Dijoux-Franca MG, Laporte F, Ghedira K, Chekir-Ghedira L (2007) Transcriptional response of genes involved in cell defense system in human cells stressed by H_2O_2 and pre-treated with (Tunisian) *Rhamnus alaternus* extracts: combination with polyphenolic compounds and classic *in vitro* assays. *Chem Biol Interact* **168**, 171–183.

Bisignano G, Tomaino A, Lo Cascio R, Ceisafi G, Uccella N, Saija A (1999) On the *in vitro* antimicrobial activity of oleuropein and hydroxytyrosol. *J Pharm Pharmacol* **51**, 971–974.

Bonham M, Arnold H, Montgomery B, Nelson PS (2002a) Molecular effects of the herbal compound PC-SPES: identification of activity pathways in prostate carcinoma. *Cancer Res* **62**, 3920–3924.

Bonham M, Galkin A, Montgomery B, Stahl WL, Agus D, Nelson PS (2002b) Effects of the herbal extracts PC-SPES on microtubule dynamics paclitaxel-mediated prostate tumor growth inhibition. *J Nat Cancer Institut* **94**, 1641–1647.

Calvano SE, Xiao W, Richards DR, Felciano RM, Bake HV *et al.* (2005) A network-based analysis of systemic inflammation in humans. *Nature* **437**, 1032–1037.

Carluccio MA, Siculella L, Ancora MA, Massaro M, Scoditti E, Storelli C, Visioli F, Distante A, De Caterina R (2003) Olive oil and red wine antioxidant polyphenols inhibit endothelial activation. Antiatherogenic properties of Mediterranean diet phytochemicals. *Arterioscler Thromb Vasc Biol* **23**, 622–629.

Caruso D, Berra B, Giavarini F, Cortesi N, Fedeli E, Galli G (1999) Effect of virgin olive oil phenolic compounds on *in vitro* oxidation of human low density lipoproteins. *Nutr Metab Cardiovasc Dis* **9**, 102–107.

Chang WS, Lee YJ, Lu FJ, Chiang HC (1993) Inhibitory effects of flavonoids on xanthine oxidase. *Anticancer Res* **13**, 2165–2170.

Chavan P, Joshi K, Patwardhan B (2006) DNA microarrays in herbal drug research. *eCAM* **3**(4), 447–457.

Coni E, Di Benedetto R, Di Pasquale M, Masella R, Modesti D, Mattei R, Carlini EA (2000) Protective effect of oleuropein, an olive oil biphenol, on low density lipoprotein oxidizability in rabbits. *Lipids* **35**, 45–54.

Cunningham ML, Bogdanffy MS, Zacharewski TR, Hines RN (2003) Workshop overview: use of genomic data in risk assessment. *Toxicol Sci*, **73**, 209–215.

Dasgupta N, De B (2007) Antioxidant activity of some leafy vegetables of India, A comparative study. *Food Chem* **101**, 471–474.

De Flora S (1998) Mechanisms of inhibitors of mutagenesis and carcinogenesis. *Mutat Res* **402**, 151–158.

De la Puerta R, Martinez-Dominguez E, Ruiz-Gutierrez V (2000) Effect of minor components of virgin olive oil on topical anti-inflammatory assays. *Z Naturforsch* **55**, 814–819.

Derocq J-M, Jbilo O, Bouaboula M, Segui M, Clere C, Casellas P (2000) Genomic and functional changes induced by the activation of the peripheral cannabinoid receptor CB2 in the promyelocytic cells HL-60. *J Biol Chem* **275**, 15621–15628.

Fleming HP, Walter Jr WM, Etchells JL (1973) Antimicrobial properties of oleuropein and products of hydrolysis from green olives. *Appl Microbiol* **26**, 777–782.

Furneri PM, Marino A, Saija A, Uccella N, Bisignano G (2002) *In vitro* antimycoplasmal activity of oleuropein. *Int J Antimicrob Agents* **20**, 293–296.

Gohil K (2002) Genomic responses to herbal extracts: lessons from *in vitro* and *in vivo* studies with an extract of *Gingko biloba*. *Biochem Pharmacol* **64**, 913–917.

Grinspoon L, Balakar J (1997) *Marijuana the Forbidden Medicine*. Yale University Press.

Hanasaki Y, Ogawa S, Fukui S (1994) The correlation between active oxygen scavenging and antioxidative effects of flavonoids. *Free Radic Biol Med* **16**, 845–850.

Hanbury D (1854) On the febrifuge properties of the olive (*Olea europa*, L.). *Pharmaceut J Provincial Trans* 353–354.

Hara A, Iizuka N, Hamamoto Y, Uchimura S, Miyamoto T, Tsunedomi R, Miyamoto K, Hazama S, Okita K, Oka M (2005) Molecular dissection of a medicinal herb with anti-tumor activity by oligonucleotide array. *Life Sci* **77**, 991–1002.

Heinrich M (2003) Ethnobotany and natural products: the search for new molecules, new treatments of old diseases or a better understanding of indigenous cultures? *Curr Top Med Chem* **3**, 141–154.

Hudson J, Altamirano M (2006) The application of DNA micro-arrays (gene arrays) to the study of herbal medicines. *J Ethnopharmacol* **108**, 2–15.

Hudson JB, Towers GHN (1999) Phytomedicines as antivirals. *Drugs Future* **24**, 295–320.

Husain SR, Cillard J, Cillard P (1987) Hydroxyl radical scavenging activity of flavonoids. *Phytochemistry* **26**, 2489–2491.

Iizuka N, Oka M, Yamamoto K, Tangoku A, Miyamoto K, Miyamoto T, Uchimura S, Hamamoto Y, Okita K (2003) Identification of common or distinct genes related to antitumor activities of a medicinal herb and its major component by oligonucleotide array. *Int J Cancer* **107**, 666–672.

Juven B, Henis Y (1970) Studies on the antimicrobial activity of olive phenolic compounds. *J Appl Bacteriol* **33**, 721–732.

Kanatt SR, Chander R, Sharma A (2007) Antioxidant potential of mint (*Mentha spicata* L.) in radiation-processed lamb meat. *Food Chem* **100**, 451–458.

Kiela PR, Midura AJ, Kuscuoglu N, Jolad SD, Solyom AM, Besselson DG, Timmerman BN, Ghishan FK (2005) Effects of *Boswellia serrata* in mouse models of chemically induced colitis. *Am J Physiol* **288**, G798–G808.

Kim SW, Choi SC, Choi EY, Kim KS, Oh JM *et al.* (2004) Catalposide, a compound isolated from *Catalpa ovata*, attenuates induction of intestinal epithelial proinflammatory gene expression reduces the severity of trinitrobenzene sulfonic acid-induced colitis in mice. *Inflamm Bowel Dis* **10**, 564–572.

Kittler JT, Grigorenko EV, Clayton Ch, Zhuang SH, Bundey SC, Trower MM, Wallace D, Hampson R, Deadwyler S (2000) Large scale analysis of gene expression changes during acute and chronic exposure to delta 9-THC in rats. *Physiol Genomics* **3**, 175–185.

Klein TW, Friedman H, Specter S (1998) Marijuana, immunity, and infection. *J Neuroimmunol* **83**, 102–115.

Lee SE, Shin HT, Hwang HJ, Kim HJ (2003) Antioxidant activity of extracts from *Alpinia katsumadar* seed. *Phytother Res* **17**, 1041–1047.

Lee SMY, Li MLY, Tse YCh, Leung SChL, Lee MMS *et al.* (2002) *Paeoniae radix*, a Chinese herbal extract, inhibit hepatoma cells growth by inducing apoptosis in a p53 independent pathway. *Life Sci* **71**, 2267–2277.

Lee-Huang S, Zhang L, Huang PL, Chang Y-T, Huang PL (2003) Anti-HIV activity of olive leaf extract (OLE) and modulation of host cell gene expression by HIV-1 infection and OLE treatment. *Biochem Biophys Res Commun* **307**, 1029–1037.

Li Y, Sarkar FH (2002) Gene expression profiles of genistein-treated PC3 prostate cancer cells. *J Nutr* **132**, 3623–3631.

Liebert M, Licht U, Böhm V, Bitsch R (1999) Antioxidant properties and total phenolics content of green and black tea under different brewing conditions. *Z Lebensm Unters Forsh A* **208**, 217–220.

Marchant GE (2002) Toxicogenomics and toxic torts. *Trends Biotechnol* **20**(8), 329–332.

Maria CB, Maximiliano T, Carlotta G, Patricia P, Analiza R, Francesco V, Ricardo B, Rebecca R (2003) Antioxydant activity of galloyl Inc. Derivatives isolated from *P. Ientiscus Ieaves*. *Free radic Res* **37**, 405–412.

Mishima S, Narita Y, Chikamatsu S, Inoh Y, Ohta S, Yoshida Ch, Araki Y, Akao Y, Suzuki KM, Nozawa Y (2005) Effect of propolis on cell growth and gene expression in HL-60 cells. *J Ethnopharmacol* **99**, 5–11.

Nagno A, Beld M, Kobayachi H (1999) Inhibition of xanthine oxidase by flavonoids. *Biosci Biotechnol Biochem* **65**, 1787–1790.

Nuwaysir EF, Bittner M, Trent J, Barrett JC, Afshari CA (1999) Microarrays and toxicology: the advent of toxicogenomics. *Mol. Carcinog* **24**, 153–159.

Oberemm A, Onyon L, Gundert-Remy U (2005) How can toxicogenomics inform risk assessment? *Tox Appl Pharm* **207**, S592–S598.

Ohinishi M, Morishita H, Iwahashi H, Shizuo T, Yoshiaki S, Kimura M, Kido R (1994) Inhibitory effects of chlorogenic acids on linoleic acid peroxidation and haemolysis, *Phytochemistry* **36**, 579–583.

Ohkawa A, Ohishi N, Yagi K (1979) Assay for lipid peroxide in animal tissues by thiobarbituric acid reaction. *Anal Biochem* **95**, 351–358.

Orphanides G, Kimber I (2003) Toxicogenetics: applications and opportunities. *Toxicol Sci* **75**, 1–6.

Owen RW, Giacosa A, Hull WE, Haubner R, Wurtele G, Spiegelhalder B, Bartsch H (2000) Olive-oil consumption and health: the possible role of antioxidants. *Lancet Oncol* **1**, 102–112.

Parmentier-Batteur S, Kunlin J, Xie L, Mao XO, Greenberg DA (2002) DNA micro-array analysis of cannabinoid signaling in mouse brain *in vivo*. *Mol Pharmacol* **62**, 828–835.

Patwardhan B (2005) Ethnopharmacology and drug discovery. *J Ethnopharmacol* **100**, 50–52.

Petroni A, Blasevich M, Salami M, Papini N, Montedoro GF, Galli C (1995) Inhibition of platelet aggregation and eicosanoid production by phenolic components of olive oil. *Thromb Res* **78**, 151–160.

Rabanal RM, Bonkanka CX, Hernandez-Perez M, Sanchez-Mateo CC (2005) Analgesic and topical anti-inflammatory activity of *Hypericum canariensis* L. and *Hypericum glandulosum* Ait. *J Ethnopharmacol* **96**, 591–596.

Raetz EA, Perkins SL, Bhojwani D, Smock K, Philip M, Caroll WL, Min DJ (2006) Gene expression profiling reveals intrinsic differences between T-cell acute lymphoblastic leukaemia and T-cell lymphoblastic lymphoma. *Ped Blood Cancer* **46**, 570–578.

Schmidt CW (2002) Toxicogenomics – an emerging discipline. *Environ Health Perspect* **110**, A750–A755.

Schultz V (2001) Incidence and clinical relevance of the interactions and side effects of *Hypericum* preparations. *Phytomedicine* **8**, 152–160.

Shon MY, Choi SD, Kahng GG, Nam SH, Sung NG (2004) Antimutagenic, antioxidant and free radical scavenging activity of ethyl acetate extracts from white, yellow and red onions. *Food Chem Toxicol* **42**, 659–666.

Smith LL (2001) Key challenges for toxicologists in the 21st century. *Trends Pharmacol Sci* **22**, 281–285.

Soleas GJ, Grass L, Josephy B, Goldberg DM, Diamandis P (2002) Comparison of the anticarcinogenic properties of four red wine polyphenols. *Clin Biochem* **35**, 119–124.

Thomas RS, Rank DR, Penn SG, Zastrow GM, Hayes KR *et al.* (2001) Identification of toxicologically predictive gene sets using cDNA microarrays. *Mol Pharmacol* **60**, 1189–1194.

Towers GHN, Page JE, Hudson JB (1997) Light-mediated biological activities of natural products from plants and fungi. *Curr Org Chem* **1**, 395–414.

Turner CE, Elsohly MA, Boeren EG (1980) Constituents of *Cannabis sativa* L. XVII, a review of the natural constituents. *J Nat Prod* **43**, 169–234.

Urquiaga I, Leighton F (2000) Plant polyphenol antioxidant and oxidative stress. *Biol Res* **33**, 55–64.

Visioli F, Galli C (1994) Oleuropein protects low density lipoprotein from oxidation. *Life Sci* **55**, 1965–1971.

Visioli F, Galli C (2001) Antiatherogenic components of olive oil. *Curr Atheroscler Rep* **3**, 64–67.

Visioli F, Galli C (2002) Biological properties of olive oil phytochemicals. *Crit Rev Food Sci Nutr* **42**, 209–221.

Visioli F, Poli A, Galli C (2002) Antioxidant and other biological activities of phenols from olives and olive oil. *Med Res Rev* **22**, 65–75.

Visioli F, Bellosta S, Galli C (1998) Oleuropein the bitter principle of olives, enhances nitric oxide production by mouse macrophages. *Life Sci* **62**, 541–546.

Vuorela P, Leinonen M, Saikku P, Tammela P, Rauha JP, Wennberg T, Vuorela H (2004) Natural products in the process of finding new drug candidates. *Curr Med Chem* **11**, 1375–89.

Wagner H (1999) New approaches in phytopharmacological research. *Pure Appl Chem* **71**, 1649–1654.

Wang Z, Gorski J, Hamman M, Huang S, Lesko L, Hall S (2000) The effects of St. John's wort on human cytochrome P450 activity. *Clin Pharmacol Ther* **70**, 317–326.

Wang Z, Du Q, Wang F, Liu Z, Li B, Wang A, Wang Y (2004) Microarray analysis of gene expression on herbal glycoside recipes improving deficient ability of spatial learning memory in ischemic mice. *J Neurochem* **88**, 1406–1415.

Watanabe CM, Wolffram S, Ader P, Rimbach G, Packer L, Maquire JJ (2001) The *in vivo* neuromodulatory effects of the herbal medicine *Gingko biloba*. *Proc Natl Acad Sci U S A* **98**, 6577–6580.

Wong M-L, O'Kirwan F, Hannestad JP, Irizarry KJL, Elashoff D, Licinio J (2004) St. John's wort and imipramine-induced gene expression profiles identify cellular functions relevant to antidepressant action. *Mol Psychiatry* **9**, 237–251.

Xie X, Hudson JB, Guns ES (2001) Tumor specific and photodependent cytotoxicity of hypericin in the human LNCaP prostate tumor model. *Photochem Photobiol* **74**, 221–225.

Yang N-S, Shyur L-F, Chen C-H, Wang S-Y, Tzeng C-M (2004) Medicinal herb extract and a single compound drug confer similar complex pharmacogenomic activities in MCF-7 cells. J Biomed Sci **11**, 418–422.

Zhuang W-J, Fong CC, Cao J, Ao L, Leung CH, Cheung HY, Xiao PG, Fong WF, Yang MS (2004) Involvement of NFkB and c-myc signaling pathways in the apoptosis of HL-60 cells induced by alkaloids of *Tripterygium hypoglaucum*. *Phytomedicine* **11**, 295–302.

Zurier RB (2003) Prospects for cannabinoids as anti-inflammatory agents. *J Cell Biochem* **88**, 462–466.

18

The Development of a Metabonomic-based Drug Safety Testing Paradigm

Muireann Coen, Elaine Holmes, Jeremy K. Nicholson and John C. Lindon

18.1 Introduction

There has been much interest in recent years in the application of 'omics' platform tech-nologies to assess drug safety (Nicholson *et al.*, 2002) and much effort has been expended on the use of toxicogenomics to study drug-induced biochemical perturbations. However, early metabonomic studies demonstrated that time-related metabolic trajectories are highly informative on the site and mechanism of toxicity (Gartland *et al.*, 1990) and more recently this approach has been developed into a more comprehensive metabonomic strategy which has the crucial advantage of being noninvasive, thus allowing individual animals or persons to be followed after drug intervention.

The metabonomic approach is defined as 'the quantitative measurement of the multipara-metric metabolic response of living systems to pathophysiological stimuli or genetic modi-fication' (Nicholson *et al.* 1999, 2002) and it has found widespread application throughout fields that range from environmental toxicology to medicine. Metabonomics involves the spectral acquisition of metabolic profiles that provide a unique and detailed 'fingerprint' for any given biofluid or tissue. This fingerprint reflects cellular status as the concentra-tions and fluxes of metabolites involved in key intermediary cellular pathways alter in response to interventions which perturb homeostasis, such as the administration of a toxin or development of a disease. It is this metabolic adjustment represented in time-related metabolic fluctuations within a spectral fingerprint that metabonomics endeavors to study, and advanced statistical methods enable models to be computed that differentiate cohorts of sample classes. Furthermore, these multivariate metabolic signatures reflect temporal response to a toxin and can be used to discover biomarkers of toxic effect, as general toxicity screening aids or to provide novel mechanistic information.

It is important to emphasize that, post metabonomic 'screening,' subsequent identification of candidate biomarkers is essential and that statistical approaches provide a powerful tool in this respect when applied in parallel with more classical molecular identification tools. In recent years, the field of metabonomics has seen fusion with metabolomic and other metabolic profiling approaches developed originally in the field of plant biology. The terms metabolomics (Fiehn *et al.*, 2000; Fiehn, 2002; Taylor *et al.*, 2002; Jenkins *et al.*, 2004) and metabonomics (Nicholson *et al.*, 1999, 2002) are widely applied to metabolic profiling studies and the terminology is often used interchangeably. However, 'philosophically' metabonomics provides a whole organism biological description of time-related multivariate metabolic response to a treatment. In addition, metabonomics aims to capture the global response of the interacting cellular metabolomes (metabolic complements for specific compartments) that are unique to each cell type in the body but are coordinated in space and time, a concept and view that has been termed the metabonome (Nicholson and Wilson, 2003).

A variety of analytical technologies are applied in metabonomic studies; however, nuclear magnetic resonance (NMR) spectroscopy and mass spectrometry (MS) are most widely used, as these technologies are capable of capturing high-resolution metabolic information in a nontargeted fashion. The majority of metabonomic publications to date deal with the applications of NMR spectroscopy; however, modern liquid chromatography (LC)–MS methods are now also being increasingly applied (Wilson *et al.*, 2005; Lenz and Wilson, 2007). In this chapter we have concentrated on the role of NMR spectroscopy in the development of rapid multivariate metabolic profiling and metabonomics with specific emphasis on toxicological applications and we have also covered the associated chemometric and statistical tools necessary for successful biomarker recovery from complex NMR spectral data sets. One of the earliest applications of metabonomics was in NMR-based toxicological studies, as the technique allowed for the metabolic consequences of treatment with a toxin to be rapidly elucidated and additionally provided metabolic insight into both the site of toxicity and the severity of response. Broad-spectrum metabolic profiling is now recognized as a powerful top-down systems biology tool that can provide a real world linkage to other 'omics' sciences (Nicholson *et al.*, 2002; Nicholson and Wilson, 2003; Lindon *et al.*, 2007).

The drug safety assessment approaches used at present in the pharmaceutical industry can still fail to prevent molecules which have no realistic chance of reaching the market from entering development and there is a need for methodologies that can pick up potential problems earlier, faster, more cheaply, and more reliably. The later that a molecule is lost from a pharmaceutical drug development pipeline, or even worse from the clinical market as a result of unknown safety issues, the higher the financial cost for the pharmaceutical company involved. A recent review highlighted that, from 1964 to 1999, approximately 8 % of the drugs approved by the Food and Drug Administration were later withdrawn from the US market (Nassar *et al.*, 2007). Minimizing attrition caused by adverse drug effects, therefore, is one of the most important aims of pharmaceutical R&D. Metabonomics and its applications in toxicology provide a novel technology that can be used to identify adverse effects at an early stage and associated biomarkers of effect, hence saving resources, and promoting safety, efficacy, and profitability. Metabonomics provides the toxicologist with information on the target organ that is affected by xenobiotic, the magnitude of the effect, and the biochemical mechanisms of adverse effects. Furthermore, the recent development of a metabonomic expert system (Ebbels *et al.*, 2007) capable of predicting and classifying response to toxins presents evidence that a metabonomic screening system is a reliable and valuable approach to include in a preclinical toxicology platform.

This chapter provides an introduction to the application of metabonomics in terms of the analytical platforms routinely used and the statistical tools that have been developed for biomarker identification. Key contributions to the development of NMR-based metabonomic and metabolic profiling tools for toxicological assessment, biomarker discovery, and toxic mechanism elucidation will be exemplified with examples derived from the Consortium on Metabonomic Toxicology (COMET) projects. The COMET projects provide examples of current large-scale application of metabonomics in toxicological screening, database building for predictive toxicology, drug safety assessment, and in providing insight into mechanisms of toxicity. We conclude this chapter with a review of recent examples of metabonomic findings and areas of study that have implications in the field of toxicology, such as 'pharmacometabonomics' and the influence of gut microflora on xenobiotic metabolism and toxic response.

18.2 Analytical Platforms Utilized for Metabonomics

18.2.1 NMR Spectroscopy

As the role of NMR spectroscopy in metabonomics has evolved, an ever-increasing range of applications has opened, enabled by the developing technology. The first applications of NMR spectroscopy for metabolic profiling of biofluids and cells followed the early advances that were made in terms of the sensitivity of the platform; for example, the introduction of Fourier transform NMR spectroscopy and superconducting magnets in the 1970s. ^{1}H NMR spectroscopy enables detection of metabolites that are present at moderate to high concentrations in biofluids or tissues; these metabolites represent the products and intermediates of many major or 'hub' pathways that are affected by toxic or disease processes. A biofluid metabolic fingerprint that reflects toxic response can be rapidly achieved without bias imposed by targeting specific metabolic changes that are expected of the toxin. Early applications of NMR spectroscopy found that quantitative changes in metabolite patterns gave information on the location and severity of toxic lesions, together with insights into the underlying molecular mechanisms of toxicity and many novel metabolic markers of organ-specific toxicity were discovered (Nicholson *et al.*, 1985a,1985b; Nicholson and Wilson, 1987, 1989). These defining studies showed for the first time that it was possible to capture site-specific, severity, and mechanistic information simultaneously with nontargeted metabolic profiling approaches.

The advent of high-field NMR impacted significantly on the biological application of metabonomics, as the much greater sensitivity that this afforded has significantly extended the NMR boundaries of detection and enabled a wider metabolic pool to be profiled. The availability of ^{1}H NMR spectrometers operating at 1 GHz or more is imminent from manufacturers such as Bruker and Varian, with 800, 900 and 950 MHz (^{1}H frequency) spectrometers now being widely available. The improvement in NMR-based technologies has resulted in enhanced biological information recovery from metabonomic studies which further extends the scope and ability of the platform. A recent innovation that has significantly improved the analytical power of NMR is the availability of cryogenically cooled probes (radio-frequency coils and preamplifier cooled to approximately 20 K with cold helium) that results in reduced electronic/thermal noise and up to a fourfold improvement in the spectral signal/noise ratio for a single scan experiment. This technology has been used to rapidly generate information-rich ^{13}C NMR profiles of urine, as the low natural abundance of ^{13}C nuclei are counteracted by the sensitivity gain (Keun *et al.*, 2002b). A typical rat urine ^{1}H NMR spectrum obtained on a 950 MHz spectrometer with a cryoprobe is shown in Figure 18.1, and the high information content of this spectrum is evident with hundreds of resonances representing metabolites from a broad spectrum of chemical classes being visible. Figure 18.2 represents three spectral acquisitions, namely a standard one-dimensional (1D) single-pulse solvent-suppressed spectrum attenuates, a Carr–Purcell–Meiboom–Gill (CPMG) spin-echo spectrum and a diffusion-edited spectrum of a human serum sample obtained on the same 950 MHz instrument with a cyroprobe. The CPMG spin-echo experiment spectral contributions from broad resonances arising from macromolecular components based on their fast relaxation times (short T_2) and, hence, improves visualization of the small molecule metabolite resonances. On the other hand,

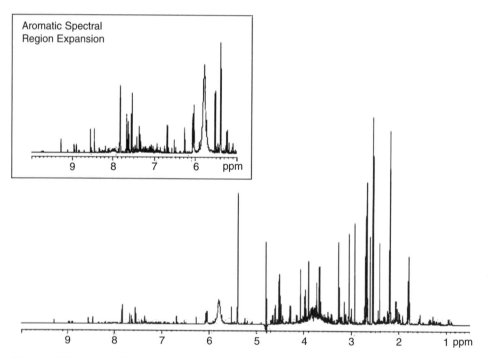

Figure 18.1 *Typical ^{1}H 950 MHz cryoprobe NMR spectrum of rat urine showing the high-resolution metabolic information obtained. This figure also includes expansion of the aromatic region of the spectrum (boxed). Author's unpublished data.*

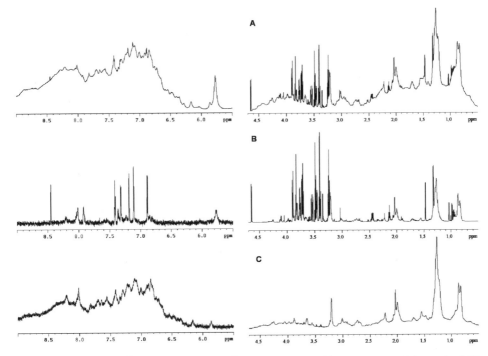

Figure 18.2 *Typical ^{1}H 950 MHz cryoprobe NMR spectra of human serum: (a) standard 1D solvent-suppressed NRM spectrum; (b) CPPMG spin-echo; (c) diffusion-edited spectrum. This figure shows both the aromatic (5.5–9.0 ppm) and aliphatic (0.2–4.7 ppm) regions of the NMR spectra. This figure represents the use of a range of NMR pulse sequences to provide metabolic profiles representative of macromolecular species, fast-diffusing small-molecule metabolites, or an overall profile representing both the above (see Section 18.2 for further detail). Author's unpublished data.*

a diffusion-edited spectrum selects macromolecules with long molecular diffusion times and, hence, provides a profile of all macromolecular species in a given sample. A standard 1D solvent-suppressed NRM spectrum provides a metabolic overview of a sample, as it represents both small- and large-molecule metabolites in a sample. Figure 18.2 exemplifies the power of NMR in metabolic profiling of a biofluid using a range of spectral pulse sequences to provide metabolic information on different molecular species without the need for chemical extraction procedures.

As automated NMR technologies are continually developed and updated, the acquisition of NMR spectra from large sample cohorts, such as those from epidemiological or population screening studies, has become feasible as high-throughput, operator-independent spectral acquisition is achievable. Such automated systems involve the use of 96-well plates, robotic sample handling, and transfer to NMR flow probes, and up to 250 ^{1}H NMR metabolic profiles can be acquired per day.

High-resolution magic-angle-spinning (MAS) NMR spectroscopy has been introduced for the determination of metabolic profiles of intact tissue samples and for investigation of the dynamics and physico-chemical properties of tissue. Prior to this, MAS NMR was principally utilized in solid-state NMR applications because it was not possible to lock the magnetic field using a ^{2}H NMR signal. The technique is based on spinning a tissue sample at the magic angle (54.7° relative to the magnetic field), which reduces anisotropic line-broadening effects and, hence, produces highly resolved spectra. It requires small amounts of tissue (~10 mg) that can be further analyzed by complementary technologies after spectral acquisition; for instance, lipid and aqueous extracts of a sample can be prepared and spectra acquired using solution-state NMR probes. Once a tissue sample has been placed in an MAS NMR rotor and is spinning at the magic angle, the conventional pulse sequences and experiments used for solution-state NMR studies can be employed to acquire MAS NMR spectra. The inclusion of MAS NMR in the traditional, solution-state NMR 'tool-box' opens up many diagnostic possibilities, since information on a variety of metabolites in different cellular environments can be rapidly obtained and together both NMR technologies provide a comprehensive and complementary set of metabolic information. In addition, specialized NMR experiments, such as those developed to measure molecular diffusion coefficients or longitudinal/transverse relaxation times, can be used to probe compartmentation within tissue. Confirmation of biochemical composition can be obtained using standard high-resolution NMR of both aqueous (protein-free) and organic solvent extracts. MAS NMR spectra, like biofluid NMR spectra, are subjected to multivariate statistical methods in order to classify the toxin-related biochemical changes in the organ under investigation and to map these time-related trajectories in multivariate space (Holmes *et al.*, 1994, 2000). Recent applications of MAS NMR include the metabolic profiling of brain (Holmes *et al.*, 2006a), intestinal tissue (Martin *et al.*, 2007), prostate (Taylor *et al.*, 2003; Swanson *et al.*, 2006), kidney (Garrod *et al.*, 1999; Waters *et al.*, 2000, 2006), and liver (Bollard *et al.*, 2000, Coen *et al.*, 2003; Duarte *et al.*, 2005; Martinez-Granados *et al.*, 2006; Yap *et al.*, 2006).

18.2.2   Mass Spectrometry

Increasingly, analytical platforms such as high-performance LC (HPLC) and ultra-performance LC (UPLC) coupled with MS are being utilized in metabonomic studies to detect and characterize a wider pool of metabolites. The development of directly coupled LC and NMR spectroscopy provides the ability to simplify complex biofluid spectra by use of a chromatographic separation step and has proved particularly advantageous for structural elucidation of drug metabolites (Gavaghan *et al.*, 2001; Connelly *et al.*, 2002; Lindon *et al.*, 1996). These hyphenated approaches involve splitting an eluted HPLC peak and analyzing it in tandem by MS and NMR platforms and can be operated in on-flow, stopped-flow and loop-storage modes to target the fractions of interest for comprehensive molecular identification. MS is generally much more sensitive than NMR spectroscopy in terms of the inherent detection limit; however, this must be balanced with problems that are commonly encountered with reproducibility, matrix effects, and ion suppression. Furthermore, it is often necessary to target chromatographic methods to a given class of chemical. However, recent developments in the applications of MS coupled with liquid chromatographic methods to metabonomic data sets have resulted in the technologies providing complementary metabolic information to NMR-based analyses, and MS is rapidly gaining acceptance as a valuable biofluid metabolic profiling tool (Want *et al.*, 2007).

UPLC is a recently developed method (Wilson *et al.*, 2005; Plumb *et al.*, 2006a) which affords greatly improved chromatographic separation via the use of small particles (1.7 μm) and high column pressures (>800 bar). The greater separation ability results in a typical three- to five-fold increase in sensitivity when compared with a conventional stationary phase and the improved resolution also reduces ion suppression problems considerably. Furthermore, the speed of data acquisition is typically increased up to 10-fold, which enables much higher throughput of samples. UPLC–MS has been applied to differentiate metabolic profiles of Zucker rats based on phenotypic differences (Plumb *et al.*, 2006b) and strain differences, namely Zucker (fa/fa) obese, Zucker lean and the lean/(fa) obese cross, and it was also possible to separate age-related effects in these animals (Granger *et al.*, 2007). A recent study by Lenz *et al.* (2007b) has used a multiplatform approach, both UPLC–MS and NMR, to profile the hepatotoxic effects of pravastatin via metabolic profiling of urine. The two analytical platforms provided complementary information, and candidate biomarkers for pravastatin toxicity were identified, such as elevated levels of taurine, creatine, and bile acids.

The power of the multiplatform metabonomic approach was further exemplified in a study of the biochemical effects of the model nephrotoxin gentamicin, which was administered daily over 7 days to rats (Lenz *et al.*, 2005). Conventional clinical chemistry urinalysis showed a significant increase in N-acetyl-β-D-glucosaminidase activity and at necropsy (9 days post-dosing) clear histological damage to the kidney was noted in all animals. ^{1}H NMR spectroscopy and HPLC coupled to time-of-flight (ToF) MS with electrospray ionization (ESI) were utilized to follow time-dependent metabolic changes as a result of gentamicin-induced toxicity and significant perturbations in the urinary metabolic profile were observed from day 7 onwards. ^{1}H NMR detected raised glucose levels and reduced trimethylamine-N-oxide (TMAO) levels, whereas HPLC–ToF-MS detected depletion of xanthurenic acid and kynurenic acid and an altered pattern of sulfate conjugation.

A further example of the complementary use of multiple analytical platforms in metabonomic studies is that of the metabolism of phenacetin in man (Nicholls *et al.*, 2006). This study involved following the metabolism of acetyl-labeled phenacetin by NMR, where the presence of nonlabeled acetaminophen glucuronide, sulfate and the N-acetyl-L-cysteinyl conjugate were detected, which confirmed the occurrence of futile deacetylation, the extent of which was calculated to be approximately 20 % for each metabolite. Further profiling was carried out using HPLC–ToF-MS and additional phenacetin metabolites were identified that included both acetylated and nonacetylated forms. This study highlights the successful application of a multiplatform approach to characterize a wide number of toxic intermediates which provides enhanced insight into analgesic nephropathy. Further examples of the combined use of NMR and HPLC–MS in metabonomics include the study of the biochemical effects of proximal tubule nephrotoxicity caused by mercury(II) chloride in the rat (Lenz *et al.*, 2004) and the monitoring of idiosyncratic hepatic response following cotreatment with ranitidine (RAN), a histamine-2 receptor antagonist, and lipopolysaccharide (LPS; Maddox *et al.*, 2006).

Gas chromatography (GC) produces highly resolved and reproducible metabolic profiles of volatile, nonpolar metabolites and has been extensively applied together with MS to the analysis of microbial and plant extracts (Fiehn, 2002; Halket *et al.*, 2005; Fiehn and Kind, 2007). The drawbacks with GC–MS include the targeted nature of the technology, the necessity for complex sample preparation procedures that involve chemical derivatization, and the collection of data over lengthy acquisition times. Evidence is gathering for the increased interest in application of GC–MS to profile biofluids, and a recent example deals with the development and optimization of an analytical protocol for the application of GC–MS to profile human plasma (O'Hagan *et al.*, 2005) and a method has also been optimized for two-dimensional (2D) chromatography of plasma (GC–GC; O'Hagan *et al.*, 2007). GC–MS has also been applied to profile serum from both human Huntington disease patients and a transgenic mouse model of Huntington disease (Underwood *et al.*, 2006). This application enabled clear discrimination of both control and disease samples in both the rat and human and a common metabolic signature of disease was seen. Furthermore, biomarkers that represented early onset of the disease were determined, and these included various markers of fatty acid breakdown, such as glycerol and malonate, and also aliphatic amino acids. A further example of the successful application of GC–MS involved profiling of plasma obtained from three strains of Zucker rats that are used as animal models of diabetes, namely Zucker lean, the lean/obese cross and the obese (fa/fa) strain (Major *et al.*, 2006). The application of pattern recognition techniques to discriminate the data from the individual strains revealed key differences between the plasma metabolic profiles of the three strains, with those of the Zucker lean and the lean/(fa) crosses being most similar to each other whilst differing from the (fa/fa) obese strain.

Capillary electrophoresis (CE) coupled to MS has also been explored as a possible technology for generation of metabolic profiles (Soga *et al.*, 2003). Metabolites are first separated by CE based on their charge and size and then selectively detected using MS; the technique is advantageous in that it requires only minimal amounts of sample and it is ideal for detection of polar metabolites. It has primarily been applied in the fields of microbial metabolomics (Mashego *et al.*, 2007) and plant metabolomics (Williams *et al.*, 2007). However, CE–MS is increasingly being applied to profile biofluids such as urine, and examples of its application in determination of the effects of nonsteroidal anti-inflammatory drug (NSAID; Suarez *et al.*, 2007) and acetaminophen (Ullsten *et al.*, 2006) exposure have been described. CE–MS has also been applied in a more targeted fashion to determine urinary nucleoside profiles (Wang *et al.*, 2007) and amino acid profiles (Mayboroda *et al.*, 2007). Recent reviews that comprehensively cover the applications of CE–MS in metabolic profiling include Servais *et al.* (2006) and Monton and Soga (2007).

18.3   Statistical Tools Applied to Model Metabonomic Data

The use of chemometric methods to analyze complex spectral data sets was perhaps the single most important development in the practical application of metabonomics and has defined the development and progression of the field ever since. The first studies that used pattern recognition to classify biofluid samples used a simple scoring system to describe the fluctuating levels of 18 endogenous metabolites in urine from rats which either were in a control group or had received a specific organ toxin which affected the liver, the testes, the renal cortex, or the renal medulla (Gartland *et al.*, 1990, 1991). These studies showed that samples corresponding to different organ toxins mapped into distinctly different regions of the pattern recognition diagrams, indicating that site-specific and severity information could be captured directly from the metabolic profile. It was also observed that capturing dynamic, temporal response to toxins provided enhanced classification of toxins and the first detailed investigation of the time course of metabolic urinary changes induced by two renal toxins involved Fisher 344 rats administered a single acute dose of the renal cortical toxin mercury(II) chloride and the renal papillary toxin 2-bromoethanamine (Holmes *et al.*, 1992a,b, 1998). The rat urine was collected for up to 9 days after dosing and was analyzed using ^{1}H NMR spectroscopy. The onset, progression, and recovery of the lesions were also followed using clinical chemistry and histopathology to provide a definitive classification of the toxic state relating to each urine sample and the geometry of the trajectory generated information relating to the mechanism and sequential targets of the toxin. The concentrations of 20 endogenous urinary metabolites were measured at eight time points after dosing and mapping methods were used to reduce the data dimensionality. These showed that the points on the plot could be related to the development of, and recovery from, the lesions.

These early pattern recognition studies on NMR spectral data involved a preselection of the metabolite signals of interest and involved detailed inspection and integration of these individual peaks. However, it is now recognized that measurement of a small set of signals does not harness the full potential of the spectral profile; although it may provide valuable information on dominant biochemical changes within the data, subtle spectral variation is often overlooked and it is difficult to envisage general effects as a function of both dose and time in a large cohort of samples with biological variability. Multivariate statistical methods provide an efficient means of analyzing and maximizing information recovery from complex NMR spectral data sets. Pattern recognition methods can be used to map the NMR spectra into a representative low-dimensional space such that any clustering of the samples based on similarities of biochemical profiles can be determined and the biochemical basis of the pattern elucidated.

The initial objective in metabonomics is thus to classify a spectral fingerprint based on identification of its inherent patterns of peaks and, second, to identify those spectral features responsible for the classification (according to physiological or pathological status), and this can be achieved using so-called supervised and unsupervised pattern recognition techniques.

The ^{1}H NMR spectra are preprocessed; typically, this involves calibration of the chemical shift scale using an internal reference standard, and phase and polynomial baseline correction are carried out. To prepare the ^{1}H NMR data for multivariate modeling, the spectra are often divided into small regions (along the chemical shift axis) whose areas are summed to provide an integral so that the intensities of peaks in such defined spectral regions are extracted, a process known as 'binning.' This results in a data matrix consisting of rows that reflect observations/samples and columns that represent variables; for example, the spectral integrals of defined 'bins' across the whole spectral width (Holmes *et al.*, 1994). Recent advances in chemometric approaches involve the utilization of full-resolution ^{1}H NMR data, where each data point in an acquired spectrum is used as a variable for modeling (Cloarec *et al.*, 2005a). This approach has many advantages; for example, the spectral structure is retained in the ensuing loadings coefficients, which enables the NMR user to identify metabolites with ease and it also avoids searching within 'bins' after data modeling to determine metabolites responsible for discrimination. Following the above preprocessing steps and the output of a data matrix consisting of samples and their associated variables (either as individual data points or summed segments of a spectrum), normalization is often applied to the rows (spectra). This adjusts spectral intensities so that concentration differences between samples are accounted for and the samples are more directly and reliably comparable. A commonly applied normalization method known as 'normalization to total area or constant sum' sets the total spectral area of each spectrum to unity so that the intensities of all data points are expressed relative to this. Scaling is the final preprocessing step typically applied to NMR spectral data prior to chemometric modeling; this is a column operation that aims to reduce the noise in the data and, hence, to improve model interpretability; for example, each column in a matrix can be set to have unit variance or a mean of zero (Craig *et al.*, 2006a).

18.3.1 Unsupervised Approaches: Principal Components Analysis

Principal components analysis (PCA; Jackson, 1991) has been widely used in metabonomic studies and is an unsupervised approach, in that it allows inherent clustering behavior of samples to be ascertained with no *a priori* knowledge of sample class membership. PCA reduces the dimensionality of a data set, as it allows multidimensional data vectors to be projected onto a hyperplane of lower dimensions (typically two or three) with this projection explaining as much of the variation as possible within the data. PC1, the first principal component (PC), is a linear combination of the original input variables and describes the largest variation in the data set; the second PC is orthogonal to PC1 and describes the next highest degree of variation in the data set. When two PCs have been defined, they constitute a plane; hence, projection of the observation vectors in the multidimensional space onto this plane enables the data to be visualized in a 2D map known as a 'scores plot.' This plot reveals any inherent clustering of groups of data, based purely on the closeness or similarity of their input coordinates. Thus, the analysis provides a convenient and objective means of reducing the complexity of the original data and of visualizing groups and classifying them. A loadings plot is used to interpret the scores plot, as it illustrates the spectral variables which contribute to the positioning of the samples on the scores plot and, hence, the variables which influence any observed separation in the data set.

18.3.2 Supervised Approaches: Partial Least Squares

Alternatively, in what are known as supervised methods, data sets can be modeled so that the class of separate samples (a validation set) can be predicted based on a series of mathematical models derived from the original data or training set. One widely used supervised method is partial least squares or projection to latent structures-discriminant analysis (PLS-DA; Wold, 1985) which relates a data matrix containing independent variables, such as spectral intensity values (an X matrix), to a matrix containing dependent variables for those samples (e.g. a Y matrix, measurements of response, such as toxicity scores or class membership). The use of PLS-DA and related technologies in metabonomics has recently been reviewed by Trygg *et al.* (2007).

Orthogonal PLS-DA (O-PLS-DA; Trygg and Wold, 2003; Eriksson *et al.*, 2004; Trygg *et al.*, 2007) is a relatively new approach that extends the traditional supervised algorithm of PLS by prefiltering classification-irrelevant, orthogonal variation from data. This method prefilters 'between-class' variation of interest and enables it to be modeled independently from 'within-class' variation; this improves the interpretability of spectral variation between classes.

A recent development (Cloarec *et al.*, 2005a) in multivariate modeling of spectral data applies O-PLS-DA to full-resolution NMR data to compute a model that is then 'back-scaled' by plotting the variables with a color scale that represents their respective correlation weights. This results in a loadings coefficient plot that is easily interpretable by a spectroscopist; it not only depicts spectral covariance, but also the variables of discriminatory importance are color-coded for ease of interpretation. O-PLS-DA modeling together with this 'back-scaling' step has been successfully applied to determine the metabolic consequences in multiple biofluid compartments of administration of the hepatotoxins galactosamine (galN) and allyl formate (Yap *et al.*, 2006; Coen *et al.*, 2007a), the renal cortical toxin mercury(II) chloride (Holmes *et al.*, 2006b), and to characterize administration of probiotic bacteria to a humanized mouse gut microflora model (Martin *et al.*, 2006). An example of such a 'back-scaled' O-PLS loadings plot that differentiates between serum from control and galN-treated animals is given in Figure 18.3. The increased and decreased levels of metabolites that discriminate between classes can clearly be identified, as the spectral structure is retained and the metabolites are color-coded with respect to discriminatory significance. The upper section of the loadings plots represents metabolites increased in the galN-treated class, whereas the lower part represents metabolites decreased in intensity.

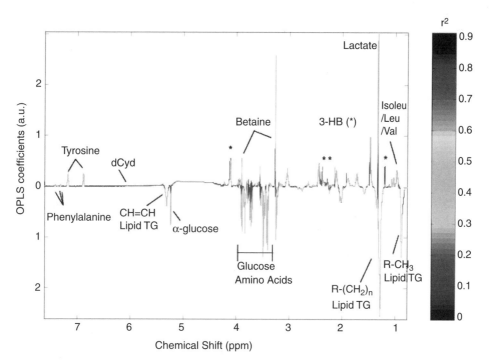

Figure 18.3 *O-PLS-DA coefficient loadings of full-resolution serum spin-echo NMR data revealing metabolites responsible for discrimination between controls and galN-treated animals (showing hepatotoxicity). The color scale represents correlation r^2 to the discriminant variable. The upper section of the loadings plots represents metabolites increased in the galN treated class, whereas the lower part represents metabolites decreased in intensity. Key: Lipid TG, lipid triglyceride; 3-HB, D-3-hydroxybutyrate; Isoleu, isoleucine; Leu, leucine; Val, valine; dCyd, 2'-deoxycytidine. Author's unpublished data.*

For example, increased levels of D-3-hydroxybutyrate, lactate, betaine, isoleucine, leucine, valine, phenylalanine, 2′-deoxycytidine, and tyrosine are seen in the galN-treated samples, together with decreased levels of glucose and lipid triglycerides.

Many other chemometric tools are used in analyzing metabonomic data, and comprehensive reviews are provided by Trygg *et al.* (2007) and Eriksson *et al.* (2004).

As chromatographic and linked mass spectrometric tools are increasingly being applied to metabonomic and metabolomic studies, data mining methods that can deal with the high information content of these data are under investigation. A typical UPLC–MS spectrum will contain tens of thousands of peaks, and this presents a new challenge in terms of biomarker information recovery, as robust methods of data handling which can reliably align and extract peaks are necessary tools prior to multivariate modeling of the data. Owing to the significantly higher sensitivity of UPLC–MS over NMR, a typical spectrum represents a much wider metabolic pool that reflects levels of both minor and major metabolites. The analytical challenge presented with analyses of UPLC–MS data is to extract the meaningful metabolic data from metabolic variation not related to class differentiation or from that caused by ionization or chromatographic retention variability. O-PLS-DA modeling has been applied to 'binned' UPLC–MS data from a study of hydrazine toxicity to reduce the data to a number of candidate biomarker peaks, i.e. those that are most significant in terms of class differentiation, these peaks then being further analyzed and structurally characterized (Crockford *et al.*, 2006a).

18.3.3   Statistical Correlation Spectroscopy: Statistical Correlation Spectroscopy and Statistical Heterospectroscopy

Recently developed statistical methods such as *statistical correlation spectroscopy* (STOCSY; Cloarec *et al.*, 2005b) have significantly enhanced information recovery from complex metabonomic data sets. STOCSY encompasses the computation of correlation statistics between the intensities of all computer points in a set of complex mixture spectra, thus generating connectivities between signals on molecules that vary in concentration between samples. This statistical method allows latent spectroscopic information of interest to be extracted from complex, highly overlapped spectra. STOCSY can be combined with supervised chemometric methods to provide linked information on those spectral features that best separate sample classes (Cloarec *et al.*, 2005a,b).

STOCSY has been used to derive assignment of biomarker metabolite NMR resonances in nephrotoxic states as a result of exposure to mercury chloride and provided an unbiased, sensitive approach to biomarker extraction and identification, and showed potential for generating novel pathway connectivities (Holmes *et al.*, 2006b). STOCSY has also been utilized for population-based identification of drug metabolites in human urine samples (Holmes *et al.*, 2007). ^{1}H NMR spectra were acquired for two groups of urine samples and the application of STOCSY to the data enabled rapid identification of the major and minor drug metabolites in common use in the population, in particular those from acetaminophen and ibuprofen. The work showed that statistical connectivities between drug metabolites could be established in routine 'high-throughput' NMR screening of human samples from participants who had self-administered pharmaceuticals. Hence, the STOCSY approach provides a powerful tool in considering interpopulation patterns of drug metabolism in epidemiological and pharmacogenetic studies. Furthermore, STOCSY has recently been applied to enhance information recovery from LC–NMR data sets (Cloarec *et al.*, 2007) and diffusion-edited NMR data sets (Smith *et al.*, 2007) arising from complex biological mixtures. The STOCSY approach is generic and can be applied to both 1D and 2D NMR spectra and to homo- or hetero-nuclear spectroscopic data to aid structural elucidation and determine pathway relationships in a given spectral sample set. Heterospectroscopic-STOCSY (HET-STOCSY) encompasses the statistical correlation of two different types of experimental data; for example, from heteronuclear NMR experiments or any given parallel combination of experimental data. The successful application of HET-STOCSY to aid metabolic biomarker assignment in a metabonomic study of galN-induced hepatotoxicity has recently been demonstrated for ^{1}H–^{31}P MAS NMR spectra of intact liver (Coen *et al.*, 2007b).

Furthermore, the recent development of Statistical HeterospectroscopY (SHY) shows the power of computation of covariance matrices for successful interrogation of multispectroscopic data sets collected in series or parallel, such as those from NMR and UPLC–MS platforms. The potential of SHY has been demonstrated for a metabonomic data set representing hydrazine toxicity (Crockford *et al.*, 2006b) where direct cross-correlation of chemical shifts (NMR) and *m/z* data (MS) has provided structural and metabolic pathway activity information. The application of SHY to MS and NMR data sets allows improved molecular biomarker identification capabilities, as not only structural information is obtained, but also higher level biological information on metabolic pathway activity and connectivities. This information is found within different levels of the NMR to MS correlation and anticorrelation matrices, and it is notable that the SHY approach is equally applicable to any two independent spectroscopic data sets.

18.4 Integrated Metabonomic and 'Multi-omic' Studies

The ability to compare biofluid and tissue NMR spectra can provide further insight into mechanisms of toxicity, as metabolic changes that occur in the target organ of toxicity can be identified. This approach provides information on biological changes within different biological matrices, information that, when considered as a whole at the biological pathway level, often provides enhanced insight on a systems response, as changes in one compartment may be reflected in another or may highlight mechanistic linkages. The combined application of solution-state and MAS NMR has been termed 'integrated metabonomics,' and a recent application involves elucidating the metabolic consequences of treatment with allyl formate, a model hepatotoxin (Yap *et al.*, 2006). Conventional solution-state NMR spectra together with MAS NMR were used to obtain metabolic profiles of urine, plasma, and liver which presented a broad systemic view of the effects of hepatic toxicity induced by allyl formate. In addition, conventional plasma clinical chemistry and histopathological assessments of liver damage were carried out and the results integrated with the metabonomic findings. In this study, inter-animal variation in response was reflected in varying degrees of liver damage, as determined from clinical chemistry measurements and histopathology; and, furthermore, this variation was also reflected in the metabolic profiles. The allyl formate-induced changes included decreased levels of liver glycogen and glucose, which suggested increased glucose utilization as a result of mitochondrial impairment and depletion of hepatic adenosine triphosphate (ATP). Increases in plasma tyrosine were also identified, and this result suggested that the hepatic ATP depletion had led to impairment in protein synthesis. Allyl formate also caused hepatic lipidosis, as evidenced from increased lipids in the MAS NMR spectra of treated animals, and the increase in liver lipids correlated with a concomitant decrease in plasma lipids. This finding suggested an impairment of hepatic-plasma lipid transport and highlights the strength of an integrated metabonomic approach in providing complementary metabolic information reflective of a systems response to a toxin.

An additional example of the utility of this integrated approach was highlighted in the study of the model hepatotoxin ethionine, and the integration of multilevel results shed new light on the mechanism of toxicity and implicated disruption of the gamma-glutamyl cycle and up-regulation of nicotinamide adenine dinucleotide (NAD) catabolism (Skordi *et al.*, 2007). MAS NMR spectra revealed increased levels of lipids, which was consistent with the histopathological observation of steatosis. ATP depletion and alteration of energy metabolism was seen through decreased urinary levels of tricarboxylic acid cycle intermediates and glucose, increased urinary levels of ketone bodies, and depleted levels of hepatic glucose and glycogen. The urinary changes identified included elevation of guanidinoacetate, which suggested impaired methylation reactions, increased 5-oxoproline and glycine that suggested disruption of the gamma-glutamyl cycle, and increased nicotinuric acid, which suggested an increase in NAD catabolism. Toxic effects on the gut microbiota were implicated, as reduced urinary levels of the microbial metabolites 3-/4-hydroxyphenyl propionic acid, dimethylamine, and tryptamine were observed. In addition, urinary markers for secondary kidney toxicity were evident in increased urinary levels of lactic acid, amino acids, and glucose at the later time-points, and this finding was supported by histopathological observation of tubular damage.

A similar approach was also applied to a compound in development (MrkA) that had been shown to induce hepatotoxicity in several animal species (Mortishire-Smith *et al.*, 2004). ^{1}H NMR spectra were acquired on urine and liver tissue samples and pattern recognition analysis of the data enabled the metabolic effects of administration of MrkA to be determined. These included a urinary depletion in tricarboxylic acid cycle intermediates and the appearance of medium-chain dicarboxylic acids. MAS NMR data revealed elevated triglyceride levels that were correlated with dicarboxylic aciduria and suggested defective metabolism of fatty acids. This metabonomic result was confirmed by subsequent *in vitro* experiments that showed that MrkA impaired fatty acid metabolism, which highlighted the potential of an integrated metabonomics approach in defining an unknown mechanism of drug-induced toxicity. This study is an example of the successful testing and validation of a metabonomic-generated hypothesis to determine a mechanism of toxicity and highlights the potential of metabonomics in the field of pharmaceutical research and development.

Metabonomics offers a complementary approach to 'omics' platforms such as genomics, transcriptomics, and proteomics, as it provides a useful connection between the 'omics' platforms and clinical chemistry or tissue histology, as real-world end points are observed and studied. Metabonomics will become increasingly important in connecting molecular events at the gene and protein level to those occurring at the macrosystem level. An example of a recent 'multi-omic' study involved the combined transcriptomic, proteomic, and metabonomic analysis of methapyrilene hepatotoxicity in the rat (Craig *et al.*, 2006b). Methapyrilene is a histamine antagonist that causes periportal liver necrosis; as the mechanism of toxicity is not fully understood, an integrated systems approach was utilized in an attempt to shed light on the mechanistic basis for the toxic outcome. Common perturbations across all three biomolecular levels representing changes in gene expression, protein expression, and global metabolic profiles were found that were related to oxidative stress responses and to changes in metabolic pathways involving lipid, glucose and choline metabolism and the urea cycle. Individual changes in each 'omic' data set were also analyzed and interpreted; for instance, animals treated with the toxic, high dose of methapyrilene were seen to excrete elevated levels of N-methylnicotinamide in the urine as detected by NMR spectroscopy, whereas associated changes in the protein or genes of the nicotinamide pathway were not seen. However, strong induction of thioesterase gene expression was observed; this, together with urinary excretion of N-methylnicotinamide, has been reported independently in the literature as a biomarker of peroxisome proliferation (Lanni *et al.*, 2002; Ringeissen *et al.*, 2003). Furthermore, two mitochondrial proteins (pyruvate carboxylase and carbamoyl phosphate synthase) were modified, which suggested covalent binding of an active methapyrilene metabolite to mitochondrial protein which had previously been implicated in the toxic mechanism. This paper presented an early framework approach to the nontrivial task of integrating complex data from parallel 'omic' platforms, each representing different levels of biomolecular organization. Such an approach has also been exemplified in the application of transcriptomics and metabonomics to acetaminophen-induced hepatotoxicity in the rat (Coen *et al.*, 2004), where the 'omic' data sets were co-interpreted in terms of common metabolic pathways. In this study, the metabolic observations were consistent with the altered levels of gene expression relating to lipid and energy metabolism in the liver which both preceded and were concurrent with the metabolic perturbations. An integrated metabonomic approach was also utilized to study the effects of peroxisome proliferator-activated receptor (PPAR) ligands on urine and plasma NMR- and HPLC-based fingerprints (Ringeissen *et al.*, 2003; Connor *et al.*, 2004) and two potential biomarkers of peroxisome proliferation in the rat were described: N-methylnicotinamide and N-methyl-4-pyridone-3-carboxamide. Both of these metabolites are end products of the tryptophan–NAD^+ pathway, suggesting alteration of this pathway as a result of peroxisome proliferation. Following this metabonomic study and the generation of a mechanistic hypothesis, a genomic study was conducted which confirmed that the relevant genes encoding two key enzymes in the NAD^+ pathway: aminocarboxymuconate-semialdehyde decarboxylase (EC 4.1.1.45) and quinolinate phosphoribosyltransferase (EC 2.4.2.19) were down-regulated (Delaney *et al.*, 2005). This example presents the power of the nontargeted metabonomic approach in generating mechanistic hypotheses that can be tested via integration of 'omic' platforms.

The statistical correlation approach that has been exemplified in terms of metabonomic data from NMR and MS platforms has also been extended to multivariate data sets derived from different biomolecular levels, such as proteomics and genomics, and has been found to enhance and aid interpretation of data from these 'multi-omic' platforms. Recent examples of statistically integrated 'multi-omic' studies include the successful integration of genome–phenotype data with metabonomic data representing a rat type II diabetes model (Dumas *et al.*, 2007). In brief, this study encompassed using plasma NMR spectroscopic metabolic fingerprints to map quantitative trait loci (QTLs) in a cross between diabetic and control rats. Candidate metabolites were proposed for the most significant QTLs, one of which included a gut microbial metabolite (benzoate) that could be explained by deletion of a uridine diphosphate glucuronosyltransferase. This data projection method allowed integration of biochemical information at a systems level and will prove useful for enhanced mechanistic understanding of *in vivo* model systems and for 'trans-omic' metabolic biomarker recovery. Related statistical cross-projection methods have also been applied to link proteomic with metabonomic data (Rantalaninen *et al.*, 2006) on data representing a human tumor xenograft mouse model of prostate cancer. Blood plasma from mice implanted with prostate tumors was profiled using both NMR spectroscopy and 2D-difference gel electrophoresis technologies. The data were integrated using O-PLS-DA modeling algorithms, and multiple correlations between metabolites and proteins were found, including associations between a serotransferrin precursor, and both tyrosine and D-3-hydroxybutyrate and reduced levels of tyrosine were correlated with increased levels of gelsolin.

18.5   COMET: A Consortium Project Using NMR-driven Metabonomics

18.5.1   Background to the Project

The development of major new paradigms for drug screening requires the engagement and resources of the pharmaceutical industry coupled to academic research groups. Such initiatives involve experimental studies, database construction, and mathematical modeling. In 2000, a pharmaceutical industry/university consortium was formed to investigate the utility of metabonomic approaches in the evaluation of xenobiotic toxicity. The main aim of the consortium was to use ^{1}H NMR spectroscopy of biofluids (and, in selected cases, tissues), and computer-based pattern recognition tools, to classify biofluids in terms of known pathological effects caused by the administration of substances causing toxic effects. Furthermore, the aim was to assess and develop methodologies to generate a metabonomic database and to build a predictive expert system for target organ toxicity. The academic part of the COnsortium for MEtabonomic Toxicology (COMET; Lindon *et al.*, 2003, 2005) was hosted at Imperial College London, UK. The original partners comprised DuPont Pharmaceuticals, Warner–Lambert, and GD Searle and Co. pharmaceutical companies, but by the time the project was underway it was sponsored by six pharmaceutical companies, namely Bristol–Myers–Squibb, Eli Lilly & Co., Hoffmann–La Roche, NovoNordisk, Pfizer Inc. and The Pharmacia Corporation (later taken over by Pfizer) with the instrument manufacturer, Bruker, also providing expertise. The COMET target compounds chosen to build this database represented an extensive area of metabolic space; hence, structures and activities were diverse, but there was an emphasis on analysis of hepato- and nephro toxins. However, other physiological stressors, such as partial hepatectomy, food restriction, and water deprivation, were included to provide positive control data sets. In addition, toxins that targeted other organs, such as the pancreas, testes, and bladder, together with multi-organ toxins were included. Furthermore, a subset of studies was carried out in mice to provide the opportunity to compare metabolic profiles reflective of toxicity across species. On completion, the COMET studies encompassed 147, 7-day, low- and high-dose toxicological and physiological studies. These studies were carried out by the sponsoring companies and typically 10 animals per group were randomly assigned to control, low-dose or high-dose treatment groups. The high-dose group, defined as a dose which exerts a clear toxic effect following a single administration, was chosen based on literature or in-house range-finding data. An acceptable low dose was one which generated a threshold response. The gold standard for the assessment of toxic response was histopathology and supporting clinical chemistry parameters. In general, these studies were acute toxicity studies employing mostly a single dose (though in some cases multiple doses were necessary to induce the toxicity). Urine samples were collected over a period of 8 days, which included a 1-day pre-dose collection, and blood was sampled at 24, 48, and 168 h post-dosing. A full list of the 147 treatments and the target organs has been published (Lindon *et al.*, 2005). A range of 1D 600 MHz ^{1}H NMR spectra were acquired for all biofluid samples (approximately 36 000), which typically involved sample preparation into 96-well plates, robotic sample-handling technologies, and the use of flow-injection NMR probes.

18.5.2 Reproducibility and Sensitivity Studies

The analytical and biological variation that might arise through the use of metabonomics was evaluated at an early stage of the COMET project and a high degree of robustness demonstrated. In the initial stages of the project, a detailed examination of the analytical and biological quality control was undertaken, with each company dosing the same frequently used model liver toxin, i.e. hydrazine, to rats using an agreed, standard protocol. This was extended to include a reproducibility assessment exercise that involved performing sample preparation and NMR data acquisition at two sites (one using a 500 MHz system and the other using a 600 MHz system) using two identical (split) sets of urine samples from an 8-day acute study of hydrazine toxicity in the rat (Keun *et al.*, 2002a,b). Despite the difference in spectrometer operating frequency, both data sets were found to be extremely similar when analyzed using PCA and gave near-identical descriptions of the metabolic responses to hydrazine treatment. These studies showed that the NMR-based analytical variation between measurement sites was very low, with a multivariate coefficient of variation below 2 %, and that those spectra that displayed intersite differences were detected as outliers in the PCA models. The hydrazine samples were further analyzed by ^{13}C NMR spectroscopy using cryogenic probe technology (Keun *et al.*, 2002b). Information-rich ^{13}C NMR spectra of rat urine were obtained in short acquisition times using a cryogenic probe, as it significantly compensates for the inherently low sensitivity of natural-abundance ^{13}C NMR spectroscopy. The significantly greater dispersion of the ^{13}C chemical shift, compared with ^{1}H, was found to present distinct advantages for metabolite identification and resonance assignment in biofluid NMR spectroscopy. The biochemical consequences of exposure to hydrazine were determined following PCA analysis of the ^{13}C NMR data, and this provided complementary metabolic information to the ^{1}H NMR data.

Multivariate statistical batch processing was also applied to model the hydrazine NMR spectral data and to establish the time-dependent metabolic variation in response to the toxicity (Antti *et al.*, 2002). Each rat was treated as an individual batch represented by a series of urine samples collected through time. A model that defined the mean metabolic status of the control samples across time was used to compare with the hydrazine-treated group, and all hydrazine-treated animals were seen to deviate from this control space. The urinary manifestations of hepatic toxicity included elevated levels of taurine, creatine, 2-aminoadipate, citrulline, and β-alanine, together with depleted levels of citrate, succinate and hippurate. As the hydrazine study was replicated at seven sites in total, the NMR data from each individual site was modeled using the statistical batch-processing approach and the reproducibility of profiles was found to be very high. The time points at which there were maximal metabolic perturbations were easily determined from the batch-processing models together with the overall profile of response. This means of modeling temporal variation inherent within metabonomic data sets may provide a powerful tool for targeting the modeling of time-dependent metabolic responses and inter-animal differences not only in response to a toxin, but also in control populations.

18.5.3   Novel Modeling Approaches and Cross-Species Comparison

Given the availability of thousands of control urine samples for both rat and mouse, the project provided an excellent opportunity to develop an understanding of the biochemical variation in these fluids by building statically based models of control urine. These models were continuously evolved as more samples became available. The control model success-fully detected outlier samples known to be deviating from normal control behavior based on clinical chemistry measurements, and also successfully detected early time-point devia-tions in animals for which a full time course was not available. Thus, valuable information was obtained on the normal metabolic variation in the rat and mouse strains used, and on the metabolic signatures that could be associated with diurnal effects and adaptation to metabolic cages.

Since various physiological factors, such as diet, state of health, age, diurnal cycles, stress, genetic drift, and strain differences, affect the metabolic composition of biofluids, part of the COMET research effort endeavored to create statistically reliable decision tools for distinguishing between physiological and pathological responses in animal models. Hence, it was deemed important to give as much consideration to optimum design of experiments (DoE) as to the subsequent data analysis. Statistical experimental design combined with partial least squares regression was proposed as an efficient approach for undertaking metabonomic studies and for analysis of the results (Antti *et al.*, 2004). The hydrazine serum NMR spectral data and associated clinical chemistry parameters collected as part of COMET were subjected to partial least squares regression to create a statistical means for screening of biomarkers in the two combined data blocks (NMR and clinical chemistry data). Partial least squares analysis was also used to reveal the correlation pattern between the two blocks of data as well as the variation within the two blocks according to dose, time, and the interaction between dose and time. The significant changes caused by dose and time could be interpreted by independent means, which suggested the DoE and partial least squares approach was able to separate changes related to time and to dose response and those that are unrelated. This combined modeling of the *J*-resolved NMR spectra and the serum clinical chemistry parameters in the same model enabled the interpretation of interblock variable correlations and covariations based on statistical significance providing greater information recovery, as well as a more detailed biochemical explanation to changes occurring due to toxic insult together with nondose-related events.

Furthermore, the COMET consortium successfully applied a host of spectral edit-ing methods, namely the Carr–Purcell–Meiboom–Gill spin-echo, diffusion-edited spec-troscopy, and skyline projections of 2D *J*-resolved spectra to profile intact liver tissue by high-resolution MAS NMR from controls and animals treated with hydrazine. The spectrally edited data were modeled using PCA and this enabled the effects of hydrazine toxicity on rat liver biochemistry (Wang *et al.*, 2003) to be determined, and included depleted levels of liver glycogen, choline, taurine, trimethylamine-N-oxide, and glucose and elevated levels of lipids and alanine. A novel metabolite, a highly unsaturated ω-3-type fatty acid, was seen in the MAS NMR data of hydrazine-treated animals for the first time. Furthermore, the use of spectral editing techniques enabled subtle fluctuations in low-intensity metabolites to be amplified and identified by PCA.

Metabonomics was also applied to model the interspecies variation between rats and mice following administration of a low and high dose of hydrazine (Bollard *et al.*, 2005) by oral gavage to male Sprague-Dawley rats (30 mg kg^{-1} and 90 mg kg^{-1}) and male B6C3F mice (100 mg kg^{-1} and 250 mg kg^{-1}). In each species, the high dose was selected to produce the major histopathological effect, namely hepatocellular lipid accumulation. The metabolites of hydrazine, namely diacetylhydrazine and 1,4,5,6-tetrahydro-6-oxo-3-pyridazine carboxylic acid (THOPC), were detected in both rat and mouse urine; however, monoacetyl hydrazine was detected only in urine samples from the rat. The absence of monoacetyl hydrazine in the urine of the mouse was attributed to a higher activity of N-acetyl transferases in the mouse than in the rat. The differential metabolic effects observed in this study between the two species included elevated urinary β-alanine, 3-D-hydroxybutyrate, citrulline, N-acetylcitrulline, and reduced trimethylamine-N-oxide excretion unique to the rat. Metabolic PC trajectories highlighted the greater degree of toxic response in the rat. A number of novel modeling strategies, such as a method to test for homothetic geometry, called scaled-to-maximum, aligned, and reduced trajectories (SMART) analysis, were developed as part of the COMET research effort (Keun *et al.*, 2004). This method facilitated scaling of the differences between metabolic starting positions and varying magnitudes of effect and, hence, enabled multivariate response similarities to be visualized from two or more sets of metabonomic measurements. Metabolic PC trajectories and SMART analysis were successfully applied to facilitate comparison of the response geometries between the rat and mouse exposed to hydrazine. This approach facilitated comparison of the response geometries between the rat and mouse. Mice followed 'biphasic' open PC trajectories, with incomplete recovery 7 days after dosing, whereas rats followed closed 'hairpin' time profiles, indicating functional reversibility. The greater magnitude of metabolic effects observed in the rat was supported by the more pronounced effect on liver pathology in the rat than with the mouse. The application of SMART analysis was further extended to encompass modeling interlaboratory variation in hydrazine response, CCl$_4$ dose–response relationships, and interspecies comparison of bromobenzene toxicity. For each of these examples, the homothetic trajectory modeling facilitated the amalgamation and comparison of the metabonomic data sets and improved the accuracy and precision of classification models based on metabolic profile data. It was suggested that such methods may be equally applicable for modeling and interpretation of multilaboratory genomic and proteomic data, as these approaches and the data generated are also influenced by interlaboratory variations, physiological variation, dose–response relationships, and interspecies differences.

18.5.4   Major Conclusions: The COMET Expert System

One of the main driving aims of the COMET consortium was to utilize the large NMR spectral database of model toxins for detection and prediction of target organ of toxicity of novel compounds. A novel modeling method developed to classify large COMET data sets utilized a density superposition approach: CLassification Of Unknowns by Density

Superposition (CLOUDS), which is a nonneural implementation of a classification technique developed from probabilistic neural networks (Ebbels *et al.*, 2003). To test this approach it was first applied to NMR spectra of rat urine from 19 different treatment groups and the data were modeled according to organ of effect with >90 % of the test samples classified as belonging to the correct group. The 19 model toxins could be classified according to the organ of toxic effect and encompassed controls, liver toxins, kidney toxins, and a group classified as other treatments, which included pancreatic and multi-organ toxins. These metabonomic data representing 19 toxins were also modeled using PCA, hierarchical cluster analysis (HCA) and *k*-nearest-neighbor (kNN) classification to reveal dose- and time-related effects (Beckonert *et al.*, 2003). PCA and HCA provided valuable overviews of the data, highlighting characteristic metabolic perturbations representative of the organ of effect and separating metabolic profiles representative of subgroups of toxins. kNN analysis of the multivariate data successfully predicted all of the different toxin classes with >85 % success rate (training/test). This work highlighted the power and reliability of the COMET approach in generating metabonomic databases representative of a wide range of toxins that could be used for predictive purposes.

The CLOUDS modeling (Ebbels *et al.*, 2003) approach was extended to construct a measure of similarity between multidimensional probability densities that represented the metabolic response to each treatment. The COMET database was used to develop this expert model for prediction and classification of drug toxicity, the details for which have recently been published (Ebbels *et al.*, 2007). It presents a significant step forward in the ability to capture and compare the characteristics of collections of metabolic trajectories representing diverse responses to nephro- and hepato toxins. The measure of similarity can be used as an estimate of the 'toxin likeness' between treatments, allowing the response to new treatments to be explained and predicted in terms of those already in the database. A subset of the COMET database representing 80 hepato- and nephro toxins was used to build this modeling system which was capable of differentiating NMR urinary metabolic profiles from controls and treated animals. The first step in the expert system involved constructing a model which captured the urinary variation of physiologically normal animals. This enabled abnormal samples to be detected at a very early stage, and those found to be abnormal were removed from the control model. Toxin-treated samples were then compared with this model and those deemed to be abnormal were then scaled and compared via the CLOUDS methodology to determine unique toxicity profiles for each toxin. A measure of similarity between the CLOUDS toxin profiles was calculated and this 'similarity matrix' could then be used to predict the 'toxin likeness' or toxic outcomes of unknown treatments. This similarity matrix is represented in Figure 18.4 for 62 treatments, where the studies are sorted by hierarchical clustering and clusters of interest are highlighted and labeled (blocks 1–5). The similarity matrix shows that studies cluster according to the treatment and target organ of toxicity; for example, blocks 1 and 2 represent the replicate hydrazine ($n = 7$) and acetaminophen ($n = 2$) studies that were carried out. Blocks 4 and 5 in this similarity matrix represent renal toxins: block 4 identifies the treatments that induced renal tubular toxicity, i.e. cisplatin, mercury(II) chloride, aurothiomalate, and hexachlorobutadiene; block 5 represents two renal papillary toxins, i.e. 2-bromoethanamine and 2-chloroethanamine (further details in Ebbels *et al.* (2007)). The sensitivity and specificity of the expert system for liver toxins was 67 % and 77 % respectively and for kidney toxins it was 41 % and 100 %

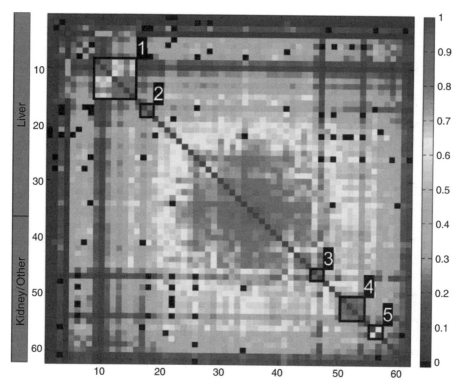

Figure 18.4 *COMET-2 expert system similarity matrix for 62 treatments sorted by hierarchical clustering; numbers 1–5 represent similarity blocks and are discussed in the text. (Ebbels, TM., Keun, HC., Beckonert, OP., Bollard, ME., Lindon, JC., Holmes, E. and Nicholson, JK. J. Proteome Res. **6**, 2007, 4407–22. Reprinted with permission of the American Chemical Society.)*

respectively, whereas the predictive ability of the system had an error rate of 8 %, where performance was tested using two blind studies and a leave-one-out cross-validation approach. This expert system uses the systems-wide window provided by metabonomic data to characterize toxic failure successfully. The overall toxic response is driven by specific mechanisms and sites of action, but the metabolic fingerprint may be derived from multiple biological sources and influences. This approach has provided the largest validation to date of the value of metabonomics in preclinical toxicology assessment and confirmed that the methodology offers practical utility and reliability for rapid *in vivo* drug toxicity screening.

18.6   The COMET-2 Project

The follow-up second COnsortium on MEtabonomic Toxicology (COMET-2) project is now under way. It extends this linkage between pharmaceutical companies (Bristol–Myers–Squibb, Sanofi–Aventis, Servier and Pfizer), a major instrument manufacturer (The Waters Corporation) and academia and aims to utilize metabonomics as a 'top-down' systems biology driver to direct research and experiments in the determination of mechanisms of toxicity, the assessment of risk factors, and an improved understanding of biological variability (Lindon *et al.*, 2004; Nicholson *et al.*, 2004). Hence, COMET-2 aims to enhance significantly the knowledge gained from the original COMET project by detailed testing of metabonomic hypotheses using NMR, LC–MS, and appropriate *in vitro* and *in vivo* labeling studies and validation of results via follow-on biochemical studies with the ultimate aim of determination of mechanisms of action of a wide range of drugs.

The initial efforts have concentrated on understanding model systems, e.g. galN hepatotoxicity, which has been utilized as a model hepatotoxin for many years. Despite extensive investigation, the mechanism of action and metabolism of galN is unclear and it also elicits hypervariable responses.

A recently published COMET-2 study investigated the protective ability of glycine in a galN-induced hepatotoxic model (Coen *et al.*, 2007a). ^{1}H NMR spectroscopy was used to investigate the metabolic effects of the hepatotoxin galN and the mechanism by which glycine protected against such toxicity. The protocol for this study included the acclimatization of rats to a 0 or 5 % glycine diet for 6 days and the subsequent administration of vehicle, galN (500 mg kg^{-1}), glycine (5 % via the diet), or both galN and glycine. In this study, urine was collected over 12 days prior to administration of galN and for 24 h thereafter. Serum and liver tissue were sampled on termination, 24 h post-dosing. The metabolic profiles of biofluids and tissues were determined using 600 MHz ^{1}H NMR and MAS NMR spectroscopy.

Representative 600 MHz ^{1}H NMR spectra of urine, serum and liver from control, glycine-treated, galN-treated, and combined galN- and glycine-treated animals (for each class, the urine, serum, and liver spectra represent samples from the same animal) are shown in Figures 18.5, 18.6 and 18.7 respectively. These figures demonstrate the level of metabolic information obtained in a typical NMR-based metabonomic study and it is evident that a wide range of metabolites can be assigned in each spectral matrix to provide complementary information on a global systems response to both galN-toxicity and glycine protection.

O-PLS-DA was applied to model the spectral data and enabled the hepatic, urinary, and serum metabolites that discriminated between control and treated animals to be determined. The urinary response to galN was found to involve elevated levels of galN, *N*-acetylglucosamine (GlcNAc) and urocanic acid, together with decreased levels of 2-oxoglutarate, *N*-methylnicotinamide, 2′-deoxycytidine and phenylacetylglycine. These metabolic differences can be seen on inspection of the representative urine NMR spectra presented in Figure 18.5. The profile representing co-treatment with galN and glycine was much closer to the control profile than that representing galN treatment. It was interesting to note that levels of GlcNAc were not significant in discriminating controls from the co-treated galN and glycine samples. In addition, lower levels of galN were present in the urine, and urocanic acid was not elevated and 2-oxoglutarate, D-3-hydroxybutyrate and phenyl-acetylglycine were not depleted. The NMR-visible level of GlcNAc correlated strongly with the degree of liver damage ascertained from the liver histopatholgy score, which suggested that it played a significant role in the hepatotoxicity of galN. Histopathological data and clinical chemistry measurements confirmed the protective effect of glycine.

Representative serum spectra for each class are given in Figure 18.6, and O-PLS-DA modeling of the full data sets revealed (Figure 18.3) that the metabolites responsible

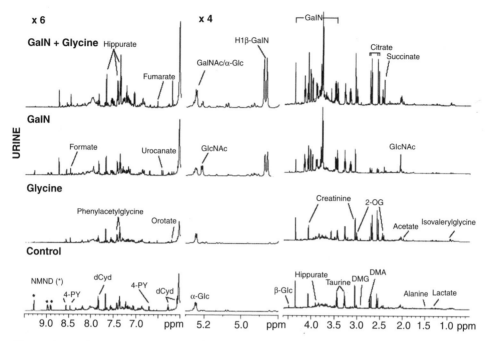

Figure 18.5 *Representative urine ¹H NMR spectra of control, galN-treated, galN- and glycine-treated animals 24 h post-dosing. Key: GalN, galactosamine; 2-OG, 2-oxoglutarate; DMG, dimethylglycine; TMAO, trimethylamine-N-oxide; DMA, dimethylamine; 4-PY, N-methyl-4-pyridone-5-carboxamide; NMND, N-methyl-nicotinamide; GlcNAc, N-acetylglucosamine; GalNAc, N-acetylgalactosamine; Glc, glucose; dCyd, 2'-deoxycytidine; Ile/Leu/Val, isoleucine/leucine/valine; 3-HB, D-3-hydroxybutyrate; Cho/PCho, choline/phosphocholine; UDP–N-acetylglucosamine (UDP–GlcNAc) and UDP–N-acetylgalactosamine (UDP–GalNAc). (Coen, M., Hong, YS., Clayton, TA., Rohde, CM., Pearce, JT., Reily, MD, Robertson, DG., Holmes, E., Lindon, JC. and Nicholson, JK. J. Proteome Res. **6**, 2007, 2711–9. Reprinted with permission of the American Chemical Society.)*

for differentiation between control and galN-treated animals were elevated levels of isoleucine, leucine, valine, tyrosine, phenylalanine, 2'-deoxycytidine, lactate, betaine, D-3-hydroxybutyrate together with reduced levels of glucose and lipids in the galN-treated samples. Conversely, the profile for treatment with galN and glycine remained much closer to that of a typical control profile, in that the changes identified were simply elevated levels of glycine and D-3-hydroxybutyrate. Elevated levels of glycine were apparent in the serum spectra following administration of glycine and no other changes were observed. The increase in tyrosine in the serum of galN-treated animals was dramatic (see Figure 18.6) and has previously been identified following administration of galN and attributed to reduced hepatic tyrosine aminotransferase activity (Clayton *et al.*, 2007).

The analysis of liver tissue by MAS NMR added a further key dimension to this study, as it provided information on liver-specific metabolic changes resulting from the treatments given. Representative MAS NMR spectral profiles of liver for each sample class are given in Figure 18.7 and marked changes in a host of metabolites are evident relative to

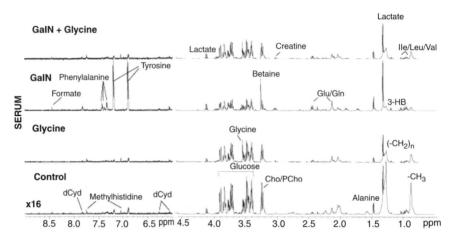

Figure 18.6 *Representative ¹H NMR serum spectra of control, galN-treated, galN- and glycine-treated animals 24 h post-dosing. See Figure 18.5 for key. (Coen M., Hong YS., Clayton TA., Rohde CM., Pearce JT., Reily MD., Robertson DG., Holmes E., Lindon JC., Nicholson JK. J. Proteome Res. **6**, 2007, 2711–9. Reprinted with permission of the American Chemical Society.)*

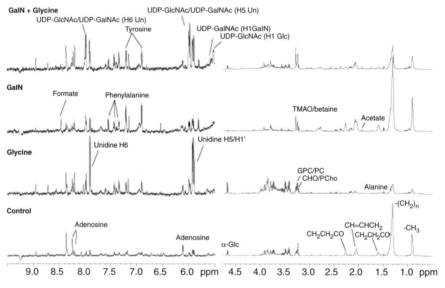

Figure 18.7 *Representative ¹H MAS NMR liver tissue spectra of control, galN-treated, galN- and glycine-treated animals 24 h post-dosing. See Figure 18.5 for key. (Coen M., Hong YS., Clayton TA., Rohde CM., Pearce JT., Reily MD., Robertson DG., Holmes E., Lindon JC., Nicholson JK. J. Proteome Res. **6**, 2007, 2711–9. Reprinted with permission of the American Chemical Society.)*

control levels following treatment with galN. The hepatic metabolic consequences of galN dosing are increased levels of lipid triglycerides and UDP–glcNAc/UDP–galNAc together with reduced levels of glucose and choline/phosphocholine. Interestingly, uridine was very clearly increased following treatment with glycine alone and the UDP–amino sugars (UDP–glcNAc and UDP–galNAc) were dramatically elevated following co-treatment with glycine and galN.

The significant increase in hepatic levels of uridine, UDP–glucose, and UDP–galactose following administration of glycine suggests that the protective role of glycine against galN toxicity involves changes in the uridine nucleotide pool. Thus, this work resulted in generation of a novel hypothesis: that administration of glycine increases the hepatic uridine nucleotide pool, which counteracts the galN-induced depletion of these pools and facilitates complete metabolism of galN. These novel data highlight the applicability of NMR-based metabonomics in elucidating multicompartmental metabolic consequences of toxicity and toxic salvage. This work demonstrated the utility of the new technologies in uncovering new metabolic facets of the mechanisms of supposedly 'well-understood' toxins. Publications are currently in preparation from recent COMET-2 studies that have involved the application of NMR- and MS-based metabonomics to investigate idiosyncratic response to galN.

Another ongoing COMET-2 area of investigation relates to bromoethanamine (BEA) and understanding the development of renal papillary necrosis (RPN) using chemical models. NSAIDs are the most widely recognized class of compounds known to cause this toxicity in rodents and in humans. However, it is difficult to study renal medullary function with noninvasive methods; hence, there are no sensitive and selective biomarkers for evaluating RPN. Furthermore, it is planned that clinical samples of RPN will be analyzed so that biomarkers relevant to the clinical setting can be differentiated from those of relevance to general physiological function.

One other aim of COMET-2 is to evaluate the strengths and weaknesses of the applicability of the LC–MS platform in metabonomic studies and more specifically in toxicological applications for drug safety evaluation. This will be achieved via a multisite reproducibility and validation study of samples from a COMET-2 toxicological study and also from samples that will be 'spiked' with known concentrations of a range of metabolites. The results of this study will enable the reliability and robustness of the LC–MS platform to be compared with the more traditional NMR platform.

18.7 Prospects for the Future

18.7.1 Pharmacometabonomics: Prediction of Responses from Pre-dose Metabolite Profiles

Metabonomics has recently been applied to modeling of pre-dose spectral fingerprints in order to predict the metabolism and toxic outcome of a drug and has been termed 'pharmacometabonomics.' This approach presents a means of understanding intersubject variability, as the metabolic fingerprint of an individual is sensitive to both genetic and environmental factors. Pharmacogenomics, on the other hand, involves determination of the genetic fingerprint of an individual which is then related to intersubject differences in the metabolism and toxicity of drugs. Pharmacometabonomics is defined as 'the prediction of the outcome (for example, efficacy or toxicity) of a drug or xenobiotic intervention in an individual based on a mathematical model of pre-intervention metabolite signatures' (Clayton *et al.*, 2006). It is suggested that pharmacometabonomics will play a major role in the development of personalized healthcare, as it may be possible to predict therapeutic outcome for individuals together with prediction of pharmaceuticals that will achieve maximal efficacy and those that will result in adverse effects. The pharmacometabonomic idea was conceived when a small-scale metabonomic study of galN hepatotoxicity found that, in terms of post-dose outcome, animals could be classified as 'responders' or 'nonresponders' as the intersubject response was so markedly varied. The pre-dose profiles of these animals were then studied to determine whether this intersubject variability in response to galN was reflected in the pre-dose fingerprint, and discrimination between responder and nonresponder animals was found. This provided an initial 'proof of principle' for the pharmacometabonomic approach and a larger scale study was subsequently carried out to determine the ability to predict the extent of liver damage sustained after acetaminophen administration from the pre-dose urinary spectral profiles. In this study, the total urinary excretion of acetaminophen and its metabolites was calculated from the post-dose spectra and the variation in this data was modeled in relation to the pre-dose urinary spectral profiles. In this study, clinical chemistry and histopathology were carried out in order to quantify the level of histopathological damage in each animal using a mean score. The variation in this mean histology score representative of post-dose outcome was also modeled in relation to the pre-dose urinary data. The predictive ability of each independent model was found to be high, and statistically significant variation in the pre-dose spectral profiles was found to relate to post-dose histopathological outcome and urinary drug metabolite excretion. Furthermore, it is encouraging that the pharmacometabonomic approach has been illustrated with studies of the toxicity and metabolism of compounds with very diverse modes of action (allyl alcohol, galN, and acetaminophen). This work suggests that the pre-existing variation in the metabolic phenotype may provide a systems-wide window to probe the metabolic basis for inter-animal variation in response to xenobiotics, dietary intervention, or external stressors and may provide insight into varied metabolic phenotypes and the associated pathological outcomes.

A later pharmacometabonomic study demonstrated a GC–MS-based approach to study two classical experimental models: the streptozotocin-induced diabetic model in Wistar rats and the high-energy, diet-induced obesity model in the Sprague-Dawley rat (Li *et al.*, 2007). Pre- and post-dose urine was collected together with serum clinical chemistry parameters and body weights so that animals could be classified via outcome as obesity prone or obesity resistant and as diabetes prone or diabetes resistant. The variation in the pre-dose/baseline urinary profiles of obesity-prone and obesity-resistant rats was found to correlate with outcome, and the discriminatory metabolites for both classes were related to gut microbial and energy metabolism. These pharmacometabonomic studies confirm that metabolic profiling methods are powerful tools for understanding disease processes and drug responses, as the baseline metabolic fingerprint of an individual can be used to predict the effects of drug interventions or the onset or susceptibility to disease.

18.7.2   Toxicological Implications of Extended Genome (Symbiotic) Interactions

The metabonomic approach has the distinct advantage that not only is the genetic status of an individual captured in the metabolic fingerprint, but so also are contributions from environmental influences such as dietary and pharmaceutical exposure, allostatic load, age, and disease status. Metabonomics has been extensively applied to model phenotypic variation in both human populations and animal models; for example, in the discrimination of diurnal variation between rats and mice (Bollard *et al.*, 2001; Gavaghan *et al.*, 2002), of strain differences (Gavaghan *et al.*, 2000, 2001; Dumas *et al.*, 2007), and of dietary intake (Stella *et al.*, 2006; Rezzi *et al.*, 2007). This inherent variation between individuals expressed in the metabolic phenotype is known to predict and influence drug metabolism and, hence, the presence and severity of adverse effects as exemplified in the pharmacometabonomic applications (Nicholson *et al.*, 2004; Clayton *et al.*, 2006).

   One further factor that modifies the overall phenotype is the activity of the gut microflora (Nicholson *et al.*, 2005), which can be thought of as a complex extension of the host genome. The gut microflora in man has been estimated to constitute over 1000 microbial species, which are poorly defined but which are known to be involved in complex interplay and coordinated response with the host. This may involve co-metabolism of substrates or the absorption by the host of gut microbial metabolites or biochemically active substrates that may induce enzymes that compete for metabolic substrates. Hence, the outcome of a drug intervention may be dependent on conditional probabilistic interactions between gut microbial metabolism and host metabolism and such interactions may be in part responsible for some idiosyncratic toxicological reactions and for adverse drug effects (Nicholson *et al.*, 2004, 2005). The study of the extended genome and its influence on pharmaceutical response will undoubtedly become an important area in the development of personalized healthcare. The variation in rodent metabolic phenotypes seen in some laboratories can be attributed to variation in animal microbiomes (Holmes and Nicholson, 2005; Robosky *et al.*, 2005; Rohde *et al.*, 2007) and, as such, may be an important and previously unsuspected source of interlaboratory and interstudy biological variation in toxic responses and metabolic fate (Martin *et al.*, 2007, 2008). Hence, it is of considerable importance in future work that the response to drug intervention is considered in the context of the fundamental phenotype (Wei *et al.*, 2008).

18.8   Concluding Remarks

We have presented an overview of the combined use of NMR spectroscopy and chemometric tools in the generation and modeling of information-rich metabolic profiles. The field of metabonomics has developed rapidly in terms of its many diverse applications, and it will continue to do so as technological advances provide means for improved biomarker discovery. As evidenced in the literature, metabonomics has been successfully and repeatedly applied to elucidate the metabolic basis of disease or toxin intervention and to determine multicompartmental metabolic consequences of therapeutic intervention. One exciting new field for further development is that of pharmacometabonomics, where the influence of the phenotype and extended genome will be modeled in baseline metabolic profiles that can be applied to predict drug response. It is believed that this work will also provide an improved mechanistic understanding of idiosyncratic toxicity. The COMET expert system for prediction of toxic response provides evidence that metabonomic screening complements and extends the traditional platforms available, and increased utilization of such a metabonomic approach in preclinical drug safety studies is envisaged. The analytical platforms used in metabonomic studies will continue to improve in terms of reliability and sensitivity and this will extend the detection limits and enable an even wider metabolic pool to be profiled. It is anticipated that there will be a significant increase in the application of MS-based analyses in the field of metabonomics, which will prove particularly useful for targeted analysis aimed at specific groups of metabolites and for the provision of metabolic information complementary to that from the NMR-based platform. The use of multiplatform approaches will also significantly extend the scope for biomarker identification and discovery. The recent efforts of the COMET-2 project have presented evidence for the value of metabonomics in mechanistic hypothesis testing; and with ongoing work and increased activity in this field, the testing and validation of many mechanistic hypotheses will become possible. The introduction of methods that allow statistical correlation of complex spectral data sets has impacted significantly on biomarker identification, and exploration and has enabled rigorous data mining and reliable data interpretation to be conducted. A major goal of the future will be the integration of 'omic' data from all levels of biomolecular organization, which will enable meaningful interpretation of a systems-level response to perturbations induced by disease or toxins and will lead to the identification of integrative systems biomarkers. This level of data integration will also facilitate thorough profiling of both genotypic and phenotypic variation between individuals, which may provide metabolic and mechanistic information related to idiosyncratic metabolism of toxins and associated adverse effects.

In summary, the examples presented herein highlight recent advances in the field of metabonomics, and particular focus has been given to toxicological applications and developments. It is anticipated that the metabonomic and metabolic profiling field will continue to gain momentum in many exciting and diverse areas, and detailed insight and understanding of the multifaceted aspects of disease states and pharmaceutical metabolism at a systems level will be achieved.

References

Antti H, Bollard ME, Ebbels TMD, Keun HC, Lindon JC, Nicholson JK, Holmes E (2002) Batch statistical processing of ^{1}H NMR derived urinary spectral data. *J Chemometr* **16**, 461–468.
Antti H, Ebbels TMD, Keun HC, Bollard ME, Beckonert O, Lindon JC, Nicholson JK, Holmes E (2004) Statistical experimental design and partial least squares regression analysis of biofluid metabonomic NMR and clinical chemistry data for screening of adverse drug effects. *Chemometr Intell Lab Syst* **73**, 139–149.
Beckonert OP, Bollard ME, Ebbels TMD, Antti H, Lindon JC, Holmes EC, Nicholson JK (2003) NMR-based metabonomic toxicity classification: hierarchical cluster analysis and *k*-nearest-neighbour approaches. *Anal Chim Acta* **490**, 3–15.
Bollard ME, Garrod S, Holmes E, Lindon JC, Humpfer E, Spraul M, Nicholson JK (2000) High-resolution ^{1}H and ^{1}H–^{13}C magic angle spinning NMR spectroscopy of rat liver. *Magn Reson Med* **44**, 201–207.

Bollard ME, Holmes E, Lindon JC, Mitchell SC, Branstetter D, Zhang W, Nicholson JK (2001) Investigations into biochemical changes due to diurnal variation and estrus cycle in female rats using high-resolution (1)H NMR spectroscopy of urine and pattern recognition. *Anal Biochem* **295**, 194–202.

Bollard ME, Keun HC, Beckonert O, Ebbels TM, Antti H, Nicholls AW, Shockcor JP, Cantor GH, Stevens G, Lindon JC, Holmes E, Nicholson JK (2005) Comparative metabonomics of differential hydrazine toxicity in the rat and mouse. *Toxicol Appl Pharmacol* **204**, 135–151.

Clayton TA, Lindon JC, Cloarec O, Antti H, Charuel C, Hanton G, Provost JP, Le Net JL, Baker D, Walley RJ, Everett JR, Nicholson JK (2006) Pharmaco-metabonomic phenotyping and personalized drug treatment. *Nature* **440**, 1073–1077.

Clayton TA, Lindon JC, Everett JR, Charuel C, Hanton G, Le Net JL, Provost JP, Nicholson JK (2007) Hepatotoxin-induced hypertyrosinemia and its toxicological significance. *Arch Toxicol* **81**, 201–210.

Cloarec O, Dumas ME, Trygg J, Craig A, Barton RH, Lindon JC, Nicholson JK, Holmes E (2005a) Evaluation of the orthogonal projection on latent structure model limitations caused by chemical shift variability and improved visualization of biomarker changes in ^{1}H NMR spectroscopic metabonomic studies. *Anal Chem* **77**, 517–526.

Cloarec O, Dumas ME, Craig A, Barton RH, Trygg J, Hudson J, Blancher C, Gauguier D, Lindon JC, Holmes E, Nicholson JK (2005b) Statistical total correlation spectroscopy: an exploratory approach for latent biomarker identification from metabolic ^{1}H NMR data sets. *Anal Chem* **77**, 1282–1289.

Cloarec O, Campbell A, Tseng LH, Braumann U, Spraul M, Scarfe G, Weaver R, Nicholson JK (2007) Virtual chromatographic resolution enhancement in cryoflow LC-NMR experiments via statistical total correlation spectroscopy. *Anal Chem* **79**, 3304–3311.

Coen M, Lenz EM, Nicholson JK, Wilson ID, Pognan F, Lindon JC (2003) An integrated metabonomic investigation of acetaminophen toxicity in the mouse using NMR spectroscopy. *Chem Res Toxicol* **16**, 295–303.

Coen M, Ruepp SU, Lindon JC, Nicholson JK, Pognan F, Lenz EM, Wilson ID (2004) Integrated application of transcriptomics and metabonomics yields new insight into the toxicity due to paracetamol in the mouse. *J Pharm Biomed Anal* **35**, 93–105.

Coen M, Hong YS, Clayton TA, Rohde CM, Pearce JT, Reily MD, Robertson DG, Holmes E, Lindon JC, Nicholson JK (2007a) The mechanism of galactosamine toxicity revisited; a metabonomic study. *J Proteome Res* **6**, 2711–2719.

Coen M, Hong YS, Cloarec O, Rhode CM, Reily MD, Robertson DG, Holmes E, Lindon JC, Nicholson JK (2007b) Heteronuclear ^{1}H–^{31}P statistical total correlation NMR spectroscopy of intact liver for metabolic biomarker assignment: application to galactosamine-induced hepatotoxicity. *Anal Chem* **79**, 8956–8966.

Connelly JC, Connor SC, Monte S, Bailey NJ, Borgeaud N, Holmes E, Troke J, Nicholson JK, Gavaghan CL (2002) Application of directly coupled high performance liquid chromatography–NMR–mass spectometry and ^{1}H NMR spectroscopic studies to the investigation of 2,3-benzofuran metabolism in Sprague-Dawley rats. *Drug Metab Dispos* **30**, 1357–1363.

Connor SC, Hodson MP, Ringeissen S, Sweatman BC, McGill PJ, Waterfield CJ, Haselden JN (2004) Development of a multivariate statistical model to predict peroxisome proliferation in the rat, based on urinary ^{1}H-NMR spectral patterns. *Biomarkers* **9**, 364–385.

Craig A, Cloarec O, Holmes E, Nicholson JK, Lindon JC (2006a) Scaling and normalization effects in NMR spectroscopic metabonomic data sets. *Anal Chem* **78**, 2262–2267.

Craig A, Sidaway J, Holmes E, Orton T, Jackson D, Rowlinson R, Nickson J, Tonge R, Wilson ID, Nicholson JK (2006b) Systems toxicology: integrated genomic, proteomic and metabonomic analysis of methapyrilene induced hepatotoxicity in the rat. *J Proteome Res* **5**, 1586–1601.

Crockford DJ, Lindon JC, Cloarec O, Plumb RS, Bruce SJ, Zirah S, Rainville P, Stumpf CL, Johnson K, Holmes E, Nicholson JK (2006a) Statistical search space reduction and two-dimensional data display approaches for UPLC–MS in biomarker discovery and pathway analysis. *Anal Chem* **78**, 4398–4408.

Crockford DJ, Holmes E, Lindon JC, Plumb RS, Zirah S, Bruce SJ, Rainville P, Stumpf CL, Nicholson JK (2006b) Statistical heterospectroscopy, an approach to the integrated analysis of NMR and UPLC–MS data sets: application in metabonomic toxicology studies. *Anal Chem* **78**, 363–371.

Delaney J, Hodson MP, Thakkar H, Connor SC, Sweatman BC, Kenny SP, McGill PJ, Holder JC, Hutton KA, Haselden JN, Waterfield CJ (2005) Tryptophan-NAD$^+$ pathway metabolites as putative biomarkers and predictors of peroxisome proliferation. *Arch Toxicol* **79**, 208–223.

Duarte IF, Stanley EG, Holmes E, Lindon JC, Gil AM, Tang H, Ferdinand R, McKee CG, Nicholson JK, Vilca-Melendez H, Heaton N, Murphy GM (2005) Metabolic assessment of human liver transplants from biopsy samples at the donor and recipient stages using high-resolution magic angle spinning ^{1}H NMR spectroscopy. *Anal Chem* **77**(17), 5570–5578.

Dumas ME, Wilder SP, Bihoreau MT, Barton RH, Fearnside JF, Argoud K, D'Amato L, Wallis RH, Blancher C, Keun HC, Baunsgaard D, Scott J, Sidelmann UG, Nicholson JK, Gauguier D (2007) Direct quantitative trait locus mapping of mammalian metabolic phenotypes in diabetic and normoglycemic rat models. *Nat Genet* **39**, 666–672.

Ebbels TMD, Keun HC, Beckonert OP, Antti H, Bollard ME, Lindon JC, Holmes EC, Nicholson JK (2003) Toxicity classification from metabonomic data usinga density superposition approach: 'CLOUDS'. *Anal Chim Acta* **490**, 109–122.

Ebbels TM, Keun HC, Beckonert OP, Bollard ME, Lindon JC, Holmes E, Nicholson JK (2007) Prediction and classification of drug toxicity using probabilistic modeling of temporal metabolic data: the Consortium on Metabonomic Toxicology Screening approach. *J Proteome Res* **6**, 4407–4422.

Eriksson L, Antti H, Gottfries J, Holmes E, Johansson E, Lindgren F, Long I, Lundstedt T, Trygg J, Wold S (2004) Using chemometrics for navigating in the large data sets of genomics, proteomics, and metabonomics (gpm). *Anal Bioanal Chem* **380**, 419–429.

Fiehn O (2002) Metabolomics – the link between genotypes and phenotypes. *Plant Mol Biol* **48**, 155–171.

Fiehn O, Kind T (2007) Metabolite profiling in blood plasma. *Methods Mol Biol* **358**, 3–17.

Fiehn O, Kopka J, Dörmann P, Altmann T, Trethewey RN, Willmitzer L (2000) Metabolite profiling for plant functional genomics. *Nat Biotechnol* **18**, 1157–61.

Garrod S, Humpfer E, Spraul M, Connor SC, Polley S, Connelly J, Lindon JC, Nicholson JK, Holmes E (1999) High-resolution magic angle spinning ^{1}H NMR spectroscopic studies on intact rat renal cortex and medulla. *Magn Reson Med* **41**, 1108–1118.

Gartland KP, Sanins SM, Nicholson JK, Sweatman BC, Beddell CR, Lindon JC (1990) Pattern recognition analysis of high resolution ^{1}H NMR spectra of urine. A nonlinear mapping approach to the classification of toxicological data. *NMR Biomed* **3**, 166–172.

Gartland KP, Beddell CR, Lindon JC, Nicholson JK (1991) Application of pattern recognition methods to the analysis and classification of toxicological data derived from proton nuclear magnetic resonance spectroscopy of urine. *Mol Pharmacol* **39**, 629–642.

Gavaghan CL, Holmes E, Lenz E, Wilson ID, Nicholson JK (2000) An NMR-based metabonomic approach to investigate the biochemical consequences of genetic strain differences: application to the C57BL10J and Alpk:ApfCD mouse. *FEBS Lett* **484**, 169–174.

Gavaghan CL, Nicholson JK, Connor SC, Wilson ID, Wright B, Holmes E (2001) Directly coupled high-performance liquid chromatography and nuclear magnetic resonance spectroscopic with chemometric studies on metabolic variation in Sprague-Dawley rats. *Anal Biochem* **291**, 245–252.

Gavaghan CL, Wilson ID, Nicholson JK (2002) Physiological variation in metabolic phenotyping and functional genomic studies: use of orthogonal signal correction and PLS-DA. *FEBS Lett* **530**, 191–196.

Granger JH, Williams R, Lenz EM, Plumb RS, Stumpf CL, Wilson ID (2007) A metabonomic study of strain- and age-related differences in the Zucker rat. *Rapid Commun Mass Spectrom* **21**, 2039–2045.

Halket JM, Waterman D, Przyborowska AM, Patel RK, Fraser PD, Bramley PM (2005) Chemical derivatization and mass spectral libraries in metabolic profiling by GC/MS and LC/MS/MS. *J Exp Bot* **56**, 219–243.

Holmes E, Nicholson J (2005) Variation in gut microbiota strongly influences individual rodent phenotypes. *Toxicol Sci* **87**, 1–2.

Holmes E, Bonner FW, Sweatman BC, Lindon JC, Beddell CR, Rahr E, Nicholson JK (1992a) Nuclear magnetic resonance spectroscopy and pattern recognition analysis of the biochemical processes associated with the progression of and recovery from nephrotoxic lesions in the rat induced by mercury(II) chloride and 2-bromoethanamine. *Mol Pharmacol* **42**, 922–930.

Holmes E, Nicholson JK, Bonner FW, Sweatman BC, Beddell CR, Lindon JC, Rahr E (1992b) Mapping the biochemical trajectory of nephrotoxicity by pattern recognition of NMR urinanalysis. *NMR Biomed* **5**, 368–372.

Holmes E, Foxall PJ, Nicholson JK, Neild GH, Brown SM, Beddell CR, Sweatman BC, Rahr E, Lindon JC, Spraul M (1994) Automatic data reduction and pattern recognition methods for analysis of ^{1}H nuclear magnetic resonance spectra of human urine from normal and pathological states. *Anal Biochem* **220**, 284–296.

Holmes E, Nicholls AW, Lindon JC, Ramos S, Spraul M, Neidig P, Connor SC, Connelly J, Damment SJ, Haselden J, Nicholson JK (1998) Development of a model for classification of toxin-induced lesions using ^{1}H NMR spectroscopy of urine combined with pattern recognition. *NMR Biomed* **11**, 235–244.

Holmes E, Nicholls AW, Lindon JC, Connor SC, Connelly JC, Haselden JN, Damment SJ, Spraul M, Neidig P, Nicholson JK (2000) Chemometric models for toxicity classification based on NMR spectra of biofluids. *Chem Res Toxicol* **13**, 471–478.

Holmes E, Tsang TM, Tabrizi SJ (2006a) The application of NMR-based metabonomics in neurological disorders. *NeuroRx* **3**, 358–372.

Holmes E, Cloarec O, Nicholson JK (2006b) Probing latent biomarker signatures and in vivo pathway activity in experimental disease states via statistical total correlation spectroscopy (STOCSY) of biofluids: application to $HgCl_2$ toxicity. *J Proteome Res* **5**, 1313–1320.

Holmes E, Loo RL, Cloarec O, Coen M, Tang H, Maibaum E, Bruce S, Chan Q, Elliott P, Stamler J, Wilson ID, Lindon JC, Nicholson JK (2007) Detection of urinary drug metabolite (xenometabolome) signatures in molecular epidemiology studies via statistical total correlation (NMR) spectroscopy. *Anal Chem* **79**, 2629–2640.

Jackson JEA (1991) *A Users Guide to Principal Components.* John Wiley & Sons, Ltd: New York.

Jenkins H, Hardy N, Beckmann M, Draper J, Smith AR, Taylor J, Fiehn O, Goodacre R, Bino RJ, Hall R, Kopka J, Lane GA, Lange BM, Liu JR, Mendes P, Nikolau BJ, Oliver SG, Paton NW, Rhee S, Roessner-Tunali U, Saito K, Smedsgaard J, Sumner LW, Wang T, Walsh S, Wurtele ES, Kell DB (2004) A proposed framework for the description of plant metabolomics experiments and their results. *Nat Biotechnol* **22**, 1601–1606.

Keun HC, Ebbels TM, Antti H, Bollard ME, Beckonert O, Schlotterbeck G, Senn H, Niederhauser U, Holmes E, Lindon JC, Nicholson JK (2002a) Analytical reproducibility in ^{1}H NMR-based metabonomic urinalysis. *Chem Res Toxicol* **15**, 1380–1386.

Keun HC, Beckonert O, Griffin JL, Richter C, Moskau D, Lindon JC, Nicholson JK (2002b) Cryogenic probe ^{13}C NMR spectroscopy of urine for metabonomic studies. *Anal Chem* **74**, 4588–4593.

Keun HC, Ebbels TM, Bollard ME, Beckonert O, Antti H, Holmes E, Lindon JC, Nicholson JK (2004) Geometric trajectory analysis of metabolic responses to toxicity can define treatment specific profiles. *Chem Res Toxicol* **17**, 579–587.

Lanni A, Mancini F, Sabatino L, Silvestri E, Franco R, De Rosa G, Goglia F, Colantuoni V (2002) *De novo* expression of uncoupling protein 3 is associated to enhanced mitochondrial thioesterase-1 expression and fatty acid metabolism in liver of fenofibrate-treated rats. *FEBS Lett* **525**, 7–12.

Lenz EM, Wilson ID (2007) Analytical strategies in metabonomics. *J Proteome Res* **6**, 443–458.

Lenz EM, Bright J, Knight R, Wilson ID, Major H (2004) A metabonomic investigation of the biochemical effects of mercuric chloride in the rat using ^{1}H NMR and HPLC-TOF/MS: time dependent changes in the urinary profile of endogenous metabolites as a result of nephrotoxicity. *Analyst* **129**, 535–541.

Lenz EM, Bright J, Knight R, Westwood FR, Davies D, Major H, Wilson ID (2005) Metabonomics with ^{1}H-NMR spectroscopy and liquid chromatography–mass spectrometry applied to the investigation of metabolic changes caused by gentamicin-induced nephrotoxicity in the rat. *Biomarkers* **10**, 173–187.

Lenz EM, Williams RE, Sidaway J, Smith BW, Plumb RS, Johnson KA, Rainville P, Shockcor J, Stumpf CL, Granger JH, Wilson ID (2007) The application of microbore UPLC/oa-TOF-MS and ^{1}H NMR spectroscopy to the metabonomic analysis of rat urine following the intravenous administration of pravastatin. *J Pharm Biomed Anal* **44**, 845–852.

Li H, Ni Y, Su M, Qiu Y, Zhou M, Qiu M, Zhao A, Zhao L, Jia W (2007) Pharmacometabonomic phenotyping reveals different responses to xenobiotic intervention in rats. *J Proteome Res* **6**, 1364–1370.

Lindon JC, Nicholson JK, Holmes E, Antti H, Bollard ME, Keun H, Beckonert O, Ebbels TM, Reily MD, Robertson D, Stevens GJ, Luke P, Breau AP, Cantor GH, Bible RH, Niederhauser U, Senn H, Schlotterbeck G, Sidelmann UG, Laursen SM, Tymiak A, Car BD, Lehman-McKeeman L, Colet JM, Loukaci A, Thomas C (2003) Contemporary issues in toxicology the role of metabonomics in toxicology and its evaluation by the COMET project. *Toxicol Appl Pharmacol* **187**, 137–146.

Lindon JC, Holmes E, Nicholson JK (2004) Metabonomics: systems biology in pharmaceutical research and development. *Curr Opin Mol Ther* **6**, 265–272.

Lindon JC, Keun HC, Ebbels TM, Pearce JM, Holmes E, Nicholson JK (2005) The Consortium for Metabonomic Toxicology (COMET): aims, activities and achievements. *Pharmacogenomics* **6**, 691–699.

Lindon JC, Holmes E, Nicholson JK (2007) Metabonomics in pharmaceutical R&D. *FEBS J* **274**, 1140–1151.

Lindon JC, Nicholson JK, Wilson ID (1996) The development and application of coupled HPLC-NMR spectroscopy. *Adv Chromatogr* **36**, 315–382.

Maddox JF, Luyendyk JP, Cosma GN, Breau AP, Bible Jr RH, Harrigan GG, Goodacre R, Ganey PE, Cantor GH, Cockerell GL, Roth RA (2006) Metabonomic evaluation of idiosyncrasy-like liver injury in rats cotreated with ranitidine and lipopolysaccharide. *Toxicol Appl Pharmacol* **212**, 35–44.

Major HJ, Williams R, Wilson AJ, Wilson ID (2006) A metabonomic analysis of plasma from Zucker rat strains using gas chromatography/mass spectrometry and pattern recognition. *Rapid Commun Mass Spectrom* **20**, 3295–3302.

Martin FP, Dumas ME, Wang Y, Legido-Quigley C, Yap IK, Tang H, Zirah S, Murphy GM, Cloarec O, Lindon JC, Sprenger N, Fay LB, Kochhar S, van Bladeren P, Holmes E, Nicholson JK (2007) A top-down systems biology view of microbiome-mammalian metabolic interactions in a mouse model. *Mol Syst Biol* **3**, 112.

Martin FP, Verdu EF, Wang Y, Dumas ME, Yap IK, Cloarec O, Bergonzelli GE, Corthesy-Theulaz I, Kochhar S, Holmes E, Lindon JC, Collins SM, Nicholson JK (2006) Transgenomic metabolic interactions in a mouse disease model: interactions of *Trichinella spiralis* infection with dietary *Lactobacillus paracasei* supplementation. *J Proteome Res* **5**, 2185–2193.

Martin FP, Wang Y, Sprenger N, Holmes E, Lindon JC, Kochhar S, Nicholson JK (2007) Effects of probiotic *Lactobacillus paracasei* treatment on the host gut tissue metabolic profiles probed via magic-angle-spinning NMR spectroscopy. *J Proteome Res* **6**, 1471–1481.

Martin FP, Wang Y, Sprenger N, Yap IK, Lundstedt T, Lek P, Rezzi S, Ramadan Z, van Bladeren P, Fay LB, Kochhar S, Lindon JC, Holmes E, Nicholson JK (2008) Probiotic modulation of symbiotic gut microbial-host metabolic interactions in a humanized microbiome mouse model. *Mol Syst Biol* **4**, 157.

Martinez-Granados B, Monleon D, Martinez-Bisbal MC, Rodrigo JM, del Olmo J, Lluch P, Ferrandez A, Marti-Bonmati L, Celda B (2006) Metabolite identification in human liver needle biopsies by high-resolution magic angle spinning ^{1}H NMR spectroscopy. *NMR Biomed* **19**, 90–100.

Mashego MR, Rumbold K, De Mey M, Vandamme E, Soetaert W, Heijnen JJ (2007) Microbial metabolomics: past, present and future methodologies. *Biotechnol Lett* **29**, 1–16.

Mayboroda OA, Neususs C, Pelzing M, Zurek G, Derks R, Meulenbelt I, Kloppenburg M, Slagboom EP, Deelder AM (2007) Amino acid profiling in urine by capillary zone electrophoresis–mass spectrometry. *J Chromatogr A* **115**(9), 149–153.

Monton MR, Soga T (2007) Metabolome analysis by capillary electrophoresis-mass spectrometry. *J Chromatogr A* **116**, 237–246.

Mortishire-Smith RJ, Skiles GL, Lawrence JW, Spence S, Nicholls AW, Johnson BA, Nicholson JK (2004) Use of metabonomics to identify impaired fatty acid metabolism as the mechanism of a drug-induced toxicity. *Chem Res Toxicol* **17**, 165–173.

Nassar AE, Talaat RE, Tokuno H (2007) Drug interactions: concerns and current approaches. *IDrugs* **10**, 47–52.

Nicholls AW, Wilson ID, Godejohann M, Nicholson JK, Shockcor JP (2006) Identification of phenacetin metabolites in human urine after administration of phenacetin–C_2H_3: measurement of futile metabolic deacetylation via HPLC/MS–SPE–NMR and HPLC–ToF MS. *Xenobiotica* **36**, 615–629.

Nicholson JK, Wilson ID (1987) High resolution nuclear magnetic resonance spectroscopy of biological samples as an aid to drug development. *Prog Drug Res* **31**, 427–479.

Nicholson JK, Wilson ID (1989) High resolution proton magnetic resonance spectroscopy of biological fluids. *Prog NMR Spectrosc* **21**, 449–501.

Nicholson JK, Wilson ID (2003) Opinion: understanding 'global' systems biology: metabonomics and the continuum of metabolism. *Nat Rev Drug Discov* **2**, 668–676.

Nicholson JK, Timbrell JA, Sadler PJ (1985a) Proton NMR spectra of urine as indicators of renal damage. Mercury-induced nephrotoxicity in rats. *Mol Pharmacol* **27**, 644–651.

Nicholson JK, Timbrell JA, Bales JR, Sadler PJ (1985b) A high resolution proton nuclear magnetic resonance approach to the study of hepatocyte and drug metabolism. Application to acetaminophen. *Mol Pharmacol* **27**, 634–643.

Nicholson JK, Lindon JC, Holmes E (1999) 'Metabonomics': understanding the metabolic responses of living systems to pathophysiological stimuli via multivariate statistical analysis of biological NMR spectroscopic data. *Xenobiotica* **29**, 1181–1189.

Nicholson JK, Connelly J, Lindon JC, Holmes E (2002) Metabonomics: a platform for studying drug toxicity and gene function. *Nat Rev Drug Discov* **1**, 153–161.

Nicholson JK, Holmes E, Lindon JC, Wilson ID (2004) The challenges of modeling mammalian biocomplexity. *Nat Biotechnol* **22**, 1268–1274.

Nicholson JK, Holmes E, Wilson ID (2005) Gut microorganisms, mammalian metabolism and personalized health care. *Nat Rev Microbiol* **3**, 431–438.

O'Hagan S, Dunn WB, Brown M, Knowles JD, Kell DB (2005) Closed-loop, multiobjective optimization of analytical instrumentation: gas chromatography/time-of-flight mass spectrometry of the metabolomes of human serum and of yeast fermentations. *Anal Chem* **77**, 290–303.

O'Hagan S, Dunn WB, Knowles JD, Broadhurst D, Williams R, Ashworth JJ, Cameron M, Kell DB (2007) Closed-loop, multiobjective optimization of two-dimensional gas chromatography/mass spectrometry for serum metabolomics. *Anal Chem* **79**, 464–476.

Plumb RS, Johnson KA, Rainville P, Smith BW, Wilson ID, Castro-Perez JM, Nicholson JK (2006a) UPLC/MS(E); a new approach for generating molecular fragment information for biomarker structure elucidation. *Rapid Commun Mass Spectrom* **20**, 1989–1994.

Plumb RS, Johnson KA, Rainville P, Shockcor JP, Williams R, Granger JH, Wilson ID (2006b) The detection of phenotypic differences in the metabolic plasma profile of three strains of Zucker rats at 20 weeks of age using ultra-performance liquid chromatography/orthogonal acceleration time-of-flight mass spectrometry. *Rapid Commun Mass Spectrom* **20**, 2800–2806.

Rantalainen M, Cloarec O, Beckonert O, Wilson ID, Jackson D, Tonge R, Rowlinson R, Rayner S, Nickson J, Wilkinson RW, Mills JD, Trygg J, Nicholson JK, Holmes E (2006) Statistically integrated metabonomic-proteomic studies on a human prostate cancer xenograft model in mice. *J Proteome Res* **5**, 2642–2655.

Rezzi S, Ramadan Z, Martin FP, Fay LB, van Bladeren P, Lindon JC, Nicholson JK, Kochhar S (2007) Human metabolic phenotypes link directly to specific dietary preferences in healthy individuals. *J Proteome Res* **6**(11), 4469–4477.

Ringeissen S, Connor SC, Brown HR, Sweatman BC, Hodson MP, Kenny SP, Haworth RI, McGill P, Price MA, Aylott MC, Nunez DJ, Haselden JN, Waterfield CJ (2003) Potential urinary and plasma biomarkers of peroxisome proliferation in the rat: identification of *N*-methylnicotinamide and *N*-methyl-4-pyridone-3-carboxamide by ^{1}H nuclear magnetic resonance and high performance liquid chromatography. *Biomarkers* **8**, 240–271.

Robosky LC, Wells DF, Egnash LA, Manning ML, Reily MD, Robertson DG (2005) Metabonomic identification of two distinct phenotypes in Sprague-Dawley (Crl:CD(SD)) rats. *Toxicol Sci* **87**, 277–284.

Rohde CM, Wells DF, Robosky LC, Manning ML, Clifford CB, Reily MD, Robertson DG (2007) Metabonomic evaluation of Schaedler altered microflora rats. *Chem Res Toxicol* **20**, 1388–1392.

Servais AC, Crommen J, Fillet M (2006) Capillary electrophoresis–mass spectrometry, an attractive tool for drug bioanalysis and biomarker discovery. *Electrophoresis* **27**, 2616–2629.

Skordi E, Yap IK, Claus SP, Martin FP, Cloarec O, Lindberg J, Schuppe-Koistinen I, Holmes E, Nicholson JK (2007) Analysis of time-related metabolic fluctuations induced by ethionine in the rat. *J Proteome Res* **6**, 4572–4581.

Smith LM, Maher AD, Cloarec O, Rantalainen M, Tang H, Elliott P, Stamler J, Lindon JC, Holmes E, Nicholson JK (2007) Statistical correlation and projection methods for improved information recovery from diffusion-edited NMR spectra of biological samples. *Anal Chem* **79**, 5682–5689.

Soga T, Ohashi Y, Ueno Y, Naraoka H, Tomita M, Nishioka T (2003) Quantitative metabolome analysis using capillary electrophoresis mass spectrometry. *J Proteome Res* **2**, 488–494.

Stella C, Beckwith-Hall B, Cloarec O, Holmes E, Lindon JC, Powell J, van der Ouderaa F, Bingham S, Cross AJ, Nicholson JK (2006) Susceptibility of human metabolic phenotypes to dietary modulation. *J Proteome Res* **5**, 2780–2788.

Suarez B, Simonet BM, Cardenas S, Valcarcel M (2007) Determination of non-steroidal anti-inflammatory drugs in urine by combining an immobilized carboxylated carbon nanotubes mini-column for solid-phase extraction with capillary electrophoresis-mass spectrometry. *J Chromatogr A* **115**, 203–207.

Swanson MG, Zektzer AS, Tabatabai ZL, Simko J, Jarso S, Keshari KR, Schmitt L, Carroll PR, Shinohara K, Vigneron DB, Kurhanewicz J (2006) Quantitative analysis of prostate metabolites using ^{1}H HR-MAS spectroscopy. *Magn Reson. Med* **55**, 1257–1264.

Taylor J, King RD, Altmann T, Fiehn O (2002) Application of metabolomics to plant genotype discrimination using statistics and machine learning. *Bioinformatics* **18** (Suppl 2), S241–S248.

Taylor JL, Wu CL, Cory D, Gonzalez RG, Bielecki A, Cheng LL (2003) High-resolution magic angle spinning proton NMR analysis of human prostate tissue with slow spinning rates. *Magn Reson Med* **50**, 627–632.

Trygg J, Wold S (2003) O2-PLS, a two-block (*X–Y*) latent variable regression (LVR) method with an integral OSC filter. *J Chemometr* **17**, 53–64.

Trygg J, Holmes E, Lundstedt T (2007) Chemometrics in metabonomics. *J Proteome Res* **6**, 469–479.

Ullsten S, Danielsson R, Backstrom D, Sjoberg P, Bergquist J (2006) Urine profiling using capillary electrophoresis–mass spectrometry and multivariate data analysis. *J Chromatogr A* **111**, 87–93.

Underwood BR, Broadhurst D, Dunn WB, Ellis DI, Michell AW, Vacher C, Mosedale DE, Kell DB, Barker RA, Grainger DJ, Rubinsztein DC (2006) Huntington disease patients and transgenic mice have similar pro-catabolic serum metabolite profiles. *Brain* **129**, 877–886.

Wang S, Zhao X, Mao Y, Cheng Y (2007) Novel approach for developing urinary nucleosides profile by capillary electrophoresis–mass spectrometry. *J Chromatogr A* **114**, 254–260.

Wang Y, Bollard ME, Keun H, Antti H, Beckonert O, Ebbels TM, Lindon JC, Holmes E, Tang H, Nicholson JK (2003) Spectral editing and pattern recognition methods applied to high-resolution magic-angle spinning ^{1}H nuclear magnetic resonance spectroscopy of liver tissues. *Anal Biochem* **323**, 26–32.

Want EJ, Nordstrom A, Morita H, Siuzdak G (2007) From exogenous to endogenous: the inevitable imprint of mass spectrometry in metabolomics. *J Proteome Res* **6**, 459–468.

Waters NJ, Garrod S, Farrant RD, Haselden JN, Connor SC, Connelly J, Lindon JC, Holmes E, Nicholson JK (2000) High-resolution magic angle spinning ^{1}H NMR spectroscopy of intact liver and kidney: optimization of sample preparation procedures and biochemical stability of tissue during spectral acquisition. *Anal Biochem* **282**, 16–23.

Waters NJ, Waterfield CJ, Farrant RD, Holmes E, Nicholson JK (2006) Integrated metabonomic analysis of bromobenzene-induced hepatotoxicity: novel induction of 5-oxoprolinosis. *J Proteome Res* **5**, 1448–1459.

Wei J, Houk AL, Zhao L, Nicholson JK, (2008). Gut microbiota: a potential new territory for drug targeting. *Nat Rev Drug Discov* **7**, 123–129.

Williams BJ, Cameron CJ, Workman R, Broeckling CD, Sumner LW, Smith JT (2007) Amino acid profiling in plant cell cultures: an inter-laboratory comparison of CE–MS and GC–MS. *Electrophoresis* **28**, 1371–1379.

Wilson ID, Plumb R, Granger J, Major H, Williams R, Lenz EM (2005a) HPLC–MS-based methods for the study of metabonomics. *J Chromatogr B Anal Technol Biomed Life Sci* **817**, 67–76.

Wilson ID, Nicholson JK, Castro-Perez J, Granger JH, Johnson KA, Smith BW, Plumb RS (2005b) High resolution 'ultra performance' liquid chromatography coupled to oa-TOF mass spectrometry as a tool for differential metabolic pathway profiling in functional genomic studies. *J Proteome Res* **4**, 591–598.

Wold H (1985) Partial least squares. In *Encyclopedia of Statistical Sciences*, S. Kotz. NL Johnson (eds). John Wiley & Sons, Ltd: New York; 581–591.

Yap IK, Clayton TA, Tang H, Everett JR, Hanton G, Provost JP, Le Net JL, Charuel C, Lindon JC, Nicholson JK (2006) An integrated metabonomic approach to describe temporal metabolic disregulation induced in the rat by the model hepatotoxin allyl formate. *J Proteome Res* **5**, 2675–2684.

19

Potential Uses of Toxicogenomic Biomarkers in Occupational Health and Risk Assessment

Paul A. Nony

19.1   Introduction

The assessment of chemical exposures in humans is concerned mainly with the measurement or estimation of both the amount of a chemical with which a person comes into contact and the resulting dose of chemical that is taken into the body (Faustman and Omenn, 2008). Chemical absorption in humans is dependent upon many factors, including the degree to which the chemical can be absorbed, the route of exposure, the concentration of chemical to which a person is exposed, and the duration of exposure. Furthermore, the resulting dose and biological effect of that dose can vary from one person to the next due to interindividual differences in metabolism and genetic susceptibility to chemical exposure. Therefore, assessing chemical exposure and resulting health outcomes in humans is a complex process with which considerable uncertainty is associated.

A reliable approach to assessing human exposure to chemicals, especially in occupational settings, is direct measurement of chemical or metabolite concentrations in biological matrices. Chemical biomonitoring is a general term that can be applied to the measurement of chemicals or their metabolites in biological matrices (i.e. hair, blood, breath, saliva, urine, whole-tissue biopsies). However, there are limitations to chemical biomonitoring that can complicate the interpretation of biomonitoring results. For example, if a chemical is short-lived in the blood due to rapid metabolism, binding to proteins, or sequestration to lipids, then a blood test to measure the chemical may not give a proper indication of the actual dose received. Conversely, if a chemical is long-lived

Toxicogenomics: A Powerful Tool for Toxicity Assessment   Edited by S. C. Sahu
© 2008 John Wiley & Sons, Ltd

in the body, then it may be difficult to determine whether a measured concentration of that chemical in a biological matrix is indicative of a recent or long-past exposure. For some chemicals for which there is a well-defined exposure index, the presence of an elevated concentration in a biological matrix is a good indication of an unusual exposure. What remains to be determined, however, is whether that observed concentration of chemical in the body results in or is associated with an observable heath effect in that individual.

For some chemicals, the concentration or concentration range in a biological matrix that produces a particular health effect is known. A good example of this is lead. Lead has known biological effects associated with defined blood-lead concentrations. In addition, the toxicokinetics of lead in children have been studied extensively and are well defined such that mathematical models have been developed for exposure and risk assessment purposes (USEPA, 2002). These models allow toxicologists and risk assessors to make reliable determinations of the levels of blood-lead, as well as the associated health effects that are likely to occur in children with potential exposures to lead-contaminated media (i.e. soil, dust, water), without performing biological monitoring. Unfortunately, there are numerous commonly encountered chemicals whose chemical properties, absorption, metabolism, and toxicity are not so well defined. Thus, even if a chemical concentration is measured in a biological matrix, there may not be sufficient information to allow the determination of whether that level is toxicologically relevant. Furthermore, models may not adequately account for interindividual variability; thus, there remains considerable uncertainty as to how an exposure in a particular individual will affect that individual.

Considerable research effort has been directed toward the identification of reliable biomarkers of chemical exposure associated with disease. As a technology aimed at the development of gene expression biomarkers, the use of toxicogenomics has already been encouraged in the risk assessment and in the drug development processes by the Environmental Protection Agency and the Food and Drug Administration respectively (Boverhof and Zacharewski, 2006). Currently, the science of toxicogenomics is somewhat limited to qualitative descriptions of alterations in gene expression and other parameters with little or no correlation of these changes with health outcomes (Boverhof and Zacharewski, 2006). However, the potential to bridge the gap between exposure assessment and health outcomes has long been a promising goal of scientists working in the field of toxicogenomics, and the technology continues to develop in that direction (Nuwaysir *et al.*, 1999; Pennie *et al.*, 2000; Morgan, 2002; Tennant, 2002; Mattes, 2006; Olden, 2006). By identifying toxicologically relevant doses at the level of gene and protein expression and/or changes in metabolism, scientists should be able to associate health outcomes with exposure levels and make better predictions of health and safety for humans, as well as for wildlife and the environment (Pennie *et al.*, 2000; Aardema and MacGregor, 2002; Kramer and Kolaja, 2002; Paules, 2003, 2006; Schmidt, 2003; Weis *et al.*, 2005; Olden, 2006). Still, much work remains to validate the technology and to ensure that the results of toxicogenomic assays are interpreted and applied properly. This chapter will explore certain areas of occupational and environmental chemical exposure assessment and risk assessment where toxicogenomics might be applied to enhance our understanding of chemical exposures and the toxicological relevance of those exposures as they relate to adverse health effects. The current limitations and potential pitfalls of the misuse of toxicogenomic data in these areas of exposure assessment will also be addressed.

19.2   Development and Application of Toxicogenomic Biomarkers

The use of biomarkers of exposure has become increasingly important in the fields of toxicology and human health (Watson and Mutti, 2004). More specifically, the application of biomarkers in human exposure assessment has been applied in population studies to establish exposure, in epidemiological studies to aid in exposure classification, and in toxicity studies to demonstrate dose at the organ, cellular, and molecular levels (Watson and Mutti, 2004). Because biomarkers can vary among individuals as a function of time as well as of individual metabolic characteristics, understanding of biomarker kinetics and the persistence of chemicals or their metabolites in the body are two important considerations when choosing a biomarker of exposure. Another important consideration is the validity of the biomarker, or how well the target effect correlates with the level of biomarker that is measured (Watson and Mutti, 2004). To date, the development of reliable biomarkers of chemical exposure and effects has not kept up with the rate of chemicals appearing in the environment and in the market place. Because thousands of chemicals have been introduced into industry and consumer products with little or no extensive toxicity testing, the chances of people being exposed to chemicals for which there is little or no toxicity information are high. Furthermore, several large data gaps remain for many chemicals due to the limitations of traditional toxicology studies. These include information on the consequences of subtle chemical effects at low doses, whether observed effects in animals are likely to occur also in exposed humans, and how exposure to a chemical might vary within the human population. Thus, the development of validated biomarkers for toxicologically relevant exposures through toxicogenomics research has the potential to fill vast toxicity information gaps for inadequately tested chemicals (Olden, 2006).

Because most toxicological processes leading to adverse health effects require alterations in gene and protein expression, it is known that similar chemicals with similar mechanisms of toxicity produce similar effects on gene expression. Thus, the application of toxicogenomics in evaluation of 'exposure and effect' is dependent on the identification of specific gene expression signatures or biomarkers that are unique to the chemical exposure at issue and have clinical relevance (Aardema and MacGregor, 2002; Benninghoff, 2007). This correlation between cause and effect has been coined 'phenotypic anchoring' (Paules, 2003, 2006). To develop biomarkers of cause and effect, studies are designed to identify alterations in gene expression profiles that can be correlated with responses observed in traditional toxicity studies (Paules, 2003). Once a specific gene expression profile can be validated as a biomarker of exposure and effect, that biomarker can then be used further to evaluate temporal and dose responses following exposure, and at lower doses (Paules, 2003). The potential applications of validated toxicogenomic biomarkers are numerous, and those related to occupational environmental exposure assessment are discussed later in this chapter.

Another goal of the development of predictive toxicogenomic biomarkers is for the assessment of effects resulting from exposures to chemical mixtures. As the gene expression changes resulting from exposure to one chemical are complex, it is reasonable to predict that the complexity in gene expression changes resulting from simultaneous exposure to more than one chemical could be considerably greater. For example, chemical mixtures may contain chemicals that will have either no effect or opposing effects on the same genes. Mixture constituents may work antagonistically to inhibit the expression of some genes and synergistically to increase the degree of expression of certain other genes (Aardema and MacGregor, 2002; Suk and Olden, 2004). Despite the added complexity, the current toxicogenomic technologies should make possible the elucidation of chemical mixture effects and the development of predictive toxicogenomic biomarkers. This will be especially important for application of biomarkers to human exposure assessment, given that humans are most often exposed to low-concentration mixtures of chemicals in the environment as opposed to single chemicals (Aardema and MacGregor, 2002). With the continued development of genomic technologies and high-throughput systems for biomarker identification in complex systems, it is reasonable to expect that the tools will eventually be available for the development and validation of predictive biomarkers of exposure to chemical mixtures. The development of these markers could initiate a revolution in the risk assessment process, in that for the first time the impact of total exposures, rather than just one chemical, can be considered to predict risk on an individual basis (Suk and Olden, 2004).

19.3   Toxicogenomics and the Identification of Interindividual Variability

The evaluation of interindividual variability in human responses and susceptibility to chemical exposures is and will continue to be a major proposed application of toxicogenomics. It has been long known that genetic variations can affect individual responses to chemical exposures (Aardema and MacGregor, 2002; Marchant, 2003b). As biomarkers of exposure and effect are validated, explanations for the historical observation that not all people respond in the same way to the same chemical exposures have and will continue to emerge in the form of specific, well-identified genotypic variations, or polymorphisms. This knowledge will lead to the development of individualized risk assessment and better understanding of the application of animal data in estimating human risk (Aardema and MacGregor, 2002; Weis *et al.*, 2005). Furthermore, the identification of chemical susceptibilities may be used as bases to remove susceptible individuals from chemical exposures. However, the ramifications of this knowledge are far reaching and present ethical, social, and legal dilemmas that must be dealt with, as discussed below.

19.4   Uses of Toxicogenomics in Occupational Health and Exposure Assessment

When validated, the use of toxicogenomic biomarkers to assess exposures and health impacts will likely force a change in the traditional scientific approach to occupational and environmental health. Currently, workplace biomonitoring serves to help industries monitor worker exposures through measurement of chemical or metabolite levels in biological matrices. Biomonitoring results can be compared with established exposure indexes based on worker populations and designed to be protective of the health of the average worker. This has been done successfully for many years for certain chemical exposures and for occupational carcinogen exposures (Watson and Mutti, 2004). However, because of interindividual differences in chemical metabolism and susceptibility, assessment of workplace safety through biomonitoring is not always capable of predicting toxicologically relevant workplace exposures for many common chemicals on an individual basis, nor can it always account for responses of chemically susceptible individuals. Through the development and validation of more sensitive toxicogenomic biomarkers, worker exposures can be evaluated on an individual basis not only to determine the level of altered gene expression caused by occupationally relevant exposures, but also the potential health impact of the exposures based on each worker's individual susceptibility.

As industries look forward to the application of toxicogenomics in occupational health, numerous ethical considerations have emerged that should be addressed prior to implementation of toxicogenomics research in the workplace (Weinstein, 2007). As the design of health and safety studies shifts from population health and safety, as in epidemiological studies, to individual health and safety, as is the potential in toxicogenomics studies, several issues emerge related to worker rights and privacy (Weinstein *et al.*, 2005). Consider the scenario in which a toxicogenomic marker of exposure and effect is employed to screen potentially exposed workers with the goal of improved worker health monitoring. Because the results of the screening could illuminate differences in biomarker responses within the worker population, there is the possibility that individuals with increased susceptibility to the effects of exposure will be singled out, possibly for removal from the workplace. It has been suggested that workers should not agree to participate in toxicogenomics research in the workplace without the implementation of job and salary protections (Weinstein, 2007). Potential solutions to the issues arising from toxicogenomic research in the workplace include the prioritization of individual privacy regarding personal toxicogenomic information and requiring employers to implement pre-employment toxicogenomic screening. These solutions may give rise to new problems of their own, as susceptible employees may choose for economic reasons to work in a job where they are at increased risk of illness from exposure, thus undermining the goal of protecting worker health. In addition, workers found to be susceptible to certain exposures may have their employment opportunities limited by the stigma of increased susceptibility to chemical exposure (Weinstein, 2007). These and other issues will confront advancements in the field of toxicogenomics as it progresses toward the development of tools aimed at improving workplace health and exposure monitoring. Still, there exists much potential for toxicogenomic research to improve the health of exposed workers through more comprehensive and individualized exposure assessment. It is likely that many of the same uses of toxicogenomic biomarkers, accompanied by the same potential misuses and pitfalls, can be applied to assessing environmental exposures in nonoccupational settings.

19.5   Applications of Toxicogenomics in Risk Assessment

Assessing chemical exposures as part of risk assessment has traditionally been an indirect process of monitoring levels of chemicals in food or environmental media (Oberemm *et al.*, 2005). Low-dose nonoccupational exposures from food and the environment are common to all individuals. Understanding the effects of low-dose chemical exposures that are most relevant to typical human exposures is a crucial aspect of the risk assessment process. However, traditional animal and *in vitro* toxicity studies are conducted at high doses to ensure that potential toxicities are not overlooked, resulting in the need for extrapolation methods to estimate the effects of the same chemicals at low doses. This creates a considerable amount of uncertainty in the risk assessment process when adapting laboratory toxicity studies to quantitative assessments of human risk. Furthermore, the need for individualized exposure assessment cannot be satisfied using information on whole-population exposure patterns. The need for application of toxicogenomic technologies to address these issues for the improvement of the risk assessment process has been widely recognized (Pennie *et al.*, 2004).

Because chemicals are expected to produce some alteration in gene expression at doses that are not sufficient to produce adverse health effects, examination of low-dose biomarker changes should greatly enhance low-dose extrapolations, aid in the identification of low-dose effects in humans, and assist in the identification of threshold doses below which no toxicity is observed (Aardema and MacGregor, 2002). In the risk assessment process, biomonitoring is most useful in the exposure assessment stage (Watson and Mutti, 2004). As toxicogenomic biomarkers of exposure and effect are developed and validated, they will likely provide more comprehensive assessments of subtle changes in gene expression at low, environmentally relevant doses. This enhanced sensitivity should reduce the inherent uncertainty of low-dose extrapolation in the quantitative risk assessment process. In addition, the use of toxicogenomics to identify individual characteristics, such as genetic susceptibility to chemical exposure, could allow the development of specific risk assessment methodologies that can be tailored to sensitive individuals. Another potential benefit of this approach is that public health decisions can be made without overly conservative and nonspecific assumptions of risk that result from the uncertainties of the current risk assessment process. Of course, as with other potential applications of toxicogenomic technologies, much work remains to be done first to validate toxicogenomic biomarkers and then to apply them properly to exposure assessment in the risk assessment process.

19.6   Limitations and Potential Misuses of Toxicogenomic Biomarkers in Exposure Assessment

Methods incorporating toxicogenomic technologies may greatly simplify the evaluation of the magnitude and potential effect of occupational and environmental chemical exposures. As alluded to previously in this chapter, and as a consequence of the numerous potential uses of toxicogenomic technologies and the wealth of new information they promise to provide, major pitfalls exist that must be recognized and properly handled to prevent the misuse of toxicogenomic data in real-world applications. One major hurdle is the elucidation of complex gene expression profiles and associated toxicological pathways in the development of biomarkers of exposure and effect. Changes in gene expression or protein levels in response to chemical exposure may be completely unrelated to toxicity (Balbus, 2005). Accordingly, exposure to a single chemical or chemical mixture may result in the activation of numerous genes that have no toxicological relevance whatsoever. Alternatively, certain gene expression patterns may be adaptive, or may even be protective or otherwise beneficial (Aardema and MacGregor, 2002). Thus, the purpose of the alteration of the expression of a gene or protein must be categorized as part of the validation of a toxicogenomic biomarker; else it might be incorrectly viewed as a toxicologically relevant event. Another area of caution in the development of toxicogenomic biomarkers for exposure and health assessment is in the identification of genetic bases for differences in individual susceptibilities to chemical exposures. Despite the potential for this technology to revolutionize occupational health and risk assessment, this information also will potentially create new labels for persons who demonstrate genetic phenotypes that make them unusually susceptible to chemical exposures. If this is not handled carefully from an ethical standpoint, then public perception could lead to a negative stigma about toxicogenomic research or, worse, discrimination or bias against individuals identified as having genetic susceptibilities (Balbus, 2005).

Toxicogenomic data may become a widespread, controversial evidentiary tool in the arena of toxic tort litigation (Marchant, 2002, 2003a; Pierce and Sexton, 2003; Redick, 2003). A key aspect of successful claims of chemical causation of disease is proof that a sufficient dose was absorbed to cause the claimed health effect; however, owing to the nature of the chemical at issue, there may be no reliable method to demonstrate exposure or quantify dose. In addition, the available science may be insufficient to support or refute claims of causation, leading to injured parties being denied fair judgment and compensation, or innocent parties being forced to pay for liability through no fault of their own. The development of toxicogenomic biomarkers of cause and effect offers the potential of reliable exposure data linked to specific health outcomes. This has great potential to assist courts in properly determining causation in toxic tort litigations. However, toxicogenomic biomarkers must be thoroughly validated for the courts to receive benefit from their use as evidence of exposure and injury. Since the current technology has not yet reached that stage of validation, the potential pitfalls are numerous and could lead to an invalid claim of chemical disease causation. For example, if the degree of alteration of a toxicogenomic biomarker that is indicative of injurious exposure is not defined, then the chances are greatly increased that any change in that biomarker could be presented and accepted as evidence that sufficient exposure occurred to produce injury. It is possible that the exposure was indeed sufficient to produce the disease and that the data properly supported the causation claim; however, it is also possible that the biomarker change was indicative of a noninjurious or unrelated exposure or change in gene expression. This example demonstrates the importance of validation of toxicogenomic biomarkers prior to their use in toxic tort litigation.

The potential power of toxicogenomic technologies is unquestionable, and the use of validated toxicogenomic biomarkers in exposure assessment will likely revolutionize the fields of occupational health and risk assessment by providing enhanced sensitivity and the development of predictive tools for linking exposures to adverse health effects. The greatest impact of this technology may be in the assessment of total exposures to environmentally relevant chemical mixtures and associated health effects. However, until the technology is validated and properly applied as a tool for exposure assessment, the potential exists for the misuse of toxicogenomic data with unwanted ethical, social, and legal consequences. The scientific and regulatory communities will be well served to take an aggressive lead in the proper interpretation and application of toxicogenomic data and technologies to ensure the realization of their full potential in exposure assessment and related fields.

References

Aardema MJ, MacGregor JT (2002) Toxicology and genetic toxicology in the new era of 'toxicogenomics': impact of '-omics' technologies. *Mutat Res* **499**, 13–25.

Balbus JM (2005) Ushering in the new toxicology: toxicogenomics and the public interest. *Environ Health Perspect* **113**, 818–822.

Benninghoff AD (2007) Toxicoproteomics – the next step in the evolution of environmental biomarkers. *Toxicol Sci* **95**, 1–4.

Boverhof DR, Zacharewski TR (2006) Toxicogenomics in risk assessment: applications and needs. *Toxicol Sci* **89**, 352–360.

Faustman EM, Omenn GS (2008) Risk assessment. In *Casarett and Doull's Toxicology: The Basic Science of Poisons*, 7th edn, Klaassen CD (ed.). McGraw-Hill: New York; 107–128.

Kramer JA, Kolaja KL (2002) Toxicogenomics: an opportunity to optimise drug development and safety evaluation. *Expert Opin Drug Saf* **1**, 275–286.

Marchant GE (2002) Toxicogenomics and toxic torts. *Trends Biotechnol* **20**, 329–332.

Marchant GE (2003a) Genomics and toxic substances: part I. Toxicogenomics. *Environ Law Reporter News Anal* **33**:10071–10093.

Marchant GE (2003b) Genomics and toxic substances: part II – genetic susceptibility to environmental agents. *Environ Law Reporter News Anal* **33**:10641–10667.

Mattes WB (2006) Cross-species comparative toxicogenomics as an aid to safety assessment. *Expert Opin Drug Metab Toxicol* **2**, 859–874.

Morgan KT (2002) Gene expression analysis reveals chemical-specific profiles. *Toxicol Sci* **67**, 155–156.

Nuwaysir EF, Bittner M, Trent J, Barrett JC, Afshari CA (1999) Microarrays and toxicology: the advent of toxicogenomics. *Mol Carcinog* **24**, 153–159.

Oberemm A, Onyon L, Gundert-Remy U (2005) How can toxicogenomics inform risk assessment? *Toxicol Appl Pharmacol* **207**, 592–598.

Olden K (2006) Toxicogenomics – a new systems toxicology approach to understanding of gene–environment interactions. *Ann N Y Acad Sci* **1076**, 703–706.

Paules R (2003) Phenotypic anchoring: linking cause and effect. *Environ Health Perspect* **111**, A338–A339.

Paules RS (2006) Toxicogenomics and environmental diseases: the search for biomarkers predictive of adverse effects. *Med Lav* **97**, 322–323.

Pennie W, Pettit SD, Lord PG (2004) Toxicogenomics in risk assessment: an overview of an HESI collaborative research program. *Environ Health Perspect* **112**, 417–419.

Pennie WD, Tugwood JD, Oliver GJ, Kimber I (2000) The principles and practice of toxigenomics: applications and opportunities. *Toxicol Sci* **54**, 277–283.

Pierce JR, Sexton T (2003) Toxicogenomics: toward the future of toxic tort causation. *North Carolina J Law Technol* **5**, 33–58.

Redick TP (2003) Twenty-first century toxicogenomics meets twentieth century mass tort precedent: is there a duty to warn of a hypothetical harm to an "eggshell" gene? *Washburn Law Journal* **42**, 547–574.

Schmidt CW (2002) Toxicogenomics: an emerging discipline. *Environ Health Perspect* **110**, A750–A755.

Suk WA, Olden K (2004) Multidisciplinary research: strategies for assessing chemical mixtures to reduce risk of exposure and disease. *Int J Occup Med Environ Health* **17**, 103–110.

Tennant RW (2002) The National Center for Toxicogenomics: using new technologies to inform mechanistic toxicology. *Environ Health Perspect* **110**, A8–A10.

USEPA (2002) *Overview of the IEUBK Model for Lead in Children*. US Environmental Protection Agency, Office of Emergency and Remedial Response: Washington, DC. EPA/540/R-99/015.

Watson WP, Mutti A (2004) Role of biomarkers in monitoring exposures to chemicals: present position, future prospects. *Biomarkers* **9**, 211–242.

Weinstein M (2007) Creating safer work environments with toxicogenomics research: worker protection, consultation and consent. *Pharmacogenomics* **8**, 209–212.

Weinstein M, Widenor M, Hecker S (2005) Health and employment practices: ethical, legal, and social implications of advances in toxicogenomics. *AAOHN J* **53**, 529–533.

Weis BK, Balshaw D, Barr JR, Brown D, Ellisman M, Lioy P, Omenn G, Potter JD, Smith MT, Sohn L, Suk WA, Sumner S, Swenberg J, Walt DR, Watkins S, Thompson C, Wilson SH (2005) Personalized exposure assessment: promising approaches for human environmental health research. *Environ Health Perspect* **113**, 840–848.

Section 4
Toxicogenomics for Regulatory Use

20

Toxicogenomics: A Regulatory Perspective

Daniel A. Casciano

20.1 Introduction

Toxicogenomics is a relatively new subdiscipline of toxicology that combines the emerging technologies of genomics, proteomics, and bioinformatics to identify and characterize mechanisms of action of known and suspected toxicants. A goal of this subdiscipline is to identify specific sets of genes that may be candidate biomarkers of specific toxic effects. These effects may result from exposure to certain chemicals or classes of chemicals or may be more related to a specific organ's response to the insult. The enormous value of these technologies, derived from the sequencing of the human genome and their application to the field of toxicogenomics, is that they provide similar tools that can be applied not only to organisms used as human surrogates, but also directly to humans. This provides toxicologists a greater degree of confidence in the use of present systems to estimate human risk of exposure to chemicals and/or allow the development of new surrogates that have greater relevance to predicting human responses.

Gene expression profiling using DNA microarrays has been widely applied to elucidate many biological processes since their introduction in 1995 (Casciano and Fuscoe, 2004). The field of toxicology was among the first to recognize the promise of this new technology to understand mechanisms of toxicity and identify biomarkers of exposure and effect, as well as to define fundamental cellular processes involved in disease. For example, Hu *et al.* (2004) and Dickinson *et al.* (2004) have described the development of *in vitro* biomarkers to distinguish between DNA-reactive (direct-acting) and DNA-nonreactive (indirect-acting) genotoxins. This distinction is particularly important in the evaluation of new pharmaceutical drugs and in the field of nutrigenomics. Early understanding of the

Toxicogenomics: A Powerful Tool for Toxicity Assessment   Edited by S. C. Sahu
© 2008 John Wiley & Sons, Ltd

biological relevance of positive genotoxic responses is expected to result in more efficient and less costly drug development and perhaps provide us with tools to determine the safety of newly developed nutriceuticals, dietary supplements, and nutrients in food. Recently, Ackerman *et al.* (2004) examined *in vitro* molecular pathways affected by DNA damage. Interestingly, they found very small effects on gene expression at low concentrations of the genotoxic polycyclic aromatic hydrocarbon benzo[*a*]pyrene diolepoxide, while robust changes occurred at doses associated with cellular toxicity and mutations. Some of the genes affected included those involved in apoptotic pathways, detoxification, and cell cycle control.

Koch-Paiz *et al.* (2004) examined the UV irradiation response in human cells in culture using microarrays. Their innovative use of an inhibitor of membrane signaling allowed these authors to assign a major role to growth factor receptor and other cytokine receptors in the UV stress response. Harris *et al.* (2004) compared the basal gene expression patterns in two promising human surrogate systems: cultured human primary hepatoctyes and the HepG2 human hepatoma cell line. Also, they investigated the effects of three genotoxic hepatocarcinogens in both model systems. Their study indicated that gene expression profiling might aid in understanding variability in drug reactions in humans, as well as showing that biomarkers of exposure may be defined despite this variability. Kier *et al.* (2004) described a microarray of rat-specific toxicology-related genes and the development of a database containing gene expression, histopathology, and clinical chemistry data on 89 compounds. Databases similar to this one, as well as those described by Waters *et al.* (2003) and Tong *et al.* (2003) show promise in exploring compound-specific expression profiles. In addition, Kier *et al.* (2004) presented data predicting human drug toxicity using primary cultured human hepatocytes as a model system.

Fielden *et al.* (2007) were interested in utilizing microarray technology to develop biomarkers that would provide early prediction and mechanistic assessment of hepatic tumor induction by nongenotoxic chemicals. They identified a novel molecular signature when they examined gene expression data from rats treated with 100 structurally and mechanistically diverse nongenotoxic hepatocarcinogens and nonhepatocarcinogens. They also showed that the gene-expression-based signature was more accurate in predicting the potential to identify nongenotoxic hepatocarcinogens than the typical alternate *in vivo* pathological biomarkers.

Desai *et al.* (2004) evaluated basal gene expression changes in the liver of rats as a function of time of day. More than 60 genes were found to be significantly altered, representing genes involved in drug metabolism, ion transport, DNA binding and regulation of transcription, signal transduction, and the immune response. Their data suggest that time of day effects need to be considered when planning *in vivo* microarray experiments. Boormann *et al.* (2005) also were interested in understanding the effect of time of day and day versus night on gene expression in the livers of the rat. Using DNA microarrays they identified differential expression in their comparisons. They found numerous periodically expressed genes including period genes, clock-controlled genes, and genes involved in metabolic pathways. They confirmed the study of Desai *et al.* (2004), demonstrating a prominent circadian rhythm in gene expression in the rat that is a critical factor in planning toxicogenomic experiments. Desai *et al.* (2007) also have developed a novel MitoChip microarray that can be used for transcriptional profiling to understand the basis of mitochondrial involvement in disease and toxicity (Desai and Fuscoe, 2007; Desai *et al.*, 2007). The array contains 542 oligonucleotides that represent genes from the mitochondrial and nuclear genomes associated with mitochondrial structure and function. They are validating the chip by measuring the expression of mitochondrial genes in the livers of p53 haplodeficient$^{(+/-)}$ and wild-type$^{(+/+)}$ C3B6F1 female mice exposed to zidovudine (AZT) and lamivudine (3TC), human drugs that are antiretroviral agents. They found that a majority of the mitochondrial genes were differentially expressed during AZT and 3TC treatment. These results confirm the present information regarding the pathology associated with rodents and humans exposed to antiretroviral agents (Poirier *et al.*, 2003).

Cornwell *et al.* (2004), Huang *et al.* (2004), and Hamadeh *et al.* (2004) examined the effects of fibric acid hepatotoxicants on gene expression in the livers of rats. In humans, fibrates are used to treat dyslipidemias. In rats and mice, these compounds induce liver toxicity and, in some cases, cancer. Corrnwell *et al.* (2004) investigated the effects of six fibric acid analogues on gene expression in rat livers and found similar changes seemingly linked to the clinical benefits of the drugs. Their analysis, however, could not distinguish the target effects from the off-target effects. Huang *et al.* (2004) studied gene expression profiles after exposure of rats to five different hepatotoxicants and found subsets of genes associated with specific types of hepatotoxicity, including inflammation, necrosis, hepatitis, and others. Their investigation suggested that gene expression profiling could be used to distinguish various types of hepatotoxicity, predict toxic endpoints, and develop hypotheses for the mechanisms of toxicity. Hamadeh *et al.* (2004) examined the gene expression changes associated with furan exposure and, importantly, related these data with clinical chemistry and pathology measures. This type of study integrating genomic information with clinical data holds great promise in the interpretation of the toxicological response. These authors also compared gene expression changes that occur in different lobes of the liver. Although similar changes in gene expression were noted, the magnitude of the change was more pronounced in the right lobe.

Yamashita *et al.* (2004), Faiola *et al.* (2004), and Sen *et al.* (2004) were interested in using gene expression profiling to understand the processes associated with the induction of tumors. Yamashita *et al.* (2004) investigated the early gene expression changes in the rat pyloric mucosae after exposure to the stomach carcinogen N-methyl-N'-nitro-N-nitrosoguanidine (MNNG) and compared these changes with the changes observed upon cessation of MNNG exposure and the changes associated with MNNG-induced stomach cancer. Many of the observed gene expression changes detected following MNNG exposure persisted in histological normal tissues and were present in the cancers. Faiola *et al.* (2004) examined the mechanism(s) of benzene-induced leukemia in the bone marrow of benzene-exposed mice by using gene expression profiling. Also, they studied a population of cells enriched for hematopoietic stem cells (HSCs) that are thought to give rise to the leukemic cells. Genes involved in apoptosis, cell cycle, and growth control were found to be altered. Interestingly, there were differences between the total bone marrow and the HSCs, with p21 being greatly induced in bone marrow but not in the HSCs. Sen *et al.* (2004) examined the effect of the Tsc2 mutation on the expression of rat kidney genes. This mutation is present in the Eker rat model that develops renal tumors with high penetrance. Many of the altered genes appeared to be directly or indirectly regulated by the P13K/Akt pathway, as well as other signaling pathways, offering new directions for the investigation of the cause of susceptibility in this model.

Rocket *et al.* (2004) used gene expression technologies to identify genes critical for male fertility. They compared gene expression in normal and abnormal human testes with those from infertility mouse models. These authors found genes associated with infertility in the mouse and the human. Intriguingly, there was little overlap in the across-species gene sets.

Sawada *et al.* (2005) utilized microarray technology to examine the molecular mechanisms that contribute to the development of phospholipidosis and to identify specific markers that might form the basis of an *in vitro* screening test. Phopholipidosis is characterized by the accumulation of phospholipids and it is well established that many cationic amphiphilic drugs have the ability to induce phospholipidosis. Sawada *et al.* (2005) treated HepG2 cells with 12 compounds known to induce phospholipidosis. They found that 17 genes showed a similar profile following treatment and were selected as candidate biomarkers. Interestingly, they found that 12 of these genes were associated with structural changes they identified using transmission electron microscopy. That is, microarray analysis revealed that factors such as alterations in lysosomal functions and cholesterol metabolism were involved in the genesis of phospholipidosis. Nioi *et al.* (2007) confirmed that the candidate genes selected by Sawada *et al.* (2005) were predictive of the phosholipidosis response and that, not surprisingly, the qualitative response was a function of dose. Based on these observations, Nioi *et al.* (2007) developed a high-throughput fluorescent-labeled phospholipids assay too identify phospholipidosis positive and negative molecular entities.

20.2   Food and Drug Administration Pharmacogenomic Guideline

The Food and Drug Administration (FDA), in anticipation of a significant increase in the use of genomic data in drug development and wishing to stimulate the use of new technologies in preclinical and clinical data submissions, issued a document: 'Guidance for Industry: Pharmacogenomic Data Submissions' (US FDA, 2005). The guideline followed workshops on pharmacogenetics/ pharmacogenomics in drug development and regulatory decision-making and it was at these workshops that the concept of 'safe harbor' was introduced as a mechanism to stimulate industry to submit data in the absence of penalties (Orr *et al.*, 2007). The guidance clarifies what type of genomic data should be submitted to the FDA and when during the drug development process. The guidance also introduces the concept of 'Voluntary Genomic Data Submissions' (VGDSs). This concept was instituted as a novel way for the regulated industry to share data with the FDA to benefit both the industry and the FDA by ensuring that the regulatory scientists become familiar with and prepared to evaluate future genomic submissions appropriately. In addition, the FDA developed a process to qualify biomarkers for use in submissions to the FDA (Goodsaid and Frueh, 2007). This pilot process covered the need identified in the 2005 guidance for qualification of 'exploratory' into 'probable valid' or 'known valid' biomarkers.

Even though the pharmaceutical industry has been utilizing genomic information in drug discovery for a number of years, they were hesitant in applying it to safety assessment because it was unclear to them how the FDA would use it in its review of investigational new drug applications. The VGDS model, a potential solution to this perceived problem, was introduced in the pharmacogenomic guidance to teach both sponsors and reviewers via multiple examples how genomic data are generated and analyzed by sponsors and how it should be reviewed by the FDA. As of 2006, the FDA had received more than 30 submissions. An Interdisciplinary Pharmacogenomic Group (IPRG) was formed to ensure that a quality review of these submissions occurred and also ensured a proper partitioning of voluntary from nonvoluntary submissions. The IPRG has representation from all of the FDA centers and is made up of scientists having significant experience in preclinical and clinical studies, as well as genomics and bioinformantics. In addition to the investigations already submitted, the FDA has had multiple consults with industry, indicating that the regulated industry is becoming comfortable sharing these types of data with the FDA. It is of interest that the concept of voluntary submissions is accepted in other geographic regions, and the first bilateral VGDS project in which both the FDA and the European Agency for the Evaluation of Medicinal Products reviewed the first submission and discussed their respective findings held a meeting with sponsors in 2005 (Orr *et al.*, 2007). The two agencies formalized this process by issuing in 2006 the 'Guiding Principles: Processing Joint FDA EMEA Voluntary Genomic Data Submissions' describing how bilateral VGDSs will be processed (US FDA, 2006).

About 60% of the VGDSs have focused on clinical design issues, while the remaining included both toxicogenomic and genomic data, including data from prototypic pharmacogenetic devices (Goodsaid and Frueh, 2007). Clinical submissions included data on oncology therapies, Alzheimer's disease, hypertension, depression, hypoglycemia, obesity, and rheumatoid arthritis. This experience has led to the identification of the need for additional information that had not been covered in the 2005 Guidance and the development of the path for qualification of exploratory biomarkers into known valid biomarkers, as well as technical recommendations on the generation and submission of genomic data.

Since the 2005 Pharmacogenomic Guideline, only the classified biomarkers 'exploratory,' 'probable valid' or 'known valid,' a description of a process by which an exploratory biomarker can be qualified as a valid biomarker, were needed. A known valid biomarker is (Goodsaid and Frueh, 2007):

> A biomarker that is measured in an analytical test system with well-established performance characteristics and for which there is wide-spread agreement in the medical or scientific community about the physiologic, toxicologic, pharmacologic, or clinical significance of the results.

The concept of biomarker qualification is needed by the FDA as a regulatory tool and to encourage accelerated identification of new biomarkers through the use of these new molecular tools. It is anticipated that the biomarkers associated with toxicity and disease discovered in animal surrogate systems may be useful in identifying similar biomarkers in humans exposed to drugs in clinical trials.

The FDA is in the process of assessing the role of pharmacogenomics in biologics (Lacaná *et al.*, 2007). One of the areas that pharmacogenomic studies could be extremely helpful is in identification of responders and nonresponders to a specific biologic challenge. Relevant pharmacogenomic information for biologics that would be useful would include variability in gene expression and how it relates to efficacy, genotype effect on efficacy, and drug–drug interactions.

20.3   ArrayTrack: Informatic Software Supporting Toxicogenomic Research and Pharmacogenomic Investigational New Drug Applications to the FDA

During the early years of the new century, the National Center for Toxicological Research (NCTR), the FDA's primary research center, was establishing core facilities for genomic, proteomic, and metabolomic technologies that used standardized experimental procedures to support centerwide toxicogenomic research. To support the large volume of data generated from the application of these new tools, the NCTR instituted a toxicoinformatic integrated system (TIS) for the purpose of fully integrating genomic, proteomic, and metabolomic data with the data in public repositories, as well as conventional *in vitro* and *in vivo* toxicology data. TIS enables curation in accordance with standard ontology and provides a rich collection of tools for data analysis and data mining. The software that was developed to accomplish this is called ArrayTrack (Tong *et al.*, 2003). ArrayTrack was recognized as a tool that was extremely suitable to be an integral part of the Center for Drug and Evaluation's (CDER's) need for an integrated bioinformatics infrastructure to support data management, analysis, and interpretation of the data submitted via the VGDS and the Pharmacogenomic Guideline to Industry.

ArrayTrack contains three integrated components: (1) Microarray DB, which stores essential data associated with a microarray experiment, including information on slide samples, treatment, and experimental results; (2) TOOL, which provides analysis capabilities for data visualization, normalization, significance analysis, clustering, and classification; and (3) LIB, which contains information from public repositories, e.g. gene annotation, protein function, and molecular and metabolic pathways. Microarray DB and LIB are used to store FDA in-house experimental results and public data, whereas TOOL provides various algorithms for data visualization and analysis. ArrayTrack is not open-source software but can be accessed at http://www.fda.gov/nctr/science/centers/toxicoinformatics/Array/Track/. Those interested in acquiring the software free of charge can do so via the website.

Microarray DB and LIB were developed based on the Oracle relational database management system. The structure was designed to accommodate the essential data associated with a microarray experiment, including the toxicogenomic experimental design, the microarray design, sample and treatment annotation, and the hybridization procedure and the experimental results. ArrayTrack is MIAME compliant with the inclusion of additional parameters related to toxicogenomics, using controlled vocabularies and is MAGE-ML compatible, allowing data in the Microarray DB to be communicated with other microarray repositories. ArrayTrack's LIB is a compilation of essential public data to facilitate annotation and interpretation of gene expression data and includes GenBank (http://www.ncbi.nlm.nih.gov; Benson *et al.*, 2003), SWISS-PROT (http://www.expasy.org/sprot and http://ebi.ac.uk/swissprot; Boeckmann *et al.*, 2003), LocusLink (http://www.ncbi.nlm.nih.gov/LocuLink; Kanehisa, 2002; Kyoto Encyclopedia of Genes and Genomes (KEGG; http://www.genome.ad.jp/kegg; Kanehisa, 2002); and Gene Ontology (GO; http://godatabaseorg/dev/database; Ashburner *et al.*, 2000).

The TOOL was designed to provide a spectrum of algorithmic tools for microarray data visualization, quality control, normalization, significant gene identification, pattern discovery, and class prediction. A quality control/quality assurance tool was developed to assist quality control of slide array results. This tool summarizes the most relevant information into one interface to facilitate the process of quality control. The investigator can determine the quality of individual microarray results through visualizing data, applying statistical measures, and viewing experimental annotation. Statistical measures are provided to determine the quality of a hybridization result based on raw data, including signal-to-signal ratio, the percentage of nonhybridized spots, etc. The experimental annotations associated with the hybridization processes, RNA extraction, and labeling are also available to the user. Also, a scatter plot of Cy3 versus Cy5, together with the original image, is available for visual inspection for quality control purposes.

Several data visualization methods are provided, two of which are ScatterPlot Viewer and VirtualImage viewer. The ScatterPlot graphs gene expression profiles of one sample versus another sample, while VirtualImage viewer displays the expression pattern in an array image format. Both of these tools permit visual identification of significant genes and hyperlink directly from the graph to additional detailed library information on any particular gene.

GeneLib, ProteinLib, and PathwayLib include general but essential information for functional genomics research. The libraries also provide a basis for linking and integrating various -omics data. For example, lists of genes, proteins, and metabolites derived from various -omics platforms can be cross-linked based on their common identifiers through these three libraries. An additional library, ToxicantLib, contains toxicological responses that can be linked to the different -omics data and provides a phenotypic anchor. Other added toxicant libraries include the endocrine disruptor knowledge base, the carcinogenicity potency base, LiverLib (genes/proteins associated with liver toxicity), and SNPsLIB (containing information on single nucleotide polymorphisms).

20.4 Microarray Quality Control Project

DNA microarray technology, a tool that can evaluate simultaneously the relative expression of thousands of genes, has developed rapidly and is transitioning to becoming the preferred technology to identify early biomarkers of toxicity and disease. The outcome of microarray studies can be affected by many technical, instrumental, computational, and interpretive factors. Indeed, a major criticism voiced about microarray studies has been the lack of reproducibility and accuracy of the derived data. To address this concern, the microarray community and regulatory agencies, led by the FDA, have developed a consortium to establish a set of quality assurance and quality control criteria to assess and assure data quality, to identify critical factors affecting data quality, and to optimize and standardize microarray procedures so that biological interpretations and decision making are not based on unreliable data. These fundamental issues are addressed by the Microarray Quality control (MAQC) project (Casciano and Woodcock, 2006).

The MAQC project aims to establish quality control metrics and thresholds for the objective assessment of the performance achievable by different microarray platforms and evaluating the merits and limitations of various data analysis methods. It is anticipated that the MAQC project will assist in improving microarray technology and foster its appropriate application in discovery, development, and review of FDA-regulated products.

The results of phase one of the MAQC project are published in a single issue of *Nature Biotechnology*, volume 24(9), September 2006. Gene expression data on four titration pools from two distinct reference RNA samples were generated at multiple test sites using a variety of microarray-based and alternative technology platforms. The data generated indicated intraplatform consistency across test sites as well as a high level of interplatform concordance in terms of genes identified as differentially expressed. The publication suggests a rich resource that represents an important first step toward establishing a framework for the use of microarrays in clinical and regulatory settings.

20.5   Conclusions

In 2004 the FDA initiated the Critical Path to Development of Medical Products (US FDA, 2004). The Critical Path refers to the scientific process through which a new medical product is transformed from a discovery or 'proof of concept' into a useful human drug or biological product or medical device. When a product is identified as a promising candidate in basic research, it then enters a product development process that consists of a series of scientific evaluations, often increasing in size and complexity, to predict whether the candidate will be safe and effective. These tests also guide a drug sponsor in choosing an appropriate dose and regimen. Unfortunately, many of the scientific tools and tests used to evaluate the safety and effectiveness of a candidate product have not kept pace with the rapid evolution occurring in other arenas and do not allow capitalization on those advances in basic science. A concerted effort to apply *new* scientific knowledge, in areas such as gene expression, analytic methods, and bioinformatics, to product development is needed. The FDA's Critical Path Initiative seeks to mobilize these new sciences in the service of product development, so we can learn more about a product before it is marketed. With better information, the pharmaceutical industry will be able to make more informed decisions; for example, about which candidate products to move to the next phase of development, what doses will most likely prove to be safe and effective, and which products will likely fail. Technologies such as genomics, proteomics, metabolomics, and bioinformatics could provide information needed to help sponsors develop safer and more effective medical products.

One of the goals of the FDA's Critical Path is to transit from practicing medicine the way we do now, namely population medicine, to personalized medicine. Personalized medicine's goal is to maximize the likelihood of therapeutic efficacy and to minimize the risk of drug toxicity for an individual patient. One of the major contributors to this concept is pharmacogenomics, a science that is concerned with interindividual genetic variation and which the FDA has promoted for the understanding of how this variation contributes to both susceptibility to diseases and responses to drugs. The FDA initiated this promotion by developing the 'Guidance for Industry: Pharmacogenomic Data Submissions' (US FDA, 2005), which introduced the concept of voluntary submissions of genomic data to facilitate the critical path to drug development.

References

Ackerman GS, Rosenzweig BA, Domon OE, McGarrity LJ, Blankenship LR, Tsai CA, Culp SJ, Macgregor JT, Sistare FD, Chen JJ, Morris SM (2004) Gene expression profiles and genetic damage in benzo(*a*)pyrene diol epoxide-exposed TK6 cells. *Mutat Res* **549**, 43–64.

Ashburner M, Ball CA, Blake JA, Botstein D, Butler H, Cherry JM *et al.* (2000) Gene ontology: tool for the unification of biology. The Gene Ontology Consortium. *Nat Genet* **25**, 25–29.

Benson DS, Karsch-Mizrachi I, Lipman DJ, Ostell J, Wheeler DL (2003) *GenBank.* Nucleic Acids Res **31**, 23–27.

Boeckmann B, Bairoch A, Apweiler R, Blatter MC, Eistreicher A, Gasteiger E *et al.* (2003) The SWISS-Prot protein knowledgebase and its supplement TREMBL. *Nucleic Acids Res* **31**, 365–370.

Boorman GA, Blackshear PE, Parker JS, Lobenhofer EK, Malarkey DE, Vallout MK, Gerken DK, Irwin RD (2005) Hepatic gene expression changes throughout the day in the Fischer rat: Implications for toxicogenomic experiments. *Toxicol Sci* **86**, 185–193.

Casciano DA, Fuscoe JC (2004) Preface to *Mutation Research* special issue on toxicogenomics. *Mutat Res* **549**, 1–4.

Casciano DA, Woodcock J (2006) Empowering microarrays in the regulatory setting. *Nat Biotechnol* **24**, 1103.

Cornwell PD, DeSouza AT, Ulrich RG (2004) Profiling of hepatic genes in rats treated with fibric acid analogs. *Mutat Res* **549**, 131–146.

Desai VG, Fuscoe JC (2007) Transcriptional profiling for understanding the basis of mitochondrial involvement in disease and toxicity using the mitochondria-specific MitoChip. *Mutat Res* **616**, 210–212.

Desai VG, Moland CL, Branham WS, Delongchamp RR, Fang H, Duffy PH, Peterson CA, Beggs ML, Fuscoe JC (2004) Changes in expression level of genes as a function of time of day in the liver of the rat. *Mutat Res* **549**, 115–130.

Desai VG, Lee T, Delongchamp RR, Moland CL, Branham WS, Fuscoe JC, Leakey JEA (2007) Development of mitochondrial-specific mouse oligonucleotide microarray and validation of data by real-time PCR. *Mitochondrion* **7**, 322–329.

Dickinson DA, Warnes GR, Quievrn G, Messer J, Zhitkovich A, Rubitski E, Aubrecht J (2004) Differentiation of DNA reactive genotoxic mechanisms using gene expression profile analysis. *Mutat Res* **549**, 29–42.

Faiola B, Fuller ES, Wong VA, Recio L (2004) Gene expression profile on bone marrow and hematopoitec stem cells in mice exposed to inhaled benzene. *Mutat Res* **549**, 195–214.

Fielden MR, Brennan R, Gollub J (2007) A gene expression biomarker provides early prediction and mechanistic assessment of hepatic tumor induction by nongenotoxic chemicals. *Toxicol Sci* **99**, 90–100.

Goodsaid F, Frueh FW (2007) Implementing the U.S. FDA guidance on pharmacogenomic data submissions. *Environ Mol Mutagen* **48**, 354–358.

Hamadeh HK, Jayedev S, Gaillard ET, Huang Q, Stoll R, Blanchard K, Chou J, Tucker CJ, Collins J, Naronpot R, Bushel P, Afshari CA (2004) Integration of clinical and gene expression endpoints to explore furan-mediated hepatotoxicity. *Mutat Res* **549**, 169–184.

Harris AJ, Dial SL, Casciano DA (2004) Comparison of basal gene expression profiles and effects of hepatocarcinogens on gene expression in cultured primary human hepatocytes and HepG2 cells. *Mutat Res* **549**, 79–100.

Hu T, Giibson DP, Carr GJ, Torontali SM, Tiesman JP, Chaney JG, Aardema MJ (2004) Identification of a gene expression profile that discriminates indirect-acting genotoxins from direct-acting genotoxins. *Mutat Res* **549**, 5–28.

Huang Q, Jin X, Gaillard ET, Knight BL, Pack FD, Stolz JH, Jayedev S, Blanchard KT (2004) Gene expression profiling reveals multiple toxicity endpoints induced by hepatotoxicants. *Mutat Res* **549**, 147–168.

Kanehisa M (2002) The KEGG database. *Novartis Found Symp* **247**, 91–101.

Kier LD, Neft R, Tang L, Suizu R, Cook T., Onsurez K, Tiegler K, Sakai Y, Ortiz M, Nolan T, Sankar U, Li AP (2004) Application of microarrays with toxicologically relevant genes (*tox* genes) for the evaluation of chemical toxicants in Sprague Dawley rats *in vivo* and human hepatocytes *in vitro*. *Mutat Res* **549**, 101–114.

Koch-Paiz CA, Amundson SA, Bittner ML, Meltzer PS, Fornace AJ (2004) Functional analysis of UV irradiation responses in human cells. *Mutat Res* **549**, 65–78.

Lacaná E, Amur S, Mummanneni P, Zhao H, Frueh FW (2007) The emerging role of pharmacogenomics in biologics. *Clin Pharmacol Ther* **82**, 466–471.

Nioi P, Perry BK, Wang EJ, Gu YZ, Snyder RD (2007) *in vitro* detection of drug-induced phospholipidosis using gene expression and fluorescent phospholipids-based methodologies. *Toxicol Sci* **99**, 162–173.

Orr MS, Goodsaid F, Amur S, Rudman A, Frueh FW (2007) The experience with voluntary genomic data submissions at the FDA and a vision for the future of the voluntary submission program. *Clin Pharmacol Ther* **81**, 294–297.

Poirier MC, Div RL, Al-Hari L, Olivero OA, Nguyen V, Walker B, Landay AL, Walker V, Charurat M (2003) Long-term mitochondrial toxicity in HIV-infected infants born to HIV-infected mothers. *J Acq Immun Def Synd* **33**, 175–183.

Rocket JC, Patrizio P, Schmid JE, Hecht NB, Dix D (2004) Gene expression patterns associated with infertility in humans and rodent models. *Mutat Res* **549**, 225–240.

Sawada H, Takami K, Asahi S (2005) A toxicogenomic approach to drug-induced phospholipidosis: analysis of its induction mechanism and establishment of a novel *in vitro* screening system. *Toxicol Sci* **83**, 282–292.

Sen B, Wolf DC, Hester SD (2004) The transcriptional profile of the kidney in Tsc2 heterozygous mutant Long Evans (Eker) rats compared to wild-type. *Mutat Res* **549**, 213–224.

Tong, W, Cao X, Harris S, Sun H, Fang H, Fuscoe J, Harris A, Hong H, Xie Q, Perkins R, Shi L, Casciano D (2003) ArrayTrack-supporting toxicogenomic research at the U.S. Food and Drug Administration National Center for Toxicological Research. *Environ Health Perspect* **111**, 1819–1826.

Tong W, Harris S, Cao X, Fang H, Shi L, Sun H, Fuscoe J, Harris A, Hong H, Xie Q, Perkins R, Casciano D (2004) Development of public toxicogenomics software for microarray data management and analysis. *Mutat Res* **549**, 241–253.

US FDA (2004) FDA Critical Path for Medical Products. http://www.fda.gov/oc/initiatives/criticalpath [May 2008].

US FDA (2005) Guidance for industry: pharmacogenomic data submissions. http://www.fda.gov/cder/guidance/6400fnl.pdf [May 2008].

US FDA (2006) Guiding principles: Processing joint FDA EMEA Voluntary Genomic Data Submissions (VGDSs) within the framework of the Confidentiality Arrangement. http://www.fda.gov/cder/genomics/FDAEMEA.pdf [May 2008].

Waters M, Boorman G, Bushel P, Cunningham M, Irwin R, Merrick A, Olden K, Paules R, Selkirk J, Stasiewicz S, Weis B, Van Houten B, Tennant R (2003) Systems toxicology and the chemical effects in biological systems (CEBS) knowledge base. *Environ Health Perspect* **111**, 15–28.

Yamashita S, Nomoto T, Abe M, Tatematsu M, Sugimura T, Ushijima T (2004) Persistence of gene expression in stomach mucosae induced by short-term *N*-methyl-*N'*-nitro-*N*-nitrosoguanidine treatment and their presence in stomach cancers. *Mutat Res* **549**, 185–194.

21

Toxicogenomics for Regulatory Use: The View from the Bench

P. Ancian, S. Leuillet, S. Arthaud, J.J. Legrand and R. Forster

21.1 Introduction

Toxicogenomic approaches are generating valuable data that cast light on the toxic actions of drugs and chemicals, aid with the prediction of potential toxicities of development candidates, and elucidate the mechanisms that underlie toxic actions. The information generated using these approaches can enrich understanding of new drugs and strengthen the conclusions of regulatory review. It is of interest therefore, to pharmaceutical companies to present toxicogenomic data in support of regulatory submissions and of interest to regulatory authorities to have access to this data and to take account of it in their evaluation of new products. This enthusiasm is reflected in the emphasis given to toxicogenomics in the Food and Drug Administration (FDA) Critical Path Initiative (FDA, 2004), which identified genomic technology as the route to a better 'safety toolkit' for future medical product development.

As a consequence, there has been great interest in the standards of performance and reporting to be required of microarray studies that are intended for regulatory submission. In this chapter we will focus on the issues that face laboratory scientists and regulatory professionals planning to perform *in vivo* gene expression studies with new drug candidates and to present the findings in a regulatory submission. The data that they provide must comply with testing standards for regulatory safety studies (including compliance with good laboratory practice (GLP)) and also the current testing standards specifically addressing toxicogenomic data generation.

Our laboratory performs toxicogenomic studies (principally in rodents and primates) intended to support the selection and development of new drugs (Bourrinet *et al.*, 2007;

Toxicogenomics: A Powerful Tool for Toxicity Assessment   Edited by S. C. Sahu
© 2008 John Wiley & Sons, Ltd

Kravtzoff *et al.*, 2007), with a throughput of 2000 microarray chips in the last 6-month period and a total of 20 000 RNA extractions in the last 24 months. Our comments below are based on our experience in the conduct of this work, and focus principally on studies of differential gene expression in whole animals (as opposed to studies in cultured mammalian cells).

21.2   Objectives of Toxicogenomics

Toxicogenomics is the study of gene expression patterns after exposure to a test item with the objective of gaining a deeper mechanistic understanding of toxic actions, and developing predictive tools to rank, select, and evaluate new drugs. Genome-based technologies such as DNA microarrays allow the simultaneous analysis of the expression of many thousands of genes in a single experiment. Whole-genome microarrays are available for several species, including humans and laboratory animal species routinely used in safety evaluation, such as the rat, mouse, primate, and pig. The study of thousands of parameters in a single sample can give new insights into the assessment of the effects which a chemical or drug can cause, whether good (pharmacology or efficacy) or bad (toxic). As significant changes in the regulation of gene expression generally occur in the early phases after exposure to the test item, this opens the way for the deployment of toxicogenomics in different ways:

- Compound classification – compounds with the same pharmacological or toxic mechanism of action generate similar changes in gene expression profiles and can be ranked accordingly.
- A better understanding of toxicological mechanisms by the study of the signaling pathways or ontologies in which proteins encoded by the modulated genes are classified.
- The discovery of new pharmacological/efficacy/toxicity biomarkers – consistently regulated genes across several compound time/dose combinations are good candidates.
- The generation of toxicity signature databases, by correlating the early gene expression profiles induced by reference compounds with traditional toxicology endpoints such as histopathology or clinical pathology findings.
- The early prediction of new compound toxicity by transcriptional profile comparisons with existing (in-house, commercially or publicly available) toxicity signature databases.
- The discovery of new pharmacological actions of existing drugs, permitting the repositioning of these drugs for new therapeutic indications.
- The different steps of a toxicogenomics study, as currently performed, are as follows:
- *In vivo* phase – generally, toxicogenomics studies are short-term studies that include several dose–time combinations in order to produce a maximal impact on gene expression. They include at least three biological replicates (animals) per dose/time combination with their matched controls. Time points for sacrifice and tissue sampling can be as short as few hours until several days or weeks following treatment. At least two test item dose-levels are used: a pharmacologically active but nontoxic dose level and a toxic dose level (usually the maximum tolerated dose). Depending on the study objectives, reference or competitor molecules can be added as comparators.
- Traditional toxicology parameter measurements (clinical signs, clinical pathology, histopathology, etc.). Characterization of the toxic actions in this way provides the so-called 'phenotypic anchoring' against which gene expression changes can be compared.

- RNA extraction, labeling and hybridization to DNA chips – RNA is fragile during laboratory handling, but the quality of RNA is of primordial importance for reliable transcriptomics data generation. After microarray processing, the fluorescence values from each probe generated at chip scanning provide the raw data for evaluation of the modulation of gene expression.
- Data analysis – depending on the study objectives, data analysis will be performed with different tools. As well as first-line statistical approaches (data filtering, clustering and discrimination), more sophisticated tools are available. These include specific genomics and statistical software, gene ontology databases (controlled vocabulary to describe gene and gene product attributes), and cellular pathway databases which can be used for the understanding of mechanisms of toxicity. Learning machines (supported vector machines for instance) may be used for the generation of toxicity signature databases in the field of predictive toxicology.

21.3 Conduct of *In Vivo* and Traditional Study Elements

The in-life part of whole-animal toxicogenomics studies is no less critical than the microarray analysis, and makes use of a different skill set and infrastructure. Know-how, experienced staff, and custom-built facilities are required, including animal rooms with sophisticated environmental control, necropsy areas, clinical pathology, and histopathology laboratories, as for the conduct of GLP-compliant regulatory toxicology studies. Location of the in-life and microarray phases of a study on the same site may not be essential, but will certainly be beneficial in permitting ready exchange and dialog between the toxicology and genomics scientists to optimize key interface phases of studies, such as necropsy and RNA extraction (see below). The skills and experience required of toxicology Study Directors are rarely found among the genomics staff; therefore, collaboration is needed to bring all of the required know-how for the conduct and interpretation of the traditional toxicology element of the study. Input from the Study Director may also help identify sources of variation that derive from animal work (for example, minimizing the impact of gene expression from phases of active growth). In general, the study designs for toxicogenomics studies will resemble standard toxicology study designs in terms of animal numbers and supporting clinical, laboratory, and histopathology analyses (Leighton *et al.*, 2006). The phenotypic correlates of treatment are essential data, and full characterization of the toxic actions must be made using traditional toxicology and histopathology approaches. The study results and data should be fully presented in a GLP toxicology report format (Leighton *et al.*, 2006; FDA, 2007).

21.4 Generation of Gene Expression Data

The use of transcriptomics data in a regulatory context can only be achieved if scientists and regulators are confident in reliable, reproducible, and validated data. To this end, sources of variation in toxicogenomics studies have to be minimized, controlled, and monitored with appropriate standards of performance. These sources of variation can be classified in three main categories: biological variation, variation deriving from microarray processing and technology, and different approaches in data analysis.

Biological variability is related to the study design, the animal strains used, the environment, and the genetic background of the host. In particular, the liver plays a central role in toxicology and is subjected to variations linked to circadian rhythm, feeding and diet, housing or fasting (Morgan *et al.*, 2005). By careful analysis of different experimental replication study designs, we and others have shown that biological variability is the most important source of variation in toxicogenomics experiments (Zakharkin *et al.*, 2005; Ancian *et al.*, 2006).

The second source of variation is that deriving from microarray processing and technology. One of the major hurdles to be overcome to permit the use of microarray data in a regulatory context was the high degree of technical heterogeneity in gene expression measurement. Huge progress has been made in this field, as a result of the efforts of providers such as Agilent, Illumina, Affymetrix, and others. Nevertheless, the use of different technical platforms introduced concerns about the transposability of gene expression data across platforms. In addition, there is a further source of variation deriving from a whole series of genomics laboratory steps, including tissue sampling, RNA extraction, processing, and labeling and microchip hybridization. Each of these steps must be rigidly standardized (in standard operating procedures (SOPs)) and monitored using appropriate quality control (QC) checks.

In terms of purity and quality, RNA samples must be as homogeneous as possible between study groups. Indeed, the quality of RNA is a determinant factor in the generation of reliable toxicogenomics data (Ancian *et al.*, 2007). Since RNA is at risk of rapid degradation in the tissues of animals during the minutes after sacrifice, modified necropsy procedures are needed to reduce the time between the moment of animal sacrifice and the onset of steps that protect the RNA (tissue snap-freezing or soaking in RNAlater and other protective solutions). The speed of RNA degradation and the difficulty of RNA extraction vary from tissue to tissue. Very different challenges are posed, for example, by bone, brain, and blood. As a consequence, tissue-specific procedures must be developed, optimized, and standardized. Those tissues with a very rapid rate of RNA degradation post-sacrifice (for example, pancreas) must be the first that are sampled at necropsy. Accordingly, a standardized necropsy sequence should be developed and adhered to.

Moreover, RNA extraction processes and RNA quality controls must be very robust. We have shown, for instance, that the RNA integrity number (RIN), an RNA quality metric developed by Agilent, is a very relevant parameter in order to define the lower limits or thresholds of RNA quality below which samples cannot be considered as reliable for analysis and should not be analyzed (Ancian *et al.*, 2007).

The sources of variation due to the technological platform must be controlled. First of all, several studies from the HESI Hepatotoxicity Working group and the Microarray Quality Control (MAQC) project have shown that the results obtained in toxicogenomics experiments are similar across microarray platforms and/or sites (Waring *et al.*, 2004; Irizarry *et al.*, 2005; MAQC Consortium, 2006). Within this context, the training of laboratory operators as well as the qualification of the microarray platform and validation of assay methods are all necessary in order to ensure the standardization of the data generated. This is not made easier by the lack of RNA analytical standards. Several approaches have been proposed in order to address these questions. Proficiency testing can be used to monitor the laboratory performances by comparing the data generated with the same starting samples. The MAQC project published (in a special issue of *Nature Biotechnology*, November

2006, made available free of charge at the MAQC website) several articles dedicated to the assessment of key factors contributing to the variability and reproducibility of microarray data. MAQC provides quality control tools and ideas to the microarray community in order to avoid procedural failures and to develop guidelines for microarray data analysis (Shippy *et al.*, 2006; Tong *et al.*, 2006). The MAQC Consortium is composed of governmental agencies, industry, and academia, and its work led to the development by the FDA of the draft guidance document 'Guidance for Industry, Pharmacogenomic Data Submissions – Companion Guidance' (FDA, 2007).

The third source of variation is that introduced during the data analysis, as there are numerous ways of data preprocessing and processing methods, with a large number of different algorithms. In that respect, appropriate data analysis procedures are essential (Shi *et al.*, 2005; Guo *et al.*, 2006) as described in the following section.

The MAQC papers, and the experience of our laboratory, show that when working with trained staff, robust SOPs, and pertinent QC tools at each step of the microarray process, the major sources of variation are controlled, giving a high level of data reproducibility and quality, suitable for regulatory use.

21.5   Bioinformatics and Interpretation

21.5.1   Data Analysis and Regulatory Requirements

The FDA Guideline on Pharmacogenomic data submissions (FDA, 2005a) does not provide any specific recommendation about the statistical and bioinformatic analysis that should be conducted on microarray raw data. The guidance document only requires that a sufficient level of information is provided to permit independent analysis of the data and verification of the results. The submitted report should contain a part describing the data analysis performed, detailing the statistical analysis, the bioinformatics tools, and software used, and the source of gene annotation or data visualization.

21.5.2   State-of-the-art Data Analysis Workflow

There is no current consensus approach to the data analysis of gene expression microarray data. The choice of the methods used depends in part on the design and aim of the toxicogenomics experiment, and whether the objective was exploratory, mechanistic, or predictive. We present here an overview of the different methods that are in current use for the interpretation of microarray data. The methods are applied successively in a series of steps.

The first step of microarray analysis consists of data preprocessing (e.g. background subtraction or log transformation) and normalization to reduce the systematic errors in experiments (due, for example, to scanner parameters, chip batches, etc.). The choice of data preprocessing/normalization algorithms is a major contributor to interlaboratory variability in results (Hoffmann *et al.*, 2002). Next, filtering of the data permits the selection of probes showing changes in transcription that are both statistically significant and meet fold-change criteria (such as twofold changes) between treatment/dose conditions. This filtering provides listings of modulated genes.

The clustering or grouping of data according to similarities across genes expression profiles provides an initial exploration of gene expression changes (Freeman, 2005). Several methods for the visualization of clustering data can be used, including hierarchical clustering, K-means clustering (a partitioning algorithm), and principal component analysis (a dimension-reducing technique).

The modulated transcripts can be associated with cellular functions (such as biochemical and signaling pathways) using standardized data resources of transcript annotation, such as Gene Ontology (GO) or the Kyoto Encyclopedia of Genes and Genomes (KEGG) (Yue *et al.*, 2005).

In order to use gene expression data in predictive mode, gene expression 'signatures' derived from known toxicants are used to predict the toxicological properties of 'unknown' candidate drugs. Such signatures can be accessed in proprietary databases (such as those of Gene Logic and Iconix). The investigator may also use supervised learning methods to identify and exploit signatures from their own bank of experimental data. Supervised learning is a machine learning technique for creating a function from training data. The methods used for classification are very varied, and include linear discriminant analysis, nearest-neighbor methods and machine learning methods, such as support vector machines and artificial neural networks (Maggioli *et al.*, 2005).

Many different bioinformatics tools, both free and proprietary, are available for the analysis of transcriptomics data. These tools are in continuous evolution, as new statistical algorithms and methods are developed. One of these tools, ArrayTrack (Tong *et al.*, 2003), a free resource developed by the National Center for Toxicoinformatics, is used by the FDA for the review of genomic data submission. ArrayTrack is a very comprehensive software package allowing the management, analysis, and interpretation of −omics data within a single tool.

21.5.3   Reference Databases

The goal of gene expression databases is to store and manage transcriptional data, together with the associated phenotypic anchoring (toxicology) data, with the purpose of facilitating data sharing among the scientific community and promoting the development of international data standards (Mattes *et al.*, 2004). Several companies (including Gene Logic, Curagen, and Iconix Pharmaceuticals) have developed proprietary toxicogenomics gene expression databases.

Some journal publishers require that data must be successfully submitted to one of the public gene expression databases repositories (Brooksbank and Quackenbush, 2006; Corvi *et al.*, 2006), such as the ArrayExpress database at the European Bioinformatics Institute (Brazma *et al.*, 2003), Gene Expression Omnibus (GEO) at the National Center for Biotechnology Information (NCBI; Edgar *et al.*, 2002), and the Center for Information Biology Gene Expression (CIBEX) database at the DNA Databank of Japan (DDBJ; Ikeo *et al.* 2003). For submission to these databases, documentation must be Minimum Information About a Microarray Experiment (MIAME) compliant (see below).

21.5.4   Bioinformatics Conclusions

Standardization of analysis and interpretation of microarray data remains a great challenge. Many statistical methods can be applied and many tools are available, which can lead the investigator to varying interpretations of results. With the development of toxicogenomics databases, supported by journal publishers, standards are currently emerging. We can hope that in the near future, by working together, the computational approaches will become more standardized, which will facilitate data sharing across scientific community.

21.6 Study Documentation

It is a tenet of preclinical drug safety that documentation of experimental work should permit reconstruction of the work performed, and this is enshrined in the GLP regulations. GLP places great emphasis on the importance of study documentation and access to study raw data. These principles apply equally to toxicogenomics studies performed for regulatory submission.

In addition to GLP requirements, further documentation needs arise from the nature of the data that is being treated. Gene expression data are highly context dependent, and without a full description of context (organism, age, organ, treatment, duration, etc.) such data have no meaning. To address this, genomics scientists have developed their own standards for documentation of microarray data in order to permit unambiguous interpretation of the data and (potentially) to permit experiments to be reproduced. The MIAME standards define these documentation requirements (Brazma *et al.*, 2001). The six principal elements of MIAME are:

1. the experimental design, including sample data relationships (e.g. which raw data file relates to which sample, which hybridizations are technical, which are biological replicates);
2. the essential sample annotation, including experimental factors and their values (e.g. compound and dose in a dose–response experiment);
3. sufficient annotation of the array (e.g. gene identifiers, genomic coordinates, probe oligonucleotide sequences or reference commercial array catalog number);
4. the raw data for each hybridization (e.g. CEL or GPR files);
5. the essential laboratory and data processing protocols (e.g. what normalization method has been used to obtain the final processed data);
6. the final processed data for the set of hybridizations in the study (e.g. the gene expression data matrix used to draw the conclusions from the study).

Both the FDA and the EMEA guidance documents on regulatory submission underline the importance of compliance with MIAME standards (EMEA, 2006b; Leighton *et al.*, 2006; FDA, 2007).

Since the original MIAME proposal, this standard has been further developed. A proposal for MIAME Version 2.0 is under discussion. An extension to MIAME adapted to toxicogenomics field has been developed by ILSI, NIEHS, NCT, NCTR and EBI (see Committee on Applications of Toxicogenomic Technologies to Predictive Toxicology and Risk Assessment, National Research Council, 2007).

The MIAME standard defines minimum data requirements, but it does not define the format in which data must be presented, or the terminologies and lexicons that must be used. The Microarray and Gene Expression Data Society (http://www.mged.org/) has made further recommendations to extend harmonization into terminology and has developed a simple spreadsheet-based, MIAME-supportive format (MAGE-TAB) for microarray data.

21.7 Regulatory Submission

In 2005, after a series of workshops and consultations, the FDA issued a guidance document on pharmacogenomics data submission (FDA, 2005a), where the term 'pharmacogenomics' covers a range of technologies, including gene expression studies and also genotyping investigations (in animals and humans) and single nucleotide polymorphism microarray studies. The document provides general guidance that applies to (toxicogenomics) gene expression studies and describes when to submit data, how to submit it, and identifies the circumstances where data submission is obligatory and or voluntary. Pharmacogenomic data which supports decision making in a clinical trial or preclinical safety study must be submitted to the FDA as part of the corresponding regulatory submission (IND or NDA). Thus, submission is obligatory when data are used as part of scientific rationale regarding a new drug, or when data refer to a well-known and valid biomarker. Pharmacogenomic data generated for other purposes (e.g. exploratory work, studies establishing the value and pertinence of a gene expression signature) may be submitted voluntarily as a Voluntary Genomic Data Submission (VGDS). A further guidance document provides examples of voluntary submissions (FDA, 2005b). Toxicogenomics studies which must obligatorily be submitted should be GLP compliant. Greater fluidity is accorded to voluntary submissions: an active IND is not required for submission, submission formats are flexible, and (most importantly) voluntarily submitted data will not be used as a basis for regulatory decisions by the agency.

The Interdisciplinary Pharmacogenomics Review Group (IPRG) was established within the FDA in order to act as the primary review body for VGDSs. For the review of (non-voluntary) submissions in the context of regulatory dossiers of new drugs, the Nonclinical Pharmacogenomics Subcommittee (NPSC) was formed within the FDA. Examples of 'mock' submissions, intended to train and test the review procedures of the NPSC, are given in Leighton *et al.* (2006). NPSC have access to the proprietary toxicogenomics database toxicity signatures and study reporting system developed by ICONIX, and examples of the value of 'contextual databases' in the interpretation of data are given in Leighton *et al.* (2006). The FDA has also set up a website portal dealing with regulatory submission (http://www.fda.gov/cder/genomics). The website provides a useful resource with access to guidance documents and explanatory material (and a tabulation of genomic biomarkers which are currently considered as 'valid biomarkers').

In a companion draft guidance document (FDA, 2007), issued last year and previously circulated as a concept paper, recommendations are presented regarding the methodological issues relating to microarray data. This document provides detailed discussion of laboratory aspects, such as ensuring the quality of RNA, labeling and hybridization steps, microarray reader settings, and data interpretation. One section of the guideline deals with proficiency testing of laboratories, for example using reference samples of mixed RNAs from the MAQC program (now commercially available) and recommends that sponsors submit data demonstrating the competence of laboratories submitting pharmacogenomic data. The guideline also comments on the submission of toxicogenomics data, describing the supporting information that should be provided. It is requested that toxicology data are submitted in SEND format (Standard for Exchange of Nonclinical Data). The guideline emphasizes the need for validated SOPs for laboratory activities and the value of inclusion of reference standards coming from the MAQC program.

The number of VGDSs to the FDA so far has been modest (about 30 at the time of writing, and experience with these submissions has been summarized in brief publications (Orr *et al.*, 2007)). The review process for nonvoluntary submissions has been tested with mock submissions as described by Leighton *et al.* (2006).

The European Medicines Evaluation Agency has issued a guidance document on pharmacogenetic briefing meetings (EMEA, 2006b). Briefing meetings are intended to allow an informal exchange on the technical, scientific, and regulatory issues that may arise for development plans and regulatory processed from the submission of pharmacogenetic/genomic data. This document also provides a format for the presentation of experimental design, statistical methods, and pharmacogenomic testing methodology.

A further EMEA guidance document describes the organization of joint FDA–EMEA voluntary genomic data submission briefing meetings (EMEA, 2006a).

21.8   Conclusions

The development of toxicogenomic data on new drugs and candidate drugs is a challenging area. Many uncertainties remain, even regarding key questions such as interpretation of the physiological significance of pharmacogenomic findings, or extrapolation of these findings and conclusions across animal species. Not only is the area technically complex, but it is also in rapid evolution. A range of standards, guidances, norms, and QC tools are available from different sources, and in order to maximize the value of data and the acceptability of data, full compliance with these technical standards is essential. The appearance of new standards is also to be expected, especially in the area of formats and submission formats.

References

Ancian P, Leuillet S, El Gana R, Arthaud S, Fisch C, de Jouffrey S, Forster R, Le Bigot JF (2006) Reproducibility of microarray analysis in toxicogenomics studies. *Toxicology Letters*, **164**(S1), S298–S299.

Ancian P, Leuillet S, Milicov A, Forster R (2007) Impact of RNA degradation on Affymetrix gene expression profiles. *Toxicol Lett* **172**(S1), S37.

Bourrinet P, Leuillet S, De Sousa A, Bonnemain B, Ancian P, Forster R (2007) Evaluation of the effects of Xenetix® and Dotarem® on hepatic and renal gene expression in rats. *Toxicologist*, **96**(S1), Abstract 2129.

Brazma A, Hingamp P, Quackenbush J, Sherlock G, Spellman P, Stoeckert C, Aach J, Ansorge W, Ball CA, Causton HC, Gaasterland T, Glenisson P, Holstege FC, Kim IF, Markowitz V, Matese JC, Parkinson H, Robinson A, Sarkans U, Schulze-Kremer S, Stewart J, Taylor R, Vilo J, Vingron M (2001) Minimum information about a microarray experiment (MIAME) – toward standards for microarray data. *Nat Genet* **29**(4), 365–371.

Brazma A, Parkinson H, Sarkans U, Shojatalab M, Vilo J, Abeygunawardena N, Holloway E, Kapushesky M, Kemmeren P, Lara GG, Oezcimen A, Rocca-Serra P, Sansone SA (2003) ArrayExpress – a public repository for microarray gene expression data at the EBI. *Nucleic Acids Res* **31**(1), 68–71.

Brooksbank C, Quackenbush J (2006) Data standards: a call to action. *OMICS J Integr Biol* **10**(2), 94–99.

Corvi R, Ahr HJ, Albertini S, Blakey DH, Clerici L, Coecke S, Douglas GR, Gribaldo L, Groten JP, Haase B, Hamernik K, Hartung T, Inoue T, Indans I, Maurici D, Orphanides G, Rembges D, Sansone SA, Snape JR, Toda E, Tong W, van Delft JH, Weis B, Schechtman LM (2006) Meeting report: validation of toxicogenomics-based test systems: ECVAM-ICCVAM/NICEATM considerations for regulatory use. *Environ Health Perspect* **114**(3), 420–429.

Edgar R, Domrachev M, Lash AE (2002) Gene expression omnibus: NCBI gene expression and hybridization array data repository. *Nucleic Acids Res* **30**(1), 207–210.

EMEA (2006a) Guiding principles: processing joint FDA EMEA voluntary genomic data submission briefing meetings. EMEA, London.

EMEA (2006b) Guideline on pharmacogenetics briefing meetings. Document ref.: EMEA/CHMP/PGxWP/20227/2004, European Medicines Evaluation Agency, London.

FDA (2004) Innovation/stagnation: challenge and opportunity on the critical path to new medical products.

FDA (March 2005a) Guidance for Industry, Pharmacogenomic Data Submissions. Food and Drug Administration.

FDA (March 2005b) Attachment to Guidance on Pharmacogenomic Data Submissions Examples of Voluntary Submissions or Submissions Required Under 21 CFR 312, 314, or 601. Food and Drug Administration.

FDA (August 2007) Guidance for Industry, Pharmacogenomic Data Submissions – Companion Guidance, Draft Guidance. Food and Drug Administration.

Freeman K (2005) From point B to point A: applying toxicogenomics to biological inference. *Environ Health Perspect* **113**(6), A388–A393.

Guo L, Lobenhofer EK, Wang C, Shippy R, Harris SC, Zhang L, Mei N, Chen T, Herman D, Goodsaid FM, Hurban P, Phillips KL, Xu J, Deng X, Sun YA, Tong W, Dragan YP, Shi L (2006) Rat toxicogenomic study reveals analytical consistency across microarray platforms. *Nat. Biotechnol* **24**(9), 1162–1169.

Hoffmann R, Seidl T, Dugas M (2002) Profound effect of normalization on detection of differentially expressed genes in oligonucleotide microarray data analysis. *Genome Biol* **3**(7), RESEARCH0033.

Ikeo K, Ishi-i J, Tamura T, Gojobori T, Tateno Y (2003) CIBEX: Center for Information Biology Gene Expression Database. *C R Biol* **326**(10–11), 1079–1082.

Irizarry RA, Warren D, Spencer F, Kim IF, Biswal S, Frank BC, Gabrielson E, Garcia JG, Geoghegan J, Germino G, Griffin C, Hilmer SC, Hoffman E, Jedlicka AE, Kawasaki E, Martínez-Murillo F, Morsberger L, Lee H, Petersen D, Quackenbush J, Scott A, Wilson M, Yang Y, Ye SQ, Yu W (2005) Multiple-laboratory comparison of microarray platforms. *Nat Methods* **2**(5), 345–350.

Kravtzoff R, Boutherin-Falson O, Forster R, Leuillet S, Arthaud S, Ancian P (2008) Predictive and mechanistic toxicology of nanoparticulate carrier by hepatic gene expression profiling in the rat. *Toxicologist* **102**(S1), Abstract 1920.

Leighton JK, Brown P, Ellis A, Harlow P, Harrouk W, Pine PS, Robison T, Rosario L, Thompson K (2006) Workgroup report: review of genomic data based on experience with mock submissions: view of the CDER Pharmacology Toxicology Nonclinical Pharmacogenomics Subcommittee. *Environ Health Perspect* **114**(4), 573–578.

Maggioli J, Hoover A, Weng L (2006) Toxicogenomic analysis methods for predictive toxicology. *J Pharmacol Toxicol Methods* **53**(1), 31–37.

Mattes WB, Pettit SD, Sansone SA, Bushel PR, Waters MD (2004) Database development in toxicogenomics: issues and efforts. *Environ Health Perspect* **112**(4), 495–505.

MAQC Consortium (2006) The MicroArray Quality Control (MAQC) project show inter- and intraplatform reproducibility of gene expression measurements. *Nat Biotechnol* **24**(9), 1151–1161.

Morgan KT, Jayyosi Z, Hower MA, Pino MV, Connolly TM, Kotlenga K, Lin J, Wang M, Schmidts HL, Bonnefoi MS, Elston TC, Boorman GA (2005) The hepatic transcriptome as a window on whole-body physiology and pathophysiology. *Toxicol Pathol* **33**(1), 136–145.

Orr MS, Goodsaid F, Amur S, Rudman A, Frueh FW (2007) The experience with voluntary genomic data submissions at the FDA and a vision for the future of the voluntary data submission program. *Clin Pharmacol Ther* **81**(2), 294–297.

Shi L, Tong W, Fang H, Scherf U, Han J, Puri RK, Frueh FW, Goodsaid FM, Guo L, Su Z *et al.* (2005) Cross-platform comparability of microarray technology: intra-platform consistency and appropriate data analysis procedures are essential. *BMC Bioinformatics* **6**(Suppl 2), S12.

Shippy R, Fulmer-Smentek S, Jensen RV, Jones WD, Wolber PK, Johnson CD, Pine PS, Boysen C, Guo X, Chudin E, Sun YA, Willey JC, Mieg JT, Thierry-Mieg D, Setterquist RA, Wilson M, Bergstrom Lucas A, Novoradovskaya N, Papallo A, Turpaz Y, Baker SC, Warrington JA, Shi L, Herman D (2006) Using RNA sample titrations to assess microarray platform performance and normalization techniques. *Nat Biotechnol* **24**, 1123–1131.

Suter L, Babiss LE, Wheeldon EB (2004) Toxicogenomics in predictive toxicology in drug development. *Chem Biol* **11**(2), 161–171.

Tong W, Cao X, Harris S, Sun H, Fang H, Fuscoe J, Harris A, Hong H, Xie Q, Perkins R, Shi L, Casciano D (2003) ArrayTrack – supporting toxicogenomic research at the U.S. Food and Drug Administration National Center for Toxicological Research. *Environ Health Perspect* **111**(15), 1819–1826.

Tong W, Bergstrom Lucas A, Shippy R, Fan X, Fang H, Hong H, Orr MS, Chu TM, Guo X, Collins PJ, Sun YA, Wang SJ, Bao W, Wolfinger RD, Shchegrova S, Guo L, Warrington JA, Shi L (2006) Evaluation of external RNA controls for the assessment of microarray performance. *Nat Biotechnol* **24**, 1132–1139.

Waring JF, Ulrich RG, Flint N, Morfitt D, Kalkuhl A, Staedtler F, Lawton M, Beekman JM, Suter L (2004) Interlaboratory evaluation of rat hepatic gene expression changes induced by methapyrilene. *Environ Health Perspect* **112**(4), 439–448.

Yue L, Reisdorf WC (2005) Pathway and ontology analysis: emerging approaches connecting transcriptome data and clinical endpoints. *Curr Mol Med* **5**(1), 11–21.

Zakharkin SO, Kim K, Mehta T, Chen L, Barnes S, Scheirer KE, Parrish RS, Allison DB, Page GP (2005) Sources of variation in Affymetrix microarray experiments. *BMC Bioinformatics*, **6**, 214.

22

Perspectives on Toxicogenomics Activities at the US Environmental Protection Agency

Karen Hamernik, Kenneth Haymes, Susan Hester and J. Thomas McClintock

22.1 Introduction and Background[1]

In June 2002, the US Environmental Protection Agency (EPA) issued its Interim Policy on Genomics which provided guidance on using genomics information in risk assessment and decision making under the various programs implemented by the agency (US Environmental Protection Agency, 2002) (http://www.epa.gov/osa/spc/genomics.htm). This Interim Policy describes genomics as the study of all the genes of a cell or tissue, at the DNA (genotype), mRNA (transcriptome), or protein (proteome) level. While genomics offers the opportunity to understand how an organism responds at the gene expression level to stressors in the environment, understanding such molecular events with respect to adverse ecological and/or human health outcomes is far from established. This Interim Policy on Genomics clearly states that genomics data alone are currently insufficient as a basis for risk assessment and management decisions. At the time of this writing, the EPA will only consider genomics information for assessment purposes on a case-by-case basis.

After the release of the EPA's Interim Policy on Genomics, the agency's Science Policy Council (SPC) created a cross-agency Genomics Task Force that was charged with

---

[1]It should be noted that this chapter is intended for informational purposes only in an evolving area subject to change. It is not intended and should not be viewed as an EPA policy, regulatory, or guideline document. In addition, this book chapter provides the reader with some highlights of the many activities associated with (toxico)genomics activities at the EPA and should not be seen as a comprehensive review.

Toxicogenomics: A Powerful Tool for Toxicity Assessment   Edited by S. C. Sahu

examining the broader implications that genomics may have on EPA programs and policies. This Genomics Task Force developed a Genomics White Paper (2004) entitled *Potential Implications of Genomics for Regulatory and Risk Assessment Applications at EPA* that identified four areas likely to be influenced by the generation of genomics information within EPA and the submission of such information to the EPA: (1) prioritization of contaminants and contaminated sites; (2) monitoring; (3) reporting provisions; and (4) risk assessment. The Task Force also identified several challenges and/or critical needs that included research, technical development, and capacity building (*e.g.*, strategic hiring practices and training; US Environmental Protection Agency, 2004 (http://www.epa.gov/osa/genomics.htm); Benson *et al.*, 2007).

The Genomics Task Force recommended that the agency charge a workgroup with developing a technical framework for analysis and acceptance criteria for genomics information for scientific and regulatory purposes. The Genomics White Paper identified issues that needed to be considered in developing this framework including the performance of genomics-based tests or assays across genomic platforms (*e.g.*, reproducibility, sensitivity, pathway analysis tools) and standardization criteria for accepting genomics data for use in risk assessment (*e.g.*, assay validity, biologically meaningful response).

In June 2004, at the request of EPA's Office of the Science Advisor, the Genomics Technical Framework and Training Workgroup (Genomics Workgroup) was established with representatives from numerous program and regional offices. The Genomics Workgroup was comprised of a Coordinating Committee, several technical genomics guidance workgroups (Performance Approach Quality Assurance Workgroup, Data Submission Workgroup, Data Analysis Workgroup, and a Data Management and Storage Workgroup), a Training Workgroup, and a Microbial Source Tracking Workgroup. The Genomics Workgroup's responsibility was to ensure that the technical framework and training activities were built upon the foundation outlined in the agency's Interim Policy on Genomics while continuing to engage other interested parties (*e.g.*, Food and Drug Administration (FDA), National Institute of Environmental Health Services (NIEHS)). The workgroup developed a guidance document entitled *Interim Guidance for Microarray-Based Assays: Data Submission, Quality, Analysis, Management, and Training Considerations* that will be used by the EPA program offices and regions to help determine the applicability of specific genomics information to the evaluation of risks under various statutes (US Environmental Protection Agency, 2007) (http://www.epa.gov/OSA/spc/pdfs/epa_interim_guidance_for_microarray-based_assays-external-review_draft.pdf). This Interim Guidance document for microarray-based assays is also intended to provide information to the regulated community and other interested parties considering submission of microarray data to the agency.

22.2   The Potential of Microarray Data

Over the next decade, the data from toxicogenomic technologies are anticipated to become an integral part of risk analysis and the regulatory decision making process at EPA (Benson *et. al.*, 2007). Currently, genomic data can be submitted voluntarily to the agency for review as part of a data submission package. However, as stated in the EPA's 2002 Interim Policy on Genomics, genomic data would be considered as supplemental information, would be evaluated on a case-by-case basis, and would not be used as the sole basis for a final regulatory decision. The EPA recognizes that, as the tools and platforms for describing biological phenomena evolve, this may in turn affect risk assessment and the regulatory decision making process.

Microarray technology has the potential for measuring large numbers of changes in gene expression and could provide information that would describe common gene pathways. This in turn could aid the elucidation of mechanisms of toxicity in the regulatory arena. Initially, this technology may be useful in augmenting currently accepted methods of toxicity screening and may also have utility in priority setting (*e.g.*, of chemicals for testing or for selecting sites for risk management and mitigation measures). A goal that is often discussed is the possible use of this technology to reduce the use of animal testing. As the technology evolves, it may be useful in developing a better understanding of the interactions and differences between cells and whole organisms. It may also provide valuable information that would aid in interspecies extrapolations in hazard assessment.

Toxicogenomic technologies have utility in numerous areas, such as investigative toxicology or the elucidation of mode of action (MOA) information or predictive toxicology and gene expression responses. Investigative toxicology seeks to discover the underlying basis of toxicity associated with a biological response (usually adverse). The resulting information could be useful in determining the MOA and provide information for the risk assessment of a particular chemical compound(s) under study. Predictive toxicology explores a compound's potential to induce a toxic response. A frequently used approach to assess both known and unknown chemical toxicity is to examine gene expression responses or profiles which reflect cell-wide alterations associated with exposure. Results may be associated with a chemical signature or molecular profile which may be unique to the chemical or the chemical classes under study. These specific signature profiles may be used to compare the gene expression profiles for untested/unknown compounds (genotoxic chemicals or compounds with unknown toxicity potential) for similarity or dissimilarity.

22.3   Limitations of Microarray Technology

The EPA is working closely with the FDA, NIEHS, the National Institutes of Health (NIH), the National Institute of Standards and Technology (NIST), academic institutions, and private industries in various consortia in the field of toxicogenomics. Some of the goals of these interactions are to help to standardize this field, develop methodologies, and work towards cross-platform comparisons of microarray data, new tools, software analysis programs, and models for data analysis. One of the identified goals that the EPA has with these collaborations is to explore the use of genomic data in risk/safety assessments and their application in regulatory decision-making.

While toxicogenomics is accepted as an important research tool, it has some inherent technological limitations. One of the limitations of gene expression analysis is that it represents a 'snapshot in time' of the toxicity or alterations that may be operative inside a cell/tissue. In addition, capturing the target cell population that is associated with the toxicity is very challenging but necessary because the gene expression signals of the target cells may be diminished by cell expression from different cell types surrounding the target cells. Another complicating issue is that the gene expression response needs to be distinguished from cellular adaptive or homeostatic responses that may have little relationship to the ongoing toxicity or may not be associated with a toxic or adverse effect. Differentiating various types of gene expression changes and interpreting their biological meaning can present real challenges to data interpretation. To understand and predict a compound's ability to induce a biologic response frequently requires many time points and different doses. Furthermore, it may be very difficult to identify a significant gene expression response if the biologic insult is mild or minimal. Another difficulty is that many chemical compounds act through multiple mechanisms, depending on many experimental variables, such as gender, age of test animals, hormonal status, beddings, diet, and lighting exposure. Each of these variables may initiate a toxic response, but in and of themselves may fail to promote the expected toxicity response. Although the assessment of genome-wide responses to chemical exposure has been shown to be useful, it should be remembered that these responses are evaluating one parameter, namely mRNA expression profiles. Additional measurements of conventionally accepted toxicology parameters, including clinical chemistry, histopathology, and biochemical assays, will still be needed. When data from these types of parameters are combined with genome-wide profiling, it is anticipated that a more comprehensive view of the exposure-related toxicity will be provided.

The development of genomic technology and its applications have progressed over the past decade and is considered more advanced and stable than when it was introduced in the early 1990s. Lagging behind is the art of microarray gene expression analysis. The gene expression analysis is analyst driven and characterized by multiple approaches, each with strengths and weaknesses. It is further complicated by the level of gene annotation, which changes frequently. Potential problems that arise are related to interpreting the array data incorrectly to generate either false-positive or false-negative data, which can lead to incorrectly identifying hazards. The EPA would like to have a modeling tool that is both predictive and quantitative in nature, and it is acknowledged that this will require the development of new algorithms. Furthermore, as the understanding of the biological and mechanistic processes of this field expands, so will our knowledge. Prior to their application, the scientific, mathematical, and statistical methods that are used for these models and analyses will need to be validated and standardized. The various models and tools used for microarray analysis, once validated, may be extremely useful for EPA in regulatory applications.

Toxicogenomics applications are continuously being refined, and journals are following Minimum Information About a Microarray Experiment (MIAME) guidelines for publication of results. Even with these advances, the development of 'gold standard' controls, such as a universal RNA, and other internal controls represent some of the main challenges to the field of toxicogenomics. In addition, standard approaches to analyze gene expression remain to be developed. As these challenges are overcome, genomic analysis will become useful to characterize toxicity responses to chemicals and environmental compounds and provide valuable information for the risk assessment and regulatory decision making process.

22.4   Microarray Data Submission Considerations

Following from the Interim Policy and the White Paper, the agency developed a document to provide guidance for the review and evaluation of genomic data by EPA scientists. When the EPA reviews a data submission, sufficient information needs to be included to allow an independent reviewer to reconstruct how the data were collected and analyzed. This will allow EPA reviewers to evaluate the quality of the genomics data and the strength of any correlations independently (*e.g.*, between a gene expression profile, a proposed toxic mechanism or MOA, and conclusions). Microarray data that are submitted to the agency are considered supplemental data and will need to be supported by comparison or integration of data from other scientifically accepted techniques.

The EPA used the MIAME standards as a basis to develop the agency's Interim Guidance document on microarray based-assays (http://www.mged.org/Workgroups/MIAME/miame.html). The agency's guidance document incorporates the EPA's Quality System and Performance Approach to Quality Measurement Systems. The EPA's Quality System requirements are described in *EPA Order 5360.1 A2, Policy and Program Requirements for the Mandatory EPA-Wide Quality System*; the *EPA Quality Manual for Environmental Programs, EPA Manual 5360 A1*; the Contracts Management Manual; and the EPA's website (http://www.epa.gov/quality). When microarray data are submitted to the agency, the submitter(s) will need to take into account the performance-related experimental design and factors related to quality management systems.

Although there is a need for standardization, the Interim Guidance document discusses the need to incorporate flexibility and innovation so that this technology can continue to evolve. Furthermore, the EPA is concerned that standards should not be burdensome, thereby discouraging data submission or slowing scientific progress. Many scientific journals have instructed that authors follow the MIAME guidelines as a standard for submission of microarray data as part of a submitted publication. The EPA has developed a slightly modified MIAME model, and this initial template is subject to change as the technology evolves. If the MIAME guidance changes, then the agency may consider modifying the Interim Guidance document as it relates to MIAME.

An in-depth description of the EPA's *Interim Guidance for Microarray-Based Assays: Data Submission, Quality, Analysis, Management, and Training Considerations* may be found at http://www.epa.gov/OSA/spc/pdfs/epa_interim_guidance_or_microarray-based_assays-external-review_draft.pdf. The following items highlight information from the Interim Guidance document that the EPA may request with the submission of microarray data. Since this is only summary information, the reader or submitter of data should refer to the Interim Guidance document for more details and is also encouraged to consult with the EPA regarding data requirements or information to support a data submission package which includes genomics data.

22.4.1   Data Submission Considerations

22.4.1.1   Abstract

The abstract should be a brief overview and provide the key highlights of the study so the reviewer will understand the source and type of data submitted, as well as the type of data evaluation performed and its final interpretation.

22.4.1.2   Experimental Design

One of the most critical aspects of the submission is a clear description of the source and nature of the data and the methods used, including standardizations/validations. Biological model system, treatment methods and doses, husbandry of animals, and cell culture information for *in vitro* systems are pertinent information needed to describe the experimental design. If whole animal models were employed, then information should be provided regarding the exposure system, exposure doses, time points, details on euthanasia, length of time between harvesting of tissues and freezing or other processing, numbers of samples utilized for DNA array analysis, and methods of RNA processing and RNA quantization.

22.4.1.3   Array Design

The submission should completely describe the platform used for transcriptional expression analysis such that the reviewer can assess the appropriateness of the analysis. The commercial platform and the specific chip used should be identified and the locations (*e.g.*, weblink) of the source of the proprietary information, allowing the reviewer to assess the data better, should be referenced. If a custom array is used, then a complete description of the production of the array needs to be included in the data submission.

22.4.1.4   Biomaterials

The physical characteristics of the biomaterials studied and information that provides details on the source properties, treatment, extract preparation, and labeling of the sample and sample controls should be stated.

22.4.1.5   Hybridization

A concise description of the procedures utilized for each hybridization should be included in the application package. Literature citations or internet material that describes the source of the hybridization protocol and materials should be included in the data submission.

22.4.1.6   Measurements

The submission should completely describe the methods used to acquire the image of the array. For example, the nature of the image (*e.g.*, Tagged Image File Format), extraction of image data into quantified image data, spreadsheets used to house the quantified data background correction, normalization methods, methods used to test usability of the raw data, and types of analytical approaches should be available to the reviewer.

22.4.2   Data Analysis

An overview of the main elements that may be considered to support the analysis and interpretation of the genomic data is described below.

22.4.2.1   Data Processing and Filtering

Data processing includes scanning the microarray to obtaining reliable estimates for the relative abundance of each gene transcript in all of the samples (*e.g.*, image analysis, quality control filtering, background correction, transformation and normalization).

22.4.2.2   Statistics

There are differences in microarray platforms, experimental variables and design (reference versus matched), levels of replication, as well as within experiment sources of variation (spot to spot, slide to slide, etc.). Other factors need to be accounted for, such as pooling, data replication, and type of replication (technical and biological). Various statistical programs may be utilized for microarray analysis, and it is up to the submitter to justify the statistical approach used.

22.4.2.3 Interpretation

Numerous approaches can be used to interpret differentially expressed genes using microarray experiments. For example, genes can be sorted by ontology (*e.g.*, gene ontology) and subsequent analyses (*e.g.*, principal component analysis, hierarchical clustering, and κ-means clustering) can be used to organize the data and help identify patterns of gene expression.

Bioinformatic tools may be used to assess expression patterns with common biological pathways and networks of co-regulated genes. The ability to link pathway and functional analyses in an effort to identify patterns will significantly help to predict adverse human and environmental effects.

22.4.2.4 Data Evaluation

A microarray-based study is a tool that can generate enormous amounts of data in a relatively short period of time. Evaluation of the raw data (type of data analysis performed, interpretation and application, and end use of the data) is an area that is still under development. The agency is also considering available resources to analyze, store, and retrieve the vast amounts of microarray data and information that could be submitted. Data generated from microarray assays would need to be confirmed using other techniques (*i.e.*, real-time quantitative polymerase chain reaction, functional enzyme assays, protein and metabolite profiles and/or linked to bioassay results) in order to be used for potential risk assessment and regulatory decision-making.

The EPA has developed a genomics Data Evaluation Record (DER) template, which can be viewed in the Interim Guidance document, that may support a regulatory application and/or submission. The DER will assist in presentation, organization, and submission of data from genomics studies (http://www.epa.gov/OSA/spc/genomicsguidance.htm).

22.5 Genomic Research Studies

The EPA relies on sound science as the basis for the development of regulations and guidance that aid in fulfilling its mission to protect both human health and the environment. The Office of Research and Development (ORD), with its diverse work force of scientists and engineers from many disciplines, provides expertise and support for the agency's mission and also conducts environmental research that can identify and solve environmental problems. ORD scientists are also available to help address scientific questions that may arise in regulatory program offices. ORD conducts research and works with program offices on research areas that include air, water, soil, ecology, and human health to reduce and prevent risk to humans and damage to the environment.

The EPA has recognized the need to develop approaches to incorporate data developed by toxicogenomic technologies into hazard and risk assessment in addition to traditional toxicological data (*e.g.*, histopathology, clinical chemistry, bioassays). In addition to the White Paper, another document, that assists in our genomic efforts is the Framework for a Computational Toxicology Research Program [EPA's National Center for Computational

Toxicology (NCCT)—http://www.epa.gov/comptox/comptoxfactsheet.html]. Ongoing research efforts exploring possible applications of toxicogenomics technologies at the EPA are presented below that use genomic approaches designed to gain insights into chemical or agent toxicity. Each of these research studies is described in terms of compound evaluated, experimental rationale, and findings of the study.

22.5.1   Study 1: Conazoles

Conazoles are antifungal agents most commonly used as pesticides in the protection of fruit, vegetable, and cereal crops (Zarn *et al.*, 2003) and are also used as pharmaceuticals in the treatment of local and systemic fungal infections in humans (Georgopapadakou and Walsh, 1996). Conazoles can be divided into two major categories according to their chemical structure. One group includes substances which contain an imidazole ring. Examples include ketoconazole and miconazole. The other group includes substances which contain a 1,2,4-triazole ring. Examples include propiconazole, triadimefon, and myclobutanil. The mechanism of antifungal action of these agents is based on their inhibition of ergosterol biosynthesis by inhibiting lanosterol 14α-demethylase (CYP51) activity (Ronis *et al.*, 1994; Debeljak *et al.*, 2003), resulting in an impairment to cholesterol biosynthesis, rendering damage to fungal elements.

In addition to their capacity to act as effective fungicides, conazoles demonstrate a wide range of toxicological properties in mammalian systems (Zarn *et al.*, 2003). For example, they can induce alterations in the liver (hepatomegaly) and of several metabolic enzymes, including cytochrome P450 (CYP) isozymes (Sun *et al.*, 2005; Juberg *et al.*, 2006; Tully *et al.*, 2006). Conazoles can act both as inducers and inhibitors of hepatic CYP activities, depending on the specific conazole. In rodents, chronic exposure to several conazoles can also induce hepatotoxicity, neurotoxicity, and tumorigenesis (Hurley, 1998; Moser *et al.*, 2001; Biagini *et al.*, 2006).

Earlier work showed that tumorigenic and nontumorigenic conazoles induced similar effects on mouse liver CYP enzyme activities and pathology. However, using conventional methods, no specific pattern of tissue responses could consistently differentiate the tumorigenic conazoles, propiconazole, and triadimefon from the nontumorigenic myclobutanil. The modes of hepatotumorigenic action of two conazoles, triadimefon and propiconazole, had not been characterized. In addition, genotoxicity studies with several of these tumorigenic conazoles provided negative results (INCHEM, 1981, 1987). Thus, a strategy was adopted to investigate early molecular changes using gene expression analyses to identify pathways that are possibly involved in conazole toxicity.

22.5.1.1   Study Rationale and Design

Historically, gene expression profiles induced by pharmaceuticals and chemicals in the livers of rodents have been used to identify potential mechanisms of toxicity (Waring *et al.*, 2001, 2004; McMillian *et al.*, 2004). A microarray study was constructed to identify alterations of gene pathway transcription which could provide clues to identify potential modes of toxicological tumorigenic action (Ward *et al.*, 2006). In a companion study employing conventional toxicological bioassays, male CD-1 mice were fed tumorigenic triadimefon and propiconazole, or the non-tumorigenic conazole myclobutanil, in a continuous oral-dose regimen for 4, 30, or 90 days (Allen *et al.*, 2006). All compounds were found to induce hepatomegaly, to induce high levels of metabolic enzymes (*e.g.*, hepatic pentoxyresorufin-*O*-dealkylase activity), to increase hepatic cell proliferation, to decrease serum cholesterol, and to increase serum triglycerides.

Evaluation of the microarray data revealed gene-pathway associations for the tumorigenic conazoles. In general, the number of altered metabolism, signaling, and growth pathways increased with time and dose and were greatest with propiconazole. All conazoles had effects on nuclear receptors, as evidenced by increased expression and enzymatic activities of a series of related cytochrome P450s (CYP). A subset of altered genes and pathways distinguished the three conazoles from each other. Both triadimefon and propiconazole altered a number of cellular pathways consistent with the observed cell proliferation, including those associated with apoptosis, oxidative stress, cell cycle, adherens junction, calcium signaling, and EGFR signaling. Propiconazole had greater effects on genes responding to oxidative stress and on the IGF/P13K/AKt/PTEN/mTor and Wnt-β-catenin pathways. In conclusion, while triadimefon, propiconazole, and myclobutanil had similar effects in mouse liver on hepatomegaly, histology, CYP activities, cell proliferation, and serum cholesterol, genomic analyses revealed major differences in their gene expression profiles.

22.5.1.2 Summary

The use of genomic technology is a relatively recent addition to an ORD program exploring the use of mechanistic information for risk assessment. The overall aim is to establish the MOA for specific tumorigenic members of the conazole family of chemicals that can induce liver or thyroid cancer and/or reproductive effects. When combined with conventional phenotypic assessments, gene expression profiling can provide a powerful tool to investigate toxicity associated with chemical exposure.

22.5.2 Study 2: Triazole Fungicides and Perfluoroalkyl Acids

Recently, a study was published that evaluated the ability of gene expression data to predict toxicity, classify chemicals, and identify relevant biological pathways and mechanisms of toxicity (Martin *et al.*, 2007). Five environmental chemicals representing two groups of chemicals, perfluoroalkyl acids (PFAAs) and triazoles, were selected based on their shared ability to induce liver toxicity. Two (PFAA) compounds, perfluorooctanoic acid (PFOA) and perfluorooctane sulfonate (PFOS) are used in industry settings as plasticizers in many products. These compounds are stable and considered to be biologically inactive in the environment. However, studies have now shown that PFOA and PFOS are detectable in the human population and animals, and preliminary studies have provided some evidence for reproductive and developmental toxicities of these chemicals in laboratory rodents (Lau *et al.*, 2004). The second group of hepatotoxicants tested, conazoles, are used as antifungals in agriculture, and each compound contains a 1,2,4-triazole ring (myclobutanil, propiconazole, triadimefon). The underlying mode or mechanism of action for both PFAAs (PFOA and PFOS) and conazoles is not fully understood. This project represents part of the Agency's research program exploring the potential of genomics for regulatory and risk assessment applications (Dix *et al.*, 2006).

22.5.2.1   Study Rationale and Design

A comparative approach was designed to query the five test chemicals against a large genomic database. The three triazole compounds and two PFAA compounds were administered by oral gavage for one, three, and five consecutive days at five different doses. In addition, other traditional toxicology endpoints accompanied the transcriptional analysis, including body and organ weights, serum hormone levels, clinical chemistry, hematology, and histopathology. These measurements were assessed to evaluate general or specific tissue/organ effects. In addition, liver gene expression profiling, using a commercially available rat expression array, was performed at all time points. The experimental liver profiles were queried against a large genomic database containing over 600 chemicals. The results of this genomic comparison were used to evaluate the ability of gene expression profiles to predict possible modes of toxicity, categorize the chemicals, and determine alterations in biological pathways.

22.5.2.2   Summary

This toxicogenomic study investigated possible MOAs for the five environmental chemicals, including xenobiotic metabolizing enzyme induction involving a number of nuclear receptors: triazoles induced the pregnane X receptor (PXR), whereas PFOA and PFOS showed effects on the peroxisome proliferator-activated receptor α (PPARα). The authors reported a concordance of gene expression profiling with the phenotypic observations, thus demonstrating that genomics could provide evidence for suggested toxic mechanisms associated with chemical treatment at a molecular level. This study serves as an example of the value of a large gene expression database coupled with traditional toxicological measures that can be used in a comparative analysis for the prediction of potential effects. The clear utility of performing toxicogenomics comparisons as demonstrated in this study suggests that such methods may be able to classify environmental chemicals, based on their 'toxicogenomic signatures.'

22.5.3   Study 3: Differentiating a Carcinogenic from a Noncarcinogenic Aldehyde

Formaldehyde exposure to humans occurs through many occupational and environmental sources, including wood products, household and personal use products, and air pollution sources. Glutaraldehyde, a compound in the same chemical class as formaldehyde, is used in sterilization processes in industry and hospital settings. Formaldehyde is a known carcinogen in the rat nose, whereas glutaraldehyde, while cytotoxic, was not shown to induce cancer. Both compounds produce a similar tissue response of inflammation and cellular damage after inhalation and, thus, cannot be distinguished by conventional toxicology endpoints after short exposures. Owing to this similarity of tissue response, and considering that the risk to humans is uncertain, a proof of principle study was undertaken at the EPA using genomics. In this study, transcriptional profiling was utilized to distinguish two aldehydes: the carcinogenic formaldehyde from the non-carcinogenic glutaraldehyde (Hester *et al.*, 2003, 2005). There is agreement in the microarray community that the power of this genomic approach is amplified when combined with conventional analytic techniques, such as descriptive histopathology (Irwin *et al.*, 2004). Using both approaches, the genetic basis for a cellular phenotype following a chemical exposure provided insights into hypothesized toxicity pathways.

22.5.3.1 Study Rationale and Design

Formaldehyde is cytotoxic and carcinogenic to the rat nasal respiratory epithelium, inducing tumors after 12 months. Glutaraldehyde is also cytotoxic, but it is not carcinogenic to nasal epithelium even after 24 months. Interestingly, both aldehydes induce similar acute and subchronic histopathology characterized by inflammation, hyperplasia, and squamous metaplasia. Early aldehyde-induced lesions are microscopically similar qualitatively. The investigation sought to determine whether transcriptional patterns using cDNA technology could explain the different cancer outcomes. Treatments included 1-, 5-, or 28-day exposure by nasal instillation of formaldehyde solution (400 mM) or glutaraldehyde solution (20 mM), doses known to induce damage and cellular proliferation (St Clair *et al.*, 1990). Animals were euthanized and the nasal respiratory epithelium removed for gene expression analysis and a subset of rats treated for 28 days was processed for microscopic examination. RNA was isolated and processed for gene expression assessment. Both aldehydes induced hyperplasia, squamous metaplasia, and inflammatory infiltrates with scattered apoptotic bodies in the epithelium covering luminal surfaces of the nasoturbinate, maxilloturbinate, and nasal septum. A subset of 80 genes that were the most variant between the treated and control groups included the functional categories of DNA repair and apoptosis. Hierarchical clustering discriminated chemical treatment effects after 5 days of exposure, with six clusters of genes distinguishing formaldehyde from glutaraldehyde.

22.5.3.2 Summary

In summary, this study provided the first known genomic comparison of a carcinogenic with a noncarcinogenic aldehyde in the rat nose. These treatments could be discriminated through generating unique gene expression profiles. The data showed that, although both aldehydes induced similar early cellular phenotypes histologically, the molecular gene expression responses could distinguish glutaraldehyde from formaldehyde. Furthermore, the observed gene expression patterns suggested that glutaraldehyde's lack of carcinogenicity may be due to its greater toxicity from lack of functional gene categories, including DNA repair, greater mitochondrial damage, and increased apoptosis.

22.5.4 Study 4: Potassium Bromate

Potassium bromate ($KBrO_3$) is a chemical oxidizing agent that has been used extensively in the food and cosmetic industries (as a component of permanent hair-wave kits). Potassium bromate is also found in drinking water as a disinfection byproduct of surface waters. Acute physiologic responses to $KBrO_3$ include damage to the kidneys and nervous system. Moreover, chronic exposures to $KBrO_3$ cause renal cell carcinomas in rats, hamsters, and mice, and thyroid and mesothelioma tumors in rats (Kurokawa *et al.*, 1990; DeAngelo *et al.*, 1998). Because of its ability to induce cancer in multiple species at multiple sites, bromate is considered to be a probable human carcinogen (Delker *et al.*, 2006).

22.5.4.1   Study Rationale and Design

Studies focusing on the mechanism of potassium bromate-induced carcinogenicity have focused on DNA damage due to oxidative stress (Chipman *et al.*, 1998; Umemura *et al.*, 1998). Potassium bromate is known to induce a number of insults to DNA, including point mutations in bacteria and chromosomal aberrations in mammalian cells (Ishidate *et al.*, 1984) and an increased incidence of kidney mutations in transgenic mice lacking the 8-oxodeoxyguanosine glycosylase (*Ogg*1) DNA repair enzyme (Nishimura, 2002). These studies have provided support for the hypothesis that $KBrO_3$ is damaging DNA through oxidative stress and that this leads to increases in mutation frequency. Through a research project developed at the EPA, gene expression analysis was utilized to investigate the molecular and cellular processes that may be involved in the toxicity and carcinogenicity of potassium bromate. A study was conducted by exposing male F344 rats to $KBrO_3$ in their drinking water, at multiple doses, from 2 to 100 weeks (Delker *et al.*, 2006). There was a substantially higher number of oxidative stress genes altered in the high-dose group compared with the low-dose group. Multiple cancer, cell death, ion transport, and oxidative stress genes were altered by potassium bromate. Multiple glutathione metabolism genes were also up-regulated in kidney following carcinogenic but not noncarcinogenic bromate exposures. A DNA repair gene, glycosylase (*Ogg1*), was up-regulated in response to bromate treatment in the kidney but not the thyroid. ^{18}O-labeled $KBrO_3$, used as a dosimeter of oxidative stress in target and nontarget organs, was administered by the oral route to a subset of rats. Tissues from exposed animals were subsequently analyzed for ^{18}O deposition. Tissue deposition of ^{18}O was observed post-$KBr^{18}O_3$ exposure, with the highest enrichment occurring in the liver, followed by the kidney, thyroid, and testes. These results provide some evidence that gene expression profiles were shown to be predictive of $KBrO_3$-induced renal toxicity after a carcinogenic dose of potassium bromate.

22.5.4.2   Summary

In conclusion, potassium bromate-induced changes in renal gene expression and ^{18}O deposition in the rat are dose dependent and provide further information regarding adverse effect level and bromate MOA. Doses below 50 mgl^{-1} only showed modest levels of ^{18}O deposition, suggesting that antioxidant defenses were effective. However, greater ^{18}O deposition in the rat kidney was observed at drinking water concentrations above 50 mg l^{-1}, which implied enhanced bromate-mediated oxidation of cellular macromolecules at higher doses. These results provide evidence of oxidative stress levels in bromate exposures.

22.5.5   Study 5: Nanoparticles

Engineered nanoparticles are widely used in medical, agricultural, industrial, manufacturing, and military settings. Nanosize titanium dioxide (TiO_2) is used in a variety of consumer products such as toothpastes, sunscreens, cosmetics, food products (Kaida *et al.*, 2004), paints and surface coatings as well as in the environmental decontamination of air, soil, and water (Choi *et al.*, 2006). The risk of potential entry through dermal, ingestion, and inhalation routes of exposure suggests that nanosize TiO_2 could pose an exposure risk to humans, livestock, and ecological species.

22.5.5.1   Study Rationale and Design

Because of their size and physical properties, nanoparticles can enter the body and cross various biological barriers and accumulate in different organs. Many *in vitro* studies have reported oxidative stress (OS)-mediated toxicity in various cell types including human bronchial cells and colon (Gurr *et al.*, 2005; Zhang and Sun, 2004), however, at the time investigations on which this study was based were conducted, little was known about TiO_2 nanoparticle exposure to the nervous system. The brain is highly susceptible to OS damage because it has low levels of antioxidants and high concentrations of macromolecular targets such as lipids, proteins, and nucleic acids. Oxidative stress in nerve tissues is related to the production of multiple reactive oxygen species (ROS) by brain microglia (Block *et al.*, 2007). In addition to the vulnerability of the brain to ROS, recent studies indicate that nano-size particles can cross the blood-brain barrier (Lockman *et al.*, 2004; Lockman *et al.*, 2003).

To investigate whether TiO_2 nanoparticles posed a potential risk to mammalian nerve tissues, the ROS response of TiO_2-exposed cell cultures of mouse brain microglia and neurons and microarray-based assessments of the microglia's response were conducted within ORD at the EPA (Long *et al.*, 2006; Long *et al.*, 2007). These nerve cultures were exposed to a commercially available nanomaterial, Degussa P25 (P25), an uncoated photoactive, largely anatase form of nanosize TiO_2. P25 is a material useful for water treatment, self-cleaning windows applications, and antimicrobial coatings and paints. This type of TiO_2 is not to be confused with the non-photoactive TiO_2 nanosize particle that is currently used in sun blocks and cosmetics.

The biochemical and morphological response to P25 exposure was investigated *in vitro* in rodent brain cultures of immortalized BV2 microglia (Blasi *et al.*, 1990) and dopaminergic (N27) neurons (Zhou *et al.*, 2000), and primary cultures of embryonic rat striatum. BV2 microglia cells were further analyzed for their genomic response to P25. These particular cell systems were chosen for study for the following reasons: BV2 microglia are derived from an immortalized mouse cell line that responds to pharmaceutical agents, particulates, and environmental chemicals with characteristic signs of OS (Block *et al.*, 2004). In addition, certain neuronal populations, such as dopaminergic (DA) neurons found in the brain striatum, are especially vulnerable to OS (Mattson, 2001, Maier *et al.*, 1994).

Throughout the study, the physicochemical properties (*e.g.*, surface area, particle shape, zeta potential and aggregate size) of P25 were described under exposure conditions that paralleled the biological response of these cells to understand how these properties affected the biological outcome.

22.5.5.2   Results

Data from these studies indicated that although P25 was not observed to be toxic to isolated N27 neurons, it did stimulate BV2 microglia to produce ROS and damaged OS-sensitive neurons in the more complex brain cultures. The genomic results were consistent with the findings of the *in vitro* cell system studies and suggested that, *in vitro*, the ROS affects genomic pathways involved in cell cycling, inflammation, apoptosis, and cell bioenergetics. These results, combined with other endpoints, could have implications for human health but require testing in other experimental systems.

22.6 Considerations to Integrate Toxicogenomics Data into Regulatory Applications

The EPA is exploring whether this technology can link biomarkers to toxic processes or biological mechanisms that may provide additional information for risk assessors. The EPA is also interested in this technology in that it may improve the toxicity screening for chemicals (*e.g.*, faster, cheaper, more accurate, greater understanding of gene expression pathways). By publishing the Interim Guidance on *Microarray-Based Assays*, the agency has attempted to clarify how this type of data may currently be used to support a registration or submission.

22.6.1 Applications and Challenges of Toxicogenomics in Risk Assessment

Genomics data may be an aid in reducing the level of uncertainty in the regulatory decision-making process and provide a means to evaluate exposure and effects. Knowledge of gene expression pathways may lead to a better understanding of potential risk and safety issues associated with the system being evaluated.

The ability to have access to the raw data from published or submitted experiments, and accompanying documentation of experimental and analysis details, is critical for the EPA to evaluate a microarray experiment properly. Even though some microarray data are available at public genomic databases such as the Gene Expression Omnibus (GEO, http://www.ncbi.nlm.nih.gov/geo/), the format is not always compatible with all monitoring or potential regulatory applications. Furthermore, many private industries have developed their own internal databases and may not be submitting data to these public databases.

Another challenge in incorporating microarray data into the risk assessment process is the plethora of computational tools available for genomic data analysis. The EPA cannot make a recommendation for a specific computer program to be utilized to analyze genomic data; however, the agency does cover in the microarray guidance document the factors that would be needed to analyze, interpret, and submit genomic data as part of a submission package for regulatory review. The agency also recognizes that, for the regulatory review process, there are other potential issues related to analysis and interpretation of genomic data. Therefore, the EPA continues to be open to suggestions on how to move forward with the use of microarray technology in support of a regulatory application. The EPA acknowledges that tremendous progress has been made in defining, testing, and validating gene expression methods for toxicogenomics. The agency is following the developments in this field closely.

22.7 Collaboration with Other Federal Agencies and Groups

The EPA is working closely with other federal agencies (*i.e.*, FDA, NIEHS), academic institutions, and stakeholders to standardize the field of microarrays and to explore the potential use of existing toxicological databases. The EPA is participating in a number of collaborations designed to develop appropriate protocols (to produce, analyze, interpret data, etc.) and methods for microarray data analysis. These include working with the FDA, NIH, NIST, and other stakeholders on the Microarray Quality Control project

(MAQC) (http://www.fda.gov/nctr/science/centers/toxicoinformatics/maqc/). The agency is also working to establish protocols for genomic data with NIEHS Chemical Effects in Biological Systems (CEBS), http://cebs.niehs.nih.gov/). The EPA's National Computational Toxicology Program has funded an EPA Office of Research and Development (ORD) National Center for Environmental Assessment (NCEA) project to develop and test an approach to using genomics data in EPA risk assessments. To test the utility of the approach that was developed, a study of dibutyl phthalate (DBP) is being performed by a multidisciplinary team of scientists at NIEHS, the Hamner Institute, the STAR Bioinformatics Center at Rutgers University, and the University of Medicine and Dentistry of New Jersey (UMDNJ), as well as EPA scientists across ORD (Euling *et al.*, 2007). Working with other federal agencies, the EPA is assessing and evaluating various computational tools available for genomic data analysis, such as the FDA's ArrayTrack database (http://www.fda.gov/nctr/science/centers/toxicoinformatics/ArrayTrack/). The EPA also participated in an international workshop on the validation of toxicogenomics-based test systems (Corvi *et al.*, 2006). It is essential for the agency to continue to collaborate with other federal agencies, academia, the regulated community, other stakeholders, and internationally in this endeavor, in order to benefit from ongoing advances in 'omics' technologies.

22.8   Future Progress on Toxicogenomics Technology

To explore the potential role of genomic techniques for immunotoxicity testing, a workshop was held in 2005 (http://www.epa.gov/nheerl/immunogen_workshop/). The discussions focused on the theoretical and practical utility of genomics techniques and whether immunotoxicogenomics approaches could actually supplement current screening procedures. These genomic approaches are in the early stages in immunotoxicology research arenas. The EPA recognizes the need to involve industry, academia, and federal agencies to focus on the development and validation of methodologies that could be beneficial in immunotoxicology research. Training on genomics and other 'omics'-related topics, developed by groups or individuals from within EPA, is currently being provided to EPA staff and other interested parties outside the agency. For example, EPA hazard and risk assessors are receiving basic and advanced training classes in 'omics' technologies and applications. Furthermore, EPA staff offered educational training classes in 'omics'-related subjects at the 2007 Annual Meeting of the Society of Toxicology held in Charlotte, NC.

The EPA is considering options to manage data, since the data affiliated with genomic (microarray) and toxicological end-points can be vast. The ability to utilize data analysis and interpretation tools has the potential to lead to the development and identification of biomarkers as well as to provide information about dose–response relationships, mechanisms of action, and predictive toxicity. Management of these data should address needs unique to scientifically based risk assessments, confidential and proprietary data security, public access, and other aspects of regulatory application. The EPA is engaged in bioinformatics research efforts (*e.g.*, EPA's Office of Research and Development's National Center for Computational Toxicology) in order to develop an agency-wide data management solution to integrate genomics, toxicological, and other key data potentially needed for regulatory applications (http://www.epa.gov/ncct/; Dix *et al.*, 2007). The EPA, by putting forth the microarray guidance document, is trying to be proactive in working with various groups to better understand and advance the field of 'omics' technology. By working closely with other federal agencies and stakeholders, the EPA hopes to overcome some of the inherent challenges of toxicogenomics and ultimately to integrate microarray data in risk assessment and the regulatory decision-making process to fulfill the agency's mission to protect human health and the environment.

References

Allen JW, Wolf DC, George MH, Hester SD, Sun G, Thai SF, Delker DA, Moore T, Jones C, Nelson G, Roop BC, Leavitt S, Winkfield E, Ward WO, Nesnow S (2006) Toxicity profiles in mice treated with hepatotumorigenic and non-hepatotumorigenic triazole conazole fungicides: propiconazole, triadimefon, and myclobutanil. *Toxicol Pathol* **34**(7), 853–862.

Benson, WH, Gallagher K, McClintock JT (2007) U.S. Environmental Protection Agency's activities to prepare for regulatory and risk assessment applications of genomics information. *Environ Mol Mutag* **48**(5), 359–362.

Biagini C, Bender V, Borde F, Boissel E, Bonnet MC, Masson MT, Cassio D, Chevalier S (2006) Cytochrome P450 expression-induction profile and chemically mediated alterations of the WIF-B9 cell line. *Biol Cell* **98**, 23–32.

Blasi E, Barluzzi R, Bocchini V, Mazzolla R, Bistoni F (1990) Immortalization of murine microglial cells by a v-raf/v-myc carrying retrovirus. *J. Neuroimmunology* **27**, 229–237.

Block ML, Wu X, Pei Z, Li G, Wang T, Qin L, Wilson B, Yang J, Hong JS, Veronesi B (2004) Nanometer size diesel exhaust particles are selectively toxic to dopaminergic neurons: the role of microglia, phagocytosis, and NADPH oxidase. *FASEB J* **18**, 1618–1620.

Block ML, Zecca L, Hong JS (2007) Microglia-mediated neurotoxicity: uncovering the molecular mechanisms. *Nat Rev Neurosci* **8**, 57–69.

Chipman J, Davies J, Parsons J, Nair O'Neill, G, Fawell J (1998) DNA oxidation by potassium bromate; a direct mechanism or linked to lipid peroxidation? *Toxicology* **126**, 93–102.

Choi H, Stathatos E, Oionysiou DO (2006) Sol–gel preparation of mesoporous photocatalytic TiO_2 films and TiO_2/Al_2O_3 composite membranes for environmental applications. *Appl Catal B Environ* **63**, 60–67.

Corvi R, Ahr H-J, Albertini S, Blakey DH, Clerici L, Coecke S, Douglas GR, Gribaldo L, Groten JP, Haase B, Hamernik K, Hartung T, Inoue T, Indans I, Maurici D, Orphanides G, Rembges D, Sansone S-A, Snape JR, Toda E, Tong W, van Delft JH, Weis B, Schechtman LM (2006) Meeting report: validation of toxicogenomics-based test systems: ECVAM-ICCVAM/NICEATM considerations for regulatory use. *Environ Health Perspect* **114**(2), 420–429.

DeAngelo A, George M, Kilburn S, Moore T, Wolf D (1998) Carcinogenicity of potassium bromate administered in the drinking water to male B6C3F1 mice and F344/N rats. *Toxicol Pathol* **26**, 724–729.

Debeljak N, Fink M, Rozman D (2003) Many facets of mammalian lanosterol 14alpha-demethylase from the evolutionarily conserved cytochrome P450 family CYP51. *Arch Biochem Biophys* **409**, 159–171.

Delker DA, Hatch G, Allen J, Crissman B, George M, Geter D, Kilburn S, Moore T, Nelson G, Roop B, Slade R, Swank A, Ward W, DeAngelo A (2006) Molecular biomarkers of oxidative stress associated with bromate carcinogenicity. *Toxicology* **221**, 158–165.

Dix DJ, Gallagher K, Benson WH, Groskinsky BL, McClintock JT, Dearfield KL, Farland WH (2006) A framework for the use of genomics data at the EPA. *Nat Biotechnol* **24**, 1108–1111.

Dix DJ, Houck KA, Martin MT, Richard AM, Setzer RW, Kavlock RJ (2007) The ToxCast program for prioritizing toxicity testing of environmental chemicals. *Toxicol Sci* **95**(1), 5–12.

Euling SY, Makris S, Sen B, Kim AS, Benson B, Gaido KW *et al* (2007) Use of toxicogenomics data in risk assessment: a case study of dibutyl phthalate and male reproductive developmental effects. *Toxicologist* **96**(1), 212 (Abstract 1023).

Georgopapadakou NH, Walsh TJ (1996) Antifungal agents: chemotherapeutic targets and immunologic strategies. *Antimicrob Agents Chemother* **40**, 279–291.

Gurr J-R, Wang AS, Chen C-H, Jan K-Y (2005) Ultrafine titanium dioxide particles in the absence of photoactivation can induce oxidative damage to human bronchial epithelial cells. *Toxicology* **213**, 66–73.

Hester SD, Benavides GB, Yoon L, Morgan KT, Zou F, Barry W, Wolf DC (2003) Formaldehyde-induced gene expression in F344 rat nasal respiratory epithelium. *Toxicology* **187**, 13–24.

Hester SD, Barry WT, Zou F, Wolf DC (2005) Transcriptomic analysis of F344 rat nasal epithelium suggests that the lack of carcinogenic response to glutaraldehyde is due to its greater toxicity compared to formaldehyde. *Toxicol Pathol* **33**, 415–424.

Hurley PM (1998) Mode of carcinogenic action of pesticides inducing thyroid follicular cell tumors in rodents. *Environ Health Perspect* **106**, 437–445.

INCHEM (1981) Triadimefon (Pesticide Residues in Food: 1981 Evaluations). FAO Plant Production and Protection Paper 42. Monograph 566: 36. www.inchem.org/documents/jmpr/jmpmono/v81pr32.htm.

INCHEM (1987) Monograph 768 Propiconazole (Pesticide residues in food: 1987 evaluations Part II Toxicology) 19 pp. www.inchem.org/documents/ jmpr/jmpmono/v87pr13.htm.

Irwin RD, Boorman GA, Cunningham ML, Heinloth AN, Malarkey DE, Paules RS (2004) Application of toxicogenomics to toxicology: basic concepts in the analysis of microarray data. *Toxicol Pathol* **32**(Suppl 1), 72–83.

Ishidate M, Sofuni T, Yoshikawa K, Hayashi M, Nohmi T, Sawada M, Matsuoka A (1984) Primary mutagenicity screening of food additives currently used in Japan. *Food Chem Toxicol* **22**, 623–636.

Juberg DR, Mudra DR, Hazelton GA, Parkinson A (2006) The effect of fenbuconazole on cell proliferation and enzyme induction in the liver of female CD1 mice. *Toxicol Appl Pharmacol* **214**, 178–187.

Kaida T, Kobayashi K, Adachi M, Suzuki F (2004) Optical characteristics of titanium oxide interference film and the film laminated with oxides and their applications for cosmetics. *J Cosmet Sci* **55**, 219–220.

Kurokawa Y, Maekawa A, Takahashi M, Hayashi Y (1990) Toxicity and carcinogenicity of potassium bromate – a new renal carcinogen. *Environ Health Perspect* **87**, 309–335.

Lau C, Butenhoff JL, Rogers JM (2004) The developmental toxicity of perfluoroalkyl acids and their derivatives. *Toxicol Appl Pharmacol* **198**, 231–241.

Lockman P, Oyewumi M, Koziara J, Roder K, Mumper R, Allen D (2003) Brain uptake of thiamine-coated nanoparticles. *J. Controlled Release* **93**, 271–282.

Lockman PR, Koziara JM, Mumper RJ, Allen DD (2004) Nanoparticle surface charges alter blood-brain barrier integrity and permeability. *J. Drug Target* **12**, 635–641.

Long TC, Saleh N, Tilton RD, Lowry GV, Veronesi B (2006) Titanium dioxide (P25) produces reactive oxygen species in immortalized brain microglia (BV2): implications for nanoparticle neurotoxicity. *Environ Sci Technol* **40**, 4346–4352.

Long TC, Tajuba J, Sama P, Saleh N, Swartz C, Parker J, Hester S, Lowry GV, Veronesi B (2007) Nanosize titanium dioxide stimulates reactive oxygen species in brain microglia and damages neurons in vitro. *Environ Health Perspect* **115**(11), 1631–1637.

Long TC, Tajuba J, Sama P, Saleh N, Swartz C, Parker J, Hester S, Lowry GV, Veronesi B (2006) Titanium dioxide (P25) produces reactive oxygen species in immortalized brain microglia (BV2): implications for nanoparticle neurotoxicity. *Environ Sci Technol* **40**(14), 4346–4352.

Maier WE, Kodavanti PR, Harry GJ, Tilson HA (1994) Sensitivity of adenosine triphosphatases in different brain regions to polychlorinated biphenyl congeners. *J Appl Toxicol* **14**, 225–229.

Martin MT, Brennan RJ, Hu W, Ayanoglu E, Lau C, Ren H, Wood CR, Corton JC, Kavlock RJ, Dix DJ (2007) Toxicogenomic study of triazole fungicides and perfluoroalkyl acids in rat livers predicts toxicity and categorizes chemicals based on mechanisms of toxicity. *Toxicol Sci* **97**, 595–613.

Mattson MP (2001) Mechanisms of neuronal apoptosis and excitotoxicity. In *Pathogenesis of Neurodegnerative Disorders*, Mattson M (ed.). Humana Press, Baltimore, MD; 1–20.

McMillian M, Nie AY, Parker JB, Leone A, Bryant S, Kemmerer M, Herlich J, Liu Y, Yieh L, Bittner A, Liu X, Wan J, Johnson MD (2004) A gene expression signature for oxidant stress/reactive metabolites in rat liver. *Biochem Pharmacol* **68**, 2249–2261.

Moser VC, Barone Jr S, Smialowicz RJ, Harris MW, Davis BJ, Overstreet D, Mauney M, Chapin RE (2001) The effects of perinatal tebuconazole exposure on adult neurological, immunological, and reproductive function in rats. *Toxicol Sci* **62**, 339–352.

Nishimura S (2002) Involvement of mammalian *Ogg*1 (MMH) in excision of the 8-hydroxyguanine residue in DNA. *Free Radic Biol Med* **10**, 225–242.

Ronis MJ, Ingelman-Sundberg M, Badger TM (1994) Induction, suppression and inhibition of multiple hepatic cytochrome P450 isozymes in the male rat and bobwhite quail (*Colinus virginianus*) by ergosterol biosynthesis inhibiting fungicides (EBIFs). *Biochem Pharmacol* **48**, 1953–1965.

St Clair MB, Gross EA, Morgan KT (1990) Pathology and cell proliferation induced by intra-nasal instillation of aldehydes in the rat: comparison of glutaraldehyde and formaldehyde. *Toxicol Pathol* **18**, 353–361.

Sun G, Thai SF, Tully DB, Lambert GR, Goetz AK, Wolf DC, Dix DJ, Nesnow S (2005) Propiconazole-induced cytochrome P450 gene expression and enzymatic activities in rat and mouse liver. *Toxicol Lett* **155**, 277–287.

Tully DB, Bao W, Goetz AK, Blystone CR, Ren H, Schmid JE, Strader LF, Wood CR, Best DS, Narotsky MG, Wolf DC, Rockett JC, Dix DJ (2006) Gene expression profiling in liver and testis of rats to characterize the toxicity of triazole fungicides. *Toxicol Appl Pharmacol* **215**, 260–273.

Umemura T, Takagi A, Sai K, Hasegawa R, Kurokawa Y (1998) Oxidative DNA damage and cell proliferation in kidneys of male and female rats during 13-weeks exposure to potassium bromate. *Arch Toxicol* **72**, 264–269.

US Environmental Protection Agency (2002) Interim Policy on Genomics. http://www.epa.gov/osa/spc/genomics.htm.

US Environmental Protection Agency (2004) Potential Implications of Genomics for Regulatory and Risk Assessment Applications at EPA. EPA 100/B-04/002. http://www.epa.gov/osa/genomics.

Ward WO, Delker DA, Hester SD, Thai SF, Wolf DC, Allen JW, Nesnow S (2006) Transcriptional profiles in liver from mice treated with hepatotumorigenic and nonhepatotumorigenic triazole conazole fungicides: propiconazole, triadimefon, and myclobutanil. *Toxicol Pathol* **34**(7), 863–878.

Waring JF, Ciurlionis R, Jolly RA, Heindel M, Ulrich RG (2001) Microarray analysis of hepatotoxins *in vitro* reveals a correlation between gene expression profiles and mechanisms of toxicity. *Toxicol Lett* **120**, 359–368.

Waring JF, Ulrich RG, Flint N, Morfitt D, Kalkuhl A, Staedtler F, Lawton M, Beekman JM, Suter L (2004) Interlaboratory evaluation of rat hepatic gene expression changes induced by methapyrilene. *Environ Health Perspect* **112**, 439–448.

Zarn JA, Bruschweiler BJ, Schlatter JR (2003) Azole fungicides affect mammalian steroidogenesis by inhibiting sterol 14 alpha-demethylase and aromatase. *Environ Health Perspect* **111**, 255–261.

Zhang AP, Sun YP (2004) Photocatalytic killing effect of TiO_2 nanoparticles on Ls-174-t human colon carcinoma cells. *World J Gastroenterol* **10**, 3191–3193.

Zhou W, Hurlbert MS, Schaack J, Prasad KN, Freed CR (2000) Overexpression of human alpha-synuclein causes dopamine neuron death in rat primary culture and immortalized mesencephalon-derived cells. *Brain Res* **866**, 33–43.

Suggested Further Reading

Andrewes P, Kitchin KT, Wallace K (2003) Dimethylarsine and trimethylarsine are potent genotoxins *in vitro*. *Chem Res Toxicol* **16**, 994–1003.

Arnold LL, Cano M, St John M, Eldan M, van Gemert M, Cohen SM (1999) Effects of dietary dimethylarsinic acid on the urine and urothelium of rats. *Carcinogenesis* **20**, 2171–2179.

Bao W, Schmid JE, Goetz AK, Ren H, Dix DJ (2005) A database for tracking toxicogenomic samples and procedures. *Reprod Toxicol* **19**(3), 411–419.

Cohen SM, Arnold LL, Uzvolgyi E, Cano M, St John M, Yamamoto S, Lu X, Le XC (2002) Possible role of dimethylarsinous acid in dimethylarsinic acid-induced urothelial toxicity and regeneration in the rat. *Chem Res Toxicol* **15**, 1150–1157.

Fostel J, Choi D, Zwickl C, Morrison N, Rashid A, Hasan A, Bao W, Richard A, Tong W, Bushel PR, Brown R, Bruno M, Cunningham ML, Dix D, Eastin W, Frade C, Garcia A, Heinloth A, Irwin R, Madenspacher J, Merrick BA, Papoian T, Paules R, Rocca-Serra P, Sansone AS, Stevens J, Tomer K, Yang C, Waters M (2005) Chemical Effects in Biological Systems – Data Dictionary (CEBS-DD): a compendium of terms for the capture and integration of biological study design description, conventional phenotypes, and 'omics data. *Toxicol Sci* **88**(2), 585–601.

Kligerman AD, Doerr CL, Tennant AH, Harrington-Brock K, Allen JW, Winkfield E, Poorman-Allen P, Kundu B, Funasaka K, Roop BC, Mass MJ, DeMarini DM (2003) Methylated trivalent arsenicals as candidate ultimate genotoxic forms of arsenic: induction of chromosomal mutations but not gene mutations. *Environ Mol Mutagen* **42**, 192–205.

Lockman PR, Koziara JM, Mumper RJ, Allen DO (2004) Nanoparticle surface charges alter blood–brain barrier integrity and permeability. *J Drug Target* **12**, 635–641.

Rockett JC, Burczynski ME, Fornace AJ, Herrmann PC, Krawetz SA, Dix DJ (2004) Surrogate tissue analysis: monitoring toxicant exposure and health status of inaccessible tissues through the analysis of accessible tissues and cells. *Toxicol Appl Pharmacol* **194**(2), 189–199.

Sen B, Wang A, Hester SD, Robertson JL, Wolf DC (2005) Gene expression profiling of responses to dimethylarsinic acid in female F344 rat urothelium. *Toxicology* **215**, 214–226.

Southgate J, Harnden P, Selby PJ, Thomas DF, Trejdosiewicz LK (1999) Urothelial tissue regulation. Unraveling the role of the stroma. *Adv Exp Med Biol* **462**, 19–30.

Thompson CJ, Ross SM, Gaido KW (2004) Di(*n*-butyl) phthalate impairs cholesterol transport and steroidogenesis in the fetal rat testis through a rapid and reversible mechanism. *Endocrinology* **145**, 1227–1237.

Thompson CJ, Ross SM, Hensley J, Liu K, Heinze SC, Young SS, Gaido KW (2005) Differential steroidogenic gene expression in the fetal adrenal gland versus the testis and rapid and dynamic response of the fetal testis to di(*n*-butyl) phthalate. *Biol Reprod* **73**, 908–917.

US Environmental Protection Agency (2007) Interim Guidance for Microarray-Based Assays: Data Submission, Quality, Analysis, Management, and Training. http://www.epa.gov/osa/spc/genomicsguidance.htm.

Wanibuchi H, Yamamoto S, Chen H, Yoshida K, Endo G, Hori T, Fukushima S (1996) Promoting effects of dimethylarsinic acid on *N*-butyl-*N*-(4-hydroxybutyl)nitrosamine-induced urinary bladder carcinogenesis in rats. *Carcinogenesis* **17**(11), 435–443.

Index

Please note: page numbers in *italics* refer to figures and those in **bold** type, to tables

Toxicogenomics: A Powerful Tool for Toxicity Assessment   Edited by S. C. Sahu
© 2008 John Wiley & Sons, Ltd